Conquering the Physics GRE

Third Edition

The Physics GRE plays a significant role in deciding admissions to nearly all US physics Ph.D. programs, yet few exam-prep books focus on the test's actual content and unique structure. Recognized as one of the best student resources available, this tailored guide has been thoroughly updated for the current Physics GRE. It contains carefully selected review material matched to all of the topics covered, as well as tips and tricks to help you solve problems under time pressure. It features three full-length practice exams, revised to accurately reflect the difficulty of the current test, with fully worked solutions so that you can simulate taking the test, review your preparedness, and identify areas in which further study is needed. Written by working physicists who took the Physics GRE for their own graduate admissions to MIT, this self-contained reference guide will help you achieve your best score.

Yoni Kahn is a theoretical physicist researching dark matter and supersymmetry. A postdoctoral research associate at Princeton University, he obtained his Ph.D. from MIT in 2015 and in 2016 received the American Physical Society's J.J. and Noriko Sakurai Dissertation Award in Theoretical Particle Physics.

Adam Anderson is an experimental physicist working at the interface between cosmology and particle physics. He received his Ph.D. from MIT in 2015 and is now a Lederman postdoctoral fellow at Fermi National Accelerator Laboratory, developing instruments for performing precision measurements of the cosmic microwave background.

Conquering the Physics GRE

Third Edition

Yoni Kahn
Princeton University, New Jersey

Adam Anderson
Fermilab, Batavia, Illinois

CAMBRIDGE
UNIVERSITY PRESS

University Printing House, Cambridge CB2 8BS, United Kingdom

One Liberty Plaza, 20th Floor, New York, NY 10006, USA

477 Williamstown Road, Port Melbourne, VIC 3207, Australia

314–321, 3rd Floor, Plot 3, Splendor Forum, Jasola District Centre, New Delhi – 110025, India

79 Anson Road, #06–04/06, Singapore 079906

Cambridge University Press is part of the University of Cambridge.

It furthers the University's mission by disseminating knowledge in the pursuit of education, learning, and research at the highest international levels of excellence.

www.cambridge.org
Information on this title: www.cambridge.org/9781108409568
DOI: 10.1017/9781108296977

Previously published by CreateSpace, 2012, 2014

Third edition 2018
Reprinted 2020

Printed in the United Kingdom by TJ International Ltd. Padstow Cornwall

A catalog record for this publication is available from the British Library.

ISBN 978-1-108-40956-8 Paperback

Chapter openings image credit: OktalStudio/DigitalVision Vectors/Getty Images

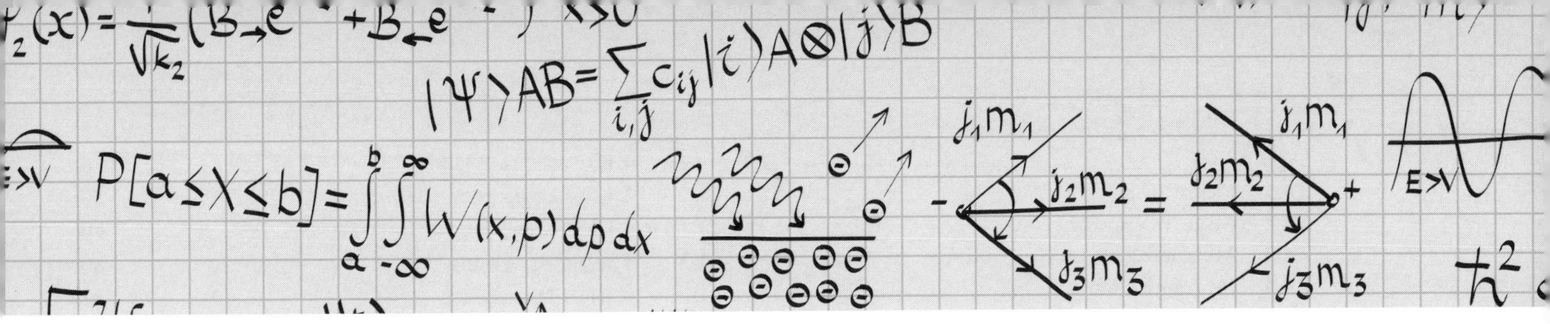

CONTENTS

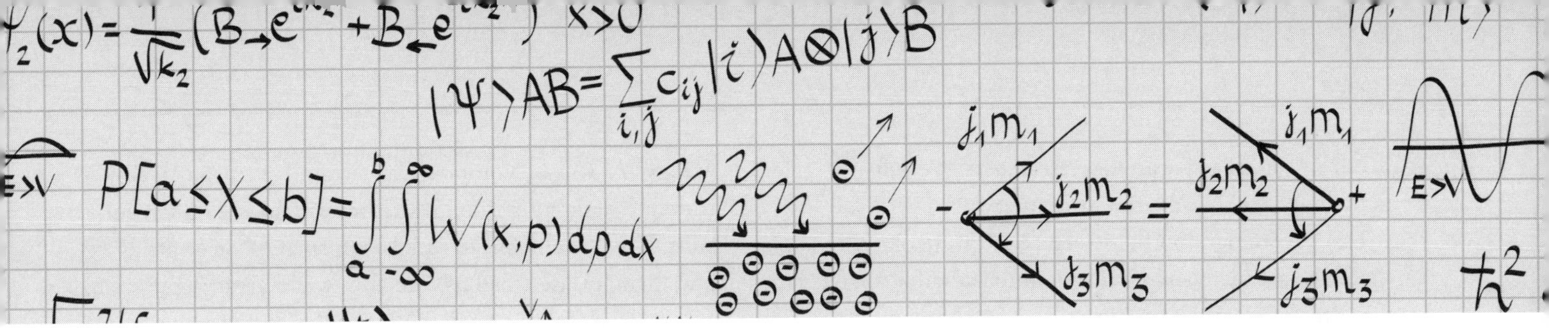

PREFACE

Conquering the Physics GRE represents the combined efforts of two MIT graduate students frustrated with the lack of decent preparation materials for the Physics GRE subject test. When we took the exams, in 2007 and 2009, we did what any student in the internet age would do – searched the various online bookstores for "physics GRE prep," "physics GRE practice tests," and so on. We were puzzled when the only results were physics practice problems that had nothing to do with the GRE specifically or, worse, GRE practice books having nothing to do with physics. Undeterred, we headed to our local brick-and-mortar bookstores, where we found a similar situation. There were practice books for every single GRE subject exam, except physics. Further web searches unearthed www.grephysics.net, containing every problem and solution from every practice test released up to that point, and www.physicsgre.com, a web forum devoted to discussing problems and strategies for the test. We discovered these sites had sprung up thanks to other frustrated physicists just like us: there was no review material available, so students did the best they could with the meager material that did exist. This situation is particularly acute for students in smaller departments, who have fewer classmates with whom to study and share the "war stories" of the GRE.

This book endeavors to fix that situation. Its main contribution is a set of three full-length practice tests and fully worked solutions, designed to be as close as possible in style, difficulty, content distribution, and format to the actual GRE exam. We have also included review material for all of the nine content areas on the Physics GRE exam: classical mechanics, electricity and magnetism, optics and waves, thermodynamics and statistical mechanics, quantum mechanics, atomic physics, special relativity, laboratory methods, and specialized topics. To our knowledge, this is the first time that reviews of standard undergraduate subjects such as classical mechanics and thermodynamics have been paired with less standard material such as laboratory methods in the same text, specifically focused on aspects of these subjects relevant for the GRE. Exam-style practice problems and worked solutions are included for each review chapter, giving over 150 additional GRE-style practice problems in addition to the 300 from the exams. The shorter chapters have review problems at the very end, while the longer ones have review problems distributed throughout the chapter.

The chapter on quantum mechanics and atomic physics is the longest, for two reasons: the combination of these two topics makes up nearly 25% of the exam, and the formalism of quantum mechanics is so different from the rest of the physics topics covered on the GRE that we felt it worthwhile to discuss a number of calculational shortcuts in detail. Unique to our book is a chapter on special tips and tricks relevant for taking the GRE as a standardized multiple-choice test. Some of the standard test-taking wisdom still applies, but we have found that the structure of the multiple answer choices often provides valuable hints on how to solve a problem: you will not find this information in any other test-prep book, because it is based on techniques such as dimensional analysis and back-of-the-envelope estimation, which most test-prep authors (who are not physicists) are simply unaware of.

Next, a brief word on what this book is *not*. This is not a detailed review of undergraduate physics: many of the more difficult subjects get an extremely abbreviated treatment, designed to highlight only those formulas and problem types relevant for the exam. We believe this will help you succeed on the Physics GRE, but if any of the standard subjects are completely unfamiliar to you, please do *not* try to teach them to yourself from our book. There are many excellent texts out there relevant for that purpose, and we have included a list of them in the Resources section following this preface. We strongly encourage you to consult these references, as we have found them useful both in writing this present text and in our careers as active physics researchers. We will often refer to them throughout the review chapters.

Last, a comment on the structure of this book. We realize that there are many, *many* equations to learn that are relevant for GRE-style physics problems. To keep the amount you feel you have to memorize to a minimum, we only assign equation numbers to equations we feel are important to remember – everything else you can safely ignore. (This is not to say that you should *memorize* every single numbered equation – Chapter 9 contains useful advice for what to memorize and what to derive.) Also, while most of the review chapters review material in roughly the order it was presented when you first learned it, Chapter 1 is structured very differently. We assume that you still remember many of the basic facts of classical mechanics from your freshman year introductory physics course, and so we focus our attention on *problem types* that are standard on the GRE, rather than on specific subtopics. We hope you will find this approach useful.

A book like this could never have been written without the help and support of other people. We especially thank Yichen Shen for his useful contributions to the condensed matter section of the Specialized Topics review. We thank Jen Sierchio and other members of the physicsgre.com community, as well as Raghu Mahajan, Nate Thomas, Jaime Varela, and Dustin Katzin at MIT, who proofread an early version of our first sample exam. Thanks also to Alex Shvonski, Kevin Satzinger, Jasen Scaramazza, Alastair Heffernan, Rizki Sharif, Benjamin Blumer, Andrew Ochoa, Ryan Janish, and especially Vinay Ramasesh for proofreading the first public versions of the sample exams and providing useful feedback. Y.K. would like to thank his advisor, Jesse Thaler, for bearing with him while working on a project that siphoned valuable time away from research. A.A. thanks Y.K. for being so accommodating and flexible toward his occasional "vanishing acts" from writing to attend to research obligations. A.A. also thanks his advisor, Enectalí Figueroa-Feliciano, and many other collaborators too numerous to name, for accepting (or at least pretending not to notice) any drag that this project caused on his research productivity.

Although we have made every effort to eliminate all factual and typographical errors from this book, the long errata lists for any physics textbook speak to the fact that this is impossible, especially in a first edition. If you find any mistakes of any kind, please email us at physics@physicsgreprep.com and let us know. Even the smallest of typos is worth fixing. We will be compiling an errata list on our website, www.physicsgreprep.com, which we will update on a regular basis. If you would like to receive information on errata as we find them, please email us. We also would greatly appreciate any feedback on this book, both positive and negative, as we strive to improve its usefulness for students everywhere.

Yoni Kahn and Adam Anderson

Preface To The Third Edition

Since *Conquering the Physics GRE* was first published, both authors have completed graduate school and gone on to careers in academia: Yoni as a theoretical particle physicist, and Adam as an observational cosmologist. If this kind of career path is what you're hoping for, this is the book for you! *Conquering the Physics GRE* remains the only comprehensive reference book specifically tailored to the topics on ETS's Physics GRE, and indeed we often refer to this book as a quick reference for key undergraduate physics topics.

The revised third edition, published by Cambridge University Press, makes numerous changes in response to comments from students and faculty who have used this book for GRE preparation. Most importantly, the three full-length sample exams have been completely reworked so that the difficulty and types of questions better match the current content of the exam. We have added an equation index, a subject index, and a problems index so you can easily look up particular terms or concepts that appear on practice problems and solutions as well as in the review material. Finally, we have made many improvements to the review chapters, including additional figures, diagrams, and practice problems; an updated Nobel Prize section; and brand-new review problems for the Tips and Tricks chapter. We hope that these changes make this book a better reference not only for the GRE but for your bookshelf in your future physics career.

We are thankful to the many people who have made this revised edition possible, including Vince Higgs, Lucy Edwards, and Esther Migueliz at Cambridge University Press, and Lia Hankla, Sean Muleady, and Ahmed Akhtar at Princeton for proofreading. We also thank the many students who submitted errata for previous editions and suggestions for topics that now appear in this book.

Yoni Kahn and Adam Anderson

HOW TO USE THIS BOOK

Studying for the GRE can be overwhelming! This book is long because it contains all the information you need to ace the exam, but not every student needs to study every chapter in equal detail. Here are some suggestions for how to use this book.

- **Only numbered equations are worth remembering.** The Physics GRE is a test of outside knowledge, so some memorization is inevitable. However, we have made a concerted effort to separate equations that are only used in specific worked examples from equations that are worth remembering for the test. Only the equations worth remembering are given equation numbers and are included in the Equation Index at the back of the book along with the page number where they appear; anything else you can safely forget for test day. This is still quite a long list, so rather than memorize each equation, check out Chapter 9 for suggestions on how to reduce your memorization workload by deriving more complex equations from more basic ones.
- **Use these sample exams as diagnostics.** ETS has released precious few actual GREs, and only the most recent (from 2001, 2008, and 2017) are representative of the current content of the test. We strongly suggest you leave the ETS exams until shortly before the actual test, where you can take them under simulated test-taking conditions. To start your studying, consider taking one of the sample exams provided in this book as a diagnostic, and note which areas you need to review the most. You can then focus on the review chapters covering these particular subject areas. Once you feel you've sufficiently filled in the gaps in your knowledge of undergraduate physics, you can take another sample exam and track your improvement, leaving the last exam for extra practice a week or two before the test, should you need it. Because we don't have access to ETS's proprietary scoring formula, we do not attempt to offer any conversion between raw score and scaled score (200–990) for our sample exams. Guessing at a formula would be extremely misleading at best, so use your score on our exams only as an estimate, but by all means use the ETS-provided conversion charts when taking the ETS exams.
- **Don't try to learn all of undergraduate physics from our book.** We have tailored the length and content of each of our review chapters to roughly follow the proportions of the GRE: 20% classical mechanics, 18% electromagnetism, 9% optics and waves, 10% thermodynamics and statistical mechanics, 22% quantum mechanics and atomic physics, 6% relativity, 6% laboratory methods, and 9% specialized topics. Our expositions of standard first- and second-year undergraduate topics are extremely brief or nonexistent, and we have given slightly more weight to more unfamiliar topics you're unlikely to find together in a single book, in order to make this book self-contained. If you find yourself totally mystified by a topic or completely unfamiliar with a formula, look it up in a more detailed reference! We've provided a list of suggested resources below.
- **Treat the end-of-chapter or end-of-section problems as subject practice rather than actual exam questions.** While our review problems follow the GRE multiple-choice format and don't require calculators, we don't intend them to exactly replicate GRE questions in style and difficulty: that's the purpose of the sample exams. Rather, the problems are there to highlight important problem types or calculational shortcuts, and as a result may feature solutions with more steps than you would see on test day. We recommend you work these problems as you're studying a particular chapter, but don't feel the need to keep to the GRE time limit of under two minutes per question.

Best of luck studying!

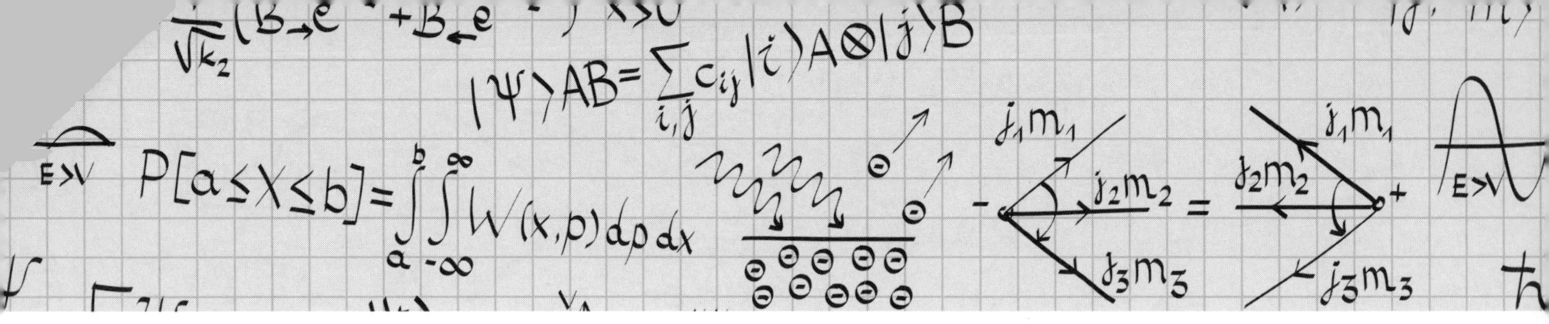

RESOURCES

Here we collect all the texts we recommend and will refer to in the review chapters. If you're wondering why books by Griffiths show up so often, it's likely because he was on the question-writing committee for the Physics GRE several years ago. Anecdotally, we know that questions are recycled *very* often (which is why so few exams have been released), so it's likely that many of the questions you'll see on your exam were written by Griffiths or consciously modeled after his books.

- **Classical Mechanics:** Whatever book you used for freshman physics should suffice here. For a more in-depth review of advanced topics, try *Classical Dynamics of Particles and Systems* by S.T. Thornton and J.B. Marion.
- **Electricity and Magnetism:** D.J. Griffiths, *Introduction to Electrodynamics*. This book covers everything you'll need to know about electricity and magnetism on the GRE, *except* for circuits. For circuits and a review of the most basic electricity and magnetism problems, which Griffiths glosses over, consult any standard freshman physics textbook. A good treatment of electromagnetic waves can also be found in R.K. Wangsness, *Electromagnetic Fields*. E. Purcell, *Electricity and Magnetism* is an extremely elegant introduction emphasizing physical concepts rather than mathematical formalism, should you need to relearn the basics of any topic. Under no circumstances should you consult Jackson! It's far too advanced for anything you'll need for the GRE.
- **Optics and Waves:** Like classical mechanics, nearly all the relevant information is covered in your freshman physics textbook. Anything you're missing can be found in the relevant chapters of *Introduction to Electrodynamics* by Griffiths.
- **Thermodynamics and Statistical Mechanics:** No overwhelming recommendation here. *Thermal Physics* and *Elementary Statistical Physics* by C. Kittel, or *Fundamentals of Statistical and Thermal Physics* by F. Reif, are decent. *Statistical Physics*, by F. Mandl has some decent pedagogy and the nice feature of many problems with worked solutions. Fermi's *Thermodynamics* is a classic for the most basic aspects of the subject.
- **Quantum Mechanics and Atomic Physics:** D.J. Griffiths, *Introduction to Quantum Mechanics*. This is really the only reference you need, even for atomic physics questions. Shankar and Sakurai are serious overkill, stay away from them for GRE purposes!
- **Special Relativity:** Chapter 12 of *Introduction to Electrodynamics* by Griffiths, and Chapter 3 of *Introduction to Elementary Particles*, also by Griffiths, for more examples of relativistic kinematics. Note that, confusingly, the two books use different sign conventions, so be careful!
- **Laboratory Methods:** For advanced circuit elements, *The Art of Electronics* by P. Horowitz and W. Hill is a classic, and used in many undergraduate laboratory courses. An excellent general reference for radiation detection is *Radiation Detection and Measurement* by G.F. Knoll. Chapter 1 covers general properties of radiation, Chapters 2 and 4 cover interactions of radiation with matter, Chapter 10 covers photon detectors, and Chapter 3 covers precisely the kind of probability and counting statistics you'll be asked about on the GRE. The rest of that book goes into far more detail than necessary, so don't worry about it. For lasers, try O. Svelto, *Principles of Lasers*, Chapters 1 and 6.
- **Specialized Topics:** The first chapter of D.J. Griffiths, *Introduction to Elementary Particles*, is a *mandatory* read. It seems that every GRE in the last several years has contained at least one question that can be answered purely by picking facts out of this chapter. The rest of the book is pretty good too, but the later chapters are almost certainly too advanced for the GRE. For condensed matter, try *Introduction to Solid State Physics* by C. Kittel, or Chapters 1–9 of *Solid State Physics* by N. Ashcroft and N. Mermin for a more advanced treatment written in a friendly and accessible style.

- **All-around:** L. Kirkby's *Physics: A Student Companion* is a nice all-around summary of a wide range of physics topics. It's geared toward students studying for exams, so it is concise and more distilled than the subject-specific books.

There are also several useful websites containing information related to the Physics GRE:

- www.grephysics.net: A compilation of the 400 problems released by ETS prior to 2011, and student-contributed solutions.
- www.physicsgre.com: A web forum for discussion of issues related to the GRE, and the grad school application process in general. Highly recommended: one of us (Y.K.) met several future colleagues on this forum before meeting them in person.
- www.aps.org/careers/guidance/webinars/gre-strategies.cfm: A webinar on Physics GRE preparation given by one of us (Y.K.) for the American Physical Society, drawing on strategies discussed in this book.

1 Classical Mechanics

Classical mechanics is the cornerstone of the GRE, making up 20% of the exam, and at the same time has the dubious distinction of being the subject that turns so many people *away* from physics. Your first physics class was undoubtedly a mechanics class, at which point you probably wondered what balls, springs, ramps, rods, and merry-go-rounds had to do in the slightest with the physics of the real world. So rather than (a) attempt the impossible task of covering your 1000-page freshman-year textbook in this much shorter reference, or (b) risk turning you away from physics before you've even taken the exam, we'll structure this chapter a little differently than the rest of the book. We're not going to review such things as Newton's laws, force balancing, or the definition of momentum; you should know these things in your sleep, or the rest of the exam will seem impossibly hard. Rather than review basic *topics*, we'll review standard *problem types* you're likely to encounter on the GRE. The more advanced topics will get their own brief treatment as well. After finishing this chapter, you will have reviewed nearly all the material you'll need for the classical mechanics section of the test, but in a format that is much more useful for the way the problems will likely be presented on the test. If you need a more detailed review of any of these topics, just open up any undergraduate physics text.

1.1 Blocks

One of the first things you learned in the first semester of freshman year physics was probably how to balance forces using free-body diagrams. Rather than rehash that discussion, which you can find in absolutely any textbook, we'll review it through a series of example problems that are GRE favorites. They involve objects, usually called "blocks," with certain masses, doing silly things like sitting on ramps, being pushed against springs, and traveling on carts. So here we go.

1.1.1 Blocks on Ramps

Here's a basic scenario (Fig. 1.1): a block of mass m is on a ramp inclined at an angle θ, and suppose we want to know the coefficient of static friction μ required to keep it in place. The usual solution method is to resolve any forces $\mathbf{F}$ into components along the ramp ($F_\parallel$) and perpendicular to the ramp ($F_\perp$). Rather than fuss with trigonometry or similar triangles, we can just do this by considering limiting cases, a theme that we'll return to throughout this book. In this case, we have to resolve the gravitational force $\mathbf{F}_g$. If the ramp is flat ($\theta = 0$), then there is no force in the direction of the ramp, so gravity acts entirely perpendicularly, and $F_{g,\parallel} = 0$. On the other hand, if the ramp is sheer vertical ($\theta = \pi/2$), then gravity acts entirely parallel to the ramp ($F_{g,\perp} = 0$), and the block falls straight down. Knowing that there must be sines and cosines involved, and the magnitude of $\mathbf{F}_g$ is mg, this uniquely fixes

$$F_{g,\parallel} = mg\sin\theta, \qquad F_{g,\perp} = mg\cos\theta.$$

For the block not to accelerate perpendicular to the ramp, we need the perpendicular forces to balance, which fixes the normal force to be $N = mg\cos\theta$. Then the frictional force is $F_f = \mu mg\cos\theta$, which must balance the component of gravity parallel to the ramp, $F_{g,\parallel} = mg\sin\theta$. Setting these equal gives

$$\mu mg\cos\theta = mg\sin\theta \implies \mu = \tan\theta.$$

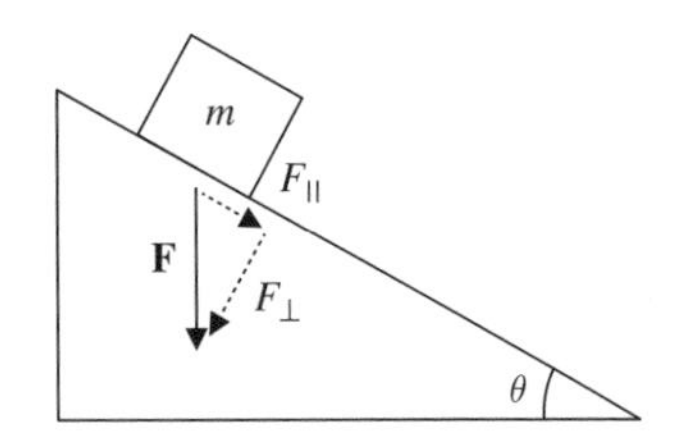

Figure 1.1 Free-body diagram of forces for a block on an inclined ramp.

Again, we can check this by limiting cases. If $\theta = 0$, then we don't need any friction to hold the block in place, and $\mu = 0$. If $\theta = \pi/2$, we need an infinite amount of friction to glue the block to the ramp and keep if from falling vertically, so $\mu = \infty$. Both of these check out.

Standard variants on this problem include applied forces and blocks attached to pulleys which hang over the side of the ramp, but surprisingly, neither the basic problem nor its variants have shown up on recent exams. Perhaps it is considered *too* standard by the GRE, such that most students will have memorized the problem and its variants so completely that it's not worth testing. In any case, consider it a simple review of how to resolve forces into components by using a limiting-cases argument, as this can potentially save you a lot of time on the exam.

1.1.2 Falling and Hanging Blocks

The next step up in complexity is to have two or more blocks interacting – for example, two blocks tied together with a rope, falling under the influence of gravity, or the same blocks hanging from a ceiling. These kinds of questions test your ability to identify precisely which forces are acting on which blocks. A foolproof, though time-consuming, method is to use *free-body diagrams*, where you draw each individual block and *only* the forces acting on it. This avoids the pitfall of double-counting, or applying the same force twice to two different objects, and ensures that you take into careful account the action/reaction balance of Newton's third law. See Example 1.1.

Sometimes, though, simple physical reasoning will suffice, especially in situations where the blocks aren't really distinct objects. For example, consider placing one block on top of another and letting them both fall under the influence of gravity. If we ignore air resistance, there is absolutely no physical distinction between the block–block system, and one larger block with the combined mass of both. In fact, a variant of precisely this argument was used in support of Galileo's discovery that the gravitational acceleration of objects was independent of their mass. We could even put a massless string between the two blocks, and the argument would *still* hold: since the whole system must fall with acceleration g, there can be no tension in the string. (Do the free-body analysis and check this yourself!) When interactions between the blocks become important, for example when they exert forces on one another through friction, then we must usually treat them as independent objects, though, as we'll see in Section 1.1.3, there are cases where the same kind of reasoning works.

EXAMPLE 1.1

A 5 kg block is tied to the bottom of a 20 kg block with a massless string. When an experimenter holds the 20 kg block stationary, the tension in the string is T_1. The experiment is repeated with the 20 kg block hanging under the 5 kg block, and the tension in the string is now T_2. What is T_2/T_1?

Our physical intuition tells us that $T_1/g = 5$ kg and $T_2/g = 20$ kg, since in both cases the function of the string is to support the weight of the lower block. So we expect $T_2/T_1 = 4$. This intuition is confirmed by a limiting-cases analysis: if the mass of the lower block is zero, then no matter the mass of the upper block, the string just dangles below the block with no tension, so the tension must be proportional to the mass of the lower block but independent of the mass of the upper one.

Let's check the intuition by doing a full free-body analysis. In order to treat both cases at once, call the mass of the top block m_1 and that of the bottom block m_2, as in Fig. 1.2. The forces on the two blocks are illustrated in Fig. 1.3. F is the force applied by the experimenter. Notice how the string tension acts up on the bottom block but down on the top block, and that the magnitude of T is the same for both blocks. For the purposes of the GRE, this is the *definition* of a massless string: it carries the same tension at every point. Setting the acceleration of m_2 equal to zero, since it is stationary, let's solve for T: $T - m_2 g = 0$, so indeed, $T = m_2 g$, the weight of the bottom block, and our intuition is correct. In this case it wasn't even necessary to consider the forces on the top block, a convenient time-saver!

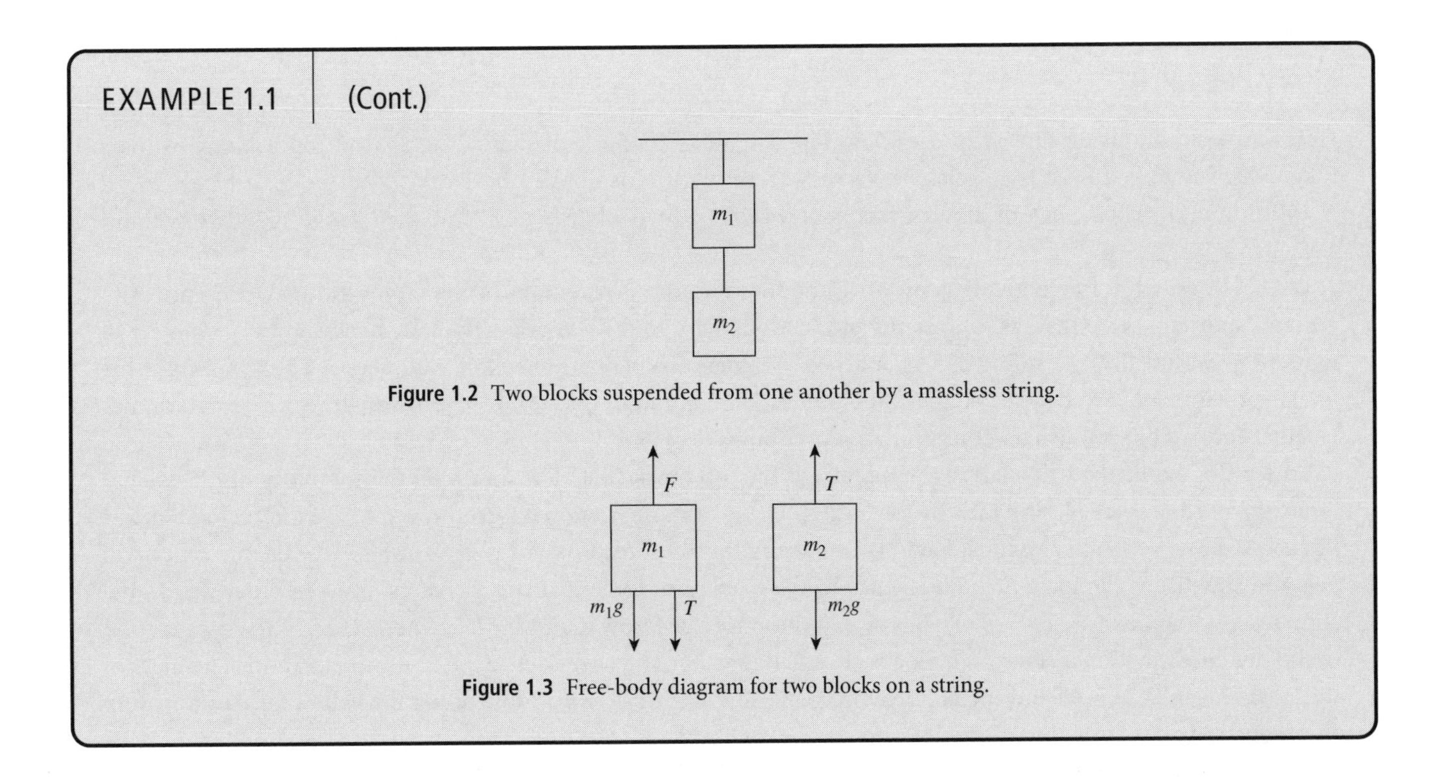

Figure 1.2 Two blocks suspended from one another by a massless string.

Figure 1.3 Free-body diagram for two blocks on a string.

1.1.3 Blocks in Contact

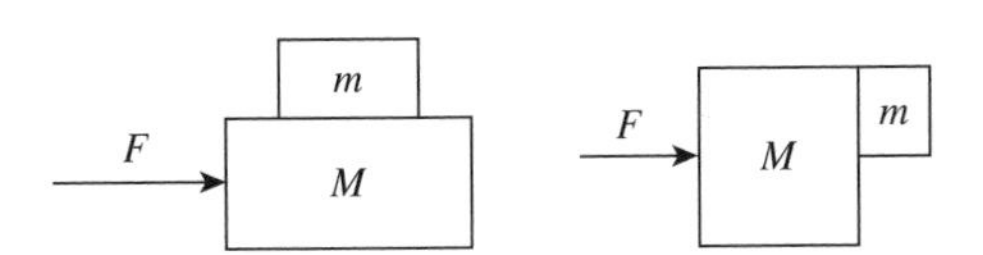

Figure 1.4 Typical setups for blocks moving together with friction.

There are two standard setups for these kinds of problems, illustrated in Fig. 1.4. Both get at all the core concepts of force balancing, Newton's second and third laws, and friction. In the second setup, you might be asked, given friction between the two blocks, what the minimum force is such that the mass m does not fall down due to gravity, or, if m is placed on the surface as well, how the force of one block on another changes depending on whether F is applied to M or m. As with the falling and hanging blocks, the key is to remember that the blocks are *independent* objects, so we must consider the forces on each independently. See Example 1.2.

1.1.4 Problems: Blocks

1. A block of mass 5 kg is positioned on an inclined plane at angle 45°. A force of 10 N is applied to the block, parallel to the ground. If the coefficient of kinetic friction is 0.5, which of the following is closest to the acceleration of the block? Assume there is no static friction.

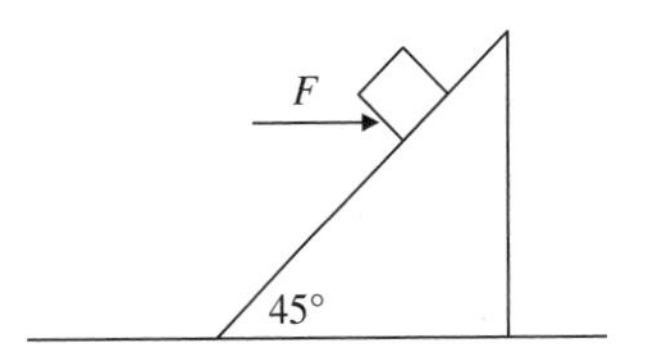

(A) $\sqrt{2}$ m/s^2 up the ramp
(B) $\sqrt{2}$ m/s^2 down the ramp
(C) $5\sqrt{2}$ m/s^2 up the ramp
(D) $5\sqrt{2}$ m/s^2 down the ramp
(E) $25\sqrt{2}$ m/s^2 down the ramp

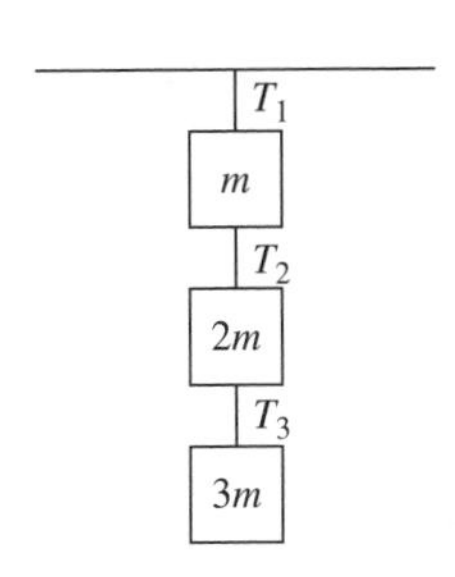

2. Three blocks of masses m, $2m$, and $3m$ are suspended from the ceiling using ropes, as shown in the diagram. Which of the following correctly describes the tension in the three rope segments, labeled T_1, T_2, and T_3?

(A) $T_1 < T_2 < T_3$
(B) $T_1 < T_2 = T_3$

EXAMPLE 1.2

Here's an example using the setup shown in Fig. 1.4 (left): A block of mass 2 kg sits on top a block of mass 5 kg, which is placed on a frictionless surface. A force of 10 N is applied horizontally to the 5 kg block. What is the minimum coefficient of static friction between the two blocks such that they move together without slipping?

We could do a full free-body diagram of all the forces in the problem, but simple physical reasoning provides a useful shortcut. Note that, as long as the blocks don't slip, the two blocks are really behaving as one object of mass $M + m$, just like the falling blocks attached by a massless string in Section 1.1.2 above. Thus we expect the final expression for μ to depend on the combination $M + m$, rather than M or m individually, since μ determines whether the two blocks stick together and act as a composite system.

To see this explicitly, let's analyze the motion of the top block first. The forces on the top block are its weight $-mg$, the normal force N_1 provided by the bottom block, and the frictional force $F_f = \mu N_1$. Since the top block is not accelerating vertically, we must have $N_1 = mg$ and the net force forward is $F_f = \mu mg$. Now the top block will begin to slip just as the force F_1 on it is equal to the maximum force that friction can supply; in other words, the slipping condition is $F_1 = F_f = \mu mg$. But by definition we also know that $F_1 = ma$, where a here is the acceleration of the two-block system – since both blocks are stuck together, they experience the same acceleration. The mass of the total system is $M + m$ and the applied force is F, so $F = (M + m)a$. Substituting the values for a and F_1 into $F_1 = ma$, we find

$$\mu mg = m\frac{F}{M+m} \implies \mu = \frac{F}{(M+m)g},$$

which as expected depends on $M+m$. Notice that we didn't ever have to do a free-body analysis of the second block alone: instead, we applied Newton's second law to the two-block system in the second step.

Of course, we can also do a free-body analysis for the block of mass M. We have the applied force F acting forwards, but there is also a force acting *backwards*, from Newton's third law: the bottom block is providing a frictional force which pushes the top block forwards, so the bottom block feels an equal force backwards. The net horizontal force is then $F - \mu mg$, where the second term is the magnitude of the friction force derived above. The acceleration of the bottom block is $a = \frac{1}{M}(F - \mu mg)$, and we want the frictional force on the top block to provide at least this acceleration, $a = F_f/m$, or the blocks will slip. Thus

$$\frac{1}{M}(F - \mu mg) = \frac{\mu mg}{m} \implies \mu = \frac{F}{(M+m)g},$$

the same answer as before. Plugging in the numbers, we find $\mu \approx 0.14$.

(C) $T_1 = T_2 = T_3$
(D) $T_1 = T_2 > T_3$
(E) $T_1 > T_2 > T_3$

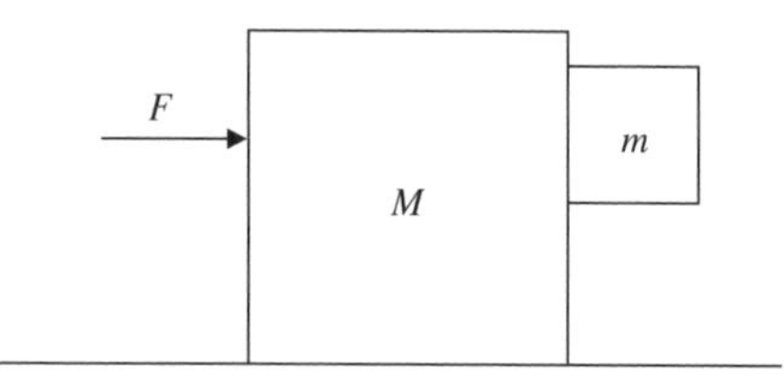

3. Two blocks of masses M and m are oriented as shown in the diagram. The block M moves on a surface with coefficient of kinetic friction μ_1, and the coefficient of static friction between the two blocks is μ_2. What is the minimum force F which must be applied to M such that m remains stationary relative to M?

(A) $\frac{\mu_1}{\mu_2}mg$
(B) $\frac{\mu_1}{\mu_2}\frac{Mm}{M+m}g$
(C) $(\mu_1 M + \mu_2 m)\, g$
(D) $\left(\mu_1 + \frac{1}{\mu_2}\right)(m + M)g$
(E) $\left(\mu_1 M + \frac{m}{\mu_2}\right) g$

1.2 Kinematics

Kinematics is the first physics that almost everyone learns, so it should be burned into the reader's mind already. For almost all problems it is sufficient to know the equations of motion for a particle undergoing constant acceleration. The primary types of problem worth reviewing are projectile motion problems and problems involving reference frames. To solve projectile motion in two dimensions, you only need the equations of motion for the x- and y-coordinates of the particle,[1]

$$x(t) = v_{0x}t + x_0, \qquad y(t) = -\frac{1}{2}gt^2 + v_{0y}t + y_0, \tag{1.1}$$

where we define coordinates such that gravity acts in the negative y-direction and $g = 10 \text{ m/s}^2$. Restricting to one dimension, there is another useful formula relating the initial and final velocities of an object, v_i and v_f, its acceleration a, and the change in position between the initial and final states Δx, if the acceleration is constant:

$$v_f^2 - v_i^2 = 2a\Delta x. \tag{1.2}$$

A two-line derivation of this formula uses the work–energy theorem, reviewed in Section 1.3.4.

For problems involving reference frames, just solve the problem in one frame, and then transform to the frame that the problem is asking about. For example, consider the situation in Fig. 1.5: a ball is thrown out of a car moving at constant velocity. Ignoring air resistance, in the frame of the car, the ball moves directly perpendicular to the road. In the frame of an observer at rest, the car is moving forwards, so the motion of the ball is the sum of the two velocities. In other words, the ball moves diagonally, both forward and away from the road. See Example 1.3.

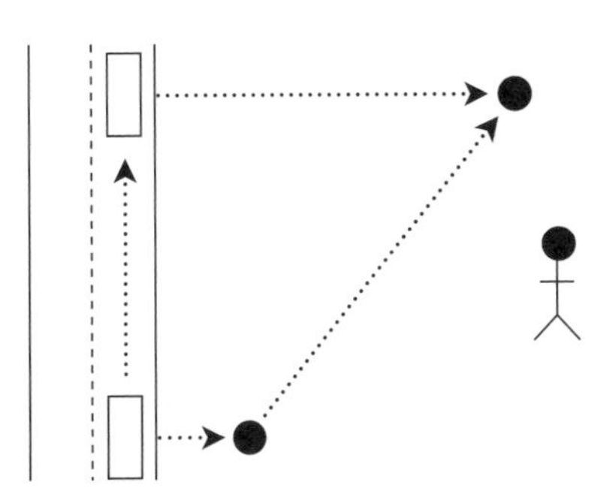

Figure 1.5 A ball thrown out of a moving car, in the frame of a stationary bystander.

[1] In this book, we use the convention of numbering *only equations describing general results worth memorizing for the exam*. We therefore numbered the kinematics formulas here, while we didn't number the equations in the previous section that applied to a specific problem involving blocks. This should help you focus on remembering the equations that actually matter for the exam. We have listed all numbered equations in the equation index at the back of the book, along with page numbers, for your convenience.

From the point of view of solving problems, however, one should avoid kinematics like the plague. It often results in having to solve quadratic equations, and although this is simple in principle, it is usually a huge waste of time. As a rule of thumb, only resort to kinematics if you need to know the explicit time dependence of a system. In nearly all other cases, the basic energy considerations discussed in Section 1.3 will be faster and computationally simpler.

1.2.1 Circular Motion

One kinematic situation that arises often on GRE questions is circular motion. We will consider this in slightly more detail in Section 1.6 when we discuss orbits. For now, consider a particle moving on a circular path. Its acceleration vector can always be decomposed into radial and tangential components. If its tangential acceleration is zero, then its tangential velocity is constant; it is moving in *uniform circular motion* about the center of the circle. But its radial acceleration is nonzero, and has value

$$a = \frac{v^2}{r}, \tag{1.3}$$

where v is the speed of the particle and r is the radius of its orbit. From this, we can immediately also infer that the force needed to keep the particle in its orbit, the *centripetal force*, is

$$F = \frac{mv^2}{r}. \tag{1.4}$$

Indeed, since the tangential acceleration is zero, it must experience some force, directed radially inwards, that keeps it moving in a circular path at a constant speed. Remember that this does not tell you what *kind* of force is acting on the body. It just tells you that if you see a body moving uniformly in a circle of radius r with *constant* speed v, then you can determine what centripetal force must be acting on it.

While uniform circular motion is perhaps the most common example, it is certainly not the most general. There are many cases of nonuniform circular motion: for example, a roller-coaster going around a circular loop-the-loop, or a vertical pendulum attached to a rigid rod with sufficient initial speed to complete a full revolution. In these cases the angle between the gravitational force vector and the velocity vector varies as the object goes around the circle, giving a varying tangential acceleration in addition to the centripetal force, and the above formulas do not apply throughout the whole orbit. However, the uniform circular motion equations *do* apply

EXAMPLE 1.3

Suppose an astronaut is on a rocket that is moving vertically at constant speed u. When the rocket is at a height h, the astronaut throws a ball horizontally out of the rocket with velocity w, as shown in Fig. 1.6. What is the speed of the ball when it hits the ground?

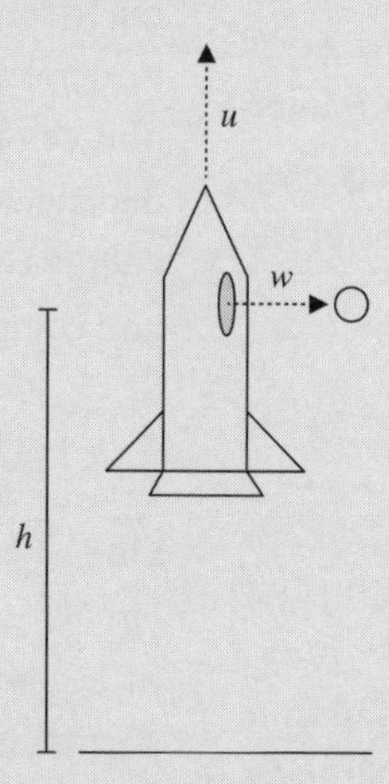

Figure 1.6 A ball is thrown horizontally at velocity w out of a rocket moving vertically upwards at constant velocity u.

In the frame of the rocket, the ball's initial y-velocity is zero, but in the ground frame its initial velocity is u, the relative velocity between the two reference frames. From our kinematic formula (1.2) above, we have for the y-component of the velocity

$$v_y = \sqrt{u^2 + 2gh}.$$

The x-component of the velocity is always the same, $v_x = w$, since no forces act in the x-direction, so we have a total speed

$$v = \sqrt{v_x^2 + v_y^2} = \sqrt{u^2 + w^2 + 2gh}.$$

at two very special places: the top and bottom of the circle, where gravity acts purely vertically, and thus radially, such that the object is instantaneously in uniform circular motion. At all other points in the orbit, other methods (such as energy conservation) must be used to find the velocity.

The centripetal force equation is not so interesting on its own, so a very common class of problems involves combining it with some other type of physics. A typical template might look roughly like this: A particle is moving in a circle. Identify the physics that is causing the centripetal force. Set the expression for this force equal to the centripetal force. Then solve for whatever quantity is requested. See Example 1.4.

1.2.2 Problems: Kinematics

1. A cannonball is fired with a velocity v. At what angle from the ground must the cannonball be fired in order for it to hit an enemy that is at the same elevation, but a distance d away?

   (A) $\arcsin(v/gd)$
   (B) $\arcsin(gd/(2v))$
   (C) $\arcsin(2gd/v)$
   (D) $(1/2)\arcsin(gd/v^2)$
   (E) $\arcsin(gd/v^2)$

2. A satellite (mass m) is in geosynchronous orbit around the Earth (mass M_E), such that its orbit has the same period as the Earth's rotation. If the Earth has angular rotational velocity ω, what is the radius of a geosynchronous orbit?

   (A) $\frac{GM_E}{\omega^2}$
   (B) $\frac{Gm}{\omega^2}$
   (C) $\left(\frac{GM_E}{\omega^2}\right)^{1/3}$
   (D) $\sqrt{\frac{GM_E}{\omega^2}}$
   (E) There is no possible geosynchronous orbit.

EXAMPLE 1.4

An electron (charge e) moves perpendicularly to a uniform magnetic field of magnitude B. If the kinetic energy of the particle doubles, by how much must the magnetic field change for the particle's trajectory to remain unchanged?

We know that the magnetic force on the electron is perpendicular to its motion (see Section 2.2.6 for a review), so it is a *centripetal* force, and the electron moves in a circle. More specifically, the forces are constant, so the electron executes uniform circular motion. Setting the magnetic and centripetal forces equal gives us

$$evB = \frac{mv^2}{r}.$$

Rearranging just a little, we can find the magnetic field

$$B = \frac{mv}{er}.$$

If the kinetic energy of the particle doubles, its velocity increases by $\sqrt{2}$, so B must increase to $\sqrt{2}B$ in order to maintain the same radius. This template occurs very frequently. Though circular motion can involve many different types of physics, identifying the centripetal force and setting it equal to mv^2/r will give you an additional equation to help solve the problem at hand.

1.3 Energy

Conservation of energy can be stated as follows:

> *If an object is acted on only by conservative forces, the sum of its kinetic and potential energies is constant along the object's path.*

Conservative forces are those for which the work done by the force is independent of the path taken between the starting and ending points, but the most useful definition (although it seems tautological) is *a force to which you can associate a (time-independent) potential energy.* The most common examples are gravity, spring forces, and electric forces. The most common example of a force that is *not* conservative, and probably the only such example you'll see on the GRE, is friction: an object traveling from point A to B and back to A will slow down due to friction the whole way through, even though the starting point is the same as the ending point.

A standard subset of GRE classical mechanics problems are most easily solved by straightforward application of conservation of energy. It's important to recognize these problems so you immediately jump to the fastest solution method, rather than fish around for the right kinematics formulas, so we'll state a general principle:

> *If you want to know how fast or how far something goes, use conservation of energy.*
>
> *If you want to know how much time something takes, use kinematics.*

It's baffling that this simple dichotomy isn't introduced in first-year physics courses. It's based on the idea that total energy is a combination of kinetic energies, which depend on velocities, and potential energies, which depend on positions. Setting $E_{\text{initial}} = E_{\text{final}}$ lets you solve for one in terms of the other, but nowhere in the equation does time appear explicitly. On the other hand, kinematics gives you explicit formulas for position and velocity as a function of time t (see equation (1.1)). Of course, some problems will require a combination of both methods, for example using conservation of energy to solve for a velocity which you then plug into a kinematics formula, but, as a very general rule, if time doesn't appear in the problem then you can leave kinematics out of the picture. However, we'll address a common exception to this rule at the end of Section 1.3.2.

1.3.1 Types of Energy

To begin with, you should know the following formulas *cold*:

$$\text{Translational kinetic energy:} \quad \frac{1}{2}mv^2 \qquad (1.5)$$

$$\text{Rotational kinetic energy:} \quad \frac{1}{2}I\omega^2 \qquad (1.6)$$

$$\text{Gravitational potential energy on Earth:} \quad mgh \qquad (1.7)$$

$$\text{Spring potential energy:} \quad \frac{1}{2}kx^2 \qquad (1.8)$$

Hopefully the standard notation is familiar to you: v is linear velocity, ω is angular velocity, m is mass, I is the moment of inertia, h and x are displacements, g is gravitational acceleration at Earth's surface (which should *always* be approximated to 10 m/s^2 on the GRE when numerical computations are required), and k is the spring constant. There are two important points to remember about potential energy:

- It is only defined up to an additive constant: we are free to choose the zero of potential energy wherever is most convenient, which is usually some physically relevant position such as the bottom of a ramp or the uncompressed length of a spring.
- It is measured from the *center of mass* of an extended object. The usefulness of the center of mass concept (see Section 1.4.4) is that it allows us to treat extended objects like point masses, with all their mass concentrated at the location of the center of mass.

There are other types of potential energy, but all can be summarized by a definition. For any conservative force $\mathbf{F}$, the change in potential energy ΔU between points a and b is

$$\Delta U = -\int_a^b \mathbf{F} \cdot d\mathbf{l}. \tag{1.9}$$

The line integral looks scary but it really isn't, since in all cases of interest the integral will be along the direction of the force vector. Probably the only time you might have to use this formula is if you can't remember the electrostatic or gravitational potential right away, so we'll do that example here. The gravitational force between two masses m_1 and m_2 is

$$\mathbf{F}_{\text{grav}} = \frac{Gm_1 m_2}{r^2}\hat{\mathbf{r}}. \tag{1.10}$$

You may have seen this equation in the form

$$\mathbf{F}_{\text{grav, 1 on 2}} = -\frac{Gm_1 m_2}{r^2}\hat{\mathbf{r}},$$

stating that the force on mass m_2 from m_1 points along the vector $\hat{\mathbf{r}}$ *from* m_1 *to* m_2, with the minus sign to indicate that the force is attractive. As we'll see, there are minus signs everywhere, so even though it's (deliberately) a bit ambiguous, we find (1.10) a more useful mnemonic for the GRE – just remember that gravity is attractive, and fill in the signs depending on which force (1 on 2 or 2 on 1) you're computing. See Example 1.5.

Alternatively, if you're given the potential, you can compute the force by inverting equation (1.9):

$$\mathbf{F} = -\nabla U. \tag{1.11}$$

Again, watch the minus sign!

1.3.2 Kinetic/Potential Problems

The simplest energy problem involves a mass on a ramp of some complicated shape, asking about its final velocity given that it starts at a certain height, or what initial height it will need to get over a loop-the-loop, or something like that. Because gravity is a conservative force, the shape of the ramp is irrelevant, as long as it's frictionless. If there's friction, then the shape of the ramp *does* matter because the work done by friction depends on the distance traveled – we'll get to that in a bit. First we'll look at a standard example.

EXAMPLE 1.5

Let's find the gravitational potential of a satellite of mass m in the gravitational field of the Earth, of mass M. The most common choice is to set the zero of potential energy at $r = \infty$, so the potential of the satellite at a finite distance r from the center of the Earth is

$$U(r) = -\int_\infty^r -\frac{GmM}{r'^2}\,dr' = -\left.\frac{GmM}{r'}\right|_\infty^r = -\frac{GmM}{r}.$$

Note the signs: the force on the satellite is directed *towards* the Earth, or in the $-\hat{\mathbf{r}}$ direction, but $d\mathbf{l} = +\hat{\mathbf{r}}\,dr'$, so the dot product is negative. The final sign makes sense because gravitational potential decreases (that is, becomes more negative) as the satellite gets closer to the Earth; in other words, it is attracted towards the Earth. Probably the most confusing part of this whole business is the signs, which the GRE *loves* to exploit. Rather than worrying about putting the signs in the right place throughout the whole problem, it may be best to just compute the unsigned quantity, then fill in the sign at the end with physical reasoning.

EXAMPLE 1.6

A block slides down a frictionless quarter-circle ramp of radius R, as shown in Fig. 1.7. How fast is it traveling when it reaches the bottom?

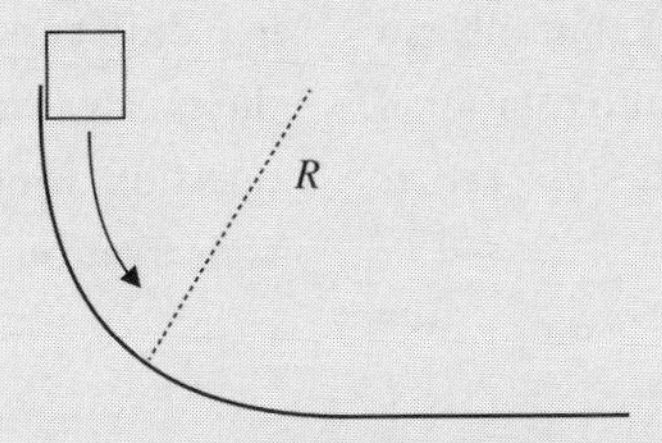

Figure 1.7 Block sliding down a quarter-circle ramp.

The quarter-circle shape is irrelevant except for the fact that it gives us the initial height: the block starts at height R above the bottom. At the top, the block is stationary, so its velocity is zero and there is no kinetic energy; all the energy is potential. Here the obvious choice is to set the zero of gravitational potential energy at the bottom of the ramp, so that the potential at the top is mgR. Wait a minute – the problem didn't tell us the mass of the block! Let's call it m, and see if we can resolve the situation as we finish the problem. At the bottom of the ramp, all the energy is kinetic, because we've defined the potential energy to be zero there. If the block's speed at the bottom is v, then its kinetic energy is $\frac{1}{2}mv^2$. We now apply conservation of energy:

$$0 + mgR = \frac{1}{2}mv^2 + 0$$
$$\implies v = \sqrt{2gR}.$$

Conveniently enough, the mass cancels out since both the kinetic and potential energies are directly proportional to m.

There are a couple things to note about Example 1.6:

- This was the very simplest version of the problem. The block could have had a nonzero speed at the top, in which case it would have had nonzero kinetic energy there. So don't automatically assume that conservation of energy is equivalent to "potential at top equals kinetic at bottom," which is *not* true in general!
- This problem can easily be extended to a kinematics problem by asking how far the block travels after it is launched off the bottom of the ramp, assuming the ramp is some height above the ground.[2] The first step of this problem would still be finding the initial velocity when it leaves the ramp, exactly as we found above.
- The fact that the mass cancels out is actually quite common in problems involving *only* a gravitational potential, since both kinetic and potential energies are proportional to m. So if the problem doesn't give you a mass, don't panic! That's actually a strong clue that the right approach is conservation of energy.

[2] Note that this is an exception to our rule about distances being associated with energy rather than kinematics: the block travels with constant horizontal speed once it leaves the ramp, so the only thing dictating how far it goes is the *time* it takes to fall vertically to the ground, which we must get from kinematics. So this is an exception only because it's actually a two-dimensional problem.

1.3.3 Rolling Without Slipping

A common variant of the above problem is a round object (sphere, cylinder, and so forth) rolling down a ramp. If the object *rolls without slipping*, then its linear velocity v and angular velocity ω are related by

$$v = R\omega, \tag{1.12}$$

where R is the radius. (Dimensional analysis dictates where to put the R so that v comes out with the correct units.) Then in addition to its kinetic energy, $\frac{1}{2}mv^2$, the object also has rotational kinetic energy $\frac{1}{2}I\omega^2$, where I is its moment of inertia. The rolling-without-slipping condition (1.12) lets you substitute v for ω and express everything in terms of v, after which you can solve for v exactly as above. Incidentally, it's *friction* that causes rolling without slipping, as friction is responsible for resisting the motion of the point of contact with the object so that it can instantaneously rotate around this pivot. In this situation friction does no work, but instead is responsible for diverting translational energy into rotational energy. Without friction, all objects would simply slide, rather than roll.

Rolling-without-slipping problems almost always boil down to the kinds of cancellations shown in Example 1.7: the kinetic energy is of the form αmv^2, with α some number that accounts for the moment of inertia. Here, α was 3/4 for the cylinder and 7/10 for the sphere. Notice that the problem didn't ask which object arrives *first*, only which object had the greater velocity at the bottom: the former is a kinematics question, which by our general principle can't be answered by conservation of energy alone.

EXAMPLE 1.7

A cylinder of mass m and radius r, and a sphere of mass M and radius R, both roll without slipping down an inclined plane from the same initial height h, as shown in Fig. 1.8. The cylinder arrives at the bottom with greater linear velocity than the sphere

(A) if $m > M$
(B) if $r > R$
(C) if $r > \frac{4}{5}R$
(D) never
(E) always

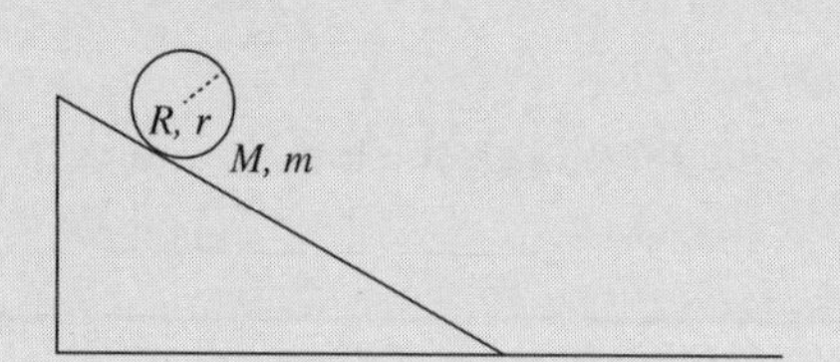

Figure 1.8 Ball or cylinder rolling down an inclined ramp.

You should immediately recognize that the mass is a red herring: since the moment of inertia is proportional to the mass, the same arguments as in Section 1.3.2 go through, and the mass cancels out of the conservation of energy equation for both objects. But let's see how this works explicitly. The moments of inertia are $\frac{1}{2}mr^2$ for the cylinder and $\frac{2}{5}MR^2$ for the sphere (neither of which you should memorize, since they're among the few useful quantities given in the table of information at the start of the test). The energy conservation equations read

$$mgh = \frac{1}{2}mv_{\text{cyl}}^2 + \frac{1}{2}\left(\frac{1}{2}mr^2\right)\omega_{\text{cyl}}^2 \qquad \text{(cylinder)},$$

$$Mgh = \frac{1}{2}Mv_{\text{sph}}^2 + \frac{1}{2}\left(\frac{2}{5}MR^2\right)\omega_{\text{sph}}^2 \qquad \text{(sphere)}.$$

As promised, we can cancel m from both sides of the first equation, and M from both sides of the second, which lets us equate the two right-hand sides. Now, substituting $\omega_{\text{cyl}} = v_{\text{cyl}}/r$ and $\omega_{\text{sph}} = v_{\text{sph}}/R$, we have

$$\left(\frac{1}{2} + \frac{1}{4}\right)v_{\text{cyl}}^2 = \left(\frac{1}{2} + \frac{1}{5}\right)v_{\text{sph}}^2.$$

The radii also cancel! So we can read off immediately that $v_{\text{cyl}} < v_{\text{sph}}$, and the cylinder *always* arrives slower, choice D.

1.3.4 Work–Energy Theorem

Since energy is conserved only if the forces acting in the problem are conservative, you might well ask how we can quantify the effects of nonconservative forces such as friction. The answer is simple: we just add a work term to one side of the energy balance equation:

$$E_{\text{initial}} + W_{\text{other}} = E_{\text{final}}, \tag{1.13}$$

where W_{other} is the work due to nonconservative forces. Because work is a signed quantity, the signs can get a little tricky, but you can usually figure them out just by reasoning logically. For example, friction always acts to oppose an object's motion, so the work done by friction is always negative, and this means that $E_{\text{final}} < E_{\text{initial}}$: the object is losing energy due to friction, as it should. You may also be used to seeing this equation in the form

$$W = \Delta\text{KE}. \tag{1.14}$$

Here, the right-hand side is the change in *kinetic* energy, while the left-hand side is the work done by *all* forces, including the conservative ones. The alternate form (1.13) simply absorbs the effect of conservative forces into the definition of the total energy by rewriting the work as a potential energy. Indeed, recall the general definition of work,

$$W = \int \mathbf{F} \cdot d\mathbf{l}, \tag{1.15}$$

which is of the same form as (1.9) up to a minus sign.

Example 1.8 works through a standard example, which can be tweaked in several ways to make it less straightforward:

- The quarter-circle ramp could have had a coefficient of friction as well. In that case, the frictional force would have varied at different points on the ramp. Note that we could *not* simply apply the formulas for uniform circular motion to determine the normal force, because the block is not moving with constant velocity (see the discussion in Section 1.2.1). This is actually a pretty interesting problem, but it requires solving an ugly differential equation for v, which is far beyond anything you'll see on the GRE.
- Similarly, the frictional surface might not have been flat, in which case the normal force at different points would also have changed.

But the problem we've solved is entirely typical of GRE problems, and illustrates possible shortcuts you should be on the lookout for. Never solve more of a problem than you absolutely must!

1.3.5 Problems: Energy

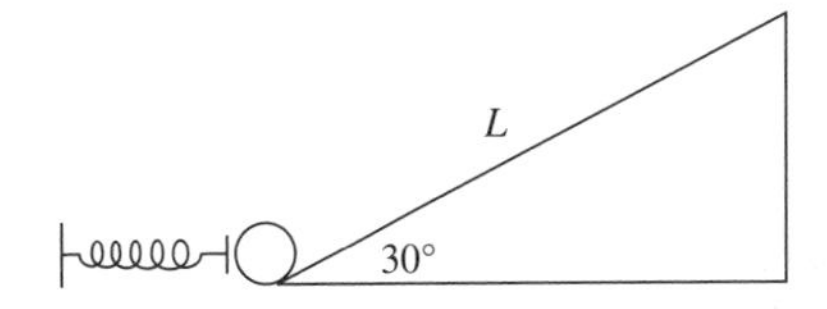

The following three questions refer to the diagram: a pinball machine launch ramp consisting of a spring of force constant k and a 30° ramp of length L.

EXAMPLE 1.8

Let's revisit the quarter-circle ramp problem (Example 1.6), but this time suppose that, after exiting the ramp, the block slides along a flat surface with coefficient of friction μ. How far down this surface does the block travel before it stops? Now, we could start where we left off, by using the speed $v = \sqrt{2gR}$ at the bottom of the ramp, then computing the kinetic energy, and continuing from there. But that would actually be too much work (no pun intended!). Instead, let's just apply the work–energy theorem directly. The block's initial energy is mgR. The frictional force is μmg along the flat surface (recall that μ is the proportionality constant between the normal force and the frictional force), and so after traveling a distance x, friction does work μmgx. When the block has stopped, its energy is zero. Applying the work–energy theorem, we have

$$mgR - \mu mgx = 0 \implies x = \frac{R}{\mu}.$$

That's it! We never even had to solve for the velocity at the bottom of the ramp. As a sanity check, we can examine the limiting cases $\mu \to 0$ and $\mu \to \infty$: as $\mu \to 0$, there is no friction, so the block never stops, and as $\mu \to \infty$, infinite friction means that the block stops right away. You can do a similar analysis for $R \to 0, \infty$.

1. You want to launch the pinball (a sphere of mass m and radius r) so that it just barely reaches the top of the ramp without rolling back. What distance should the spring be compressed? You may assume friction is sufficient that the ball begins rolling without slipping immediately after launch.

   (A) $\sqrt{\frac{2mgL^2}{5kr}}$

   (B) $\sqrt{\frac{mgL}{k}}$

   (C) $\sqrt{\frac{2mgL}{k}}$

   (D) $\sqrt{\frac{mgr}{k}}$

   (E) $\sqrt{\frac{2mgr}{k}}$

2. What is the ball's speed immediately after being launched?

   (A) $\sqrt{gL}$

   (B) $\sqrt{\frac{2}{5}gL}$

   (C) $\sqrt{\frac{5}{7}gL}$

   (D) $\sqrt{\frac{7}{10}gL}$

   (E) $\sqrt{\frac{10}{7}gL}$

3. Now suppose the ramp is waxed, so there is no friction. What is the distance the spring should be compressed this time?

   (A) $\sqrt{\frac{2mgL^2}{5kr}}$

   (B) $\sqrt{\frac{mgL}{k}}$

   (C) $\sqrt{\frac{2mgL}{k}}$

   (D) $\sqrt{\frac{mgr}{k}}$

   (E) $\sqrt{\frac{2mgr}{k}}$

1.4 Momentum

If you have gotten this far in physics, then you don't need a refresher on the physics of conservation of momentum. Some of the problems, however, take a bit of practice to learn to solve quickly. In general, you just need to remember that

Momentum is always conserved in a system in the absence of external forces.

This caveat about external forces is sometimes important: for example, if two balls collide in mid-air, the total horizontal momentum is conserved, but not the total vertical momentum because gravity is acting in that direction. In fact, the vertical momentum will continually increase in the downward direction according to Newton's second law $\mathbf{F} = \dot{\mathbf{p}}$. The trick with momentum problems is just to be sure that you are counting all types of momenta – linear and angular – and writing down the correct conservation equations.

1.4.1 Linear Collisions

This class of problem involves point particles that undergo collisions or explosions: for the purposes of the GRE,

If things are colliding, try conservation of momentum first.

Collisional forces can be arbitrarily complicated, but because they are all internal among the colliding particles, the total momentum is conserved as long as there are no additional external forces such as gravity. You just need to set the initial momentum equal to the final momentum and solve for the necessary variables. A special case is when the initial and final energies of the system are the same – this is known as an *elastic* collision, and imposing conservation of energy can give you an additional equation to solve (to find outgoing velocities, for example). Don't assume a collision is elastic unless you are explicitly told so, as this can lead to many trap answers. See Example 1.9.

Solving momentum conservation problems like this one invariably reduces to solving systems of linear equations. This often gets complicated, and if you're like most people, it is easy to make algebraic errors. Don't do it! Exhaust all limiting cases and dimensional analysis arguments before doing algebra. After this, if you think the algebra is easy, then do it. If you think it will be messy, just skip it and come back later. Since you may not even have time to finish the exam, triage is essential.

1.4.2 Rotational Motion and Angular Momentum

Like linear momentum problems, the game here is always to write the angular momentum in the initial state and in the

EXAMPLE 1.9

A ball of mass M strikes another ball of mass m initially at rest. The ball of mass M scatters at an angle θ relative to its initial direction. Suppose the ball of mass M initially has speed V, and both balls have a final speed v. What is the scattering angle ϕ of the ball of mass m, as defined in Fig. 1.9?

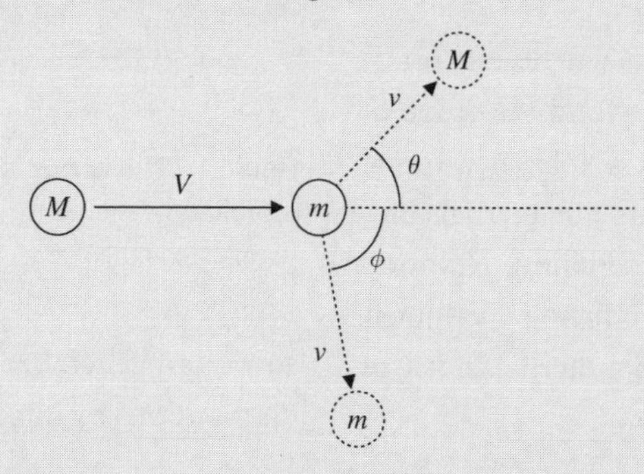

Figure 1.9 Collision of two balls in two dimensions.

Applying conservation of momentum in two dimensions, we get the two equations

$$MV = Mv\cos\theta + mv\cos\phi \qquad \text{(parallel to initial direction)},$$
$$0 = Mv\sin\theta + mv\sin\phi \qquad \text{(perpendicular to initial direction)}.$$

We know M, m, V, v, and θ here and we are solving for ϕ. Thus all we actually need is the second of the two equations, which gives us the result that

$$\phi = \arcsin\left(-\frac{M}{m}\sin\theta\right).$$

The minus sign makes good physical sense: if θ is positive, the mass-M ball goes up, giving a negative ϕ. The ball of mass m goes down, conserving momentum perpendicular to the initial direction. For practice, do a limiting-case analysis for the M and m dependence as well.

final state, then set the two equal. The angular momentum of a point particle of linear momentum $\mathbf{p}$ is defined by

$$\mathbf{L} = \mathbf{r} \times \mathbf{p}, \tag{1.16}$$

where $\mathbf{r}$ is the vector from a chosen reference point to the particle. Remember that rotational motion is always defined with respect to a reference point or axis. For an extended body we also have

$$\mathbf{L} = I\boldsymbol{\omega}, \tag{1.17}$$

where I is the moment of inertia and $\boldsymbol{\omega}$ is the angular velocity vector. Conceptually, I plays the same role as the mass m in the definition of linear angular momentum $\mathbf{p} = m\mathbf{v}$. Extending the metaphor, the analogue of force $\mathbf{F}$ for rotational motion is the torque

$$\boldsymbol{\tau} = \mathbf{r} \times \mathbf{F}. \tag{1.18}$$

Classic problems include merry-go-rounds and spinning disks. For instance, if a person jumps onto a spinning disk with a known moment of inertia, how does the rotational velocity change? Just equate $\mathbf{L} = I\boldsymbol{\omega}$ in the initial and final states.

We wrote angular momentum and torque in their vector form for completeness above. Note, however, that the vector form is only really needed for the definitions of $\mathbf{L}$ and $\boldsymbol{\tau}$. The analogues of the equations $\mathbf{p} = m\mathbf{v}$ and $\mathbf{F} = d\mathbf{p}/dt$ are only used on the GRE in their scalar forms:

$$L = I\omega, \tag{1.19}$$

$$\tau = \frac{dL}{dt}. \tag{1.20}$$

Problems involving angular momentum can also be conceptual, asking for the configuration of momentum, velocity, and acceleration vectors for a system involving rotational motion. The key point to remember is that the angular momentum vector **L** is generally *parallel*[3] to the angular velocity vector $\boldsymbol{\omega}$, which points along the axis of rotation, just like an object's linear momentum is parallel to its velocity. The direction of **L** is determined by the right-hand rule: curl the fingers of your right hand in the direction of rotation, and your thumb gives the direction of **L**.

More advanced classical mechanics texts will discuss rotating reference frames, which are mostly beyond the scope of the GRE. All you need to know is that a reference frame rotating at constant angular velocity $\boldsymbol{\Omega}$ is *not* inertial, but, nonetheless, one can write a formula resembling Newton's second law $F = ma$ at the price of introducing "fictitious" forces, which only appear because of the noninertial choice of coordinates:

$$F_{\text{centrifugal}} = -m\Omega^2 r, \tag{1.21}$$

$$F_{\text{Coriolis}} = -2m\boldsymbol{\Omega} \times \mathbf{v}. \tag{1.22}$$

The centrifugal force (which we emphasize once again is *not a real force*!) is the apparent force on an object in a uniformly rotating frame that pushes it away from the axis of rotation. The Coriolis force vanishes if the object is stationary in the rotating frame, but often appears in the context of the Earth's rotation, which defines a rotating frame. Unless the motion of the object is defined with respect to a rotating frame, in which case you typically need to use the Coriolis force, we recommend sticking with inertial frames to avoid confusion.

1.4.3 Moment of Inertia

As we've seen, an object's moment of inertia is analogous to its mass in the context of rotation, but, unlike mass, it depends on the distance from the center of rotation. Let's start with a point particle of mass m: the moment of inertia scales with the radius as

$$I = mr^2, \tag{1.23}$$

which accounts for the fact that, at fixed rotational frequency, the particle will have a higher linear velocity at higher radii. Thankfully, the moment of inertia of a system of many particles is just the sum of the individual moments of inertia,

[3] Strictly speaking, this is only true if the system is rotating about one of its principal axes. But these correspond to various axes of symmetry, and practically all the rotating objects you'll see on the GRE are symmetric to some extent and rotating about their axes of symmetry, so to the best of our knowledge you can ignore this subtlety for GRE purposes. The exception is in problems involving *precession*, which we haven't yet seen appear on the GRE.

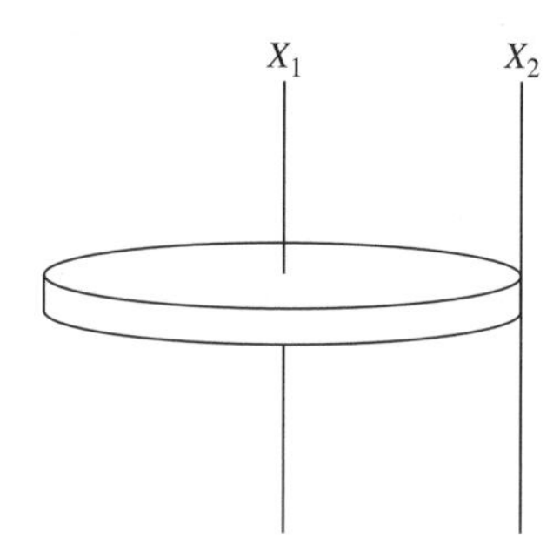

Figure 1.10 A penny is rotated about two axes X_1 and X_2. The moments of inertia for rotation about each axis are related by the parallel axis theorem.

so we can generalize to extended objects with arbitrary mass distributions by integrating:

$$I = \int r^2 \, dm, \tag{1.24}$$

where dm is an infinitesimal mass element (which can depend on position) and the integration is taken over the entire system. Conceptually, objects with more mass further from the axis of rotation are "harder" to rotate and have a larger moment of inertia.

Typically, the integral for moment of inertia is solved by changing integration variables to spatial variables, using the density $dm = \rho dV$. Note that, if you're given the density, for example $\rho = Ar^3$ for a sphere, you actually have to do *two* integrals: one to set the total mass m equal to $\int \rho dV$ to eliminate the constant A, and the other to compute the moment of inertia. The GRE does provide the formulas for the moments of inertia for rods, disks, and spheres on the equations page at the beginning of the test. Typically this is sufficient, though it is useful practice to compute these formulas.

The *parallel axis theorem* is a fast and frequently invaluable tool for computing the moment of inertia of systems built out of smaller pieces whose moments of inertia are known. If we know the moment of inertia I of a system of mass M rotating about an axis through its center of mass (CM), then its moment of inertia about any axis parallel to the CM axis is given by

$$I = I_{\text{CM}} + Mr^2, \tag{1.25}$$

where r is the distance between the CM axis and the parallel axis. For instance, the moment of inertia of a penny rotating about an axis perpendicular to the center of one of its faces is given on the GRE formula page as $I = (1/2)MR^2$. The moment of inertia of the penny when rotating about an axis that passes through the edge of the penny (see Fig. 1.10) would just be $I = (1/2)MR^2 + MR^2 = (3/2)MR^2$.

EXAMPLE 1.10

Consider a rod of mass M and length L whose density varies quadratically, $\rho(x) = Ax^2$, where A is a constant and x is the distance from the left end of the rod, as shown in Fig. 1.11. What is the position of the center of mass of the rod?

Figure 1.11 A rod of length L with a position-dependent density $\rho(x)$.

The total mass is

$$M = \int_0^L \rho(x)\,dx = \frac{1}{3}AL^3,$$

so $A = 3M/L^3$. The center of mass is then

$$x_{\text{CM}} = \frac{1}{M}\int_0^L x\rho(x)\,dx = \frac{1}{M}\frac{3M}{L^3}\int_0^L x^3\,dx = \frac{3}{L^3}\left(\frac{1}{4}L^4\right) = \frac{3}{4}L,$$

to the right of the center of the rod, which makes sense because the density increases from left to right.

1.4.4 Center of Mass

For reference, the center of mass of an extended object of mass M can be calculated similarly to the moment of inertia using

$$\mathbf{r}_{\text{CM}} = \frac{\int \mathbf{r}\,dm}{M}. \tag{1.26}$$

Instead of weighting the mass by the square of the distance from the axis of rotation, the center of mass weights the mass by the *displacement* from the origin. In principle, equation (1.26) is actually three integrals, one for each coordinate. The corresponding formula for a system of point masses just replaces the integral by a sum

$$\mathbf{r}_{\text{CM}} = \frac{\sum_i \mathbf{r}_i m_i}{M}. \tag{1.27}$$

In particular, for a single mass m at position $\mathbf{r}$, the center of mass is just $\mathbf{r}$, as it should be. See Example 1.10.

1.4.5 Problems: Momentum

1. What is the moment of inertia, about an axis through its center, of a sphere of radius R, mass M, and density varying with radius as $\rho(r) = Ar$?

   (A) $\frac{4}{3}MR^6$

   (B) $\frac{4}{9}MR^2$

   (C) $\frac{2}{5}MR^2$

   (D) $\frac{2\pi}{3}MR^2$

   (E) $\frac{4\pi}{5}MR^2$

2. A block explodes into three pieces of equal mass. Piece A has speed v after the explosion, and pieces B and C have speed $2v$. What is the angle between the directions of piece A and piece B?

   (A) π

   (B) $\pi - \arccos(1/2)$

   (C) $\pi - \arccos(1/3)$

   (D) $\pi - \arccos(1/4)$

   (E) 0

3. A disk of mass M and radius R rotates at angular velocity ω_0. Another disk of mass M and radius r is dropped on top of the rotating disk such that their centers coincide. Both disks now spin at a new angular velocity ω. What is ω?

   (A) $r^2\omega_0/(R^2 + r^2)$

   (B) $R^2\omega_0/(R^2 + r^2)$

   (C) $(R^2 + r^2)\omega_0/r^2$

   (D) $(R^2 + r^2)\omega_0/R^2$

   (E) ω_0

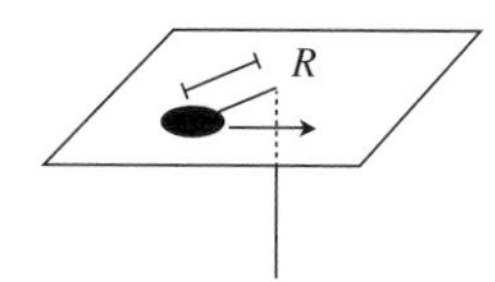

4. A small puck of mass M is attached to a massless string that drops through a hole in a platform, as shown in the diagram above. The puck rotates at radius R when the tension in the string is T. The string is pulled downwards until the radius of rotation is $r < R$. What is the change in energy of the puck when the radius is decreased? You may assume the puck is a point mass.

   (A) $(1/2)TR(R^2/r^2 - 1)$
   (B) $(1/2)TR(r^2/R^2 - 1)$
   (C) $(1/2)Tr(R^2/r^2 - 1)$
   (D) $(1/2)Tr(r^2/R^2 - 1)$
   (E) 0

1.5 Lagrangians and Hamiltonians

As you probably know, Lagrangian and Hamiltonian mechanics provide an elegant way of rewriting the results of classical mechanics, which are most useful for formal results and for dealing with systems with strange constraints, such as a particle confined to the surface of a sphere. Lagrangians and Hamiltonians are fascinating in their own right, and the basis for much of quantum mechanics and quantum field theory, but, as we'll emphasize throughout this book, almost none of this is relevant for the GRE. Instead, Lagrangian and Hamiltonian questions fall into just two categories:

- Write down the Lagrangian or Hamiltonian function.
- Write down the Lagrangian or Hamiltonian equations of motion.

Of course, there are sub-topics within each of these, most importantly conceptual questions dealing with conserved quantities, but these two topics cover all the important bases. Note in particular that *you don't have to solve the equations of motion.* While Lagrangians and Hamiltonians make it easy to write down the equations, they're typically horrible coupled differential equations with no easy solutions. This is sort of the idea of this formulation of mechanics: the actual coordinates of the particle as a function of time aren't so important, but instead one is concerned with properties of the motion (such as energy, momentum, and time dependence) which are easy to see in this framework.

Warning: what follows will be a *drastically simplified* version of Lagrangian and Hamiltonian mechanics. We are leaving out many subtleties and special cases, which are covered in standard treatments, but are not important for the GRE.

1.5.1 Lagrangians

The Lagrangian L of a system is a scalar function described by this absurdly simple formula:

$$L(q, \dot{q}, t) = T - U. \tag{1.28}$$

Here T is the total kinetic energy of the system, U is the potential energy, and q is the collection of all the coordinates describing the degrees of freedom of the system. Note the minus sign! The Lagrangian is *not* the total energy of the system. Note also that the Lagrangian is not only a function of the coordinates q, but also of the *velocities* $\dot{q}$. It is a peculiarity of the Lagrangian formalism that the coordinates and their time derivatives are considered as *independent* variables.

Let's discuss the coordinates in more detail, since that's where almost all of the difficulty of Lagrangians comes in. The power of Lagrangian mechanics lies in being able to choose coordinates to describe *only* the directions in which the system is allowed to move. For example, consider a particle of mass m attached to the end of a massless rod of length ℓ, which is free to rotate in a plane about a pivot (Fig. 1.12). The particle is not allowed to take on any old Cartesian coordinates (x, y), but is forced to move on a circle of radius ℓ. So the most convenient (read "correct") coordinate to use is the angular coordinate θ. But what is the kinetic energy in terms of θ? Here is where things get tricky. We'll now give a recipe for computing the correct expression for T for any question you'll see on the GRE:

- Write down expressions for the *Cartesian coordinates* in terms of your chosen coordinates q.
- Differentiate the Cartesian coordinates (x, y, z) with respect to time to get $(\dot{x}, \dot{y}, \dot{z})$, paying careful attention to the chain rule.
- Form the expression for the Cartesian kinetic energy, $T = \frac{1}{2}m(\dot{x}^2 + \dot{y}^2 + \dot{z}^2)$ for a point particle or $T = \frac{1}{2}I\omega^2$ for an extended object, as appropriate. For the latter, you'll need to express ω in terms of the velocities $\dot{q}$, but this is typically easy because you will have chosen coordinates such that $\dot{q}$ *is* ω.

EXAMPLE 1.11

Applying the Lagrangian recipe to Fig. 1.12, we first define our coordinates carefully: choose the origin of Cartesian coordinates (x, y) to be the pivot, so that $\theta = 0$ corresponds to the rod hanging straight down, with the mass at $(0, -\ell)$. We have

$$x = \ell \sin\theta,$$
$$y = -\ell \cos\theta,$$

and taking time derivatives, we get

$$\dot{x} = \ell \cos\theta\, \dot{\theta},$$
$$\dot{y} = \ell \sin\theta\, \dot{\theta}.$$

Notice how the chain rule gets applied to θ. Finally, we form the kinetic energy:

$$T = \frac{1}{2}m\left(\dot{x}^2 + \dot{y}^2\right) = \frac{1}{2}m(\ell^2 \cos^2\theta\, \dot{\theta}^2 + \ell^2 \sin^2\theta\, \dot{\theta}^2) = \frac{1}{2}m\ell^2\dot{\theta}^2.$$

Not surprisingly, we reproduce exactly the expression for the rotational kinetic energy of a point mass, $\frac{1}{2}I\omega^2$.

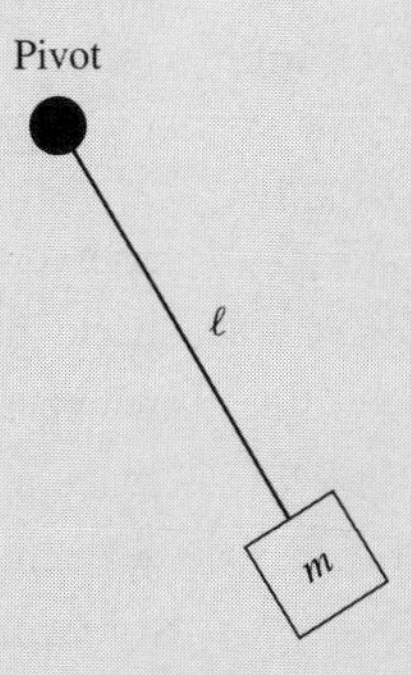

Figure 1.12 A mass m on the end of a rigid rod of length l that rotates about a pivot.

Example 1.11 is probably a simpler example than you'll see on the real exam, so you might have been able to write down the answer right away, but we can't emphasize strongly enough the importance of following this recipe. For example, if the pivot was sliding with velocity v, the total kinetic energy would *not* simply be the sum of the rotational and translational kinetic energies, but would contain an additional cross term $m\ell v \cos\theta\, \dot{\theta}$. You'll see examples of this in our practice tests and in those released by ETS.

1.5.2 Euler–Lagrange Equations

The Lagrangian is a useful quantity because one can derive the equations of motion directly from it. Unlike in Newtonian mechanics, where the equations of motion are the vector differential equations $\mathbf{F} = m\mathbf{a}$, in Lagrangian mechanics the equations of motion are *scalar* equations derived from the scalar quantity L. These equations are known as the Lagrangian equations of motion, or more commonly, the *Euler–Lagrange equations*,

$$\frac{d}{dt}\frac{\partial L}{\partial \dot{q}} = \frac{\partial L}{\partial q}, \tag{1.29}$$

one equation for each coordinate q.

There are several important things to note about the Euler–Lagrange equations. The first is that signs are *very easy* to mix up, so be careful! A good way to check is to make sure that these equations reduce to $F = ma$ for a particle moving in one dimension x in a potential $U(x)$. In that case, the Lagrangian is

$$L = \frac{1}{2}m\dot{x}^2 - U(x),$$

and we have

$$\frac{\partial L}{\partial \dot{x}} = m\dot{x}, \qquad \frac{\partial L}{\partial x} = -U'(x),$$

which gives

$$m\ddot{x} = -U'(x) = F,$$

as expected. (Of course, you still have to remember that $\mathbf{F} = -\nabla U$ with the correct sign, but you should know that already!) The second thing to note is that d/dt is a *total* time derivative, not a partial derivative, which is why it gives $\ddot{x}$ when applied to $\dot{x}$. There are extra terms where the kinetic term happens to have an explicit time dependence – it's unlikely you'll run across something like this on the GRE, but it's good to be careful in any case.

Writing down the correct Lagrangian is typically the hardest part of the problem. Once you have the Lagrangian, the equations of motion are easy, provided you're careful about taking derivatives. Speaking of derivatives, we already know from classical mechanics that quantities whose time derivatives are zero are important – these are *conserved* quantities. Looking at (1.29), we see that if the right-hand side $\partial L/\partial q$ is zero, the quantity $\partial L/\partial \dot{q}$ is conserved! Whether or not this quantity is conserved, it is so important it is given its own name:

$$p_i \equiv \frac{\partial L}{\partial \dot{q}} : \textit{momentum conjugate to } q. \tag{1.30}$$

This name arises because the conjugate momentum is *usually* some kind of momentum (linear or angular), but not always. To reiterate,

> *Iff the Lagrangian is independent of a coordinate q, the corresponding conjugate momentum $\partial L/\partial \dot{q}$ is conserved.*

(Here, "iff" means "if and only if," a shorthand reminder that the converse of the statement is true as well.) In this case, the coordinate q is called *cyclic*. Questions about conserved quantities in Lagrangian mechanics are very common on the GRE, so we've included several representative problems at the end of this section.

1.5.3 Hamiltonians and Hamilton's Equations of Motion

There is an alternative formulation of Lagrangian mechanics, called *Hamiltonian mechanics*, which on the surface is nothing more than a change of variables. As the name implies, the formalism depends on a quantity called the Hamiltonian, derived from the Lagrangian as follows:

$$H(p, q) = \sum_i p_i \dot{q}_i - L. \tag{1.31}$$

Here, i runs over all the coordinates q_i, and p_i is the momentum conjugate to q_i, as defined above in equation (1.30). Note that this relation can be inverted, to give $\dot{q}_i$ as a function of p_i and q_i. To construct H, we solve for $\dot{q}_i$ in this way and plug back into both terms on the right-hand side of (1.31), so that the final result is a function of the momenta p_i *rather than the velocities* $\dot{q}_i$.

OK, that was the textbook definition. On the GRE, if you're only asked for the Hamiltonian, you'd prefer not to take the time to first find the Lagrangian, solve for all the momenta, and only then construct H. With two slight restrictions, there is a much simpler definition:

$$H = T + U \quad \text{(if } U \text{ does not depend explicitly on velocities or time).} \tag{1.32}$$

So for all potentials U that only depend on coordinates, the Hamiltonian *is* the total energy, albeit expressed in terms of the funny position and momentum variables. For a simple example, let's consider the particle moving in one dimension again. As we derived above, $p_x = \partial L/\partial \dot{x} = m\dot{x}$, so

$$\dot{x} = \frac{p_x}{m}.$$

Assuming the potential is time independent, we have

$$H = T + U = \frac{1}{2} m \left(\frac{p_x}{m}\right)^2 + U(x) = \frac{p_x^2}{2m} + U(x),$$

an expression we will meet again in quantum mechanics. The tricky part about this formalism is once again the kinetic term, which *usually* takes the form of a momentum squared over twice a mass. In the case of *angular* coordinates, we usually see a moment of inertia in the denominator: you can work out for yourself that the Hamiltonian for a free particle moving in two dimensions in polar coordinates is

$$H = \frac{p_r^2}{2m} + \frac{p_\theta^2}{2mr^2},$$

with the promised moment of inertia mr^2 showing up in the denominator of the angular momentum term. But for complicated examples, you should still go through the first couple of steps of the Lagrangian construction, carefully identifying the kinetic terms. As long as the potential is velocity and time independent, as is true for all ordinary potentials, there is no need to construct the rest of the Lagrangian in order to calculate the momenta.

As with the Lagrangian, the Hamiltonian is a single scalar function encoding the equations of motion. But this time, we get a system of coupled *first*-order differential equations, as opposed to the second-order Euler–Lagrange equations. These are *Hamilton's equations*:

$$\dot{p} = -\frac{\partial H}{\partial q}, \qquad \dot{q} = \frac{\partial H}{\partial p}. \tag{1.33}$$

Again, the signs are tricky, but again, the same simple example of a particle in a one-dimensional potential will fix them for you. The first equation reduces in that case to

$$\dot{p}_x = -U'(x),$$

once again reproducing Newton's second law because p_x is precisely the linear momentum of the particle.

Finally, one can derive conservation laws from Hamilton's equations as well. Looking at the first equation, the momentum p is constant if $\partial H/\partial q = 0$. So we have the result:

Iff the Hamiltonian is independent of a coordinate q, the corresponding conjugate momentum p is conserved.

Incidentally, this tells you that if the Lagrangian is independent of q, so is the Hamiltonian, since the conjugate momentum is conserved in both cases.

1.5.4 Problems: Lagrangians and Hamiltonians

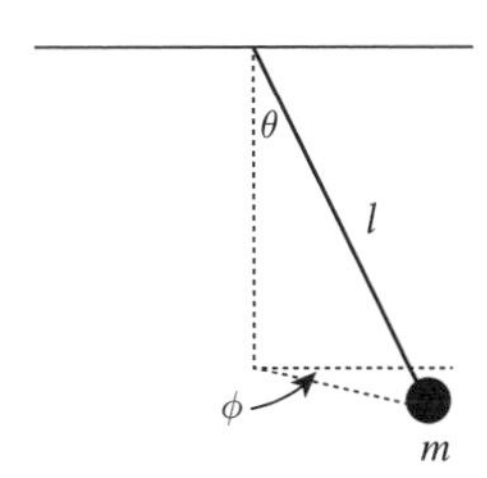

The following four questions all refer to a mass m suspended from a rigid massless rod of length l, but free to rotate otherwise (a spherical pendulum). One can take generalized coordinates θ and ϕ as shown in the figure.

1. Which of the following is a possible Lagrangian for the system?

   (A) $\frac{1}{2}ml^2\left(\dot{\phi}^2 + \dot{\theta}^2\right) - mgl\cos\theta$
   (B) $\frac{1}{2}ml^2\left(\dot{\phi}^2 + \sin^2\phi\dot{\theta}^2\right) + mgl\cos\theta$
   (C) $\frac{1}{2}ml^2\left(\dot{\theta}^2 + \sin^2\theta\dot{\phi}^2\right) - mgl\sin\theta$
   (D) $\frac{1}{2}ml^2\left(\dot{\theta}^2 + \sin^2\theta\dot{\phi}^2\right) + mgl\cos\theta$
   (E) $\frac{1}{2}ml^2\left(\dot{\theta}^2 + \sin^2\theta\dot{\phi}^2\right) - mgl\cos\theta$

2. Which of the following is a conserved quantity for the system?

   (A) $ml^2\dot{\phi}$
   (B) $ml^2\sin^2\theta\dot{\phi}$
   (C) $ml^2\dot{\theta}$
   (D) $mgl\cos\theta$
   (E) $mgl\sin\theta$

3. Which of the following is a possible Hamiltonian for the system?

   (A) $\dfrac{p_\theta^2}{2ml^2} + \dfrac{p_\phi^2}{2ml^2\sin^2\theta} - mgl\cos\theta$
   (B) $\dfrac{p_\theta^2}{2ml^2} + \dfrac{p_\phi^2}{2ml^2} - mgl\cos\theta$
   (C) $\dfrac{p_\theta^2}{2m} + \dfrac{p_\phi^2}{2ml^2} - mgl\cos\theta$
   (D) $\dfrac{p_\theta^2}{2m\sin\phi^2 l^2} + \dfrac{p_\phi^2}{2ml^2\sin^2\theta} - mgl\cos\theta$
   (E) $\dfrac{p_\theta^2}{2ml^2} + \dfrac{p_\phi^2}{2ml^2\sin^2\theta}$

4. Suppose the pendulum is confined to the plane $\phi = 0$. What is the Euler–Lagrange equation for θ?

   (A) $\dot{\theta} = \frac{g}{l}\sin\theta$
   (B) $\ddot{\theta} = -\frac{g}{l}\theta$
   (C) $\ddot{\theta} = -\frac{g}{l}\sin\theta$
   (D) $\ddot{\theta} = gl\sin\theta$
   (E) $\ddot{\theta} = -gl\cos\theta$

1.6 Orbits

The two fundamental forces of classical mechanics, gravity and electromagnetism, are remarkably similar: they both have $1/r^2$ force laws, but, perhaps more importantly, they are *central forces*, meaning that the force vector points along the line between the two interacting bodies. Without exception, these are the forces that appeared on recent GREs in the context of orbit problems, so we will confine our discussion of orbits to one where all forces are central and spherically symmetric. This means that the force can be derived from a potential function $U(r)$, which only depends on the radial distance between the two bodies; this is known as a *central potential.*

While one can discuss orbits quite straightforwardly using only the language of forces and Newtonian dynamics, our discussion will simplify *immensely* if we throw in a bit of Lagrangian mechanics. As a result, we strongly urge you to study Section 1.5 on Lagrangians and Hamiltonians carefully before reading this section: the material presented there should be more than sufficient to understand our treatment here.

1.6.1 Effective Potential

The fact that our potential has the form $U(r)$ immediately gives us conservation laws which we can put right to use. First, let's write down the Lagrangian for a particle of mass m moving in the potential U: after writing x, y, and z in spherical coordinates, we find

$$L = \frac{1}{2}m\dot{r}^2 + \frac{1}{2}mr^2\dot{\theta}^2 + \frac{1}{2}mr^2\sin^2\theta\dot{\phi}^2 - U(r).$$

(The polar angle θ shows up in the kinetic energy roughly for the same reason that it shows up in the spherical volume element $r^2 \sin\theta$.) Reverting to Newtonian reasoning for a bit, conservation of angular momentum implies the conservation of a whole *vector* $\mathbf{L}$ (whose magnitude is l), and the fact that the *direction* of this vector is constant means that *the particle is confined to a plane.* By spherical symmetry, we can choose this plane to be at $\theta = \pi/2$; the second term (involving $\dot{\theta}$) vanishes since θ is constant, and $\sin(\pi/2) = 1$ means the third term simplifies as well. We are left with the restricted form:

$$L = \frac{1}{2}m\dot{r}^2 + \frac{1}{2}mr^2\dot{\phi}^2 - U(r), \tag{1.34}$$

which we will use from now on. Now, since $U(r)$ is independent of the azimuthal angle ϕ, so is the Lagrangian, and that gives us conservation of the conjugate momentum to ϕ, which we identify as the ordinary angular momentum l:

$$l = mr^2\dot{\phi}. \tag{1.35}$$

The radial behavior of the orbit is of course described by the Euler–Lagrange equation for the radial coordinate:

$$\frac{d}{dt}(m\dot{r}) = mr\dot{\phi}^2 - U'(r).$$

Substituting for $\dot{\phi}$ in terms of l using (1.35), we get

$$m\ddot{r} = \frac{l^2}{mr^3} - U'(r).$$

First of all, we have reduced a complex system of partial differential equations in three dimensions to a single ordinary differential equation, which we may have some hope of understanding. And secondly, this looks suspiciously like Newton's second law of motion. We can improve the resemblance by "factoring" the derivative on the right-hand side to find

$$m\ddot{r} = -\frac{d}{dr}\left(\frac{l^2}{2mr^2} + U\right).$$

This is now in exactly the same form as Newton's second law, except that the "potential" now has an additional term depending on l. We call the expression in parentheses the *effective potential:*

$$V(r) = \frac{l^2}{2mr^2} + U(r). \tag{1.36}$$

Now we can draw potential energy graphs just as we would with an ordinary one-dimensional problem, remembering in the back of our minds that we're really dealing with an entire orbit, which also has some angular dependence. The effective potential formalism is most useful for dealing with *shapes* of orbits, $r(\phi)$, rather than time dependences, $r(t)$ and $\phi(t)$. Happily, the GRE only cares about orbit shapes, with one simple exception which we'll discuss when we come to Kepler's laws.

We should briefly mention a subtlety of the most common application of this formalism, namely two bodies orbiting each other under the influence of gravity. In that case, the bodies orbit about their mutual center of mass, but instead of dealing with two separate orbits, one can perform a coordinate transformation to describe the *relative* motion. In doing so, the mass m gets replaced by the *reduced mass,*

$$\mu = \frac{m_1 m_2}{m_1 + m_2}. \tag{1.37}$$

This is a great one to remember by dimensional analysis and limiting cases. The numerator has to contain the product in order for the whole thing to have dimensions of mass, and in the limit that $m_2 \to \infty$, we have $\mu \approx m_1$, corresponding to a center of mass that is very near the heavy body, m_2, which barely moves at all. So in a two-body problem, replace all instances of m by μ (that is, in *both* the kinetic energy and the effective potential), and you're good to go. The most common situation is the limit $m_2 \gg m_1$ just noted, so $\mu \approx m_1$ is usually a good approximation – but not always!

1.6.2 Classification of Orbits

Let's exploit conservation laws in order to learn something about possible orbits of objects in a central force. Now that we have defined the effective potential, we can define the total energy of the orbit:

$$E = T + V = \frac{1}{2}m\dot{r}^2 + \frac{l^2}{2mr^2} + U(r), \tag{1.38}$$

which is conserved if $U(r)$ is time independent. The most interesting part of this formula is the l-dependent term in the effective potential, whose $1/r^2$ dependence acts as a "centrifugal barrier," which imposes an infinite energy cost to get to $r = 0$ if the body has a nonzero angular momentum. To learn more, let's consider a sample shape for $V(r)$, as shown in Fig. 1.13, under the assumptions that $U(r) \to 0$ as $r \to \infty$ and that the centrifugal term dominates as $r \to 0$ (meaning that $U(r)$ must have a smaller power of r in the denominator than $1/r^2$, as is the case for gravity, where $U(r) \sim 1/r$).

Three representative orbit energies are marked, E_1, E_2, and E_3. An orbit with energy $E_1 > 0$ is unbound: the body comes in from infinity, "strikes" the centrifugal barrier, and "reflects" back out to infinity. An orbit with energy E_2 is bound, and has two "turning points," with a minimum distance r_1 and a maximum distance r_2; the body is always stuck between them.

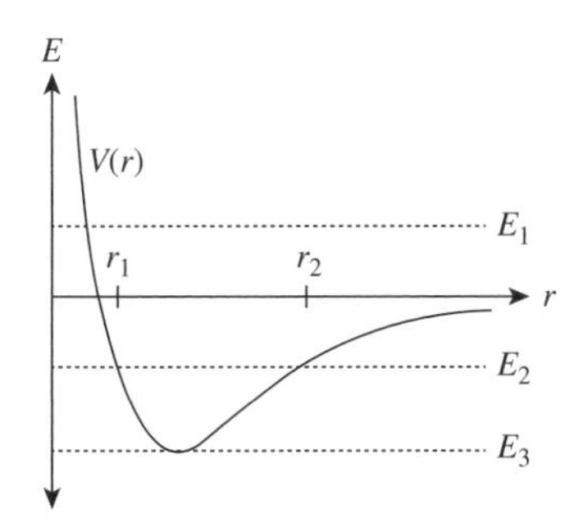

Figure 1.13 Effective potential with some representative orbit energies.

That is *not* to say that the orbit must be periodic: all it means is that, for a general potential, the body's orbit shape is enclosed within a ring-shaped region bounded by the circles of radii r_1 and r_2. The orbit E_3 is a special case, where we sit exactly at the minimum of the effective potential: then, there is not enough energy to change the value of r, so *the minimum of* $V(r)$ *corresponds to circular orbits*. To find the radius of these orbits, just solve $V'(r) = 0$ for r. You should also check for stability by ensuring that $V''(r) > 0$, otherwise we'd be sitting at an unstable *maximum*. Similarly, for more general orbit energies, we can read off the distance of closest approach by solving $E = V(r)$ for r.

We can be much more specific about orbit shapes in the case $U(r) = k/r$, as would be the case for a gravitational potential. Without getting into the details of the derivation, the results are

- $E > 0$: hyperbolic orbit
- $E = 0$: parabolic orbit
- $E < 0$: elliptical orbit

As in the more general discussion, since $U(r)$ falls off at large r, $E < 0$ corresponds to bound orbits, and for a $1/r$ potential these happen to take an elliptical shape. For $E = V_{\min}$, we have the special case of a circular orbit, which has the lowest possible energy.

1.6.3 Kepler's "Laws"

Yes, those scare quotes are there for a reason – Kepler's three "laws" are not really laws at all, in the sense of Newton's laws, which are *always* true (in the context of classical mechanics). Rather, they're three completely logically independent rules of thumb: one of them is almost trivial and totally general, and two of them are far too specific and only approximately true! In any case, here they are, in the traditional order:

I. Planets move in elliptical orbits with one focus at the Sun.
II. Planetary orbits sweep out equal areas in equal times.
III. If T is the period of a planetary orbit, and a is the semi-major axis of the orbit, then $T = ka^{3/2}$, with k the same constant for all planets.

Let's start with the first law. The first part is trivial: planets have bound orbits by definition, and, as we've seen above, a gravitationally bound orbit is either elliptical or circular, where a circular orbit can be considered a limiting case of an elliptical orbit where the two foci coincide. The second part is fairly difficult to derive, and, strictly speaking, it's not even true! Remember that two bodies orbit about their mutual center of mass – the precise statement is that the Sun and planet both undergo elliptical orbits, with a common focus of both ellipses located at their mutual center of mass. The Sun only sits still at the focus under the approximation that it is much heavier than any of the planets, such that the reduced mass is nearly equal to the planetary mass. We can even be a little sloppier, and say that as long as the center of mass of the Sun–planet system lies inside the Sun, the Sun is "at the focus" throughout its motion. Unfortunately, the GRE has been known to ignore this subtlety from time to time, and a question from a 2008 GRE suggested that the statement "the Sun is at one focus" is exactly true.

Kepler's second law is also known as *conservation of areal velocity*, and means that if you drew vectors from an orbiting planet to the Sun, at equal time intervals along the planet's orbit, the orbit would be sliced up into equal-area segments. In Fig. 1.14, if the two shown portions of the orbit are traversed in equal times, the regions marked A and B have equal areas.

In fact, this law is completely general, for *any central potential*, not just gravitational force laws, since it follows immediately from conservation of angular momentum. Recall the definition of l:

$$l = mr^2 \frac{d\phi}{dt}$$

$$\implies \frac{l}{m} dt = r^2 \, d\phi.$$

The expression on the right-hand side is precisely the area element in polar coordinates (up to a factor of 2), and l/m is constant, so integrating both sides gives us the second law.

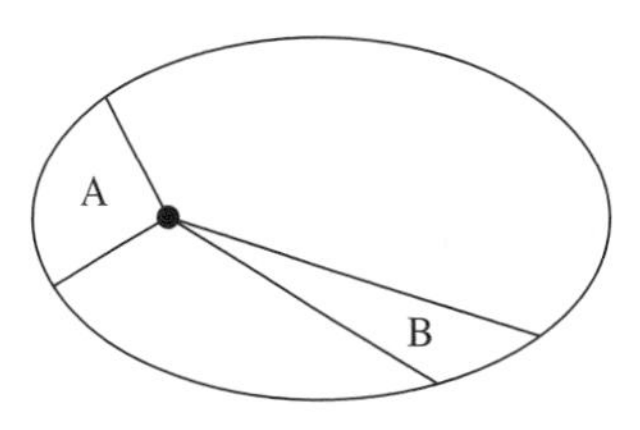

Figure 1.14 Areal sections illustrating Kepler's second law.

The third law is rather tricky to derive, so best to just memorize it. The proportionality between T and $a^{3/2}$ is exactly true for a pure $1/r$ potential. You probably won't need it, but the proportionality constant k is

$$k = \frac{2\pi}{\sqrt{G(m_{\text{planet}} + m_{\text{Sun}})}}.$$

Like the first law, the statement that k is the same for all planets is only true in the approximation that the Sun is infinitely massive; otherwise, the m_{planet} term in the denominator of k matters, and k then varies from planet to planet.

1.6.4 Problems: Orbits

1. A particle of mass m is attached to one end of a spring with spring constant k in a zero-gravity environment; the other end of the spring is attached to a fixed pivot. The spring's equilibrium position is fully compressed. The spring is stretched to a length r and the particle is given an initial angular momentum l. Which of the following is a possible value of r so that the spring stays at a constant extension r throughout the entire motion of the particle?

   (A) $(l^2/mk)^{1/4}$
   (B) $(l^2/mk)^{1/2}$
   (C) $\sqrt{mkl^2}$
   (D) ml^2/k
   (E) $(mk/l)^{1/3}$

2. Suppose a new planet were discovered orbiting the Sun, whose orbital period was exactly twice that of Mars. Assuming the new planet's mass is much smaller than the Sun's mass, which of the following must be true?

   I. The new planet's mass is exactly twice that of Mars.
   II. The major axis of the new planet's orbit is smaller than that of Mars.
   III. The major axis of the new planet's orbit is bigger than that of Mars.

   (A) I only
   (B) II only
   (C) III only
   (D) I and II
   (E) I and III

3. An asteroid of mass m orbits the Sun (mass M) on a parabolic trajectory. Which of the following relates its distance of closest approach d to its orbital velocity v at the point of closest approach? You may assume m is negligibly small compared to M.

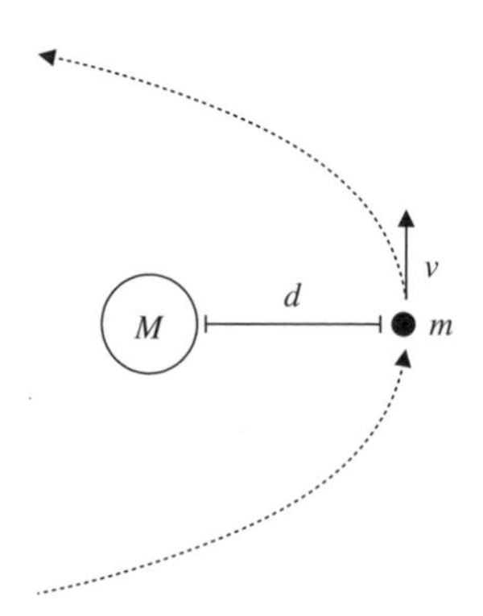

   (A) $d = \dfrac{GM}{v^2}$
   (B) $d = \dfrac{2GM}{v^2}$
   (C) $d = \dfrac{Gm}{v^2}$
   (D) $d = \dfrac{2Gm}{v^2}$
   (E) $d = \dfrac{GM}{2v^2}$

1.7 Springs and Harmonic Oscillators

Spring problems appear in many different forms, though they tend to use only a few basic facts. Obviously Hooke's law is the starting point for most problems:

$$F = m\ddot{x} = -kx, \tag{1.39}$$

where x is the displacement of the spring from equilibrium, and k is the spring constant. This is an ordinary differential equation describing a *harmonic oscillator* whose solutions are of the form $x(t) = A\cos(\omega t + \phi)$. The angular frequency is given by

$$\omega = \sqrt{\frac{k}{m}}. \tag{1.40}$$

Note that the amplitude A is *not* determined by Hooke's law, but is instead a constant of integration fixed by the initial conditions. The phase ϕ is the second constant of integration. Since complex exponentials

$$x(t) = Ae^{i\omega t}, \tag{1.41}$$

with A allowed to be complex, also satisfy Hooke's law, it is often easier to write the solutions in terms of complex numbers and take the real part at the end of the problem. This convenient shorthand will be elaborated in the chapter on waves, Section 3.1.2.

The potential energy of a spring of spring constant k at displacement x is given by

$$U = \frac{1}{2}kx^2. \qquad \text{(p.7)} \qquad (1.8)$$

You can prove this to yourself just by integrating the force from Hooke's law over the displacement of the spring to obtain the potential energy.

In fact, this is all that you really need to know for solving spring problems! Like much of the rest of mechanics, spring problems – at worst – just reduce to solving second-order ordinary differential equations. Simple as it sounds, this can sometimes be a time-consuming task, so you should use the potential energy considerations whenever possible. To summarize, the order of operations for spring problems should be:

1. Try limiting cases, dimensional arguments, and symmetry.
2. Try conservation of energy.
3. Try writing down a differential equation and solving it.

Even this last method should not be too bad for problems on the GRE, but consider it a last resort.

1.7.1 Normal Modes

Now suppose that, instead of having one body attached to a spring, we have two bodies, and we want to solve for the motion of both. This is complicated because we have a set of two coupled differential equations with a large family of solutions. It turns out that all of the solutions, however, are just superpositions of two basic solutions, called *normal modes*, which have the usual sinusoidal form of a harmonic oscillator. One very common GRE question is to ask for the normal modes of a system.

The mathematical setup is as follows. Consider the general case of n masses attached to springs, whose displacements[4] are q_k. Then the equations of motion for the entire system can be written

$$\sum_k (A_{jk}q_k + m_{jk}\ddot{q}_k) = 0.$$

This is just the most general linear combination of coordinates that we can form out of acceleration terms and forces from Hooke's law.

With great foresight, use the ansatz

$$q_k(t) = a_k e^{i\omega t}, \qquad (1.42)$$

whose form is inspired by (1.41). For this guess, $\ddot{q}_k = -\omega^2 q_k$, so the equations of motion reduce to

$$\sum_k (A_{jk} - \omega^2 m_{jk})a_k = 0.$$

This is just a matrix equation for the coefficients a_k, and linear algebra teaches us that in order for it to have a nontrivial solution the determinant must vanish:

$$\det(A_{jk} - \omega^2 m_{jk}) = 0. \qquad (1.43)$$

This *secular equation* defines a polynomial whose solutions give n frequencies ω_i, which are called the *normal frequencies* of the system. The normal modes alluded to above are just the solutions at which the *entire* system oscillates collectively at a fixed normal frequency, with the motion in time given by equation (1.42). A linear combination of solutions to an ordinary differential equation is still a solution, so we can build up more complicated motion by taking linear combinations of normal mode solutions with various coefficients.

Often, it's important to determine what kind of motion normal frequencies correspond to. For Example 1.12, we can immediately tell that the normal mode with frequency $\sqrt{k/m}$ will correspond to the two blocks moving in sync with fixed separation. In this case, the middle spring has no effect on the system, and the motion reduces to a single block–spring system with total mass $2m$ and total spring constant $2k$, thus giving $\omega = \sqrt{2k/2m} = \sqrt{k/m}$. (This should remind you of the discussion of blocks stuck together in Section 1.1.3.) We might guess that the other normal mode corresponds to motion of the blocks in exactly opposite directions. The frequency of this motion should be higher than $\sqrt{k/m}$ because the middle spring is now exerting an additional restoring force on the two blocks. Depending on answer choices given in the problem, these observations may be sufficient to pick the correct answer.

Solving for the normal modes of a system is a very common problem, in both quantitative and qualitative contexts. For a quantitative solution, the recipe is exactly as above:

- Write down the equations of motion for the system.
- Determine the matrices A_{jk} and m_{jk} and write down the secular equation (1.43).
- Solve to find the normal frequencies ω_i.

In many cases, this machinery is overkill, especially as most GRE problems are designed to be solved in about one minute. For most problems, you should ask yourself:

- What are the simplest ways that the system can oscillate at a fixed frequency? Drawing a picture is often helpful.

[4] In this section, by "displacement" we will always mean "displacement from equilibrium."

EXAMPLE 1.12

Two blocks of mass m are coupled to each other and to two walls by springs of spring constant k, as shown in Fig. 1.15. What are the normal modes of the system?

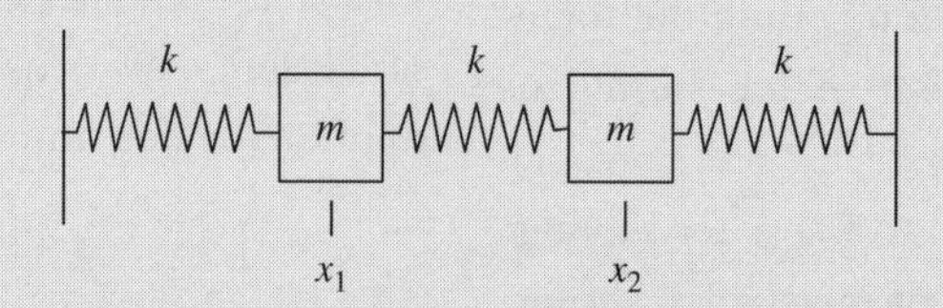

Figure 1.15 Two blocks and three springs.

We start by writing the equations of motion. Let x_1 and x_2 denote the displacements of the left and right blocks, respectively. The force on the left block is the force due to each of the adjacent springs, giving an equation of motion

$$m\ddot{x}_1 = -kx_1 - k(x_1 - x_2).$$

Similarly, for the right block, we have

$$m\ddot{x}_2 = -kx_2 - k(x_2 - x_1).$$

Using our guess (1.42) for x_1 and x_2, we find that the equations of motion become

$$m\omega^2 a_1 = 2ka_1 - ka_2,$$
$$m\omega^2 a_2 = 2ka_2 - ka_1.$$

Writing this explicitly in matrix form we have

$$\begin{pmatrix} 2k - m\omega^2 & -k \\ -k & 2k - m\omega^2 \end{pmatrix} \begin{pmatrix} a_1 \\ a_2 \end{pmatrix} = 0.$$

Notice that we didn't actually have to plug in the exponential guess solution to find this matrix equation; with practice, you can determine the matrices A_{jk} and m_{jk} just by staring at the equations of motion for x_1 and x_2, and jump straight to the secular equation (1.43). Taking the determinant of the matrix and setting to zero gives

$$(2k - m\omega^2)^2 - k^2 = 0.$$

Luckily, we don't even need the quadratic equation to solve this: rearranging and taking square roots gives

$$2k - m\omega^2 = \pm k,$$

handing us immediately the two solutions

$$\omega = \sqrt{\frac{k}{m}}, \sqrt{\frac{3k}{m}}.$$

- What is the normal frequency corresponding to the entire system oscillating together? A general rule of thumb is that *symmetric* motion will have lower frequency than asymmetric motion, with the most symmetric mode (corresponding to collective oscillations, as in the case above of the blocks moving in sync) having the lowest frequency. This is often $\sqrt{K/M}$, where K and M are the effective spring constant and mass of the entire system ($2k$ and $2m$ in the example above).
- Can this information be used to eliminate incorrect answers?

1.7.2 Damping, Driving, and Resonance

What happens when there is an oscillatory system with a damping force? One example would be a block–spring system placed underwater, where drag forces act to oppose the block's motion. As always in classical mechanics, our strategy is to write down the equations of motion of the system and find solutions. Damping terms such as drag or air resistance usually appear as a force proportional to the *velocity* of a particle, $F_{\text{damp}} = -b\dot{x}$, so the general equation of motion for an oscillator with damping is

$$m\ddot{x} + b\dot{x} + kx = 0. \tag{1.44}$$

There are three qualitative types of solutions: underdamped, critically damped, and overdamped. These are most conveniently expressed in terms of a damping parameter $\beta = b/2m$ and the natural frequency $\omega_0 = \sqrt{k/m}$. The underdamped solution, for $\beta^2 < \omega_0^2$, corresponds to motion in which the oscillations follow an exponentially decaying envelope. The equation of motion takes the convenient form

$$x(t) = Ae^{-\beta t}\cos(\omega_1 t - \delta),$$

where $\omega_1^2 = \omega_0^2 - \beta^2$. It is not critical to remember all of the constants in this expression, but it is valuable to remember that underdamped oscillations are a sinusoid with an exponentially decaying envelope. This should make intuitive sense: a small amount of damping will produce the same behavior as free oscillation, but with gradually decreasing energy (and thus amplitude) because of the damping force.

The overdamped solution, for $\beta^2 > \omega_0^2$, corresponds to motion in which the damping is so strong that no oscillation occurs. While the solutions to the underdamped case are sinusoidal functions (i.e. *complex* exponentials), the solutions to the overdamped oscillator are *real* exponentials, with the oscillator returning exponentially to its equilibrium position. The intermediate case between these two is called critical damping, and it corresponds to $\beta^2 = \omega_0^2$.

A harmonic system can also be driven by some external sinusoidal force. In this case, the equation of motion picks up an additional term due to the driving; in the most general case including possible damping,

$$\ddot{x} + 2\beta\dot{x} + \omega_0^2 x = A\cos\omega t.$$

The solutions to this equation of motion can be found using elementary methods for solving inhomogeneous ordinary differential equations. Without going into details, the important point is that, unlike the case of the oscillator with no driving force, the amplitude of the steady-state solution (that is, after a long time has elapsed) is not a free parameter but instead depends on the coefficient A and the frequency ω of the driving force. For a given A, the amplitude is maximized when the driving frequency equals the so-called *resonant frequency*:

$$\omega_R = \sqrt{\omega_0^2 - 2\beta^2}. \tag{1.45}$$

It is also possible to calculate the amplitude of oscillation when the driving frequency is different from resonance. It's extremely unlikely that you would ever need the exact expression for the GRE, but it could be worth remembering the scaling relation in the absence of damping. The amplitude D of an undamped oscillator of natural frequency ω_0 subject to a driving force at frequency ω is proportional as follows:

$$D \propto \frac{1}{|\omega_0^2 - \omega^2|}.$$

Note that this expression diverges at a driving frequency $\omega = \omega_0$! Of course, we don't see infinite amplitude in real-life oscillators because of small damping forces such as friction. But this proportionality should hold well near resonance in the weak damping limit.

To summarize, we have three different types of motion with three different characteristic frequencies, in order of increasing generality:

- Free oscillation: $\omega^2 = k/m$
- Damped oscillation:
  - overdamping
  - critical damping
  - underdamping, with characteristic frequency $\omega_1^2 = \omega_0^2 - \beta^2$
- Driven oscillation: $\omega_R^2 = \omega_0^2 - 2\beta^2$

1.7.3 Further Examples

There are many additional examples of the methods presented in this section. We show a few common examples. The basic unifying feature of all the examples in this section is that the behavior of the system can be described with a system of second-order linear ordinary differential equations with constant coefficients.

- **Pendulums.** In the limit of small displacements, the equation of motion of a pendulum of length L is described by simple harmonic motion. The full equation of motion for a pendulum with angular displacement θ is (see problem 4 in Section 1.5.4)

  $$mL\ddot{\theta} = -mg\sin\theta.$$

  For small displacements, this becomes

  $$m\ddot{x} = -mgx/L, \tag{1.46}$$

  which describes simple harmonic motion of angular frequency

  $$\omega = \sqrt{\frac{g}{L}}. \tag{1.47}$$

A similar equation holds for an extended object of mass m swinging on a massless rod of length R, which has moment of inertia I about the pivot. More precisely, R is the distance between the center of mass of the object and the pivot. The oscillation frequency, easily derived using Lagrangian mechanics, is now

$$\omega = \sqrt{\frac{mgR}{I}}, \tag{1.48}$$

which is easy to remember using dimensional analysis once you know equation (1.47).

- **Circuits.** The structure of the differential equations describing damped and driven oscillations is identical to the differential equations describing electrical circuits under the replacements in Table 1.1. All differential equations remain perfectly valid for electrical systems after these substitutions, with the technical caveat that the quantities in the table refer to the effective quantities for the *whole* circuit or system, not individual circuit elements or springs. This shouldn't be too surprising since, while one can add resistors in series or in parallel, it doesn't make much sense to "add damping resistance in parallel." In particular, while it is true that $1/k_{\text{eq}} = C_{\text{eq}}$, one *cannot* just replace all individual capacitors by springs, though there is a simple rule for equivalent spring constants for springs in series and parallel that mirrors the rule for capacitors. See below for more details.
- **Parallel and series springs.** Pursuing the electrical analogy, we can consider springs attached to a block of mass m in both parallel and series configurations; see Example 1.13.

EXAMPLE 1.13

When the springs are in parallel (Fig. 1.16, left), the equation of motion is

$$m\ddot{x} = -kx - kx = -2kx,$$

so the effective spring constant is the sum of the two spring constants $k + k = 2k$. When the springs are in series (Fig. 1.16, right), the situation is slightly more complicated. Call the displacement of the (massless) joint between the springs x_1, and call the displacement of the block x_2. To determine the equations of motion, we can use a trick: pretend a very small mass is attached to the joint between the two springs. In this case, we just have a two-mass/two-spring system, very similar to the one we already solved in Section 1.7.1! Simply copying down the first equation of motion and sending the small mass to zero gives

$$0 = -kx_1 - k(x_1 - x_2).$$

Notice that this is no longer a differential equation, but an algebraic constraint equation, $x_1 = x_2/2$. This enforces the condition that there must be zero force acting on the joint between springs; otherwise, since the joint is massless, the acceleration would be infinite. The block just experiences a force due to the spring touching it, so its equation of motion is

$$m\ddot{x}_2 = -k(x_2 - x_1).$$

Plugging in the constraint, we get

$$m\ddot{x}_2 = -\frac{k}{2}x_2,$$

giving an effective spring constant of $k/2$.

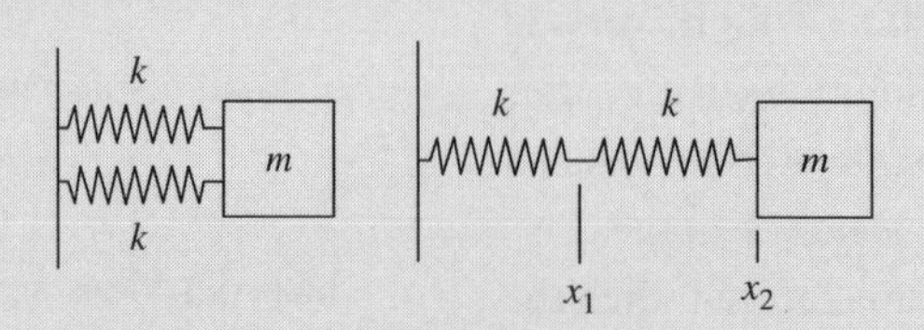

Figure 1.16 Springs in series and parallel.

Table 1.1 Correspondence of quantities in the analogy between electrical and mechanical systems.

Mechanical system	Circuit system
x displacement	q charge
$\dot{x}$ velocity	I current
m mass	L inductance
b damping resistance	R electrical resistance
$1/k$ spring stiffness	C capacitance
F amplitude of driving force	V amplitude of driving voltage

If you go through the derivation in Example 1.13 with two different spring constants k_1 and k_2, you can prove to yourself that the rule for computing equivalent spring constants is $1/k_{eq} = 1/k_1 + 1/k_2$, the same as the rule for computing equivalent capacitances for capacitors in series, $1/C_{eq} = 1/C_1 + 1/C_2$.

This leads to a subtle pitfall: the electrical analogue of our series spring system above is two capacitors in series. If we were to replace each spring k by a capacitor C, the rule of Table 1.1 gives $1/k \to C$. But the equivalent capacitance of this system is $C/2$, and the equivalent spring constant is $k/2$, so the rule $1/k_{eq} \to C_{eq}$ becomes $2/k \to C/2$, or $1/k \to C/4$! So the electrical–mechanical analogue is not as simple as just replacing $1/k$ by C everywhere in the circuit. To keep things straight, it's best to remember the two rules as logically distinct:

- The rule for adding springs k in series and parallel is the same as adding capacitors C in series and parallel.
- The electrical analogue to a mechanical system with equivalent spring stiffness $1/k_{eq}$ is an equivalent capacitor C_{eq}.

1.7.4 Problems: Springs

1. A block of mass M is attached to two springs, both with spring constants k, in series. Another block of mass m is attached to three springs, each of spring constant k, in parallel. What is the ratio of the oscillation frequency of the block of mass M to the frequency of the block of mass m?

   (A) $\sqrt{3m/(2M)}$
   (B) $\sqrt{2M/m}$
   (C) $\sqrt{3M/m}$
   (D) $\sqrt{m/(6M)}$
   (E) $\sqrt{6m/M}$

2. A ball of mass m is launched at 45° from the horizontal by a spring of spring constant k which is depressed by a displacement d. What horizontal distance x does the ball travel before returning to its height at launch?

   (A) $2mgd/k$
   (B) mgd/k
   (C) $kd^2/(gm)$
   (D) $kd^2/(2gm)$
   (E) $2kd^2/(gm)$

3. Suppose a motor drives a block on a spring at a frequency ω, and the natural frequency of the spring–block system is ω_0. If damping is negligible, by what factor does the amplitude of oscillation change when the driving frequency is increased from $\omega = 2\omega_0$ to $\omega = 3\omega_0$?

   (A) 4/9
   (B) 3/8
   (C) 2/3
   (D) 9/64
   (E) 9/4

1.8 Fluid Mechanics

In general physics courses, fluid dynamics problems typically appear only in two simple forms. The first is in applications of Bernoulli's principle – essentially a reformulation of conservation of energy. The second uses the concept of buoyant forces.

1.8.1 Bernoulli's Principle

Consider a fluid that is traveling through some pipe. The pipe may go up and down, and it may change diameter. Regardless, the following quantity is constant along a streamline of the fluid moving through the pipe:

$$\frac{v^2}{2} + gz + \frac{p}{\rho} = \text{constant}, \tag{1.49}$$

where v is the velocity of the fluid, g is gravitational acceleration, z is the height of a point along the streamline, p is the

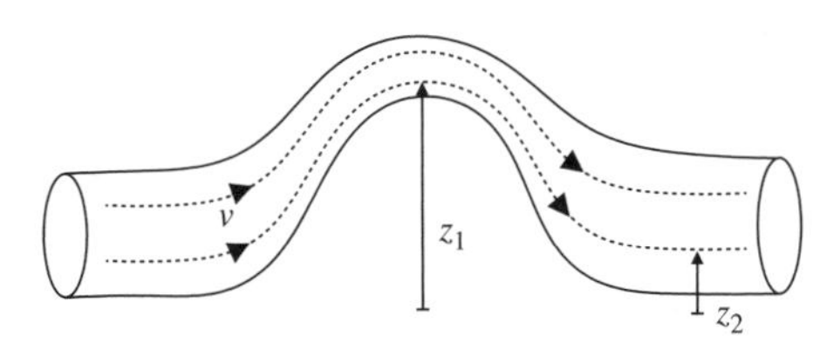

Figure 1.17 General setup for the Bernoulli equation, describing a fluid traveling through a pipe of variable size and height. Dashed lines represent streamlines.

EXAMPLE 1.14

To see how Bernoulli's principle works in context, consider a horizontal square pipe with two sections: one of side length a, and another of side length b (Fig. 1.18). The cross-sectional areas of the two pipes are a^2 and b^2, respectively. The amount of fluid flowing past a point in the first section in a time Δt must equal the amount of fluid flowing past a point in the second section, since the fluid can't appear or disappear. Mathematically, we require the following *fluid conservation equation* to hold:

$$\rho v_1 A_1 \Delta t = \rho v_2 A_2 \Delta t,$$

or just

$$v_1 A_1 = v_2 A_2, \tag{1.50}$$

where A_1 and A_2 are the cross-sectional areas of the two sections of pipe. Note that the density entering into both sides of the above equation is the same because of our assumption that the fluid is incompressible. It doesn't matter what the pressure is – an incompressible fluid will always have the same density. In our specific case, we have

$$v_1 a^2 = v_2 b^2.$$

From Bernoulli's principle, we know that we must have

$$\frac{v_1^2}{2} + gz_1 + \frac{p_1}{\rho} = \frac{v_2^2}{2} + gz_2 + \frac{p_2}{\rho}. \tag{1.51}$$

Since the pipe is horizontal (except in the region between the two pipe sections), streamlines are horizontal and $z_1 = z_2$. Substituting equation (1.50) into equation (1.51), we find that

$$\frac{v_1^2}{2} + \frac{p_1}{\rho} = \frac{v_1^2 a^4}{2b^4} + \frac{p_2}{\rho}.$$

If we know the pressure and velocity in the first part of the pipe, we can now calculate the pressure in the second part of the pipe:

$$p_2 = \frac{\rho v_1^2}{2}\left(1 - \frac{a^4}{b^4}\right) + p_1.$$

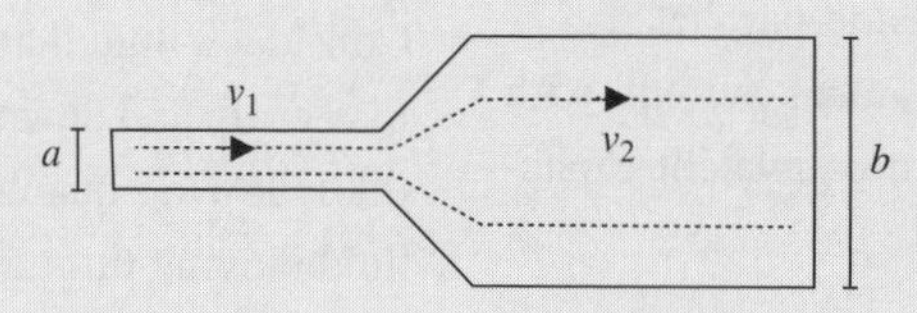

Figure 1.18 Fluid flows through two horizontal segments of square pipe, with side lengths a and b.

pressure, and ρ is the density of the fluid. Nearly all common fluids are *incompressible*, in the sense that ρ is constant for any reasonable range of pressures p. You can remember this equation by noting its relation to conservation of energy. The first term is a form of kinetic energy, the second term is a form of gravitational potential energy, and the third term is an energy associated with pressure. Since the first two terms are just the usual kinetic and potential energy terms with the mass divided out, the units of the constant on the right-hand side of (1.49) must be energy per unit mass, a fact that can help you remember the form of the third term. Problems involving Bernoulli's principle typically just require applying this relation to two different points along a streamline, often in conjunction with a fluid conservation equation.

From Example 1.14, we can abstract the general strategy for problems involving Bernoulli's principle:

- Write down the Bernoulli equation at all relevant points for the system.
- Write down the equation for fluid conservation.
- Solve these equations for the desired variables.

Remember that not all problems need both the fluid conservation and the Bernoulli equation. If you need to know something about pressure, for example, you'll certainly need Bernoulli, but maybe not the conservation equation. If you need to know something about the velocity of the fluid, the reverse could be true.

1.8.2 Buoyant Forces

Consider a block that floats in water (Fig. 1.19). Obviously the force of gravity pushes down on it, yet it does not sink. This is, of course, because of the buoyant force of the water that the block displaces. If the block displaces a volume of V of water when it is floating, then the buoyant force pushing back up on it will be

$$F = \rho V g, \tag{1.52}$$

where ρ is the density of water (or whatever fluid the block is floating in). You can think of this as the weight of the displaced water pushing up on the submerged object. The mass of the displaced water is ρV, and so its weight is $\rho V g$. Problems involving floating and submerged blocks can usually be solved by simply assigning all forces as usual, and then adding in the buoyant force which pushes upward on the object.

For example, suppose you blow all the air out of your lungs and sink to the bottom of a pool. How much do you weigh underwater? Suppose you weigh 60 kg and your volume is 50 L. You are displacing 50 kg of water, so there is about 500 N of buoyant force pushing up on you. But your weight is 600 N, so your net weight is only 100 N. Quite an effective weight-loss program!

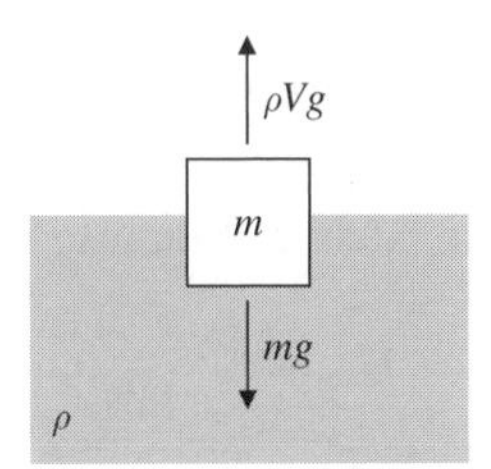

Figure 1.19 Example of buoyant forces. A block floating in water displaces a mass ρV of water, whose gravitational force pushes up against the weight of the block.

1.8.3 Problems: Fluid Mechanics

1. A diver in water picks up a lead cube of side length 10 cm. How much force is needed to lift the cube? The density of lead is approximately 11 g/cm^3.

   (A) 10 N
   (B) 11 N
   (C) 50 N
   (D) 100 N
   (E) 110 N

2. A vertical cylindrical tube has radius 1 cm. The tube is plugged with a stopper at the bottom end and filled with water so that the top of the water is 1 m above the bottom stopper. What is the frictional force required to hold the stopper in the tube?

   (A) 3.1 N
   (B) 2.4 N
   (C) 1.0 N
   (D) 0.62 N
   (E) 0 N

3. An aqueduct consisting of a pipe filled completely with water passes up a hill that is 10 m high. At the bottom of the hill, a flowmeter measures the speed of the fluid to be 2.0 m/s. At the top of the hill, a flowmeter measures the speed of the fluid to be 1.0 m/s. Which is closest to the difference in the fluid pressure between the bottom and top of the hill?

   (A) 1.50×10^3 Pa
   (B) 9.85×10^4 Pa
   (C) 1.00×10^5 Pa
   (D) 1.02×10^5 Pa
   (E) 9.85×10^7 Pa

1.9 Solutions: Classical Mechanics

Blocks

1. B – The happy fact that the plane is at a 45° angle, and $\sin 45^\circ = \cos 45^\circ$, means that we don't have to be especially careful about decomposing forces since the sines and cosines will always be equal. We know that the applied force contributes a normal force of $5\sqrt{2}$ N and a force *up* the ramp of $5\sqrt{2}$ N, and, similarly, gravity (approximating $g \approx 10$ m/s^2) contributes a normal force of $25\sqrt{2}$ N and a force *down* the ramp of $25\sqrt{2}$ N. So before considering

friction, the net force is $20\sqrt{2}$ N *down* the ramp. Now, friction contributes a force $(0.5)(30\sqrt{2}) = 15\sqrt{2}$ N opposing the block's motion, which means up the ramp in this case. The total net force is then $5\sqrt{2}$ N down the ramp. Dividing by the mass to find the acceleration, we have $\sqrt{2}$ m/s^2 down the ramp, choice B.

2. E – As explained in the text, we could either do separate free-body diagrams for each of the three blocks and solve for the tensions one by one, or we could use some physical intuition and realize that each rope segment must support the full weight of all the blocks below it. So, as long as the blocks have nonzero masses, $T_1 > T_2 > T_3$, *always*.
3. D – As always, we begin by separating the two blocks and drawing free-body diagrams as above. Here, F_{bb} is the block-on-block force, which is equal in magnitude for the two blocks by Newton's third law. Note that the normal force, which provides the friction keeping the mass m stationary, is the block-on-block force, *not* its weight! The plan is to solve for F_{bb} using the second block, then plug that in and solve for F using the first block. For the vertical forces to balance, we need $\mu_2 F_{bb} = mg$, so $F_{bb} = mg/\mu_2$. The acceleration of the second block is $a = F_{bb}/m = g/\mu_2$. This must equal the acceleration of the first block for the two to move together. Now, the normal force on the first block is $Mg + \mu_2 F_{bb} = (M+m)g$, where the $\mu_2 F_{bb}$ term is from the action–reaction pair of the frictional force on m. The net force on the first block is then $F - F_{bb} - \mu_1(m+M)g = F - mg/\mu_2 - \mu_1(m+M)g$, so we set its acceleration equal to g/μ_2 and solve:

$$\frac{F - mg/\mu_2 - \mu_1(m+M)g}{M} = \frac{g}{\mu_2}$$
$$\Longrightarrow F = \left(\mu_1 + \frac{1}{\mu_2}\right)(m+M)g.$$

The limiting cases check out: if $\mu_2 \to 0$, $F \to \infty$ since a vanishing frictional force between the two blocks means that they slip no matter what, and if $\mu_1 \to \infty$, $F \to \infty$

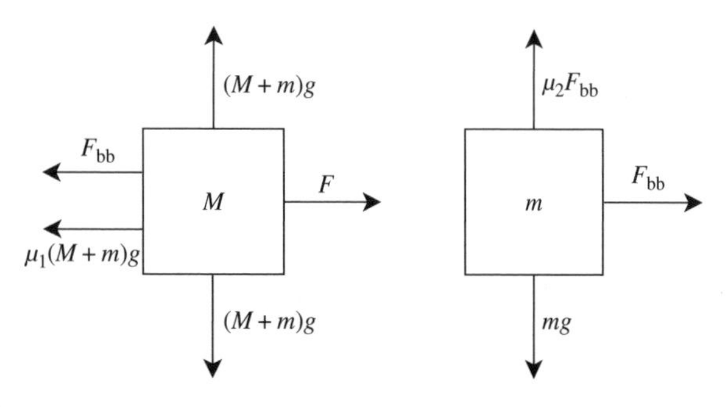

Figure 1.20 Solution for block problem 3.

since the strong kinetic friction prevents block M from accelerating enough and providing a large enough F_{bb}.

Alternatively, by the reasoning of Section 1.1.3, the fact that the two blocks don't slip means that they move as a composite system with mass $m + M$, and so the total force required to make them move together should depend on the combination $m + M$, which singles out choice D. The GRE may not always be this kind to you, but if you see a major simplification that allows you to solve a problem without drawing free-body diagrams, take it! If you have time at the end of the test, of course, come back to the problem and do it the long way to make sure that you didn't oversimplify.

Kinematics

1. D – The projectile will land on the ground at a time given by the solution to
$$0 = -\frac{1}{2}gt^2 + vt\sin\theta.$$
The solution is
$$t = \frac{2v\sin\theta}{g}.$$
The distance away at time t is given by
$$vt\cos\theta = d.$$
Substituting the result above, we find that
$$\cos\theta\sin\theta = \frac{gd}{2v^2}.$$
Using the double angle formula, we conclude that
$$\theta = \frac{1}{2}\arcsin\frac{gd}{v^2}.$$
2. C – In a geosynchronous orbit, the satellite will orbit with the same angular velocity as the Earth, which is constant, so its orbital speed is constant and we can use the uniform circular motion formulas. Just equate the centripetal force to the gravitational force to obtain
$$\frac{GM_E m}{r^2} = \frac{mv^2}{r}.$$
The velocity of the satellite at radius r is $v = r\omega$. Thus, we have
$$\frac{GM_E m}{r^2} = mr\omega^2.$$
Solving for r, we find that
$$r = \left(\frac{GM_E}{\omega^2}\right)^{1/3}.$$

Energy

1. B – From the 30 – 60 – 90 triangle, the ramp has height $L/2$. Setting the zero of gravitational potential at the bottom of the ramp, the initial energy is all potential energy of the spring, $\frac{1}{2}kx^2$. At the top of the ramp, we have potential $mgh = mgL/2$ and *that's it*: if it just barely reaches the top of the ramp, its kinetic energy (translational *and* rotational) must be zero there. So we have simply

$$\frac{1}{2}kx^2 = \frac{mgL}{2} \implies x = \sqrt{\frac{mgL}{k}}.$$

2. C – Now there is no change in potential energy immediately after being launched, so the ball's energy is all kinetic. Because it rolls without slipping, we have to take into account translational and rotational energies, using the appropriate moment of inertia for a sphere:

$$T = \frac{1}{2}mv^2 + \frac{1}{2}I\omega^2 = \frac{7}{10}mv^2.$$

(See the calculation at the end of Section 1.3.3 for more details.) We may be tempted to set this equal to the potential energy of the spring, but that's more work than necessary: since energy is conserved everywhere, we can feel free to use the simpler expression $mgL/2$ for the energy at the top. This gives

$$\frac{7}{10}mv^2 = \frac{mgL}{2} \implies v = \sqrt{\frac{5}{7}gL}.$$

3. B – Without doing any work, we know the answer here must be the same as the answer to problem 1. While it's true that, during the trip up the ramp, the ball's kinetic energy will no longer be shared between rotational and translational, this is irrelevant once it gets to the top of the ramp, when all the energy is potential. So the spring is compressed by precisely the same distance. Once again, the *time* it takes the ball to get up the ramp will be different than in the case with friction, but the problem doesn't ask about that, so we need not worry about it.

Momentum

1. B – It is important to keep the notation straight in this problem. The r in the mass density $\rho(r)$ refers to the distance between the origin and a point in the sphere, but when we compute the moment of inertia, the argument of the integral is the square distance from *a point in the sphere to the axis of rotation* – in this case, we can choose coordinates to make it the z-axis. If we use $s = r\sin\theta$ to denote this distance, then the moment of inertia is

$$\begin{aligned} I &= \int s^2 dm \\ &= \int r^2 \sin^2\theta \, \rho(r) \, dV \\ &= \int_0^{2\pi}\int_0^{\pi}\int_0^{R} Ar^3 \sin^2\theta \, r^2 \sin\theta \, dr \, d\theta \, d\phi \\ &= \frac{1}{6}AR^6 \int_0^{2\pi}\int_0^{\pi} \sin^3\theta \, d\theta \, d\phi \\ &= \frac{\pi}{3}AR^6 \int_0^{\pi} \sin^3\theta \, d\theta \\ &= \frac{4\pi}{9}AR^6. \end{aligned}$$

As usual, we get rid of A by expressing it in terms of the total mass of the sphere. Since the density depends only on radius, we can integrate spherical shells of thickness dr and mass $4\pi r^2 \, dr\rho(r)$, so the only integral is the radial one:

$$M = 4\pi \int_0^R Ar(r^2 \, dr) = \pi AR^4 \implies A = \frac{M}{\pi R^4}.$$

Plugging this in, we get $I = \frac{4}{9}MR^2$, which has the correct units for a moment of inertia.

2. D – Since the momentum must sum to zero, and since all the masses of the pieces are equal, the velocity vectors must also sum to zero. This means that the velocity vectors can be arranged as the sides of an isosceles triangle. Solving the rest of the problem is just basic geometry. The angle between the long and short sides of the isosceles triangle is given by $\cos\theta = (v/2)/(2v) = 1/4$, so $\theta = \arccos(1/4)$. The angle between the outward-going velocity vectors of the exploding fragments is then $\pi - \arccos(1/4)$.
3. B – Straightforward conservation of angular momentum. The initial angular momentum is $L_i = \frac{1}{2}MR^2\omega_0$. The final angular momentum is $L_f = \frac{1}{2}M(R^2 + r^2)\omega$. Solving for ω, we find choice B.
4. A – Call the initial angular velocity ω_0 and the final angular velocity ω. Angular momentum is conserved because tension acts radially and hence provides no torque, so we have the constraint

$$MR^2\omega_0 = Mr^2\omega,$$

so

$$\omega = \frac{R^2}{r^2}\omega_0.$$

The initial tension must be equal to the centripetal force, so

$$T = MR\omega_0^2.$$

The change in energy is just

$$\begin{aligned}\Delta E &= \frac{1}{2}M\left(r^2\omega^2 - R^2\omega_0^2\right) \\ &= \frac{1}{2}M\left(\frac{TR^3}{Mr^2} - \frac{TR}{M}\right) \\ &= \frac{1}{2}TR\left(\frac{R^2}{r^2} - 1\right).\end{aligned}$$

Note that this is a case in which energy is not conserved, but angular momentum is. Work is required to pull the string downwards and change the radius of the rotation, hence the energy increases.

Lagrangians and Hamiltonians

1. D – Following the recipe outlined in Section 1.5.1, we define the origin of coordinates to be at the pivot, giving

$$\begin{aligned}x &= l\sin\theta\cos\phi, \\ y &= l\sin\theta\sin\phi, \\ z &= -l\cos\theta.\end{aligned}$$

Note the minus sign in front of the z-coordinate! These are not quite spherical coordinates because of the way the angle θ was defined by the problem. This is conventional for pendulums though, so it's good to get used to it. Now, taking time derivatives, we have

$$\begin{aligned}\dot{x} &= l\cos\theta\cos\phi\,\dot\theta - l\sin\theta\sin\phi\,\dot\phi, \\ \dot{y} &= l\cos\theta\sin\phi\,\dot\theta + l\sin\theta\cos\phi\,\dot\phi, \\ \dot{z} &= l\sin\theta\,\dot\theta.\end{aligned}$$

We now form the kinetic energy:

$$T = \frac{1}{2}m(\dot{x}^2 + \dot{y}^2 + \dot{z}^2) = \frac{1}{2}ml^2\left(\dot\theta^2 + \sin^2\theta\dot\phi^2\right).$$

The algebra really isn't all that bad once you realize that the cross term cancels in $\dot{x}^2 + \dot{y}^2$, and all the rest collapse using the trig identity $\sin^2\alpha + \cos^2\alpha = 1$. The potential energy is all gravitational,

$$U = mgz = -mgl\cos\theta,$$

so we have

$$L = T - U = \frac{1}{2}ml^2\left(\dot\theta^2 + \sin^2\theta\dot\phi^2\right) + mgl\cos\theta.$$

Note that this is not the unique Lagrangian, since we can always add a constant to U, and hence to L, without changing the physics. But it matches one of the answer choices, so we can move on.

2. B – We notice that the Lagrangian is independent of the coordinate ϕ, which means the conjugate momentum p_ϕ is conserved:

$$p_\phi = \frac{\partial L}{\partial\dot\phi} = ml^2\sin^2\theta\dot\phi.$$

Happily, this matches answer choice B. When looking for conserved quantities in problems like this, it's a *much* better idea to find them yourself and check them against the answer choices, rather than trying to ascertain whether each of the answer choices individually is conserved. Incidentally, the only other conserved quantity in this problem is the total energy, since L is independent of time.

3. A – Here we can apply the trick mentioned in Section 1.5.3: rather than compute the Hamiltonian using the Legendre transform, just use the fact that $H = T+U$ since there is no time dependence. We've already computed one canonical momentum, so we need the other:

$$p_\theta = ml^2\dot\theta.$$

This gives us

$$\begin{aligned}\dot\theta &= \frac{p_\theta}{ml^2}, \\ \dot\phi &= \frac{p_\phi}{ml^2\sin^2\theta},\end{aligned}$$

and plugging into T gives

$$T = \frac{p_\theta^2}{2ml^2} + \frac{p_\phi^2}{2ml^2\sin^2\theta}.$$

Adding, rather than subtracting, the potential energy this time gives us the Hamiltonian:

$$H = T + U = \frac{p_\theta^2}{2ml^2} + \frac{p_\phi^2}{2ml^2\sin^2\theta} - mgl\cos\theta.$$

4. C – Restricting to a constant value of ϕ means that we can drop the $\dot\phi$ term from the Lagrangian:

$$L_{\phi=0} = \frac{1}{2}ml^2\dot\theta^2 + mgl\cos\theta.$$

Now we compute the two quantities we need for the Euler–Lagrange equations:

$$\begin{aligned}\frac{\partial L}{\partial\dot\theta} &= ml^2\dot\theta, \\ \frac{\partial L}{\partial\theta} &= -mgl\sin\theta.\end{aligned}$$

Applying the Euler–Lagrange equation (1.29), we have

$$ml^2\ddot\theta = -mgl\sin\theta,$$

or canceling ml from both sides and rearranging,

$$\ddot\theta = -\frac{g}{l}\sin\theta.$$

Expanding $\sin\theta \approx \theta$ for small θ, you should recognize this as the simple harmonic oscillator equation. Indeed, this is just an ordinary pendulum, with the correct angular frequency $\sqrt{g/l}$.

Orbits

1. A – Don't let the complicated-seeming problem statement fool you: this is just an application of the orbit formalism to the central potential $U(r) = \frac{1}{2}kr^2$ for a spring. We're looking for the radius of a circular orbit, so we solve $V'(r) = 0$:

$$\frac{-l^2}{mr^3} + kr = 0 \implies r = (l^2/mk)^{1/4}.$$

We can check that $V''(r) = k + 3l^2/mr^4 > 0$, so this is indeed a stable circular orbit. Incidentally, since all of the answer choices have different units, this would have been a perfect chance to practice using dimensional analysis.

2. C – This is a straightforward application of Kepler's third law. In the limit where planetary masses are small compared to that of the Sun, we have

$$\frac{T_{\text{new}}}{T_{\text{Mars}}} = \left(\frac{a_{\text{new}}}{a_{\text{Mars}}}\right)^{3/2} \implies a_{\text{new}} = a_{\text{Mars}} \times 2^{2/3}.$$

$2^{2/3} > 1$, so only III is true. In this limit the planet masses don't show up explicitly in Kepler's third law, so I is not necessarily true.

3. B – A parabolic orbit means that the total energy of the orbit is zero. Because m is negligibly small compared to M, we can use m instead of the reduced mass μ in the formulas for angular momentum and effective potential. Right at the point of closest approach, $\dot{r} = 0$, so all the motion is tangential and we can write $l = mvd$. Substituting into the expression for orbital energy (1.38), we have

$$E = \frac{(mvd)^2}{2md^2} - \frac{GMm}{d} = \frac{1}{2}mv^2 - \frac{GMm}{d}.$$

Setting this equal to zero and solving for d, we find choice B.

Springs and Harmonic Oscillators

1. D – Computing equivalent spring constants is discussed in Section 1.7.3. The equivalent spring constant for the block of mass M attached to the series springs is

$$k_{\text{eq}} = \left(\frac{1}{k} + \frac{1}{k}\right)^{-1} = \frac{k}{2}.$$

The equivalent spring constant for the block of mass m attached to the three parallel springs is

$$k_{\text{eq}} = k + k + k = 3k.$$

The ratio of the frequencies is therefore

$$\frac{\sqrt{k/(2M)}}{\sqrt{3k/m}} = \sqrt{\frac{m}{6M}}.$$

2. C – The kinetic energy of the ball at launch is equal to the potential energy stored in the spring, so

$$\frac{1}{2}mv^2 = \frac{1}{2}kd^2,$$

and

$$v = \sqrt{\frac{k}{m}}d.$$

The velocity in both the vertical and horizontal directions is therefore

$$v_0 = \frac{\sqrt{2}}{2}\sqrt{\frac{k}{m}}d.$$

The horizontal displacement is $x = v_0 t$, and the vertical displacement is $y = -(1/2)gt^2 + v_0 t$. Solving the latter equation for t at $y = 0$, we find

$$t = \frac{2v_0}{g}.$$

Substituting this into the equation for the horizontal displacement, we find

$$x = \frac{2v_0^2}{g} = \frac{kd^2}{gm}.$$

3. B – Recall that the amplitude of oscillation of an undamped driven oscillator scales as

$$A \sim \frac{1}{|\omega_0^2 - \omega^2|}.$$

Thus, as we increase ω from $2\omega_0$ to $3\omega_0$, the amplitude is multiplied by a factor of 3/8.

Fluid Mechanics

1. D – The mass of the lead cube is 11 kg, so the weight of the cube is 110 N. The volume of the cube is 10^3 cm^3, so the mass of the displaced water is 1 kg, since the density of water is 1 g/cm^3. The force required to lift the cube is just the weight minus the buoyant force from the displaced water: $F_{\text{lift}} = 110\text{ N} - 10\text{ N} = 100\text{ N}$.

2. A – By the Bernoulli equation, the pressure exerted by the fluid on the bottom stopper must be $p = \rho g y$, where y

is the level of the water line. (Atmospheric pressure acts equally at the top and at the bottom, so it cancels out.) The frictional force is therefore $F = pA = \pi \rho g y r^2$. Plugging numbers we have

$$F = \pi (10^3 \text{ kg/m}^3)(10 \text{ m/s}^2)(1 \text{ m})(10^{-2} \text{ m})^2 \approx 3.1 \text{ N}.$$

3. B – By Bernoulli's principle, we have

$$\frac{1}{2}\rho v_t^2 + \rho g y_t + p_t = \frac{1}{2}\rho v_b^2 + \rho g y_b + p_b,$$

where the left-hand side contains quantities measured at the top of the hill, and the right-hand side contains quantities measured at the bottom of the hill. We can rearrange and solve for the difference between the pressures at the top and bottom:

$$p_b - p_t = \rho g(y_t - y_b) + \frac{1}{2}\rho(v_t^2 - v_b^2).$$

Plugging in $y_t - y_b = 10$ m, $\rho = 1000$ kg/m^3, $g = 10$ m/s^2, $v_t = 1$ m/s, and $v_b = 2$ m/s, we find the result 9.85×10^4 Pa.

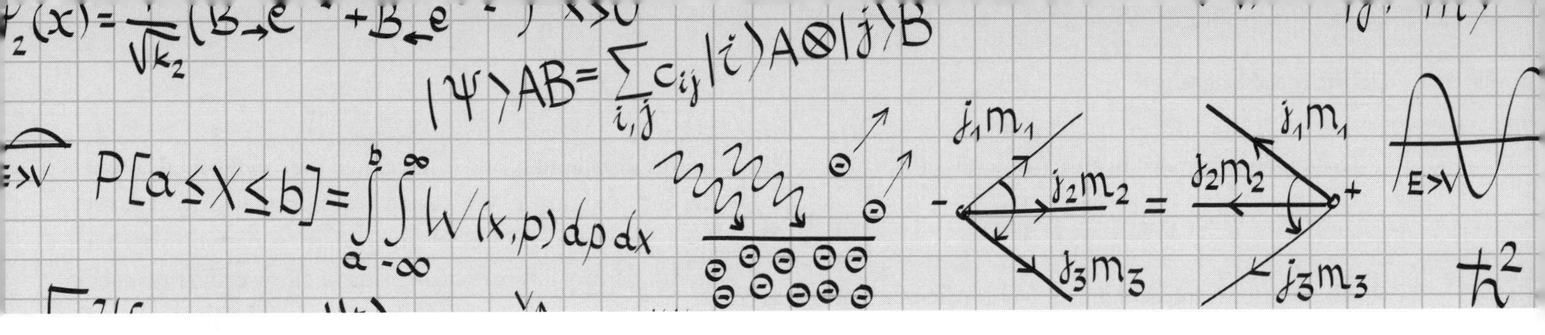

2 Electricity and Magnetism

According to ETS, approximately 18% of the exam covers electromagnetism. Given the format of the exam, the questions tend to emphasize a broad set of topics in electricity and magnetism rather than deep theoretical issues or analysis of complicated charge configurations. You will, for example, find plenty of AC and DC circuits in addition to the more traditional topics of electro- and magnetostatics, induction, Maxwell's equations, and electromagnetic waves.

The general philosophy that you should take away from this chapter is that electromagnetism is conceptually simple. There are only a few key concepts, such as symmetry and boundary conditions, most of which are buried somewhere in Maxwell's equations. The vast majority of the work lies in figuring out how to apply these basic ideas to specific configurations. The key to success on the electromagnetism problems on the GRE is to develop enough intuition with a few specific classes of problems that you can quickly deploy the necessary equations to solve them. We'll try to outline the general ideas concisely and try to illustrate how to choose the fastest methods for the common classes of problems.

2.1 Electrostatics

Electro*statics* refers to problems where charges and fields are not moving or changing in time. If you have a configuration of charges that does not change in time, electrostatics lets you calculate resulting electric fields and forces. We'll generalize this to include time dependence in Section 2.3.

2.1.1 Maxwell's Equations for Electrostatics

The tools needed for analyzing electrostatics problems are very simple. In fact, it just boils down to the first two of Maxwell's equations in free space in the absence of magnetic fields:

$$\nabla \cdot \mathbf{E} = \frac{\rho}{\epsilon_0}, \tag{2.1}$$

$$\nabla \times \mathbf{E} = 0 \text{ (electrostatics).} \tag{2.2}$$

The first equation actually holds true regardless of whether a magnetic field or time dependence is present, but the second is only true in electrostatics. These two equations tell you how to build the electric field $\mathbf{E}$ from a charge distribution ρ. To figure out how a test charge q moves in response to an electric field,[1] you need one more equation,

$$\mathbf{F}_E = q\mathbf{E}. \tag{2.3}$$

That's it! In principle, you can solve for any static electric field with these equations and then compute the motion of a particle in the field, given knowledge of a charge distribution. As you're probably well aware, however, partial differential equations are almost always quite nasty to solve, so we use a few additional tools to calculate the electric field in practice.

2.1.2 Electric Potential

The first tool is the scalar electric potential. Under some fairly general conditions, the fact that $\nabla \times \mathbf{E} = 0$ implies that there is a *scalar potential* field $V(\mathbf{r})$, which we also call the *electric potential* such that

$$\mathbf{E} = -\nabla V. \tag{2.4}$$

[1] This is still electrostatics, since the *field* isn't moving, only the test charge, whose own field we don't care about in this context.

This can be integrated to give V in terms of $\mathbf{E}$,

$$V(b) = -\int_a^b \mathbf{E} \cdot d\mathbf{l}, \tag{2.5}$$

which hopefully looks familiar from potential energy in mechanics. Here the point a has been defined as the zero of potential; as is the case with the potential energy in classical mechanics, electric potential must always be defined relative to some reference location. For electrostatics problems, unless otherwise specified, the reference point is often taken to be infinitely far from the location of interest. As with potential energy in mechanics: *mind the sign!* There is a relative minus sign between the electric field and the electric potential, which is easy to forget. Also note that V is *not* a potential energy; rather, it's a potential energy *per unit charge*, and the real potential energy of a particle of charge q in a region with electric potential V is $U = qV$. Because it is proportional to the true potential energy, the electric potential is also only defined up to a constant, so changing V by a constant value does not change the value of the electric field. Thus it is only *differences* in potential that we measure. Finally, since the potential $V(\mathbf{r})$ is a scalar quantity, not a vector like the electric field, it is sometimes easier to calculate the potential for a particular charge configuration and then convert to the electric field. For practical purposes, both contain equivalent information.

At this point we should mention the fact that both $\mathbf{E}$ and V are *additive*: Maxwell's equations are linear differential equations, so to find $\mathbf{E}$ or V of several charges, you just add up (using vector addition in the first case, and ordinary addition in the second) the corresponding $\mathbf{E}$ and V of the individual charges.

The second tool for solving electrostatics problems is the direct solution of Maxwell's equations. If we plug (2.4) into (2.1), we obtain the single scalar equation

$$\nabla^2 V = -\frac{\rho}{\epsilon_0}, \tag{2.6}$$

known as *Poisson's equation*. Often we are asked to solve this equation in a region where the charge density ρ is zero, in which case it reduces to *Laplace's equation*,

$$\nabla^2 V = 0 \qquad \text{(empty space)}. \tag{2.7}$$

We can write the general solution to Poisson's equation as an integral over the charge distribution $\rho(\mathbf{r})$:

$$V(\mathbf{r}) = \frac{1}{4\pi\epsilon_0} \int \frac{\rho(\mathbf{r}')}{|\mathbf{r} - \mathbf{r}'|} d^3\mathbf{r}'. \tag{2.8}$$

Note the variables carefully: $\mathbf{r}$ labels the point where you're measuring the potential, but $\mathbf{r}'$ labels the position of the charge distribution, and is integrated over so the final answer depends only on $\mathbf{r}$. We should mention that (2.8) is rarely used in practice, since the integral is so nasty, unless the charge density ρ is particularly simple. It is worth remembering mostly as a sanity check for the case of a point charge, where $\rho(\mathbf{r}) \propto \delta^3(\mathbf{r})$. We'll come back to this case below.

2.1.3 Integral Form of Maxwell's Equations

The third useful tool is Maxwell's equations themselves, though in a slightly different form. Using Gauss's theorem and Stokes's theorem, we can rewrite Maxwell's equations in integral form:

$$\oint_S \mathbf{E}(\mathbf{r}) \cdot d\mathbf{S} = \frac{Q_{\text{enc}}}{\epsilon_0}, \tag{2.9}$$

$$\oint_C \mathbf{E}(\mathbf{r}) \cdot d\mathbf{l} = 0 \ \text{(electrostatics)}, \tag{2.10}$$

where Q_{enc} is the charge enclosed by the closed surface S, and C is some closed curve. The first equation is known as *Gauss's law*, and the left-hand side is defined as the *electric flux*. The important implication of Gauss's law is that regions that have no net flux in or out can enclose no net charge. Problems on the GRE that involve actually solving for realistic charge configurations always require the integral forms of Maxwell's equations. The differential forms, if they are ever needed, are mainly involved in conceptual questions or questions that just require you to evaluate the divergence or curl, given some electric field.

There are always problems on the GRE that involve simply solving for the electric field or electric potential of a charge configuration. If a problem has a high degree of symmetry (e.g. spherical, cylindrical, or planar), the fastest route is to use the integral form of Gauss's law (2.9). The recipe to calculate the field is simple and probably familiar from an electromagnetism course.

1. **Figure out the "symmetry" of the problem:** Should the electric field point radially outward from a single point (spherical symmetry), radially outward from a central axis (cylindrical symmetry), or away from a plane (planar symmetry)? This can usually be deduced from the shape of the charge configuration.
2. **Find a Gaussian surface:** Visualize a "Gaussian" surface S such that the electric field $\mathbf{E}$ is always either (a) perpendicular to S with constant magnitude, or (b) parallel to S – this is where the symmetry is important, as you usually *guess* where this surface is, based on the symmetry of the problem.

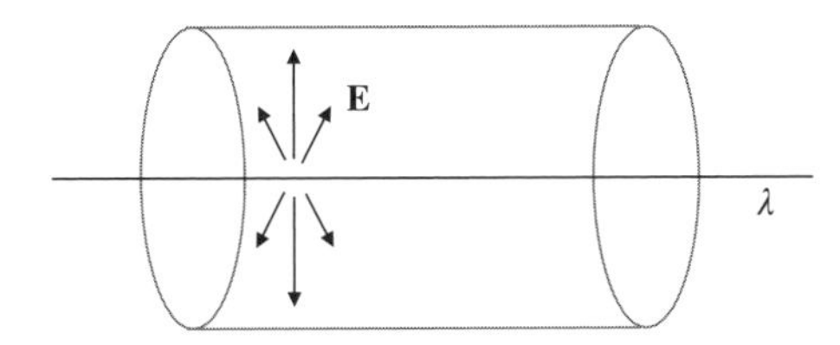

Figure 2.1 Example of a Gaussian surface over which the electric field from the charged line λ is both constant and perpendicular.

3. **Solve for the field:** The dot product $\mathbf{E} \cdot d\mathbf{S}$ then vanishes whenever $\mathbf{E}$ is parallel to S, and is equal to $|\mathbf{E}|dS$ over the rest of S. The constant magnitude $|\mathbf{E}|$ can be pulled out of the integral, and (2.9) reduces to

$$|\mathbf{E}| \oint_{S_\perp} dS = \frac{Q_{\text{enc}}}{\epsilon_0},$$
$$|\mathbf{E}|\, A = \frac{Q_{\text{enc}}}{\epsilon_0},$$

where A is the area of the Gaussian surface $S_\perp$ over which $\mathbf{E}$ is perpendicular to the surface.

We'll treat several more standard applications of Gauss's law in the following Section 2.1.4.

2.1.4 Standard Electrostatics Configurations

As we discussed at the beginning of this chapter, the hard part about electromagnetism is applying the ideas, not understanding them. In this section, we'll therefore summarize the important charge configurations that frequently appear on the exam.

- **Point charges.** The force, electric field, and potential due to a collection of point charges can always be found by summing up the Coulomb terms for each charge in the system. Problems generally fall into two types. In the first type, you are given an arbitrary arrangement of charges and you need to find the force, electric field, or potential due to the charges at some nearby point in space. It is always best to start with symmetry considerations to limit the calculations needed. For example, if the point charges are at the vertices of a regular polygon, then the field and force at the center will be zero because all components will cancel. Only after using as much symmetry as possible should you start writing down terms. If possible, you want to find the potential first and only then calculate the field by taking a derivative, since scalar addition is much simpler than vector addition.

EXAMPLE 2.1

The simplest possible application of all these ideas is to "derive" Coulomb's law. Hopefully you do not need a reminder about Coulomb's law, but this is still a nice simple setting to see how these ideas work without the complications of strange charge configurations. Let's say that we have a point charge q; that is, a charge distribution that is just proportional to a delta function,

$$\rho(\mathbf{r}) = q\delta^3(\mathbf{r}).$$

We know that the electric field due to a point charge is directed radially outward. What surface has a normal vector that points radially outward? A sphere, of course. So we integrate over a sphere of radius r: the enclosed charge is just q since the delta function will integrate to 1 over any region containing the origin, and we get

$$|\mathbf{E}|\,(4\pi r^2) = \frac{q}{\epsilon_0},$$

from which we deduce *Coulomb's law*,

$$\mathbf{E}(\mathbf{r}) = \frac{q}{4\pi\epsilon_0 r^2}\hat{\mathbf{r}}. \tag{2.11}$$

We can deduce the potential from a similar procedure starting with equation (2.8). Or we can simply obtain the potential by integrating the electric field from some reference point, where we set V equal to zero. Here, we pick $r = \infty$ to be our reference point, so we obtain a potential

$$V(\mathbf{r}) = \frac{q}{4\pi\epsilon_0 r}, \tag{2.12}$$

known as the Coulomb potential.

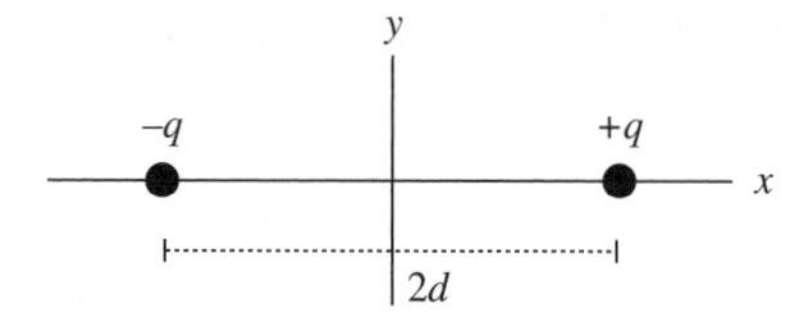

Figure 2.2 Electric dipole.

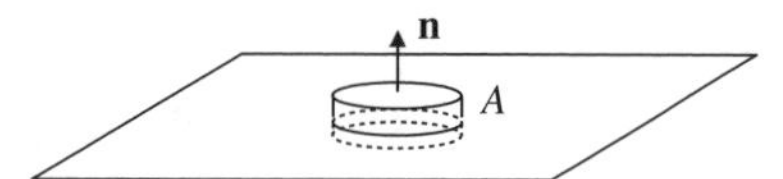

Figure 2.3 Gaussian pillbox for calculating the field of an infinite charged sheet.

In a second common class of problems, you are asked to find the electric field or force of some configuration at a point very far from the origin. A standard example (which we'll treat in a different context in Section 2.4) is two point charges separated by some distance. Suppose we have a charge q and a charge $-q$ separated by a distance $2d$, and choose coordinates such that q is at $(d, 0, 0)$ and $-q$ is at $(-d, 0, 0)$ (Fig. 2.2).

The potential at some point $(x, 0, 0)$ on the x-axis is

$$V(x) = \frac{1}{4\pi\epsilon_0}\left(\frac{q}{|x-d|} - \frac{q}{|x+d|}\right).$$

Taylor expanding this expression for $x \gg d$ tells you approximately how the field behaves far away. Note that the lowest-order terms in the Taylor expansion cancel, but the next nonzero term in the Taylor series does not vanish and gives a much better approximation.

- **Planes.** The classic example here is an infinite flat sheet of charge with surface charge σ. By symmetry, the field must be constant in magnitude: the sheet is infinite, so there is no local measurement you can do to tell you how close you are to it, and it looks the same from every distance. Again by symmetry, the field must point perpendicularly *away* from the sheet, since that is the only preferred direction in the problem. Now we draw a Gaussian "pillbox" surrounding part of the sheet, which cuts out an area A (Fig. 2.3).

  The thickness of the pillbox doesn't matter, because, as we argued, $\mathbf{E}$ has constant magnitude. The integral in Gauss's law is $\oint \mathbf{E} \cdot d\mathbf{S} = 2|\mathbf{E}|A$ (one from each of the two opposite faces of the pillbox), the charge enclosed is $Q = \sigma A$, and solving for $\mathbf{E}$ we find

$$\mathbf{E} = \frac{\sigma}{2\epsilon_0}\hat{\mathbf{n}}, \qquad (2.13)$$

  where $\hat{\mathbf{n}}$ is a unit normal pointing away from the plane. This particular result shows up so often that it's best to memorize it, so we've given it an equation number. You'll see below in Section 2.1.5 how precisely this same argument can be used to get the boundary conditions for $\mathbf{E}$ at a surface.

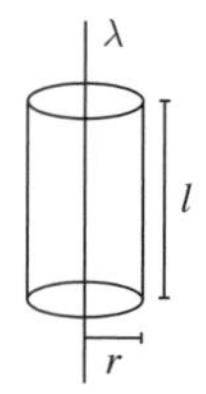

Figure 2.4 Cylindrical Gaussian surface for calculating the field of an infinite line charge.

- **Line charges and cylinders**. Problems involving infinite line charges and cylinders are usually solved with cylindrical Gaussian surfaces. For example, consider the electric field due to an infinite line of charge with charge per unit length λ (Fig. 2.4).

  The field of a point charge points radially outward, but here we have an infinite line charge; by symmetry, the field can't have a component along the line, and by the Maxwell equation $\nabla \times \mathbf{E} = 0$, the only option is for it to point in the $\hat{\mathbf{r}}$ direction in cylindrical coordinates. From Gauss's law applied to a cylinder of height l and radius r surrounding the line, we have

$$|\mathbf{E}|2\pi rl = \frac{\lambda l}{\epsilon_0},$$
$$|\mathbf{E}| = \frac{\lambda}{2\pi\epsilon_0 r},$$
$$\mathbf{E} = \frac{\lambda}{2\pi\epsilon_0 r}\hat{\mathbf{r}}.$$

  Notice that the height l of the cylinder cancels out when we express the result in terms of the linear charge density λ.

- **Spherical surfaces**. Concentric spherical surfaces are another common geometry in electrostatics problems. Similar to the case of the cylinder, solutions can usually be obtained by using a sphere as the Gaussian surface to compute the electric field at each radius. See Examples 2.1 and 2.2.

2.1.5 Boundary Conditions

In the previous section, we learned everything needed to calculate electric fields and potentials in the bulk of regions – such as inside or outside of a sphere. But what about on the boundary *between* regions? Figuring out how fields behave on boundaries between regions of space is so important that it deserves its own section. Luckily, like everything else, it too follows from Maxwell's equations.

Let's say that we have a continuous surface surrounded by vacuum, and we zoom in so close that the surface appears flat. We now have a plane, which we can use to define $\mathbf{E}^{\parallel}$, the two

EXAMPLE 2.2

Find the electric field created by a solid sphere of radius a with charge density $\rho(r) = \alpha r$ and a spherical cavity of radius b at its center (Fig. 2.5).

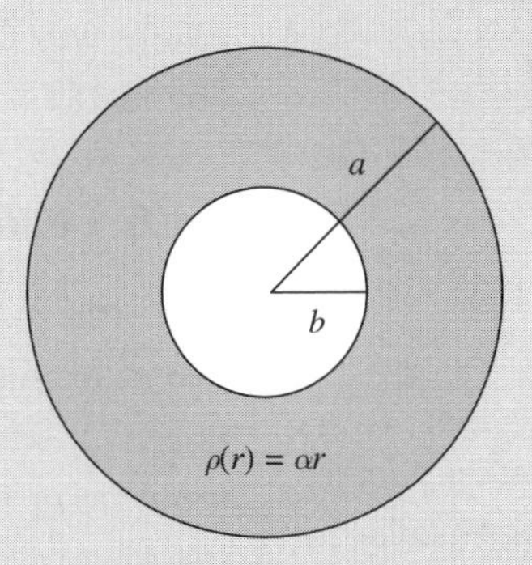

Figure 2.5 A spherically symmetric charge distribution with a cavity at the center.

For $r < b$, there is no electric field because a spherical Gaussian surface lying entirely inside the cavity contains no charge. This is an important result in its own right: *the electric field inside an empty spherically symmetric cavity is always zero.* In the solid region $b < r < a$, we need to use Gauss's law with a sphere as our Gaussian surface, so we can use the usual trick of pulling out $|\mathbf{E}|$ from the surface integral. The electric field is given by

$$\begin{aligned}
\oint_S \mathbf{E} \cdot d\mathbf{S} &= \frac{1}{\epsilon_0} \int_b^r \rho(r') d^3\mathbf{r}', \\
|\mathbf{E}|\,(4\pi r^2) &= \frac{4\pi}{\epsilon_0} \int_b^r \alpha r'^3 dr', \\
|\mathbf{E}|\,(r^2) &= \frac{\alpha}{4\epsilon_0}(r^4 - b^4), \\
\mathbf{E} &= \frac{\alpha}{4\epsilon_0} \frac{r^4 - b^4}{r^2} \hat{\mathbf{r}}.
\end{aligned}$$

In the outermost region $r > a$, the electric field is just given by Coulomb's law for the total charge in the sphere. The total charge in the sphere is just the integral of the charge density:

$$\begin{aligned}
Q &= \int_b^a \rho(r') d^3\mathbf{r}' \\
&= 4\pi \int_b^a \alpha r'^3 dr' \\
&= \pi\alpha(a^4 - b^4),
\end{aligned}$$

so the electric field is given by

$$\mathbf{E} = \frac{\alpha(a^4 - b^4)}{4\epsilon_0 r^2} \hat{\mathbf{r}}.$$

Notice that the electric field is everywhere continuous: at $r = b$ our expression for the field in the region $b < r < a$ gives 0, as did our argument for the field in the region $r < b$. Similarly, at $r = a$, both expressions give $\alpha(a^4 - b^4)/(4\epsilon_0 a^2)$ for the magnitude of the field. As we will see in Section 2.1.5, the continuity of $\mathbf{E}$ is due to the fact that there are no surface charges in this problem, only volume charge densities.

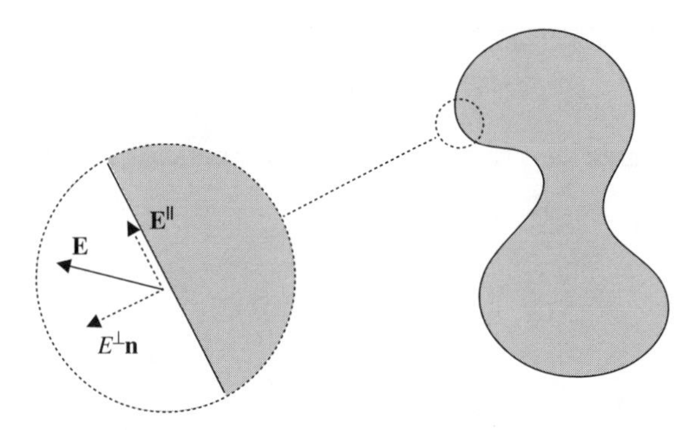

Figure 2.6 Parallel and perpendicular components of **E** at a boundary.

components of the three-component vector **E** that are parallel to this plane, and $E^{\perp}$, the third component perpendicular to the plane. Mind the notation: $\mathbf{E}^{\|}$ is bold because it's a (two-component) vector, but $E^{\perp}$ is a one-component scalar. Next, let's integrate over a narrow rectangular path with long edges parallel to the surface and short edges perpendicular to the surface. Using the second of Maxwell's equations in integral form, equation (2.10), the line integral implies that

$$\mathbf{E}^{\|}_{\text{out}} - \mathbf{E}^{\|}_{\text{in}} = 0, \tag{2.14}$$

where $\mathbf{E}_{\text{out}}$ is the electric field just outside the surface, and $\mathbf{E}_{\text{in}}$ is the field just inside the surface. This is true since the lengths of the paths perpendicular to the surface can be taken to be arbitrarily small, and the other two opposite sides of the rectangle contribute equally and opposite in the integral.

Next, let's consider the same surface, but integrate over the surface of a Gaussian pillbox and use Gauss's law, equation (2.9). If we take the height of the box to be extremely small, then, using the same arguments as in the plane symmetry discussion in the previous section, we end up with the condition that

$$E^{\perp}_{\text{out}} - E^{\perp}_{\text{in}} = \frac{\sigma}{\epsilon_0}, \tag{2.15}$$

where σ is the surface charge density. Equations (2.14) and (2.15) let you patch together electric fields in different regions of space across surfaces, where strange things may be happening. A simple, classic example of why this is useful is the case of conductors, which we discuss below in Section 2.1.6.

Before we do that, however, we should talk about how the potential behaves at boundaries, since working with the potential is almost always easier than working with the electric field. It's not difficult to show that equations (2.14) and (2.15) and the fact that $\mathbf{E} = -\nabla V$ imply that

V is always continuous.

Similarly, the first derivative of V is constrained so that

Derivatives of V are continuous, except at charged surfaces.

These rules for the potential and electric field boundary conditions will allow you to match solutions across any type of boundaries.

2.1.6 Conductors

While the GRE might throw a question or two your way that relies on knowing formal aspects of boundary conditions, you are most likely to use these ideas when solving problems involving conductors. An ideal conductor is a material where charge can flow freely without resistance: usually "metal" and "conductor" are synonymous for the purposes of the GRE. There is only one fact you really need to know about conductors:

V is constant throughout a conductor.

From this fact, you can derive four important corollaries:

- The electric field inside a conductor is zero.
- The net charge density inside a conductor is zero.
- Any net charge on a conductor is confined to the surface.
- The electric field just outside a conductor is perpendicular to the surface.

These properties are a direct consequence of the fact that an ideal conductor has no resistance to the movement of charges. If there *were* any bulk electric fields, free charges would experience a force until they had arranged themselves to cancel out any electric fields.

On the GRE, you will occasionally work with *grounded* conductors. Usually a conductor is said to be grounded if it is connected to the reference for the electric potential. In other words, something is grounded if it is connected to $V = 0$. You can think of this as setting the constant reference scale for the potential. The other important property of the idealized ground in electrostatics problems is that it is an infinite sink and source of charge. So, if you put a charge near a grounded conductor, some charge will be induced on the grounded conductor with no cost in energy. See Example 2.3.

2.1.7 Method of Images

The examples that we have discussed so far have been simplified by totally spherical, cylindrical, or planar geometry which allowed us to use simple arguments with Gauss's law.

EXAMPLE 2.3

As an example of a problem involving conductors, consider the simple example of a thick, uncharged, conducting shell of inner radius r_1 and outer radius r_2, with a point charge q in the center, shown in Fig. 2.7. Further suppose that the potential at infinity is zero, $V(r = \infty) = 0$. What is the potential everywhere in space?

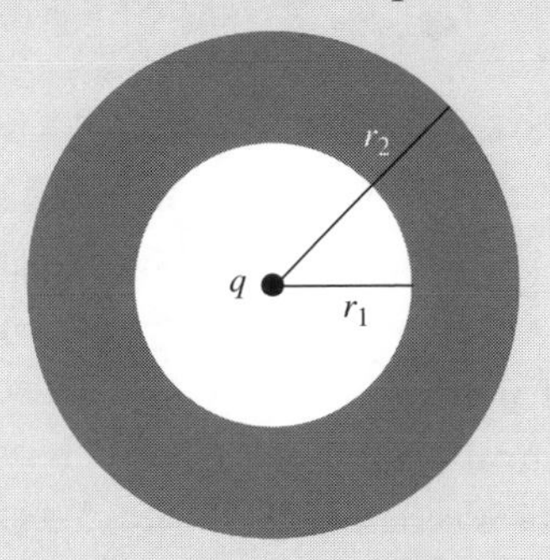

Figure 2.7 Conducting shell with point charge q at center.

From Coulomb's law, the field in the region $0 < r < r_1$ is simply

$$\mathbf{E}(r) = \frac{q}{4\pi\epsilon_0 r^2}\hat{\mathbf{r}}.$$

Inside the conductor, the electric field must be zero for $r_1 < r < r_2$:

$$\mathbf{E}(r) = 0.$$

Outside of the conductor ($r > r_2$), we can use Gauss's law to find the electric field. The conductor is uncharged, so the enclosed charge is just q. Thus, the electric field is

$$\mathbf{E}(r) = \frac{q}{4\pi\epsilon_0 r^2}\hat{\mathbf{r}}.$$

If we specify that the potential at infinity is 0, then the potential outside the conductor is

$$V(r) = \frac{q}{4\pi\epsilon_0 r}, \quad r > r_2.$$

Inside the conductor, we demand that V be constant. By continuity at $r = r_2$, we must have

$$V(r) = \frac{q}{4\pi\epsilon_0 r_2}, \quad r_1 < r < r_2.$$

Finally, inside the shell ($r < r_1$), the potential is of the form

$$V(r) = \frac{q}{4\pi\epsilon_0 r} + \text{const.}$$

Requiring continuity at $r = r_1$, we determine the value of the constant and find that the potential is

$$V(r) = \frac{q}{4\pi\epsilon_0 r} + \frac{q}{4\pi\epsilon_0}\left(\frac{1}{r_2} - \frac{1}{r_1}\right), \quad r < r_1.$$

Since the electric field is zero inside the conductor, we conclude from Gauss's law, after integrating over a sphere of radius $r_1 < r < r_2$, that

$$Q_{\text{enc}} = 0.$$

Since there is a charge q at the center, the charge on the inner surface of the conductor must be $-q$. This charge is said to be *induced* by the charge at the center. Since the conductor has zero net charge, there must be a corresponding charge of $+q$ uniformly distributed on the outer surface of the conductor. This gives us exactly the field structure we have shown.

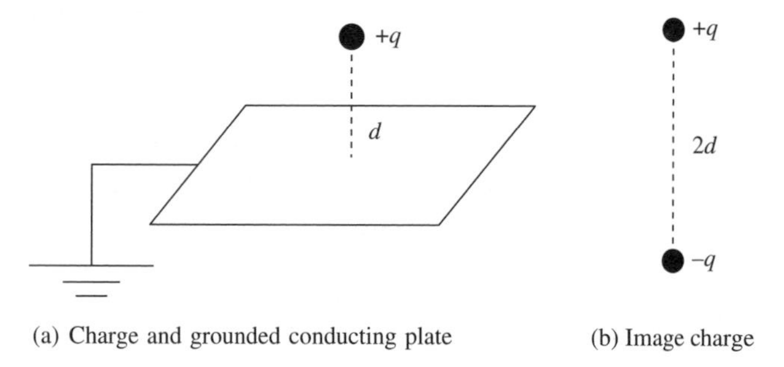

(a) Charge and grounded conducting plate (b) Image charge

Figure 2.8 Setup for the method of images.

As you might imagine, solving for the potential and induced charges of a configuration without this kind of total symmetry can be extremely complicated. In general, this is certainly true, but there are many tricks for systematically dealing with more complex geometries, particularly involving conductors. The only one of these methods that you are likely to need for the GRE is the so-called "method of images," which relies on a property of Laplace's equation (2.7), the *uniqueness of solutions*: if you can guess the potential V and it satisfies the boundary conditions, it must be *the* correct answer.

Consider the case of a point charge a distance d above an infinite, grounded conducting plane (Fig. 2.8(a)). Naïvely, this looks complicated! The point charge induces some complicated surface density of charge on the conductor. The arrangement does not have spherical or planar symmetry so we cannot rely on Gauss's law easily. But note that the potential on the surface must be constant because the surface is a conductor, and it must be zero because the conductor is grounded. Now switch gears for a moment and imagine a totally different charge distribution, that of the original charge q and a so-called "image" charge $-q$ at $z = -d$ (Fig. 2.8(b)). The potential on the plane $z = 0$ is exactly the same (namely $V = 0$) for these two charges as in the case of the conductor! Furthermore, if we restrict ourselves to the region $z \geq 0$, the sources for Poisson's equation are also the same, namely a single point charge q at $z = d$ in both cases. Therefore, by uniqueness, the potential V produced by the charge and the image charge *must* be the solution to the original problem with the conductor, since it satisfies the same boundary conditions at $z = 0$ with the same sources.

The essence of the method of images is this: whenever we have a conductor, if we can find an arrangement of point charges that exactly reproduces the potential on the surface, then the potential everywhere will be the same as if there were no conductor. But you must beware to place image charges only *below* the conductor, in the region that would be inaccessible in the original setup; if you placed them in the same region as the original charge, you would be adding different sources to Poisson's equation and solving a different problem. The only example you are likely to encounter on the GRE is the case of a grounded, conducting, infinite plane with some charges on one side. In this case, for each charge, just put an opposite mirror charge on the other side of the plane at the same distance away and you will have the potential immediately. While it is good to know this prescription, understanding why it works more generally is important and might come in handy for other problems.

There is one very important subtlety about the method of images. When calculating anything involving work or energy, remember that

There is no energy cost to moving an image charge.

The image charge is just a fake construct, a trick to get the correct potential; in reality, there is no field below the conductor, and so no work can be done on anything below the conductor. This can get confusing because the position of the image charge depends on the position of the real charge. If you move the real charge in towards the conducting plate, the image charge moves with it, but work is only done *on the real charge*. A useful mnemonic that lets you forget about this subtlety is to calculate the electric *field* directly from the image configuration. In other words, do *not* calculate the electric field by taking the derivative of V, because doing so will implicitly count the motion of the image charges as contributing to the energy. A more detailed discussion of this subtlety can be found in Griffiths, Section 3.2.3.

2.1.8 Work and Energy in Electrostatics

As in classical mechanics, using work and energy considerations wherever possible often saves time and energy (no pun intended) when problem solving. The same is true in electrostatics. Let's start with point charges and then generalize to other configurations. The work required to put together n point charges is

$$W = \frac{1}{2}\sum_{i=1}^{n} q_i V(\mathbf{r}_i), \tag{2.16}$$

where q_i is the charge of each point charge and $V(\mathbf{r}_i)$ is the potential due to *all* of the charges, but evaluated at the location of the ith point charge. The intuition for this formula is that each of the $q_i V(\mathbf{r}_i)$ terms gives us the potential energy between one charge and every other charge. When we sum up

EXAMPLE 2.4

Consider a point charge at the center of a thin, grounded, conducting, spherical shell of radius a (Fig. 2.9(a)). The shell is removed and taken to infinity (Fig. 2.9(b)). How much work is done during this process?

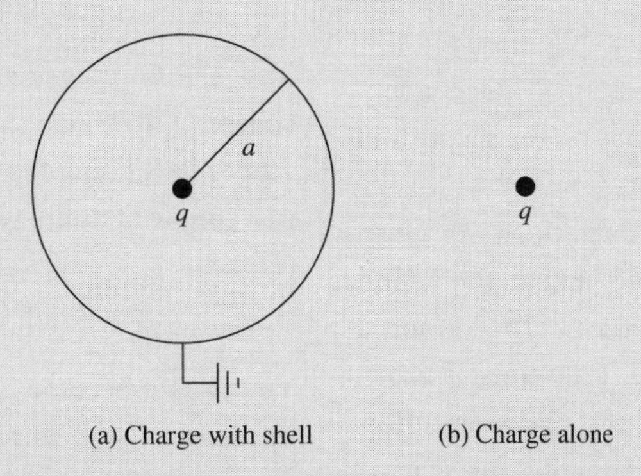

(a) Charge with shell (b) Charge alone

Figure 2.9 Change in energy due to a conducting shell.

We can just compute the change in energy for the two configurations. When the shell is in place, there is no field outside the shell, so the energy is just

$$U_1 = \frac{\epsilon_0}{2}\int_0^a |\mathbf{E}|^2\, 4\pi r^2 dr.$$

After the shell is gone, the energy is

$$U_2 = \frac{\epsilon_0}{2}\int_0^\infty |\mathbf{E}|^2\, 4\pi r^2 dr.$$

The work done must be

$$W = U_2 - U_1 = \frac{\epsilon_0}{2}\int_a^\infty |\mathbf{E}|^2\, 4\pi r^2 dr.$$

Using Coulomb's law, this quantity is easy to evaluate. But beware! If we had actually tried to evaluate U_1 or U_2 explicitly before taking the difference, we would have found them to be infinite! This is an important point: the total energy due to a point charge is infinite. The reasons for this are rather complicated and deep, so we will avoid discussing them. Nevertheless, this example shows that the *difference* between formally infinite quantities still has meaning in electromagnetism.

all of these contributions, we double count the energy so we need to divide by 2 to get the actual work. It should not be too mysterious to see that when we're not dealing with point charges, the work becomes an integral:

$$W = \frac{1}{2}\int \rho(\mathbf{r})V(\mathbf{r})d^3\mathbf{r}. \tag{2.17}$$

We showed that it takes work to move charge around, but there's also energy stored in the *fields* themselves. The energy is just

$$U_E = \frac{\epsilon_0}{2}\int |\mathbf{E}|^2\, d^3\mathbf{r}. \tag{2.18}$$

This simple formula can be extremely useful, as shown in Example 2.4.

2.1.9 Capacitors

Suppose you have two conductors with different net charges – for concreteness, give one $+Q$ and the other $-Q$. This gives rise to an electric field between the conductors, which in turn puts them at different potentials, say 0 and V. (This is a well-defined concept because potential is constant throughout a conductor.) This arrangement is known as a *capacitor*. In many practical cases, Q and V are proportional, and the constant of proportionality is called the *capacitance* C:

$$Q = CV. \tag{2.19}$$

You'll often hear the statement "C only depends on the geometry of the problem." All this means is that the proportionality

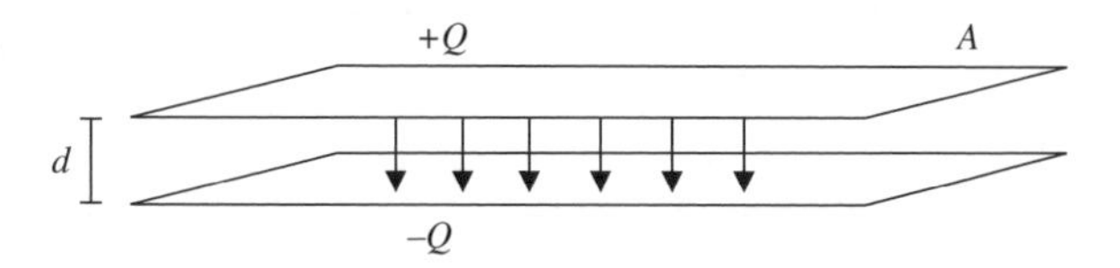

Figure 2.10 Parallel-plate capacitor.

between Q and V holds with the same constant C independent of how the charge got there; it depends only on the shapes and relative orientation of the two conductors.[2]

Finding the capacitance is usually straightforward: given some conductors, imagine putting charges $\pm Q$ on them, find the potential between them, and extract C. The standard example is two parallel plates of area A a distance d apart. We'll assume that $A \gg d^2$ such that the plates are effectively infinite and the field between them is constant. Placing charges $\pm Q$ on the plates gives them surface charge densities $\sigma = \pm Q/A$. A straightforward application of Gauss's law in a plane geometry gives the electric field of each plate as $\mathbf{E} = \frac{\sigma}{2\epsilon_0}\hat{\mathbf{n}}$, where $\hat{\mathbf{n}}$ is a unit vector pointing away from the surface. The fields of the two plates cancel each other outside, but reinforce each other between the plates, such that the total field is $|\mathbf{E}| = \frac{Q}{A\epsilon_0}$ pointing from the positive to the negative plate. Doing the line integral along a straight line between the two plates to get V gives $V = \frac{Qd}{A\epsilon_0}$, or rearranging,

$$Q = V \times \frac{A\epsilon_0}{d}.$$

As promised, Q is proportional to V, so we extract the capacitance:

$$C = \frac{\epsilon_0 A}{d} \quad \text{(parallel-plate capacitor).} \tag{2.20}$$

This result is common enough and important enough that we give it an equation number. Note that it only depends on the geometric constants A and d, along with ϵ_0, which comes along for the ride in any electrostatics problem. By the way, you should reach the point where every step of the derivation is intimately familiar and obvious: the application of Gauss's law to an infinite sheet of charge, integrating $\mathbf{E}$ to find the potential, and checking that the boundary conditions for $\mathbf{E}$ are satisfied at the surfaces of the conductors are all part of classic GRE problems.

The device we have just described is called a *parallel-plate capacitor*. Along with the resistor and the inductor, it is one of the three building blocks of elementary circuits. It has the interesting property that it produces a strong, *uniform* electric field in a limited region of space: the field is $Q/A\epsilon_0$ between the plates, and zero everywhere else. This lets it store electrical energy; indeed, using (2.18), we find that the energy stored in the field of a charged capacitor is

$$U_C = \frac{\epsilon_0}{2}\left(\frac{Q^2}{A^2\epsilon_0^2}\right)(Ad) = \frac{1}{2}\frac{Q^2 d}{A\epsilon_0} = \frac{1}{2}\frac{Q^2}{C}.$$

This can be interpreted as the energy it takes to remove a charge Q from one (initially neutral) plate and put it on the other plate. Using the relation $Q = CV$ this can be expressed in a couple of useful ways:

$$U_C = \frac{1}{2}\frac{Q^2}{C} = \frac{1}{2}CV^2. \tag{2.21}$$

You should become intimately familiar with capacitors: they are completely defined by (2.19)–(2.21), so learn those equations well. There are other arrangements of conductors that act as capacitors: in such cases, both (2.19) and (2.21) still hold, but you have to derive the analogue of (2.20). One example is two concentric spherical conducting shells, and another example is treated in the problems below.

2.1.10 Problems: Electrostatics

1. A point charge q is brought to a distance d from a grounded conducting plane. What is the magnitude of the force on the plane from the point charge?

   (A) $q^2/(16\pi\epsilon_0 d^2)$
   (B) $q^2/(8\pi\epsilon_0 d^2)$
   (C) $q^2/(4\pi\epsilon_0 d^2)$
   (D) $q^2/(\epsilon_0 d^2)$
   (E) 0

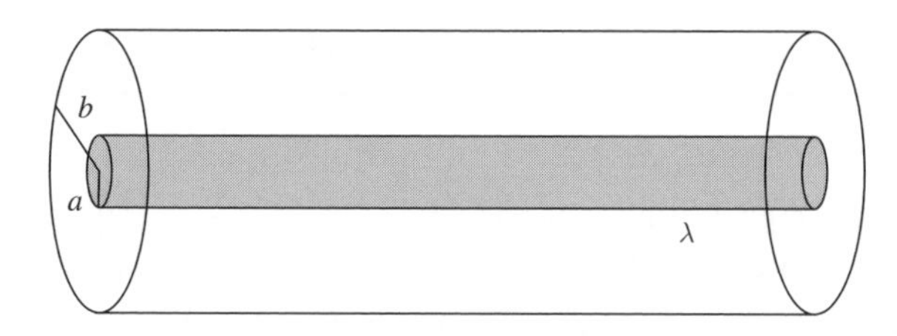

2. A cylindrical wire of charge of radius a and charge per unit length λ is at the center of a thin cylindrical conducting shell of radius b. What is the capacitance per unit length of this configuration?

   (A) ∞
   (B) $2\pi\epsilon_0 ab/(b^2 - a^2)$
   (C) $2\pi\epsilon_0\lambda/\ln(b/a)$
   (D) $2\pi\epsilon_0/\ln(b/a)$
   (E) 0

[2] Note that C is an intrinsically *positive* quantity, so you often don't have to worry too much about sign conventions for V.

3. A capacitor is formed from two square plates of edge length a and separation d, with $d \ll a$. If all linear dimensions of the capacitor are tripled, by what factor does the capacitance change?

   (A) 1/3
   (B) 1
   (C) 3
   (D) 9
   (E) 27

4. What is the work needed to assemble four point charges q into a regular tetrahedron of side length a?

   (A) $q^2/(4\pi\epsilon_0 a)$
   (B) $q^2/(2\pi\epsilon_0 a)$
   (C) $q^2/(\pi\epsilon_0 a)$
   (D) $3q^2/(2\pi\epsilon_0 a)$
   (E) $3q^2/(\pi\epsilon_0 a)$

5. The electric field inside a sphere of radius R is given by $\mathbf{E} = E_0 z^2 \hat{\mathbf{z}}$. What is the total charge of the sphere?

   (A) $\frac{\pi}{2}\epsilon_0 E_0 R^4$
   (B) $\pi\epsilon_0 E_0 R^3$
   (C) $2\pi\epsilon_0 E_0 R^4$
   (D) $4\pi\epsilon_0 E_0 R^3$
   (E) 0

2.2 Magnetostatics

So far, we have discussed configurations with static electric fields. There is a very similar story for static *magnetic* fields, which are produced by constant currents of charge. This is a slight abuse of terminology, since *moving* charges (not static ones) are what create currents, but we're assuming that there are enough individual charges all moving together that the net current they create is constant in time; this is known as the steady-current approximation.

2.2.1 Basic Tools

The fundamental equations for these problems are the other two of Maxwell's equations, in the absence of changing electric fields:

$$\nabla \cdot \mathbf{B} = 0, \tag{2.22}$$

$$\nabla \times \mathbf{B} = \mu_0 \mathbf{J} \qquad \text{(magnetostatics)}. \tag{2.23}$$

The first equation is simply the statement that there are no magnetic monopoles, which is true in general, not just in magnetostatics. The second equation describes how currents act as sources for magnetic fields. As with Maxwell's equations for electrostatics, we can write these equations in integral form:

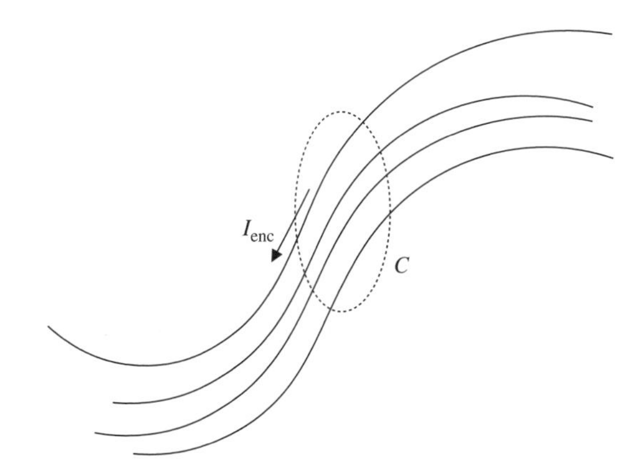

Figure 2.11 Example of a curve enclosing a current as in Ampère's law.

$$\oint_S \mathbf{B} \cdot d\mathbf{S} = 0, \tag{2.24}$$

$$\oint_C \mathbf{B} \cdot d\mathbf{l} = \mu_0 I_{\text{enc}} \qquad \text{(magnetostatics)}, \tag{2.25}$$

where S is a closed surface and C is a closed curve, just as in equations (2.9)–(2.10), and I_{enc} is the current piercing the surface defined by C. Equation (2.24) is sometimes referred to as *Gauss's law for magnetism*, and, just like the electric case, the left-hand side is known as the *magnetic flux*. We should emphasize that S is a *closed* surface, which means that it doesn't have any boundary: just picture a floppy ball. This distinction will be important when we discuss magnetic flux through surfaces that are *not* closed in Section 2.3.2. Equation (2.25) is known as *Ampère's law* and will be discussed in more detail in the following Section 2.2.2.

Similarly to the case of electrostatics, we can construct a potential for $\mathbf{B}$. However, since $\nabla \times \mathbf{B} \neq 0$, we cannot use a scalar potential. We are instead forced to use a *vector* potential, which has the defining property that

$$\nabla \times \mathbf{A} = \mathbf{B}. \tag{2.26}$$

Since the vector potential has three components (just like $\mathbf{B}$), it isn't as useful as the scalar potential V for calculations. The vector potential shows up so rarely on the GRE that it's not even worth discussing further apart from its defining equation.

In addition to Maxwell's equations, we can completely determine the effects of fields on test charges with the Lorentz force law. This gives the force on a test charge q due to a magnetic field:

$$\mathbf{F}_B = q\mathbf{v} \times \mathbf{B}. \tag{2.27}$$

Generalizing this to the force on a wire carrying current I, we have

$$d\mathbf{F}_B = Id\mathbf{l} \times \mathbf{B}. \tag{2.28}$$

The direction of the force is often more important than the magnitude on the GRE, so now is a good time to mention the famed *right-hand rule*: to evaluate the cross product, put the fingers of your right hand in the direction of the wire, curl them around toward the direction of $\mathbf{B}$, and your thumb will point in the direction of $\mathbf{F}_B$. Alternatively, you can use the rules appropriate to right-handed coordinate systems: $\hat{\mathbf{z}} \times \hat{\mathbf{r}} = \hat{\boldsymbol{\phi}}$ in cylindrical coordinates, $\hat{\boldsymbol{\theta}} \times \hat{\boldsymbol{\phi}} = \hat{\mathbf{r}}$ in spherical coordinates, and so on.

Problems in magnetostatics almost always fall into three general classes:

- Finding the field due to a configuration of currents
- Finding forces on wires or charges due to fields
- Finding energies of fields

The first class can generally be tackled with Ampère's law or the Biot–Savart law (discussed below). The second class can usually be solved with some variant of the Lorentz force law. The final class can be solved by integrating to find the energy in a field configuration. Obviously, these are just general guidelines, but they help to put these topics in perspective.

2.2.2 Ampère's Law and the Biot–Savart Law

Suppose that we have some collection of wires carrying currents. What is the magnetic field produced by the wire configuration? This question can be answered by Ampère's law and the Biot–Savart law.

Ampère's law is generally only useful for configurations that possess a high degree of symmetry, and can be thought of as analogous to Gauss's law for electrostatics. Referring back to equation (2.25), we want to pick the closed curve C such that the magnetic field is parallel to the path and constant. This allows us deduce the magnetic field by

$$|\mathbf{B}| \oint_C dl = \mu_0 I_{\text{enc}},$$

$$|\mathbf{B}| = \frac{\mu_0 I_{\text{enc}}}{L},$$

where L is the length of the curve C (for example, $2\pi r$ for an Amperian loop at a distance r from a current-carrying wire). This method relies crucially on being able to choose a path along which the magnetic field is constant, which is why it only works in highly symmetric problems. When we can use it, however, it dramatically simplifies our work.

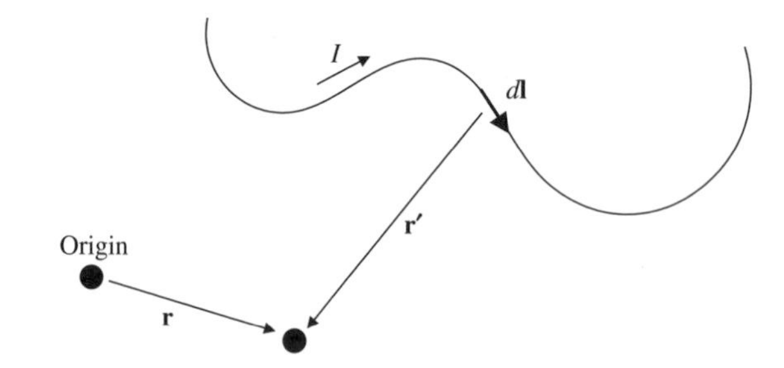

Figure 2.12 Calculating magnetic field from the Biot–Savart law.

In cases that are not so symmetric, we can use the Biot–Savart law instead, integrating over all of the current elements in a configuration. The Biot–Savart law reads[3]

$$\mathbf{B}(\mathbf{r}) = \frac{\mu_0 I}{4\pi} \int \frac{d\mathbf{l} \times \hat{\mathbf{r}}'}{r'^2}. \tag{2.29}$$

The notation in this expression can be a bit cryptic and deserves some explanation. Referring to Fig. 2.12, $\mathbf{r}$ is the point where the field is evaluated, the integration is over the entire wire producing the magnetic field, $d\mathbf{l}$ is the line element along the wire, $\mathbf{r}'$ is the vector pointing from the line element to $\mathbf{r}$, and I is the current carried in the wire. This is similar to (2.8), which gives the potential sourced by a charge distribution; there is an analogous expression for the electric field as an integral over a charge configuration, but it's rarely used since we have the advantage of the scalar potential in the electrostatics case. The Biot–Savart law is clearly a good deal more complicated than Ampère's law, so avoid it unless you have no choice. Chances are good that it will arise at least once in a simple form on the GRE; if it does, the line element $d\mathbf{l}$ will likely take some simple form such as a square or a circle. You'll see an example in the problems at the end of this section.

2.2.3 Standard Magnetostatics Configurations

As in the case of electrostatics, there are a few magnetostatic configurations that seem to come up over and over again. This tends to occur because there just are not very many configurations that one can solve analytically in a reasonable amount of time. This is lucky for you! If you can master the following configurations, you should have a good general intuition for most problems that the GRE will throw at you. Our discussion will be extremely brief, because you probably covered these examples at least twice: once in your freshman physics course,

[3] This form of the equation assumes that the current is confined to a wire and has constant magnitude I, which can be pulled outside the integral. As far as we know this will always be the case for GRE problems.

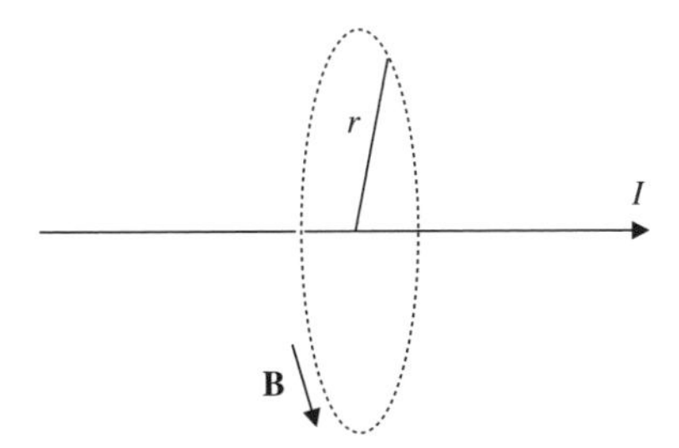

Figure 2.13 Circular Amperian loop for calculating the magnetic field of an infinite wire.

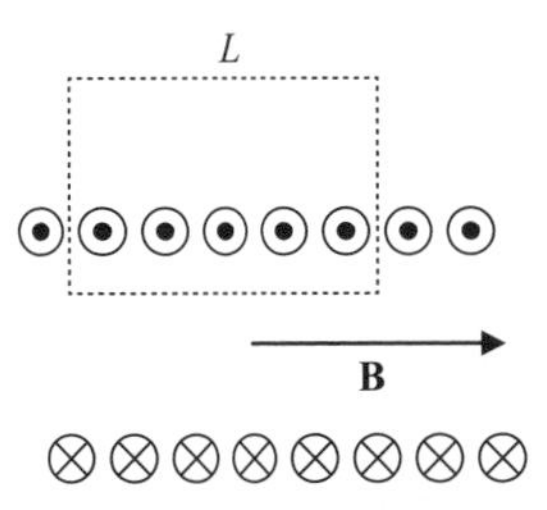

Figure 2.14 Rectangular Amperian loop for calculating the magnetic field of a solenoid.

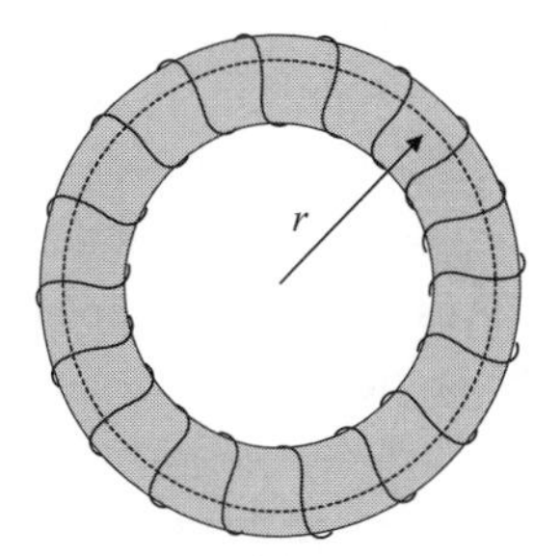

Figure 2.15 Example of a toroid. The dashed circle inside the toroid is a typical Amperian curve which can be used to solve for the magnetic field inside the toroid.

and then again in an advanced electromagnetism course. We recommend going back and reviewing this material in detail, just to convince yourself that the symmetry arguments make sense; but once you do this, you can forget them and just remember the problem-solving techniques for the GRE.

- **Wires.** Since there are no point charges in magnetostatics, the wire is the simplest example. Consider an infinite wire along the z-axis carrying a current I in the positive $\hat{\mathbf{z}}$ direction (Fig. 2.13). Draw an Amperian loop of radius r around the wire, so the current enclosed is I. By (cylindrical) symmetry, the **B**-field must only be a function of the distance r from the wire. Since we know **B** has a curl, we can guess that it "curls" around the wire in a direction dictated by the right-hand rule; in this case, the $\hat{\boldsymbol{\phi}}$ direction.

  This means that it is always parallel to the Amperian loop, so we can use the trick described above and pull out **B** from the integral in Ampère's law:

$$|\mathbf{B}|(2\pi r) = \mu_0 I \implies \mathbf{B} = \frac{\mu_0 I}{2\pi r}\hat{\boldsymbol{\phi}}. \tag{2.30}$$

  This result shows up very often and should be memorized.

  A common generalization is a thick wire with a volume current density that changes with radius, such that the current enclosed changes with radius inside the wire. Note that the wire has to be *infinite* for these symmetry arguments to work: to find the field of a finite wire, you have to go all the way back to the Biot–Savart law.

- **Solenoids.** The previous example had a straight line of current and a circular Amperian loop surrounding it. Now imagine the opposite scenario: take a bunch of tightly wound circular coils ("turns") of wire carrying current in the $\hat{\boldsymbol{\phi}}$ direction, stack the coils in a cylinder, and draw a rectangular Amperian loop with one vertical side of length L inside the cylinder and the other outside (Fig. 2.14).

  Various symmetry arguments, which you can find in Griffiths, tell you that the field must point along the axis of the cylinder and be constant inside. If there are n turns per unit length, then the only nonzero term in Ampère's law is from the one inside the cylinder, which gives you BL. The current enclosed is InL, so setting these equal gives

$$B = \mu_0 n I \text{ (solenoid)}. \tag{2.31}$$

  The direction of the field can be found, as usual, with the right-hand rule: curl your fingers in the direction of the current going around the coils, and your thumb points in the direction of **B**. This device, with the coils stacked in a cylinder, is called a *solenoid*, and we will discuss it in much more detail in Section 2.3.3. By the way, applying the same arguments to an Amperian loop *outside* the cylinder tells you that the field is identically zero outside, so, just like a capacitor, the solenoid confines a strong uniform field to a limited volume.

- **Toroids.** If you bend the solenoid cylinder around into a circle like a donut, you get a *toroid* (Fig. 2.15). Once again, symmetry arguments tell you that the field still points along the axis of a cylinder, which has now been bent around into the $\hat{\boldsymbol{\phi}}$ direction. Drawing a circular Amperian loop in the plane of the tube, just as in the wire example above, gives you

$$B = \frac{\mu_0 N I}{2\pi r} \text{ (toroid)}, \tag{2.32}$$

  where N is the total number of turns. Note that the field is *no longer constant*, but depends on r, the distance from the center of the toroid to the point inside where we are measuring the field. Notice also that this result is independent of the cross-sectional shape of the toroid, so long as it is constant. Just as with a solenoid, the field vanishes outside the volume enclosed by the loops of wire.

2.2.4 Boundary Conditions

We can also study the boundary conditions for magnetostatics, in complete analogy to the boundary conditions for electrostatics. They are slightly more complicated, but luckily they're not needed as frequently. As in Section 2.1.5, let's zoom into the surface of a boundary so that it can be taken to be flat, and define $\mathbf{B}^{\|}$ as the two-component vector parallel to the surface, and $B^{\perp}$ as the component normal to the surface (Fig. 2.16). If we put a cylinder across the surface and evaluate using equation (2.24), we find that the normal component of the magnetic field must be continuous:

$$B^{\perp}_{\text{out}} - B^{\perp}_{\text{in}} = 0. \tag{2.33}$$

Using equation (2.25), we can integrate over a narrow rectangle around the surface. To be completely general, let's suppose that there is some current density **K** on the surface, and let's orient the plane of our loop to be perpendicular to this surface current. If we take the sides of the rectangle perpendicular to the surface to be arbitrarily small, then we can neglect its contribution to the integral, and we are left with the boundary condition for the parallel component of the magnetic field:

$$\mathbf{B}^{\|}_{\text{out}} - \mathbf{B}^{\|}_{\text{in}} = \mu_0 \mathbf{K} \times \hat{\mathbf{n}}, \tag{2.34}$$

where **K** is the surface current density and $\hat{\mathbf{n}}$ is a unit vector pointing perpendicular to the surface, from "in" to "out."

A good mnemonic to remember these equations is to notice that they're sort of the reverse of the analogous electrostatic boundary conditions (2.14) and (2.15): normal **B** becomes parallel **E** and vice versa, and ϵ_0 goes in the denominator while μ_0 goes in the numerator. Also, note that σ (the surface charge density) is a *scalar*, while **K** (the surface current density) is a *vector*; this means that σ must be related to the scalar $E^{\perp}$, and **K** is related to the vector $\mathbf{B}^{\|}$.

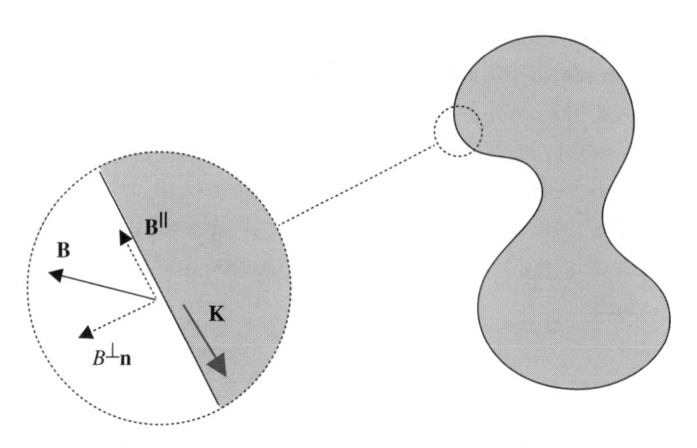

Figure 2.16 Parallel and perpendicular components of **B** at a boundary where surface current **K** flows along the boundary.

2.2.5 Work and Energy in Magnetostatics

Unlike electric fields, *magnetic fields do no work.* This can be seen immediately from the Lorentz force law (2.27): the cross product $\mathbf{v} \times \mathbf{B}$ means that the force is always perpendicular to the velocity, the speed of a particle cannot increase and **B** does no work. This is the cause of much conceptual confusion, as there are many standard problems where work is being done by *something* (usually, an external source or gravity), but it's not the magnetic field. However, magnetic fields store energy just like electric fields:

$$U_B = \frac{1}{2\mu_0} \int |\mathbf{B}|^2 \, d^3\mathbf{r}. \tag{2.35}$$

This is identical to the analogous expression (2.18) for electric fields except for the prefactor.

2.2.6 Cyclotron Motion

Even though magnetic fields do no work, they can still change the *direction* of a charged particle's motion. The most common situation is the motion of a charge q with mass m in a uniform magnetic field $\mathbf{B} = B\hat{\mathbf{z}}$. Suppose the particle moves with velocity $\mathbf{v} = v\hat{\mathbf{y}}$. Then evaluating the cross product,

$$\mathbf{F} = qvB(\hat{\mathbf{y}} \times \hat{\mathbf{z}}) = qvB\hat{\mathbf{x}}.$$

The force is constant and in the $\hat{\mathbf{x}}$-direction, which is perpendicular to the particle's motion. As the particle begins to accelerate in the $\hat{\mathbf{x}}$-direction, the force will remain perpendicular thanks to the cross product, so what we have is *uniform circular motion*: a charged particle in a constant magnetic field will move in a circle confined to the plane with normal parallel to the **B** field. Actually, if the particle has some initial velocity *parallel* to the magnetic field, this velocity component will be unchanged because of the cross product. Thus the most general motion in a constant magnetic field is a *helix*, with the velocity perpendicular to the magnetic field $\mathbf{v}^{\perp}$ playing the role of v above.

By using the uniform circular motion formula from Section 1.2.1, you can easily work out the radius of the circle:

$$R = \frac{mv}{qB}, \tag{2.36}$$

and the angular frequency, known as the *cyclotron frequency*:

$$\omega = \frac{qB}{m}. \qquad (2.37)$$

Amusingly, the formula for the cyclotron radius remains the same even at relativistic velocities, provided we replace the numerator mv with the relativistic momentum p (see Chapter 6).

2.2.7 Problems: Magnetostatics

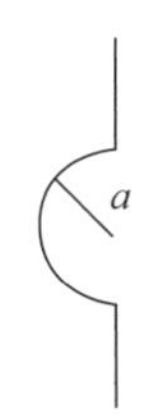

1. A wire consists of a half circle whose ends extend perpendicular to the circle as shown above. If current I flows downward through the wire, what is the magnitude of the magnetic field at the center of the circle?

   (A) $\mu_0 I/(4a^2)$
   (B) $\mu_0 I/(4a)$
   (C) $\mu_0 I/a$
   (D) 0
   (E) $\mu_0 I/(4\pi a)$

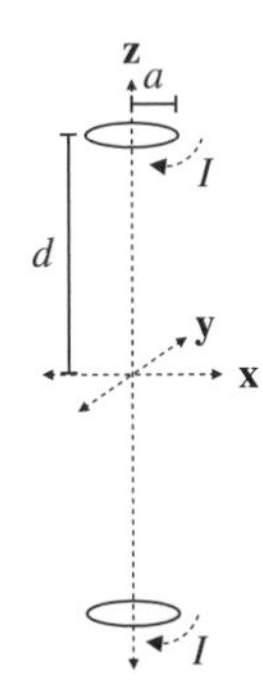

2. Two circular loops of wire, both with radius a, are oriented parallel to the xy-plane with their centers at $(0, 0, -d)$ and $(0, 0, d)$, where $d \gg a$. If both wires carry clockwise currents, which best describes the direction of the force from the loop at $z = d$ on electrons in the loop at $z = -d$?

   (A) Radially inward
   (B) Radially outward
   (C) In the $+\hat{\mathbf{z}}$ direction
   (D) In the $-\hat{\mathbf{z}}$ direction
   (E) There is no force.

3. What is the magnetic force per unit length between two parallel wires, separated by a distance d, each carrying current I in the same direction?

   (A) $\mu_0 I/(2\pi d)$, attractive
   (B) $\mu_0 I/(2\pi d)$, repulsive
   (C) $\mu_0 I^2/(2\pi d)$, attractive
   (D) $\mu_0 I^2/(2\pi d)$, repulsive
   (E) $\mu_0 I^2/(2\pi d^2)$, attractive

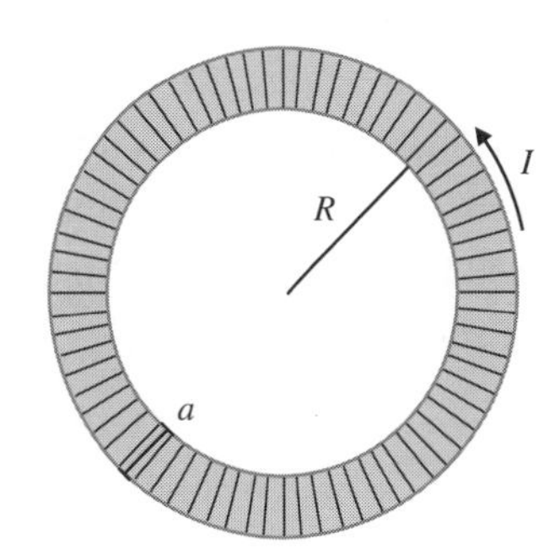

4. What is the magnetic energy stored in a toroid of wire with a square cross section of side length a, N total winds, inner radius R, and current I?

   (A) $\dfrac{\mu_0 N^2 I^2 a}{4\pi} \ln\left(\dfrac{R+a}{R}\right)$

   (B) $\dfrac{\mu_0 N^2 I^2 R}{4\pi} \ln\left(\dfrac{R+a}{R}\right)$

   (C) $\dfrac{\mu_0 N^2 I^2 a}{2\pi}$

   (D) $\dfrac{\mu_0 N^2 I^2 a}{4\pi}$

   (E) $\dfrac{\mu_0 N^2 I^2 R}{4\pi}$

2.3 Electrodynamics

So far we have given a treatment of how static electric fields and magnetic fields behave. The story becomes more complicated as we fill in the final piece of the puzzle and ask what happens when charges and currents – the sources of electromagnetic fields – move and change *in time*.

2.3.1 Maxwell's Equations

This story is summarized by the complete Maxwell's equations:

$$\nabla \cdot \mathbf{E} = \frac{\rho}{\epsilon_0}, \qquad (2.38)$$

$$\nabla \cdot \mathbf{B} = 0, \qquad (2.39)$$

$$\nabla \times \mathbf{E} = -\frac{\partial \mathbf{B}}{\partial t}, \tag{2.40}$$

$$\nabla \times \mathbf{B} = \mu_0 \mathbf{J} + \mu_0 \epsilon_0 \frac{\partial \mathbf{E}}{\partial t}. \tag{2.41}$$

The only changes from the static case are the time derivative terms in equations (2.40) and (2.41) which describe how a changing magnetic field produces an electric field, and how a changing electric field produces a magnetic field. The former is called inductance, and the latter is called the displacement current, also known as Maxwell's correction to Ampère's law. These are the main new effects that we must account for when we deal with electro*dynamics*. As it turns out, the displacement current is almost always too small an effect to be measured, so we will focus our attention on induction, which is much more practically important.

2.3.2 Faraday's Law

Integrating both sides of equation (2.40) over a surface S and using Stokes's theorem, we find the integral form of the equation:

$$\int (\nabla \times \mathbf{E}) \cdot d\mathbf{S} = \oint_C \mathbf{E} \cdot d\mathbf{l} = -\frac{d\mathbf{\Phi}_B}{dt},$$

where $\mathbf{\Phi}_B = \int_S \mathbf{B} \cdot d\mathbf{S}$ is the magnetic flux through the (not closed!) surface S with boundary C (a *closed* curve). In most cases of interest, the curve C is a loop of current such as a wire. The surface S can be any surface with C as the boundary; see Fig. 2.17 for an example. The middle term of this expression is just the electric potential around the loop (up to a sign), and the right-hand side is the change in magnetic flux. This expression is telling us that a changing magnetic flux through a loop of wire sets up a potential (and therefore a current) through the wire, much like a battery would. The electric potential in this context is often called by the unfortunate name *electromotive force* (emf) and denoted by $\mathcal{E}$. Most

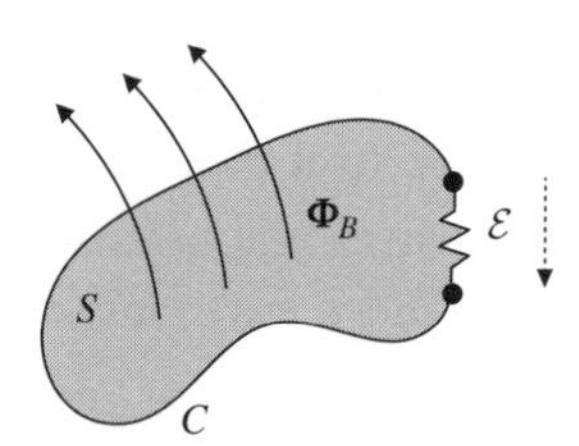

Figure 2.17 Example geometry described by Faraday's law. A wire along a curve C encloses a surface S, through which a changing magnetic flux Φ_B penetrates. A segment of the wire contains a resistor, across which there is a voltage or emf $\mathcal{E}$ in response to a changing magnetic flux. Note that the sign of the voltage must obey Lenz's law: the current induced by the emf must generate a magnetic field that opposes the external change in flux.

often, it is only the emf $\mathcal{E}$ that matters, so you can stick to the easier-to-remember form,

$$\mathcal{E} = -\frac{d\mathbf{\Phi}_B}{dt}, \tag{2.42}$$

provided you remember that (despite the notation) $\mathcal{E}$ is the potential, *not* the electric field.

Equations (2.40) and (2.42) are collectively referred to as Faraday's law. There is, however, one critical additional twist concerning the minus sign on the right-hand side of equation (2.42). This minus sign is often referred to as Lenz's law. The subtlety arises from the following problem. Imagine a wire loop with an increasing magnetic flux from some external source; Faraday's law implies there is a current induced in this wire. But the induced current in the wire also sets up a magnetic field itself. If both the external magnetic field and the magnetic field from the wire point in the same direction then there will be a runaway increase in magnetic field and current, violating conservation of energy! Clearly, energy conservation requires that these two fields point in opposite directions. This is the origin of the minus sign and the essential content of Lenz's law:

Induced currents always oppose changes in magnetic flux.

If you calculate everything correctly and use a minus sign in equations (2.40) and (2.42), then you should obtain the correct answer. But *always* check to make sure that the direction of current in your final solution does not end up violating conservation of energy.

2.3.3 Inductors

When two current loops are positioned close to each other, a changing current in one produces a time-varying magnetic field that can influence the other and vice versa. The flux Φ_{21} through loop 2 is proportional to the current I_1 in loop 1 via

$$\Phi_{21} = M_{12} I_1, \tag{2.43}$$

where M_{12} is a constant entirely dependent on geometry and known as the mutual inductance. It turns out that $M_{12} = M_{21}$, so this relationship is symmetric: $\Phi_{12} = M_{21} I_2 = M_{12} I_2$.

While mutual inductance rarely appears on the GRE, it is related to a much more common quantity that almost certainly will appear in some form on every exam: *self-inductance.* The self-inductance (or simply *inductance*) is generally defined to be the constant L in the expression

$$\Phi_B = LI, \tag{2.44}$$

in which the magnetic flux through an arrangement of wires, where the field is produced *by the wires themselves*, is proportional to the current carried by the wires. This immediately implies that

$$\mathcal{E} = -L\frac{dI}{dt}, \tag{2.45}$$

where I is the current. The self-inductance of an arrangement of wires is the magnetic analogy of the capacitance of an arrangement of conductors, and, as with capacitance, the self-inductance is purely determined by the geometry of the arrangement. If you ever need to calculate it directly, there is a set of simple guidelines:

- Calculate the magnetic field through the current loop and integrate it to obtain the magnetic flux. This can be done with either the Biot–Savart law or Ampère's law, but obviously Ampère's law is always preferable when the geometry permits.
- Plug the result into Faraday's law, and identify the coefficient L.

In the same way that "capacitor" usually refers to the particular parallel-plate model, "inductor" usually means a solenoid with a large number of turns carrying a current. Just like a capacitor, it produces a strong uniform field in a limited spatial region, and therefore is able to store energy. The inductance of this solenoid is

$$L = \frac{\mu_0 N^2 A}{l} \text{ (solenoid)}, \tag{2.46}$$

where N is the total number of turns, l is the length, and A is the cross-sectional area. Rather than derive this for you, we will leave the derivation as an exercise. The factor of N^2 is peculiar, so we recommend memorizing this expression so you don't get tripped up on the exam.

Solenoids store magnetic field energy just as capacitors store electric field energy. From equations (2.32) and (2.35), we can compute the total stored energy in terms of the inductance (2.46) and the current, which works out to

$$U_L = \frac{1}{2}LI^2. \tag{2.47}$$

2.3.4 Problems: Electrodynamics

1. A circular loop of wire of radius a and resistance R is oriented in the xy-plane. A uniform magnetic field of magnitude B points in the $+\hat{\mathbf{z}}$-direction. If the loop of wire is rotated about the x-axis with an angular frequency ω, what is the average power dissipated by Joule heating in the loop?

   (A) $\pi^2 a^4 B^2 \omega^2/2R$
   (B) $\pi^2 a^4 B^2 \omega^2/R$
   (C) $\pi^2 a^4 B^2/2R$
   (D) 0
   (E) $\pi a B\omega/R$

2. What is the inductance of a toroid with N winds, circular cross section of radius a, and overall radius R (from the center of the torus to the center of the circular cross section)? You may assume that $a \ll R$.

   (A) $\mu_0 a^2 N/(2R)$
   (B) $\mu_0 a N^2/2$
   (C) $\mu_0 a^2 N^2/(2R)$
   (D) $\mu_0 a N/2$
   (E) $\mu_0 R^2 N/(8a)$

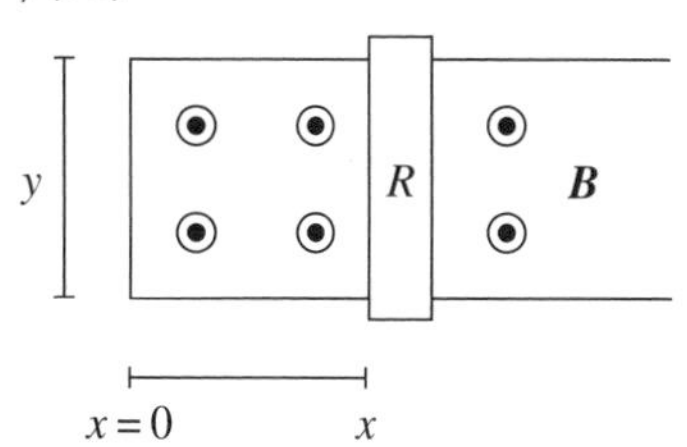

3. A rod of mass m and resistance R is attached to frictionless rails in the presence of a magnetic field of magnitude B pointing out of the page, as shown in the diagram above. The rod and rails form a closed electrical circuit. If the rod is launched from $x = 0$ with velocity v_0 to the right, at what time t is the velocity of the rod v_0/e? Assume that the rails have negligible resistance and neglect the self-inductance of the circuit.

   (A) $my/(R^2B^2)$
   (B) $my/(2R^2B^2)$
   (C) $mR/(y^2B^2)$
   (D) $mR/(2y^2B^2)$
   (E) The rod never reaches this speed because it travels at constant velocity v_0.

4. A capacitor made from two circular parallel plates of area A and separation d is connected in series to a voltage supply which maintains a constant current I in the circuit, charging the capacitor. What is the magnitude of the magnetic field between the parallel plates, as a function of r, the distance from the central axis of the plates? Assume r is smaller than the radius of the plates.

   (A) $\mu_0 r I/(2A)$
   (B) $\mu_0 I/(2\pi r)$

(C) $\mu_0 I/(2d)$
(D) $\mu_0 Id/(2\pi r^2)$
(E) 0

2.4 Dipoles

There is one final parallel between electricity and magnetism that is worth exploring. There are many cases when we are confronted with a situation in which two opposite charges are located very close to each other and we want to find the field far away. (If you remember your chemistry from high school, then you should know that a salt is a good example of a dipole.) Since there are no magnetic monopoles, most configurations of magnetic fields are dipoles. Although the mathematical formalism is identical in both cases, we'll examine them one by one.

2.4.1 Electric Dipoles

If we have two opposite charges q and $-q$ located at positions $\mathbf{r}_1$ and $\mathbf{r}_2$ respectively (Fig. 2.18), then we can define the dipole moment as

$$\mathbf{p} = q\mathbf{r}_1 - q\mathbf{r}_2 = q\mathbf{d}, \tag{2.48}$$

where $\mathbf{d}$ is the vector *from* the negative *to* the positive charge.[4] The dipole moment is a vector, so the dipole moment for several dipoles is the vector sum of the individual dipole moments. This leads to a trick for finding the dipole moments of funny charge configurations: pair them up into dipoles, and add up the vectors. More generally, for an arbitrary collection of point charges we have

$$\mathbf{p} = \sum_i q_i \mathbf{r}_i.$$

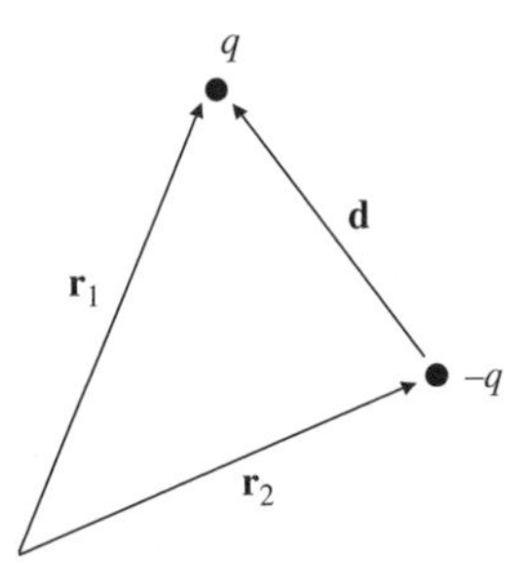

Figure 2.18 Two charges of opposite sign have a dipole moment proportional to their charge and the displacement $\mathbf{d}$ between them. The direction of the dipole moment points from the negative charge to the positive charge.

[4] Watch this sign convention! This is the opposite of the more familiar case where electric fields point from positive to negative charges.

Or, most generally, for a charge density $\rho(\mathbf{r})$, we have a dipole moment

$$\mathbf{p} = \int \mathbf{r}\rho(\mathbf{r})d^3\mathbf{r}. \tag{2.49}$$

You can recover the expression for the dipole moment of two opposite point charges $\pm q$ by just substituting delta functions for the charge density (try it yourself!). Dipoles have a nonzero electric field, which we will not dwell on here because knowing the precise form doesn't seem to be important for actually doing GRE problems. The potential is much easier to work with, and is simply

$$V(\mathbf{r}) = \frac{1}{4\pi\epsilon_0}\frac{\mathbf{p}\cdot\hat{\mathbf{r}}}{r^2}. \tag{2.50}$$

Notice that the potential of a dipole goes as $1/r^2$, which is *not* the same as the $1/r$ potential of a point charge! This comes from the Taylor expansion mentioned in Section 2.1.4, but it's an important enough fact that it's worth remembering on its own. Since $\mathbf{E}$ is related to V by a derivative, this means that the electric field of a dipole goes as $1/r^3$, rather than $1/r^2$.

Electric dipoles tend to align themselves with electric fields because the negative part gets pulled in one direction and the positive part gets pulled in the opposite direction. But since pure dipoles have net charge zero, they are not accelerated by electric fields – they just experience a torque. This torque is given by

$$\mathbf{N} = \mathbf{p} \times \mathbf{E}. \tag{2.51}$$

Since there is a torque, there is also a potential energy for a dipole in an electric field:

$$U = -\mathbf{p}\cdot\mathbf{E}. \tag{2.52}$$

2.4.2 Magnetic Dipoles

Magnetic dipoles behave quite similarly to electric ones, with one key conceptual difference. Magnetic dipoles are *not* composed of two opposite charges because there are no magnetic monopoles. Magnetic dipoles are pure and irreducible: if you chop a magnetic dipole in half, you only get two magnetic dipoles, not two monopoles.

Since we cannot build magnetic dipoles from monopoles like the electric case, we typically build them out of current loops. The magnetic dipole moment associated with a current loop carrying current I is given by

$$\mathbf{m} = I\mathbf{A}, \tag{2.53}$$

where $\mathbf{A}$ is a vector pointing normal to the surface subtended by the current loop, with magnitude equal to the area of the surface. As always in this business, the direction of the normal is fixed by the right-hand rule, the direction your right thumb points when your fingers curl around the direction of the current. The torques and potential fields due to a dipole in a magnetic field are analogous to the electric case:

$$\mathbf{N} = \mathbf{m} \times \mathbf{B}, \tag{2.54}$$

$$U = -\mathbf{m} \cdot \mathbf{B}. \tag{2.55}$$

And, just as with electric dipoles, the magnetic field of a magnetic dipole falls off as $1/r^3$.

2.4.3 Multipole Expansion

The idea of a dipole can be generalized with a tool known as the *multipole expansion*, a series that gives a quantitative measure to how "lumpy" a charge distribution is. If it has a net charge, the first term in the series is nonzero; if it is neutral but has a separation of charge within it, the second term is nonzero; and so on. While it is unlikely that you will have to compute anything with a multipole expansion, understanding it will help you guess correct answers. The potential due to an arbitrary charge configuration is given by equation (2.8). By expanding the fraction inside the integral in terms of Legendre polynomials, we arrive at a general series expression:

$$V(\mathbf{r}) = \frac{1}{4\pi\epsilon_0}\sum_{n=0}^{\infty}\frac{1}{r^{n+1}}\int (r')^n P_n(\cos\theta')\rho(\mathbf{r}')d^3\mathbf{r}'.$$

This looks complicated, but the idea is that the first few Legendre polynomials are quite simple, so it will be easy to evaluate the first few terms of this series. Since the first few terms dominate, this will often be a good enough approximation for the problem we wish to solve. In addition, *all* the dependence on $\mathbf{r}$ is now contained in the power series $1/r^{n+1}$, so the integral that remains is "easy" in the sense that it only depends on the coordinates $\mathbf{r}'$ of the charge distribution.

The first two terms are very simple, in fact. The first term is the monopole term, which is just given by the total charge of the configuration:

$$V_0(\mathbf{r}) = \frac{1}{4\pi\epsilon_0}\frac{Q}{r}.$$

As promised, the second term is the dipole term:

$$V_1(\mathbf{r}) = \frac{1}{4\pi\epsilon_0}\frac{\mathbf{p}\cdot\hat{\mathbf{r}}}{r^2},$$

where $\mathbf{p}$ is defined by (2.49). While this discussion only dealt with the scalar potential V, you could play the same game with the vector potential $\mathbf{A}$; in that case, by Gauss's law for magnetism, you would find that the monopole moment of a current distribution automatically vanishes, and the first term would be the dipole term (2.53).

2.4.4 Problems: Dipoles

1. Consider a pure electric dipole of moment $\mathbf{p} = p\hat{\mathbf{z}}$. A small test particle is located at $(0, 0, z)$ and experiences an electric field of magnitude E. What is the magnitude of the electric field experienced by a test particle at $(0, 0, 2z)$ from a dipole of moment $2\mathbf{p}$?

   (A) $E/4$
   (B) $E/2$
   (C) E
   (D) $2E$
   (E) $4E$

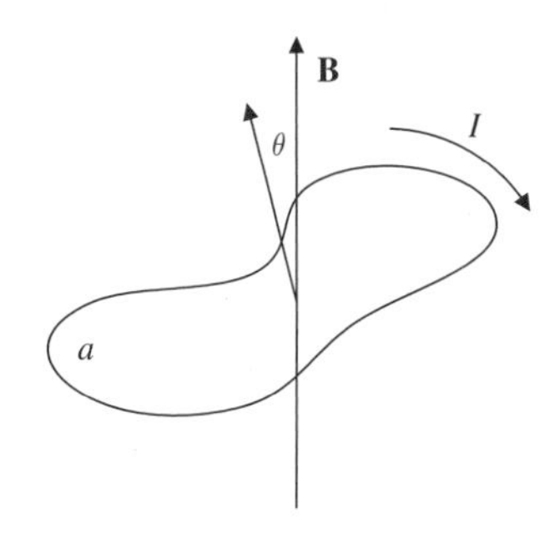

2. Suppose that a current loop of area a carrying current i, with moment of inertia I is placed in a uniform magnetic field of magnitude B. The normal to the loop is initially misaligned from the direction of the magnetic field by a small angle θ. When the loop is released, what is the period of oscillation?

   (A) $2\pi\sqrt{I/(iaB)}$
   (B) $2\pi\sqrt{i/(IaB)}$
   (C) $\sqrt{IaB}$
   (D) $2\pi/(IaB)$
   (E) $1/(IaB)$

2.5 Matter Effects

In everything that we have done so far with electric and magnetic fields, we have implicitly assumed that we are working in vacuum. This is rarely the case in the real world. Matter, for example, is often full of microscopic dipoles that can align themselves to slightly cancel out electric fields. Similar effects occur with magnetic fields in many materials. The behavior of electric and magnetic fields in matter used to be *much* more important on the GRE ten or twenty years ago, but these questions have gradually fallen out of fashion, paralleling a similar development in undergraduate physics curricula.

You are unlikely to see more than one question on your exam related to matter effects, so our treatment here will be even briefer than usual. In fact, the three most recently released exams from ETS had only one question about matter effects each, both times about capacitors and/or dielectrics.

2.5.1 Polarization

The primary effect of electric fields in matter is the dielectric effect: small dipoles in a material become slightly polarized by the presence of an external electric field. This can be described by a quantity called the *polarization* **P**, which is the *electric dipole moment per unit volume.* On the GRE, the most you will be asked to do is calculate the electric field, given the polarization. This is straightforward: **P** gives rise to effective surface and volume charge densities, known as *bound* charges:

$$\sigma_b = \mathbf{P} \cdot \hat{\mathbf{n}}, \tag{2.56}$$

$$\rho_b = -\nabla \cdot \mathbf{P}. \tag{2.57}$$

Here, $\hat{\mathbf{n}}$ is the outward-pointing normal to the surface of the polarized object. To calculate the electric field, just apply Gauss's law as usual to the bound charges σ_b and ρ_b, exploiting whatever symmetry is appropriate to the problem.

The situation becomes considerably more complicated in the presence of external electric fields, which will back-react on the polarization. There are also analogous bound currents for magnetized materials. These scenarios are treated in any advanced electrodynamics textbook, but are likely too advanced for the GRE.

2.5.2 Dielectrics

Dielectrics are materials, such as insulators, that can be polarized in an applied field, and thus slightly *cancel* the applied electric field. This effect can be parameterized by making the substitution

$$\epsilon_0 \mapsto \epsilon = \kappa\epsilon_0 \tag{2.58}$$

in all formulas (potential, electric field, etc.), where κ is known as the *dielectric constant*. The most common situation is when a dielectric is placed between the two plates of a parallel-plate capacitor, and the capacitance becomes

$$C = \frac{\epsilon A}{d} = \kappa\frac{\epsilon_0 A}{d}. \tag{2.59}$$

This is a common way to increase the capacitance of a capacitor.

2.5.3 Problems: Matter Effects

1. What is the work needed to insert a dielectric with dielectric constant $\kappa = 2$ into a parallel-plate capacitor with capacitance C that is maintained at a constant voltage V?

   (A) $\frac{1}{2}CV^2$
   (B) CV^2
   (C) $2CV^2$
   (D) $\frac{1}{2}\frac{C}{V^2}$
   (E) $\frac{C}{V^2}$

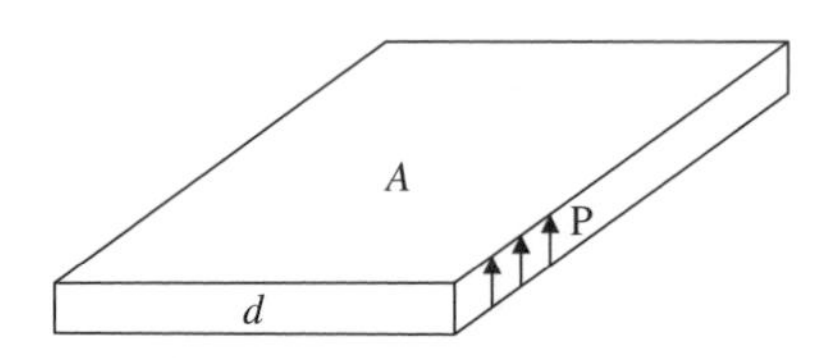

2. A thin slab of material of area A and thickness d carries uniform polarization **P**, as shown in the diagram. What is the magnitude of the electric field just above the slab, assuming $d^2 \ll A$?

   (A) $|\mathbf{P}|/(2\epsilon_0)$
   (B) $|\mathbf{P}|/\epsilon_0$
   (C) $2|\mathbf{P}|\epsilon_0$
   (D) $|\mathbf{P}|\epsilon_0$
   (E) 0

2.6 Electromagnetic Waves

Back in the early twentieth century, there was a famous conflict in physics between experiments such as diffraction, which indicated that light was a wave, and other experiments such as the photoelectric effect, which indicated that light was a particle. Now, of course, we know that light is both a particle and a wave in some sense. Here we will see the argument for light being a wave by showing how classical electrodynamics implies the existence of waves that travel at the speed of light.

2.6.1 Wave Equation and Poynting Vector

Start by taking the curl of equation (2.40). Using the identity from vector calculus that $\nabla \times (\nabla \times \mathbf{E}) = \nabla(\nabla \cdot \mathbf{E}) - \nabla^2\mathbf{E}$, and assuming that we are in vacuum where $\rho = 0$ and $\mathbf{J} = 0$ we have

$$\nabla(\nabla \cdot \mathbf{E}) - \nabla^2\mathbf{E} = -\frac{\partial}{\partial t}(\nabla \times \mathbf{B}),$$

$$\nabla^2\mathbf{E} = \mu_0\epsilon_0\frac{\partial^2\mathbf{E}}{\partial t^2}.$$

By a similar argument,

$$\nabla^2 \mathbf{B} = \mu_0 \epsilon_0 \frac{\partial^2 \mathbf{B}}{\partial t^2}.$$

As we will see in Chapter 3, these are just wave equations with velocity

$$c = 1/\sqrt{\epsilon_0 \mu_0}. \tag{2.60}$$

If you plug in the numbers, you really do find that $1/\sqrt{\epsilon_0 \mu_0}$ gives the correct speed of light. So, in vacuum, the electric and magnetic fields have wavelike solutions that travel at the speed of light. The wave solutions have the explicit form

$$\tilde{\mathbf{E}}(\mathbf{r}) = \tilde{E}_0 e^{i(\mathbf{k}\cdot\mathbf{r}-\omega t)} \hat{\mathbf{n}}, \tag{2.61}$$

$$\tilde{\mathbf{B}}(\mathbf{r}) = \frac{1}{c} \tilde{E}_0 e^{i(\mathbf{k}\cdot\mathbf{r}-\omega t)} (\hat{\mathbf{k}} \times \hat{\mathbf{n}}). \tag{2.62}$$

Here, $\hat{\mathbf{k}}$ is the propagation vector, describing the direction the wave travels in, and $\hat{\mathbf{n}}$ is the polarization vector. Note that when discussing electromagnetic waves, the polarization refers to the direction of the *electric* field only; the magnetic field is polarized in a perpendicular direction, as you can see from the cross product in equation (2.62). The three vectors $\hat{\mathbf{k}}$, $\hat{\mathbf{n}}$, and $\hat{\mathbf{k}} \times \hat{\mathbf{n}}$ form a right-handed coordinate system; the fact that the electric and magnetic field vectors are both perpendicular to the propagation vector means that electromagnetic waves in vacuum are *transverse.*

We are using some extremely convenient, but potentially confusing, notation here, where we are representing the electric and magnetic fields as complex exponentials. The magnitude $\tilde{E}_0$ may also be complex. There is nothing imaginary about the fields, but it simplifies the algebra considerably. The *physical* part of the fields is just the real part: $\mathbf{E} = \mathrm{Re}(\tilde{\mathbf{E}})$, and similarly for $\mathbf{B}$. So the rule for working in this notation is to calculate everything in the complex formalism as normal, and then take the real part at the end. This is straightforward when dealing with superpositions of waves, since the real part of a sum is the sum of the real parts, but for products it requires some care in definitions, as we'll see below. For a slightly more in-depth treatment of this material, see the following chapter, Section 3.1.

Since electromagnetic fields have energy, it should be perfectly natural to expect the wave solutions to transport energy. This is described by the Poynting vector:

$$\mathbf{S} = \frac{1}{\mu_0} (\mathbf{E} \times \mathbf{B}), \tag{2.63}$$

which gives the flux of energy of the wave (energy per unit area per unit time, or power per unit area). This expression

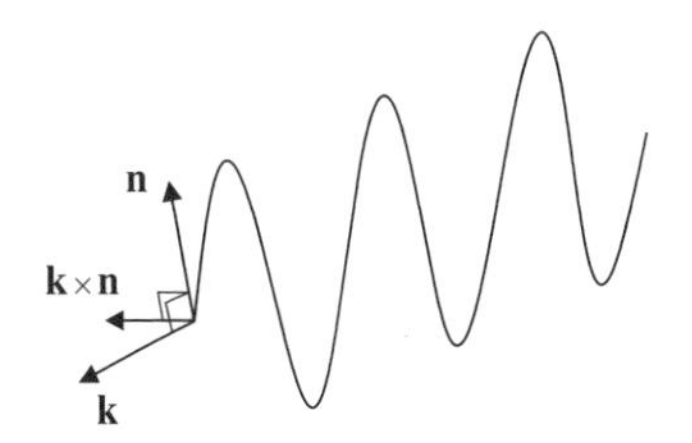

Figure 2.19 Vectors describing propagation and component fields for an electromagnetic wave. The wave propagates in the direction **k**, the electric field **E** is proportional to the vector **n**, and the magnetic field **B** is proportional to the vector **k** × **n**.

is in terms of the physical (real) fields: if you prefer to use complex notation, use the definition

$$\mathbf{S} = \frac{1}{2\mu_0} \mathrm{Re}(\tilde{\mathbf{E}} \times \tilde{\mathbf{B}}^*) \tag{2.64}$$

instead. Since electromagnetic waves have extremely high frequencies, it's often more useful to average the magnitude of the Poynting vector over one complete cycle in time, which gives the *intensity* $\langle S \rangle$, the *average* power per unit area. Using the helpful fact that the average of $\sin^2$ or $\cos^2$ over one cycle is $1/2$, we obtain

$$I = \langle S \rangle = \frac{1}{2} c \epsilon_0 E_0^2. \tag{2.65}$$

When we are analyzing electromagnetic waves, we often want to know what happens when we pass across an interface of different materials. This is the case in optics, for example, where it is possible to use electrodynamics to derive all of the formulas in geometric optics. We won't do this, but if you want a little practice, it's a nice way to check whether you truly understand all of this material. Snell's law is a good place to begin. The crucial trick when analyzing electromagnetic waves at boundaries is to use the boundary conditions that we described in detail earlier in this chapter. The schematic approach is:

- Figure out the generic form of the field on either side of the interface.
- Write the boundary conditions for the fields at the interface.
- Match the fields using the boundary conditions.

Conductors are good example of this process. The fields inside a perfect conductor must be zero. We also have the boundary condition that $\mathbf{E}^{\parallel}_{\mathrm{out}} = \mathbf{E}^{\parallel}_{\mathrm{in}} = 0$. So, if an electromagnetic wave is normally incident on a conductor, the electric component of the reflected wave points in the opposite direction to the incident wave so as to cancel off the parallel electric field just outside the conductor. Working through the cross products, the magnetic fields of the incident and reflected waves therefore point in the same direction.

2.6.2 Radiation

Since moving electric charges can create electromagnetic waves and these electromagnetic waves carry energy, it is possible to produce electromagnetic radiation far away from the charges. Indeed, an accelerating point charge radiates a total power

$$P = \frac{q^2 a^2}{6\pi\epsilon_0 c^3} = \frac{\mu_0 q^2 a^2}{6\pi c}, \tag{2.66}$$

where q is the charge of the oscillating particle, and a is its acceleration. The two forms are related to each other by equation (2.60). This formula, known as the Larmor formula, only holds when the charge is moving at small velocities, $v \ll c$. The prefactors are not so important, what really matters is the q^2 and a^2 dependence.

An oscillating *dipole* will also radiate, and in fact this situation is more common because most molecules in nature are dipoles rather than free charges. Let the dipole have dipole moment $\mathbf{p}(t) = p_0 \cos(\omega t)\hat{\mathbf{z}}$. There are two useful formulas here. The first is the intensity:

$$\langle S \rangle = \left(\frac{\mu_0 p_0^2 \omega^4}{32\pi^2 c} \right) \frac{\sin^2\theta}{r^2}, \tag{2.67}$$

where θ has its usual meaning in spherical coordinates. As usual, don't worry about the numerical factors, but what is important is the p_0^2 dependence (the same as the q^2 dependence in the Larmor formula), the frequency dependence ω^4, the fact that $\langle S \rangle$ falls off as $1/r^2$, and the $\sin^2\theta$ term which means that *no radiation occurs along the dipole axis.* Integrating (2.67) over a sphere of radius r gives the total power

$$\langle P \rangle_E = \frac{\mu_0 p_0^2 \omega^4}{12\pi c}, \tag{2.68}$$

which has the same p_0^2 and ω^4 dependence. By the way, the reason we deal with dipole radiation rather than monopole radiation (for example, a sphere of charge with an oscillating radius) is the curious fact that *monopoles do not radiate.* This actually follows from Gauss's law, which says the field outside a spherically symmetric charge distribution is independent of the size of the sphere.

The above formulas applied to electric dipoles only. There is an analogous formula for *magnetic* dipole radiation:

$$\langle P \rangle_B = \frac{\mu_0 m_0^2 \omega^4}{12\pi c^3}, \tag{2.69}$$

where m_0 is the average magnetic dipole moment. The only other change is the additional factor of $1/c^2$, which represents an *enormous* suppression compared to the electric case because c is so large. This means that electric dipole radiation will dominate unless the system is contrived to eliminate an electric dipole moment: for example, when current is driven around a wire loop.

2.6.3 Problems: Electromagnetic Waves

1. An AC current is driven around a loop of wire. Suppose the amplitude and frequency of the current are both doubled. By what factor does the power radiated by the antenna increase?

   (A) 4
   (B) 8
   (C) 16
   (D) 32
   (E) 64

2. A perfectly conductive plate is placed in the yz-plane. An electromagnetic wave with electric field $\mathbf{E} = E_0 \cos(kx - \omega t)\hat{\mathbf{y}}$ is incident on the conductor. If the wave strikes the plate at $t = 0$, what are the directions of the electric and magnetic fields of the reflected wave immediately after reflection?

   (A) $\mathbf{E} \propto -\hat{\mathbf{x}}$, $\mathbf{B} \propto -\hat{\mathbf{y}}$
   (B) $\mathbf{E} \propto \hat{\mathbf{y}}$, $\mathbf{B} \propto \hat{\mathbf{z}}$
   (C) $\mathbf{E} \propto -\hat{\mathbf{y}}$, $\mathbf{B} \propto -\hat{\mathbf{z}}$
   (D) $\mathbf{E} \propto -\hat{\mathbf{y}}$, $\mathbf{B} \propto \hat{\mathbf{z}}$
   (E) There is no reflected wave.

3. What is the speed of light in a medium with a permeability of $2\mu_0$ and a permittivity of $3\epsilon_0$?

   (A) c
   (B) $c/\sqrt{3}$
   (C) $c/2$
   (D) $c/\sqrt{6}$
   (E) $c/6$

2.7 Circuits

Depending on the flavor of your undergraduate education, your knowledge of circuits might be a little rusty. One of us actually never learned circuits in an undergraduate course, possibly because they were deemed too practical and not of fundamental importance! Apparently the GRE does not share this opinion, so it behooves you learn this material

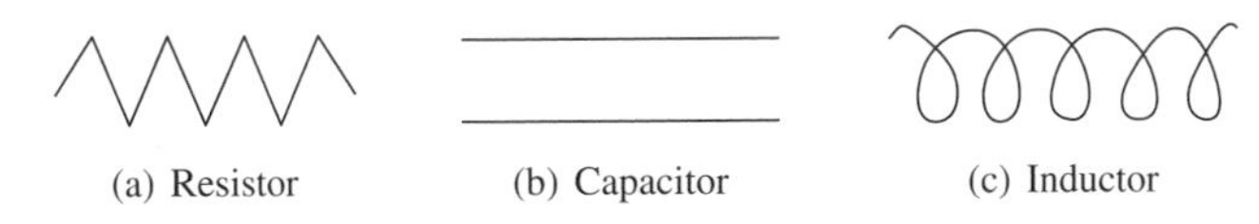

Figure 2.20 Symbols for three fundamental circuit elements.

well. Thankfully it is not too difficult at the level the GRE tests it.

2.7.1 Basic Elements

There are three fundamental circuit elements: resistors, capacitors, and inductors. Their icons in circuit diagrams are shown in Fig. 2.20. The voltages across each element are given by

$$V_R = IR, \tag{2.70}$$

$$V_C = \frac{Q}{C}, \tag{2.71}$$

$$V_L = L\frac{dI}{dt}, \tag{2.72}$$

where R is the resistance (units of ohms), C is the capacitance (units of farads), and L is the inductance (units of henries). When confronted with a circuit to analyze, one typically wants to find the current or voltage across some element. For simple circuits, this is done by subdividing the circuit into "equivalent" blocks and computing the total resistance, capacitance, or inductance of each section. The practical formulas for adding elements in series are

$$R_{\text{eq}} = \sum_i R_i \text{ (series)}, \tag{2.73}$$

$$\frac{1}{C_{\text{eq}}} = \sum_i \frac{1}{C_i} \text{ (series)}, \tag{2.74}$$

$$L_{\text{eq}} = \sum_i L_i \text{ (series)}. \tag{2.75}$$

In parallel, we have the rules

$$\frac{1}{R_{\text{eq}}} = \sum_i \frac{1}{R_i} \text{ (parallel)}, \tag{2.76}$$

$$C_{\text{eq}} = \sum_i C_i \text{ (parallel)}, \tag{2.77}$$

$$\frac{1}{L_{\text{eq}}} = \sum_i \frac{1}{L_i} \text{ (parallel)}. \tag{2.78}$$

These rules come from the fact that circuit elements in series see the same current, while circuit elements in parallel see the same voltage.

While capacitors and inductors are created by very particular geometries of conductors and wires, resistance is a property of all materials. More precisely, we define a material-dependent, intrinsic property called resistivity. The resistance is the impedance to electrical current for a particular geometrical configuration of that material. The general relation between the resistivity ρ of a material and the resistance of a geometry is given by

$$R = \frac{\rho\ell}{A}, \tag{2.79}$$

where A is the cross-sectional area of the resistor, and ℓ is the length.

2.7.2 Kirchhoff's Rules

A more general method for solving for currents and voltages in a circuit is to use Kirchhoff's rules. This method is more systematic for larger circuits, but it leads to a system of linear equations which can be cumbersome to solve quickly on the GRE. There are two rules, which are consequences of conservation of charge and energy, respectively:

1. The sum of currents flowing into every node must be zero:

$$\sum_k I_k = 0. \tag{2.80}$$

2. The sum of the voltages across elements around any closed loop must be zero:

$$\sum_k V_k = 0. \tag{2.81}$$

The strategy is then to write an equation for every node and loop in a circuit, and then solve them all simultaneously for the desired current or voltage. Since they are so systematic, Kirchhoff's rules are good to use when you are uncertain how to solve a problem using the series and parallel circuit rules above.

2.7.3 Energy in Circuits

An important distinction between the three circuit elements is between dissipative elements and elements that conserve energy. Resistors *dissipate* energy according to the famous rule

$$P = IV = \frac{V^2}{R} = I^2R, \tag{2.82}$$

and this power usually shows up as heat, which raises the temperature of circuit elements (which is why the back of your computer gets hot). In contrast, capacitors and inductors do not dissipate energy, but *store* it in electric and magnetic fields, as we have previously discussed. We'll repeat the formulas here for convenience:

$$U_C = \frac{1}{2}CV^2, \qquad \text{(p. 44)} \qquad (2.21)$$

$$U_L = \frac{1}{2}LI^2. \qquad \text{(p. 51)} \qquad (2.47)$$

2.7.4 Standard Circuit Types

Circuits generally have two kinds of behavior: transient, which describes charges and currents that die off quickly with time, and steady-state, which describes the state of the circuit after a sufficiently long time has passed. When analyzing circuits that contain a combination of different elements, the transient (as opposed to steady-state) behavior of the circuit is very important. The general approach for analyzing such circuits is to write down the voltage around some section of the loop (e.g. using Kirchhoff's rules), and then use the relations in equations (2.70)–(2.72) to produce a differential equation in I or Q. You can then solve the ODE with exponentials to extract the time dependence. For a more qualitative, but very useful, discussion of the time behavior of circuit elements, see Section 7.3.1.

- **RL circuits.** Consider the example of a circuit with a resistor and inductor in series with a voltage source. The circuit satisfies the equation

$$V = IR + L\frac{dI}{dt}.$$

If the voltage source is suddenly switched on, the current is

$$I = \frac{V}{R}\left(1 - e^{-t/\tau_{RL}}\right).$$

The time constant

$$\tau_{RL} = L/R \qquad (2.83)$$

is the characteristic response time of an RL circuit. All transient behavior of RL circuits with DC voltages is exponential with this characteristic time.

- **RC circuits.** The situation with RC circuits is very similar to RL circuits. An RC circuit is a circuit containing a resistor and a capacitor in series, and possibly a voltage source as well. The equation of current of an RC circuit with no voltage source is

$$0 = R\frac{dQ}{dt} + \frac{1}{C}Q.$$

The discharging of a capacitor in such a circuit (or the charging of a capacitor in a circuit with a voltage supply) is again exponential with characteristic time constant

$$\tau_{RC} = RC. \qquad (2.84)$$

- **RLC circuits.** Combining all three basic circuit elements gives the most interesting elementary circuit, the RLC circuit. Happily, the mathematics of this situation are *identical* to a problem we've already treated, namely masses and springs with friction. Just refer to Table 1.1 in Section 1.7.3 and make the appropriate substitutions. Note that an LC circuit can be considered a special case where the resistance R goes to zero. In that case, the resonant frequency of the circuit is

$$\omega_0 = \frac{1}{\sqrt{LC}}. \qquad (2.85)$$

2.7.5 Problems: Circuits

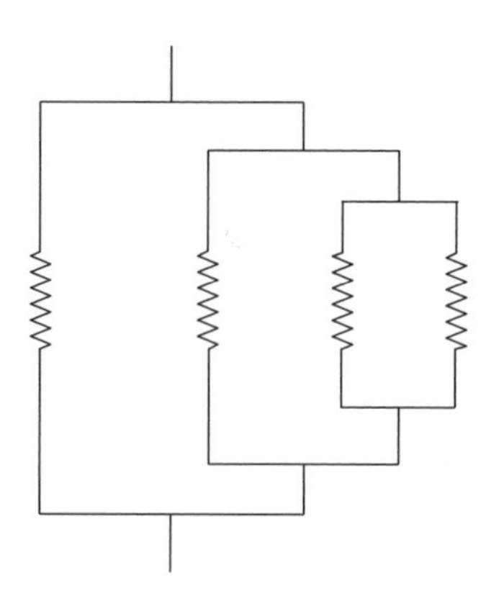

1. What is the equivalent resistance of the network above, if all resistors have resistance R?
   (A) R
   (B) $R/4$
   (C) $R/2$
   (D) $3R/4$
   (E) $4R/3$

2. A capacitor C is in series with a resistor R. The capacitor is initially charged, and a switch is closed at time $t = 0$ to complete the circuit. After what time t has the resistor dissipated half of the energy originally stored in the capacitor?
   (A) RC
   (B) $(RC \ln 2)$
   (C) $(RC \ln 2)/2$
   (D) $(RC)/2$
   (E) $(RC \ln 2)/4$

2.8 Solutions: Electricity and Magnetism

Electrostatics

1. A – By the method of images, the configuration is equivalent to two point charges, each a distance d from the putative conducting plane. The force on the image charge, and thus the magnitude of the force on the conducting plane is just given by the Coulomb force law:
$$F = \frac{q^2}{16\pi\epsilon_0 d^2}.$$
2. D – As we have calculated previously, the electric field due to a line of charge is given by
$$\mathbf{E} = \frac{\lambda}{2\pi\epsilon_0 r}\hat{\mathbf{r}}.$$
The potential between the line and the cylindrical shell is given by
$$V = \int_a^b \frac{\lambda}{2\pi\epsilon_0 r}dr = \frac{\lambda}{2\pi\epsilon_0}\ln\frac{b}{a}.$$
The capacitance per unit length is consequently
$$\frac{C}{\ell} = \frac{\lambda}{V} = \frac{2\pi\epsilon_0}{\ln(b/a)}.$$
3. C – The factor of 3 just comes by scaling the equation for the capacitance of a parallel-plate capacitor:
$$C = \frac{\epsilon_0 A}{d} = \frac{\epsilon_0 a^2}{d}.$$
4. D – The work needed to assemble the configuration is just given by the potential between each pairwise combination of vertices of the tetrahedron, i.e. the number of edges of a tetrahedron times the potential between two point charges. Each point charge is separated by a distance a and there are six edges on a tetrahedron, so the work needed is
$$W = \frac{3q^2}{2\pi\epsilon_0 a}.$$
5. E – To find the total charge enclosed, we can use the first of Maxwell's equations to find the charge density and then integrate it over the sphere. From the first Maxwell equation, we have
$$\rho = \epsilon_0 \nabla\cdot\mathbf{E} = \epsilon_0\frac{\partial}{\partial z}\left(E_0 z^2\right) = 2\epsilon_0 E_0 z.$$
The total enclosed charge is therefore (recalling $z = r\cos\theta$ in spherical coordinates)
$$Q = \int \rho(r)d^3\mathbf{r} = 2\epsilon_0 E_0 \int_0^R\int_0^\pi\int_0^{2\pi}(r\cos\theta)r^2\sin\theta\, dr\, d\theta\, d\phi = 0.$$
In hindsight, this makes sense: the electric field always points in the same direction throughout the sphere, so, thinking in terms of field lines, there is no net charge for the field lines to start or end on. This is equally apparent from the charge distribution $\rho = 2\epsilon_0 E_0 z$, which is positive for $z > 0$ and negative for $z < 0$, with the same magnitude on both sides of the plane $z = 0$.

Magnetostatics

1. B – We can use the Biot–Savart law to find the field. Note that there is no force from either straight portion of the wire, since $d\mathbf{l}$ and $\hat{\mathbf{r}}'$ are parallel and the cross product vanishes. Applying Biot–Savart to the semicircle, we have
$$|\mathbf{B}| = \frac{\mu_0}{4\pi}\int_0^\pi \frac{Ia d\phi\left|\hat{\boldsymbol{\phi}}\times(-\hat{\mathbf{r}})\right|}{a^2} = \frac{\mu_0 I}{4a}.$$
2. B – The clockwise current of the upper loop produces a magnetic field in the $-\hat{\mathbf{z}}$-direction. The electrons moving in the lower wire are moving in the $+\hat{\boldsymbol{\phi}}$-direction, and therefore feel a force
$$\mathbf{F} = q\mathbf{v}\times\mathbf{B} \propto -e\hat{\boldsymbol{\phi}}\times(-\hat{\mathbf{z}}) \propto \hat{\mathbf{r}},$$
in the outward radial direction in cylindrical coordinates. Beware: by convention current is the flow of positive charges, so negative charges move in the direction opposite to the "direction of current." So a clockwise current is a counterclockwise flow of electrons!
3. C – This is a classic problem and well worth remembering. The field from one wire is $\mathbf{B} = \frac{\mu_0 I}{2\pi r}\hat{\boldsymbol{\phi}}$. Letting the z-axis run along that wire, the force on a length dl of the other wire is then
$$d\mathbf{F} = (Idl\,\hat{\mathbf{z}})\times\frac{\mu_0 I}{2\pi d}\hat{\boldsymbol{\phi}} = \frac{\mu_0 I^2}{2\pi d}(-\hat{\mathbf{r}})dl.$$
Thus, the force per unit length $d\mathbf{F}/dl$ is $\mu_0 I^2/2\pi d$, towards the first wire. This can easily be remembered as "like currents attract."

4. A – Recall that the energy stored in a magnetic field $\mathbf{B}$ is given by equation (2.35),

$$U_B = \frac{1}{2\mu_0}\int |\mathbf{B}|^2 d^3\mathbf{r}.$$

Using the expression for the magnetic field of a toroid, we have (using cylindrical coordinates)

$$\begin{aligned} U &= \frac{1}{2\mu_0}\int_0^a\int_0^{2\pi}\int_R^{R+a}\frac{\mu_0^2N^2I^2}{4\pi^2r^2}r\,dr\,d\phi\,dz \\ &= \frac{\mu_0N^2I^2a}{4\pi}\int_R^{R+a}\frac{dr}{r} \\ &= \frac{\mu_0N^2I^2a}{4\pi}\ln\frac{R+a}{R}. \end{aligned}$$

Electrodynamics

1. A – The magnetic flux through the loop of wire is given by

$$\Phi_B = \pi a^2 B\cos\omega t.$$

By Faraday's law, we have

$$\mathcal{E} = \pi a^2 B\omega\sin\omega t.$$

The power dissipated by Joule heating is therefore

$$P = \frac{\mathcal{E}^2}{R} = \frac{\pi^2a^4B^2\omega^2\sin^2\omega t}{R},$$

or on average (using the fact that $\langle\sin^2\rangle = 1/2$),

$$\langle P\rangle = \frac{\pi^2a^4B^2\omega^2}{2R}.$$

2. C – To calculate the inductance we just calculate the magnetic field and flux through the toroid and then identify the coefficient of the current term. From Ampère's law, the magnetic field is simply

$$\mathbf{B} = \frac{\mu_0NI}{2\pi r}\hat{\boldsymbol{\phi}}.$$

Since the field is approximately constant inside the toroid (because $a \ll R$), we can just multiply by the cross-sectional area and the number of turns to obtain the magnetic flux:

$$\begin{aligned} \Phi_B &= N\pi a^2\frac{\mu_0NI}{2\pi R} \\ &= \frac{\mu_0a^2N^2I}{2R}. \end{aligned}$$

In this problem it is easy to miss the factor of N that must multiply the magnetic flux through one of the turns. Remember that, at a fixed magnetic field in the toroid, as the number of turns grows, the total flux must increase. The self-inductance is then just the coefficient of the current term:

$$L = \frac{\mu_0a^2N^2}{2R}.$$

3. C – The magnetic flux through the loop formed by the rod and rails is $\Phi_B(t) = x(t)yB$. The emf is given by

$$\mathcal{E} = -\dot{x}(t)yB.$$

Note that we have neglected the "back emf" from the magnetic field of the induced current. The power dissipated is

$$P = \frac{\mathcal{E}^2}{R}.$$

Using the work–energy theorem, the work performed on the rod, as a function of the velocity, is given by

$$W = \frac{1}{2}mv_0^2 - \frac{1}{2}m\dot{x}^2.$$

Setting power equal to the time derivative of the work, we find that

$$\frac{dv}{dt} = -\frac{vy^2B^2}{mR}.$$

The solution to this differential equation is just

$$v(t) = v_0\exp\left(-\frac{y^2B^2t}{mR}\right),$$

So the time needed to reach a velocity of v_0/e is given by

$$t_0 = \frac{mR}{y^2B^2}.$$

If this problem seems rather involved, note that E can be eliminated by common sense because the problem clearly involves induction, and A and B can be eliminated by dimensional analysis. It's a tough call between C and D, but one could plausibly guess that a factor of 2 should not enter into the solution.

4. A – While 0 is a tempting answer because there is no current flowing across the capacitor, it is not correct. In fact, a charging capacitor is about the only time you'll have to invoke the concept of displacement current. To find the answer carefully, note that the electric field between the two plates is approximately constant with magnitude

$$E = \frac{Q}{\epsilon_0A},$$

where Q is the charge on the plates. The charge on the plates is equal to the constant current times the time since the beginning of charging, so

$$E = \frac{It}{\epsilon_0A}.$$

From the integral form of Ampère's law with Maxwell's correction, we have

$$\oint \mathbf{B} \cdot d\mathbf{l} = \mu_0 \epsilon_0 \frac{d\Phi_E}{dt}.$$

Pick a circular loop between the parallel plates, centered at the axis of the plates, of radius r. The electric flux through this loop is

$$\Phi_E = \frac{\pi r^2 It}{\epsilon_0 A}.$$

The magnitude of the magnetic field is therefore

$$B = \frac{\mu_0 r I}{2A},$$

which you can think of as the field sourced by the fictitious "displacement current" between the capacitor plates. Note that without Maxwell's correction to Ampère's law, the answer *would* be zero!

Dipoles

1. A – Recall that the potential of a pure dipole scales as $V \sim \hat{\mathbf{r}} \cdot \mathbf{p}/r^2$. The electric field along the z-direction is simply the z-derivative of the potential V, so the electric field will scale as p/z^3 and will be reduced by a factor of 4 when p and z are both doubled.
2. A – The torque on the dipole is given by

$$\mathbf{N} = \mathbf{m} \times \mathbf{B}.$$

For small oscillations, we have (using Newton's second law in the form $N = I\ddot{\theta}$)

$$\ddot{\theta} = -\frac{iaB}{I}\theta.$$

The angular frequency is therefore

$$\omega = \sqrt{\frac{iaB}{I}},$$

and the period is

$$T = 2\pi\sqrt{\frac{I}{iaB}}.$$

Matter Effects

1. A – The potential energy stored in the capacitor before the dielectric is

$$U_i = \frac{1}{2}CV^2.$$

After the dielectric is introduced, the effective capacitance doubles, since $C' = \kappa C$. So the new energy stored is

$$U_f = CV^2,$$

and the total work done must be

$$W = \frac{1}{2}CV^2.$$

2. E – Since the polarization is uniform, the bound volume charge vanishes. The polarization is normal to the area A, so the bound surface charge is $\sigma_b = \pm|\mathbf{P}|$ on the upper and lower surfaces, respectively, and zero on the sides. Just above the surface, then, the slab looks like an infinite parallel-plate capacitor, where the fields from the two plates reinforce each other between the plates (here, inside the slab), but cancel outside. Thus, the field just above the slab is zero. This picture makes sense if we think of the polarized slab as a bunch of dipoles collected in a rectangular area, which will create an effective sheet of positive charge at the top and negative charge at the bottom.

Electromagnetic Waves

1. E - The total power radiated is proportional to $\omega^4 m_0^2$. Doubling the amplitude of the current doubles m_0, and doubling the frequency doubles ω, so overall the power changes by a factor of $2^6 = 64$.
2. D – Since the plate is a perfect conductor the electric field inside must be identically zero:

$$\mathbf{E}_{\text{inc}} + \mathbf{E}_{\text{refl}} = \mathbf{E}_{\text{transmitted}} = 0.$$

Since $\mathbf{E}^{\parallel}_{\text{out}} = \mathbf{E}^{\parallel}_{\text{in}}$ by the EM boundary conditions, the parallel component of the incident and reflected waves must cancel. At $t = 0$ the electric field vector of the incident wave is $\mathbf{E} = E_0\hat{\mathbf{y}}$, so the reflected wave must be polarized in the $-\hat{\mathbf{y}}$-direction. Since the propagation vector of the reflected wave is $-\hat{\mathbf{x}}$, the magnetic field of the reflected wave is in the $(-\hat{\mathbf{x}}) \times (-\hat{\mathbf{y}}) = \hat{\mathbf{z}}$-direction.
3. D – The speed of an electromagnetic wave in vacuum is

$$c = \frac{1}{\sqrt{\mu_0 \epsilon_0}}.$$

Making the substitutions $\mu_0 \mapsto 2\mu_0$ and $\epsilon_0 \mapsto 3\epsilon_0$, we find answer D.

Circuits

1. B – The rightmost parallel section has a resistance of $R/2$. Adding the next branch to the left we find a resistance $(1/R + 2/R)^{-1} = R/3$. Finally, adding the final section, we get a total resistance of $R/4$.
2. C – After the switch is closed, the charge on the capacitor decreases as

$$Q(t) = Q_0 \exp\left(-t/(RC)\right).$$

The energy initially stored in the capacitor is

$$U = \frac{Q_0^2}{2C},$$

so we want to solve for the time such that

$$\frac{Q(t)^2}{2C} = \frac{1}{2}\frac{Q_0^2}{2C},$$

$$\exp\left(-\frac{2t}{RC}\right) = \frac{1}{2},$$

$$t = \frac{RC \ln 2}{2}.$$

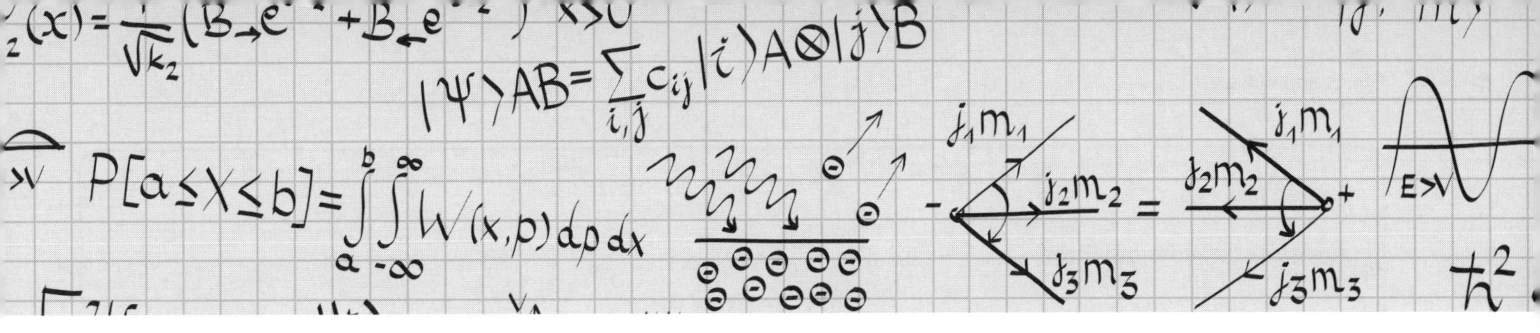

3 Optics and Waves

The Optics and Waves part of the GRE weighs in at 9% of the test, and contains a mix of some very basic material and some rather advanced material. Optics is a part of any standard freshman physics course, while waves appear in all areas of physics, and their treatment can vary greatly in difficulty and sophistication. For the purposes of the GRE, "optics" refers to geometric optics (lenses, mirrors, and so on), while "waves" refers to properties such as interference and diffraction as well as some more advanced topics such as Rayleigh scattering. We'll first discuss general properties of waves, including behavior such as diffraction and interference that can occur with any type of general wave, then we'll go over specific examples involving light waves, finishing with geometric optics. Many of the equations required for solving optics problems arise from fairly technical calculations that are outside the scope of the exam. It is therefore worth memorizing the key equations in this chapter and knowing the situations where they can be applied.

3.1 Properties of Waves

3.1.1 Wave Equation

Roughly speaking, a wave is a disturbance that propagates in time. More precisely, a wave (in one dimension) is any solution to the *wave equation*,

$$\frac{\partial^2 f}{\partial t^2} = v^2 \frac{\partial^2 f}{\partial x^2}. \tag{3.1}$$

It turns out that, for *any* function $f(x)$, the related functions $f(x \pm vt)$ solve the wave equation. We interpret these as disturbances of fixed shape, given by the function $f(x)$, which propagate either to the left or to the right with constant speed v. Now, the crucial property of the wave equation is its linearity in the function f, which leads to the *principle of superposition*: for any two solutions $f(x, t)$ and $g(x, t)$ that solve the wave equation, the function $f+g$ *also* solves the wave equation. This makes analyzing wave behavior quite easy, since we can always break up any complicated wave profile into a sum of simpler pieces.

The waves described above, $f(x \pm vt)$, are known as *traveling waves*: just as $f(x - a)$ represents the graph of $f(x)$ translated to the right by a units, so does $f(x - vt)$ represent the shape $f(x)$ translated to the right by $\Delta x = vt$ units after time t. In other words, the wave travels to the right. If, instead, a solution to the wave equation looks like $f(x, t) = A(x)B(t)$, we say that it represents a *standing wave*. Indeed, there is no longer any translation in time, and the shape at $t = 0$, given by $A(x)$ (up to a constant factor $B(0)$), is modulated in time by the function $B(t)$. See Example 3.1.

3.1.2 Nomenclature and Complex Notation

Often, instead of being given the wave equation (and the associated constant v, which represents the speed of the wave), we are simply given the functional form of the wave: for example,

$$f(x, t) = A\cos(kx - \omega t + \delta). \tag{3.2}$$

It turns out that all solutions to the wave equation can be written as sums of functions of this form, with different values of the constants A, k, ω, and δ; so without loss of generality, whenever we say "wave" we will often mean (3.2). Let's now give the constants names:

A: amplitude

k: wavenumber

EXAMPLE 3.1

Consider the standing wave

$$f(x,t) = \cos x \cos vt$$

for various values of t. At $t = 0$, the wave has a shape given by $\cos x$, but at time $t = \pi/2v$, it disappears entirely since the second term $\cos vt$ vanishes. At $t = \pi/v$, it now has the shape $-\cos x$, and so on. But the essential point is that, while the shape oscillates in time, it doesn't *go* anywhere, hence the term "standing wave." However, using the product-to-sum formula for cosines, we can write

$$f(x,t) = \frac{1}{2}\left(\cos(x+vt) + \cos(x-vt)\right),$$

so in fact the standing wave is secretly the sum of a left-moving wave and a right-moving wave, with equal amplitudes. In fact, we can decompose *any* solution of the wave equation into the sum of a left-moving and a right-moving wave, possibly of different shapes and different amplitudes.

ω: angular frequency

δ: phase

The amplitude A is the maximum value of the function $f(x,t)$, and represents the "size" of the wave. It is a general fact that the energy carried by waves is proportional to the *square* of the amplitude, A^2. In the very common case of light waves, the energy (up to some constants) is called the *intensity*.

The wavenumber k is related to the period of $\cos x$ as a function of x: since cosine has period 2π, $\cos kx$ has period $2\pi/k$. In other words, the wave starts repeating after a distance $\lambda = 2\pi/k$, called the *wavelength*. Up to a factor of 2π, the wavenumber counts the number of wave crests contained within a distance x, but the more useful quantity is often the wavelength. Similarly, ω is related to the period T by $T = 2\pi/\omega$, and to the (ordinary, not angular) frequency[1] f by $\omega = 2\pi f$. Summarizing these definitions,

$$\lambda = \frac{2\pi}{k}, \qquad T = \frac{2\pi}{\omega}, \qquad \omega = 2\pi f. \tag{3.3}$$

Finally, the phase represents the "offset" of the shape of the cosine function from the usual one centered at $x = 0$. You've likely seen all these names before, since, with the exception of wavenumber, the notation and nomenclature are identical to those of the classical simple harmonic oscillator. These same concepts also show up in quantum mechanics, with k related to momentum by the de Broglie relation $p = \hbar k$ and frequency related to energy by the Einstein relation $E = \hbar\omega$.

The above information is conveniently represented in terms of complex numbers: we can write

$$f(x,t) = \mathrm{Re}(Ae^{i(kx-\omega t)}),$$

where the amplitude A is now allowed to be complex, $A = |A|e^{i\delta}$. This allows the phase to be absorbed in the amplitude, and by working with complex numbers through the whole calculation and taking the real part at the end, we can avoid using some of the more annoying trig identities.[2]

The generalization to waves traveling in three dimensions is quite simple. We just replace kx with $\mathbf{k}\cdot\mathbf{r}$, and call $\mathbf{k}$ the *wavevector*. Its magnitude $|\mathbf{k}|$ is the wavenumber, and its direction is the direction in which the wave propagates. Such a wave solves the wave equation in three dimensions in Cartesian coordinates, and is called a *plane wave*. This is in contrast to a spherical wave, whose oscillations take the form $e^{i(kr-\omega t)}$, where r is the distance from the origin in spherical coordinates. Note the distinction from the case of plane waves: a spherical wave has a wavevector that is always parallel to the position vector $\mathbf{r}$, which is *not* constant.

[1] In many physics texts, the frequency is denoted by ν, the Greek letter "nu." But this looks too much like v, which we've used for wave velocity, so to avoid any notational confusion we'll use f for frequency wherever possible, which is the standard convention in engineering. Also, be careful not to confuse f and ω! The first quantity f is the frequency in of units of Hz or cycles per second. The second quantity is the rate at which the argument of the sine or cosine functions change, in units of radians per second. When working with waves, it's easy to forget the factors of 2π that connect these quantities.

[2] Caveat: expressing quantities that are *products* of waves in complex notation requires tweaking some definitions, since the real part of a product of complex numbers is not the product of the real parts. See the discussion of the Poynting vector in Section 2.6.1. However, complex notation is entirely straightforward for linear superpositions of waves. You probably won't need complex notation for the GRE, but it's good to be aware of it in case it helps you solve a problem faster.

3.1.3 Dispersion Relations

For waves satisfying the wave equation (3.1), the relation

$$\omega = vk \tag{3.4}$$

holds: you can check this by simply plugging (3.2) into (3.1). If we modify the wave equation, for example to incorporate phenomena such as frictional dissipation or changes in the density of the medium, we lose this property, but we can still specify the behavior of the wave by a relation between ω and k, known as a *dispersion relation*. The special case $\omega(k) = vk$ is known as linear dispersion, and is the one obeyed by light in vacuum, with $v = c$ the speed of light. You may have seen it written in the more familiar form $\lambda f = c$. From the dispersion relation, we can calculate two quantities with the dimensions of velocity:

$$\text{Phase velocity: } \frac{\omega}{k}, \tag{3.5}$$

$$\text{Group velocity: } \frac{d\omega}{dk}. \tag{3.6}$$

For linear dispersion, these quantities turn out to be the same, but this is a special case.

In general, phase velocity measures the velocity of an individual wave crest, while group velocity measures the velocity with which a whole bunch of waves (for example, a wave packet of various wavelengths centered around a certain wavelength λ_0) move together. For light waves traveling in a medium (for example, radio waves carrying an FM signal or light signals in a fiberoptic pipe), the group velocity is the speed at which information is transmitted. In fact, there are cases where the phase velocity can be *greater* than the group velocity! If the phase velocity turns out to be greater than the speed of light c, you shouldn't be worried, since information is only transmitted at the group velocity and there is no conflict with special relativity. The fact that the phase velocity can in general be a function of k implies that waves of different wavelengths travel at different speeds; in other words, a wave packet is *dispersed*. This is exactly what happens when white light is passed through a prism.

3.1.4 Examples of Waves

We can see some of these general considerations in action by looking at a few common instances of waves. The amplitude y of waves on a uniform vibrating string, for example, is described by the equation

$$\frac{\partial^2 y}{\partial t^2} = \frac{T}{\mu}\frac{\partial^2 y}{\partial x^2},$$

where μ is the mass density per unit length and T is the tension in the string. This is exactly the same as equation (3.1), with the replacement

$$v = \sqrt{\frac{T}{\mu}} \tag{3.7}$$

for the velocity of the wave. Memorizing the wave equations for every possible type of wave would obviously be a bit of a nuisance. Thankfully this is unnecessary, since you should usually be able to reconstruct velocities such as equation (3.7) from other physical constants by dimensional analysis. This expression also makes physical sense: increasing the tension in the string makes the restoring force on a displaced segment of string greater, and so the oscillation will occur more rapidly. Similarly, increasing the mass density of the string means a displaced segment of string will accelerate more slowly, reducing the speed of propagation.

Another example is sound waves. By analogy with the string, we might suspect that the parameters that would determine the speed of sound waves in air would be some measure of inertia, such as the density ρ (units of kg/m^3), and some measure of "stiffness," such as the bulk modulus K (units of pressure, Pa or N/m^2). The only combination of these factors with the units of speed is

$$c_s = \sqrt{\frac{K}{\rho}}.$$

Using numbers appropriate for air at STP, this is approximately 340 m/s, as expected.

3.1.5 Index of Refraction

The optical properties of a medium can be roughly summarized by a single number n called the *index of refraction*, which is the factor by which the speed of light is reduced in that medium. In the language of dispersion relations

$$\omega/k = c/n \quad \text{(for light waves)}. \tag{3.8}$$

Consider a beam of light with frequency ω and wavenumber k in vacuum, incident on a medium with index of refraction n. Inside the medium, the speed of light changes to c/n, so clearly either ω or k (or both) must change to satisfy (3.8). An analysis of Maxwell's equations at the boundary of the medium shows that the *frequency* of the light in the medium does *not* change. Rather, the wavelength λ is modified to

$$\lambda \to \frac{\lambda}{n} \tag{3.9}$$

so that (3.8) holds. By the way, real materials, such as glass, do *not* have a constant index of refraction; rather, $n = n(k)$ is a function of wavenumber, hence dispersion of white light. But in nearly every application on the GRE you can consider n to be a constant that represents the reduction of the speed of light in that medium, or, equivalently, the factor by which the wavelength is reduced.

3.1.6 Polarization

So far we have written waves as real-valued (or sometimes complex-valued) functions. But electromagnetic waves are *vector*-valued because $\mathbf{E}$ and $\mathbf{B}$ are both vectors. The only new ingredient in the mathematical description of the wave is the polarization unit vector $\hat{\mathbf{n}}$. For example, we can write the electric field component of an EM wave as

$$\mathbf{E}(x, t) = E_0 \cos(\mathbf{k} \cdot \mathbf{r} - \omega t + \delta)\hat{\mathbf{n}},$$

so while the amplitude is E_0, the vectorial information is contained in $\hat{\mathbf{n}}$. If a source of light emits EM waves with constant $\hat{\mathbf{n}}$, we say the light is *polarized* in the direction $\hat{\mathbf{n}}$, as we have already discussed in Section 2.6.1. If the oscillating vector is polarized perpendicular to the wavevector (as is the case for EM waves in vacuum), the wave is called *transverse*; if the vector is parallel to the wavevector (as can occur for EM waves in waveguides), the wave is *longitudinal*.

There are two very common types of applications of polarization that show up on the GRE:

- **Malus's law.** Suppose we have a device, called a polarization filter or polarizer, for which all light exiting the device is polarized in a certain direction $\hat{\mathbf{n}}_0$ (Fig. 3.1). Then for incident light of intensity I_0, which is polarized at an angle θ with respect to $\hat{\mathbf{n}}_0$ (i.e. $\hat{\mathbf{n}} \cdot \hat{\mathbf{n}}_0 = \cos\theta$), the intensity I of the transmitted light is given by Malus's law:

$$I = I_0 \cos^2\theta. \tag{3.10}$$

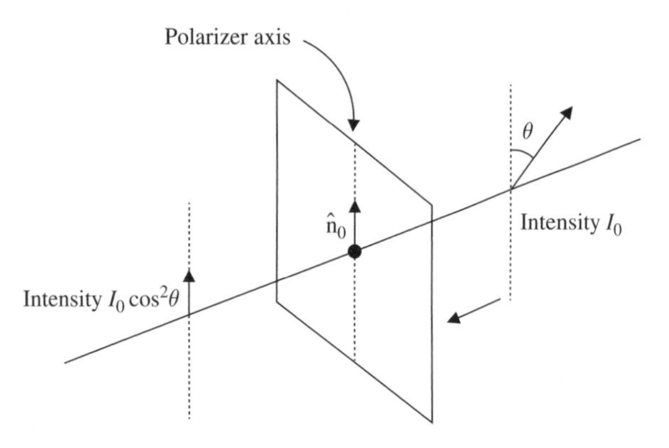

Figure 3.1 Polarized light of intensity I_0 is incident on a polarizer. If the light is polarized at an angle θ with respect to the polarizer axis $\hat{\mathbf{n}}_0$, then the light emerging from the polarizer will have an intensity $I_0 \cos^2\theta$ according to Malus's law, and it will be polarized in the direction $\hat{\mathbf{n}}_0$.

This implies some curious properties of polarizers. For example, if we have an arrangement of two polarizers oriented at 90 degrees with respect to one another, the intensity of transmitted light will be *zero*, independent of the initial polarization: light will hit the first filter, emerge polarized with some reduced intensity, then be promptly absorbed by the second filter, since $\cos(90°) = 0$. However, if we place a third filter in between the first two, oriented at 45 degrees with respect to the filters on either side, the emitted intensity of incident light polarized parallel to the first filter is $I_0 \cos^4(45°) = I_0/4$, where I_0 is the intensity of light emerging from the first filter. So in this case, placing an extra filter *increases* the intensity of transmitted light, despite the usual intuition that a filter removes light rather than augments it. Another important application is when *unpolarized* light is shined on a polarizer: the transmitted intensity is then the average of $I_0 \cos^2\theta$ over all angles θ, which works out to be $I_0/2$. So unpolarized light incident on a polarizer of arbitrary orientation comes out parallel to the polarizer axis, with intensity reduced by half.
- **Brewster's angle.** Suppose we have two media with indices of refraction n_1 and n_2, joined at an interface. Some formidable manipulation of Maxwell's equations implies that there exists an angle θ_B at which incident unpolarized light going from n_1 to n_2 will reflect off the interface and emerge *completely polarized* perpendicular to the plane formed by the incident ray and the normal to the surface (the incident plane) (Fig. 3.2). θ_B is known as Brewster's angle, and is given by

$$\theta_B = \arctan\left(\frac{n_2}{n_1}\right). \tag{3.11}$$

By the same reasoning, light polarized *parallel* to the incident plane incident at θ_B will not be reflected at all. The polarization properties of reflected light are what make Polaroid sunglasses useful: even if the incident angle is not exactly θ_B, the reflected light will still be mostly polarized in one direction, so sunglasses whose polarization filters are perpendicular to this direction will block most of the reflected light, reducing glare off a road or off water. Indeed, picturing this scenario helps to make concrete the rather abstract and confusing term "incident plane." If the surface is a flat road, the normal will be a vertical line pointing towards the sky, and if the light beam is coming right

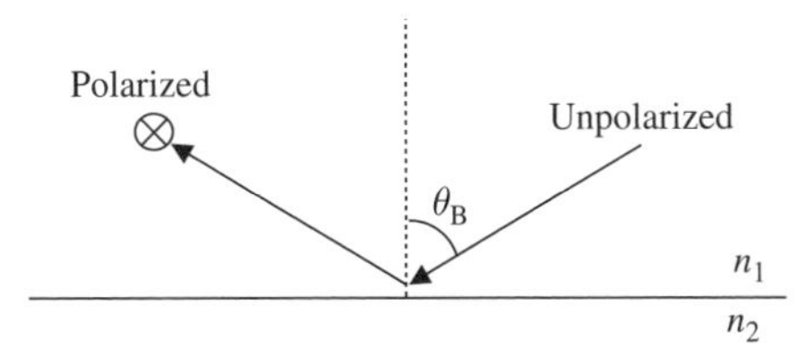

Figure 3.2 Unpolarized light travels in a medium with index n_1 and reflects off an interface with a medium of index n_2. If the angle of incidence (relative to normal incidence) is Brewster's angle θ_B, then the reflected light will emerge linearly polarized in a direction perpendicular to the page.

at you, the polarization of the reflected beam is perfectly horizontal.

3.2 Interference and Diffraction

One of the more striking consequences of the superposition principle for waves is *interference*. Consider two waves:

$$f(x,t) = A\cos(kx - \omega t),$$
$$g(x,t) = A\cos(kx - \omega t + \pi).$$

Both f and g are perfectly acceptable solutions to the wave equation, differing only by their phase δ, but their sum, $f + g$, is identically zero! The phase of g is $\delta = \pi$, so at every (x,t), $g(x,t) = -f(x,t)$, and the two waves cancel identically at every spatial point and time. This is known as *destructive* interference, and occurs whenever the phase difference is an odd multiple of π. Adding 2π to the phase changes nothing because 2π is the period of cosine, so the same destructive interference argument works for π, 3π, etc. Similarly, if the phase difference were $\delta = 0$, the two waves would be identical and would simply add to give a wave of amplitude $2A$; this is known as *constructive* interference. Once again, this argument also holds for phase differences of 2π, 4π, etc. To summarize:

$$\text{Constructive interference} \Longleftrightarrow \text{phase difference of } 2m\pi, \tag{3.12}$$

$$\text{Destructive interference} \Longleftrightarrow \text{phase difference of } (2m+1)\pi, \tag{3.13}$$

where m can be any integer. The classic examples of interference are when a phase difference arises from a difference in path length, or from traveling between media with different indices of refraction. The former is exemplified by double-slit interference and single-slit diffraction, while the latter arises in situations involving thin films.

3.2.1 Double-Slit Interference

The setup for Young's classic double-slit experiment is a point source of monochromatic light (wavelength λ), shining on a barrier with two narrow slits cut in it a distance d apart, with a screen at a distance L behind the slits (Fig. 3.3).

The waves from the two slits interfere because, at a given point on the screen, light from the two slits will have traveled a different distance; let's call that path difference Δx. This leads to a phase shift

$$\delta = k\Delta x. \tag{3.14}$$

To see why, note that one wave will arrive at the screen with functional form $A\cos(kx - \omega t)$, but the other wave will arrive as $A\cos(k(x+\Delta x) - \omega t) = A\cos(kx - \omega t + k\Delta x)$, where the last term in the cosine can be interpreted as a phase. For constructive interference, we need $\delta = 2m\pi$, so $\Delta x = 2m\pi/k = m\lambda$, and for destructive interference, we need $\delta = (2m+1)\pi$, so $\Delta x = (2m+1)\pi/k = (m+1/2)\lambda$. In words,

Constructive interference occurs when the path difference is an integral multiple of the wavelength, and destructive interference occurs when the path difference is a half-integral multiple of the wavelength.

On the screen, we will see bright bands at the locations of constructive interference, and dark bands at the locations of destructive interference. The bright bands and the dark bands are called interference maxima and minima, respectively.

If the screen distance L is much larger than the distance between the slits d, we can find the positions of the interference maxima and minima using a little geometry. When $L \gg d$, the paths of the two light rays to a fixed point on the screen are very nearly parallel, so the path difference to a point at an angle θ from the center is shown from Fig. 3.4 to be $d\sin\theta$.[3]

The maxima and minima then satisfy:

$$\text{Maxima: } d\sin\theta = m\lambda, \tag{3.15}$$
$$\text{Minima: } d\sin\theta = (m+1/2)\lambda. \tag{3.16}$$

Here, m is any integer, positive, negative or zero. Conventions differ, but the problem statement will always be unambiguous with regards to counting maxima and minima. Rather than having to draw the diagram over and over, it's best to just memorize these relations. An easy way to remember which

[3] In Fig. 3.3, θ is measured from the top slit, but in the limit $L \gg d$, θ is the same whether it is measured from the top slit, the bottom slit, or between the two slits. This is *always* the approximation that will be used in double-slit interference, and is known as the Fraunhofer or far-field regime.

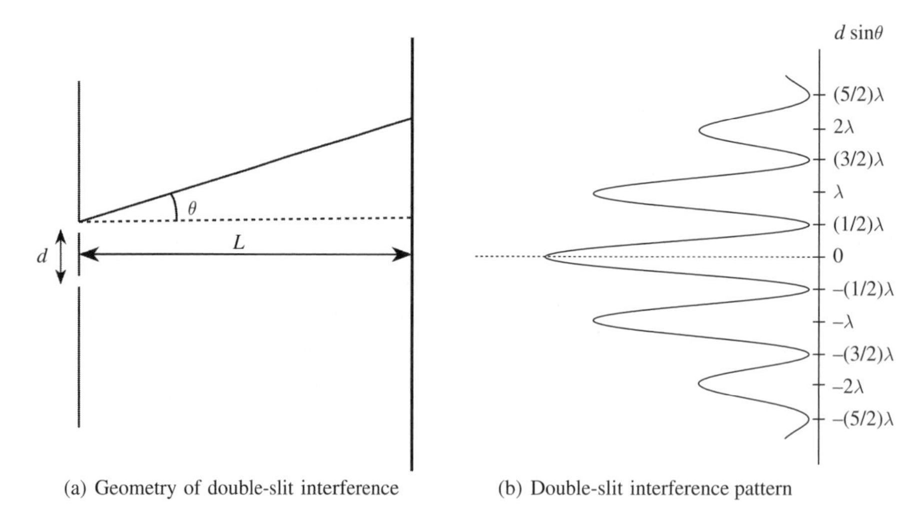

(a) Geometry of double-slit interference (b) Double-slit interference pattern

Figure 3.3 The interference pattern produced behind a double slit has a maximum when $d \sin\theta$ is an integer multiple of the wavelength λ and a minimum when $d \sin\theta$ is a half-integer multiple of the wavelength.

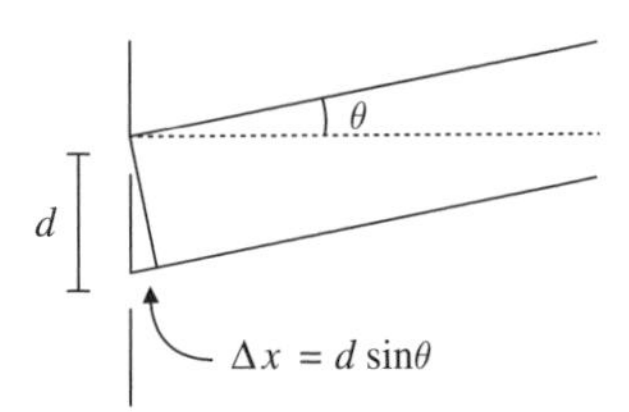

Figure 3.4 Double-slit interference arises because of the difference in path lengths $d \sin\theta$ between light passing through each slit. The path-length differences produce interference maxima and minima as a function of the angle θ relative to the horizontal.

formula applies to maxima is to note that $\theta = 0$ must be a maximum since both waves travel the same distance: this means the right-hand side must be an integer rather than a half-integer, giving (3.15).

3.2.2 Single-Slit Diffraction

We now consider a modification of the above setup, replacing the two thin slits with a single slit of width a (Fig. 3.5).

Once again, light that passes through different parts of the slit travels different lengths on the way to the screen, but the derivation of the path difference involves a lot of handwaving and is totally useless for the GRE. Instead we will just write down the answer: interference *minima* occur when

$$a \sin\theta = m\lambda, \qquad m = 1, 2, \ldots \tag{3.17}$$

Fortunately this is easy to remember because of the similarity with the double-slit formula, but don't confuse the two equations! The *minima* for single-slit diffraction occur at nearly the same condition as the *maxima* for double-slit diffraction. There is no simple formula for the position of the diffraction maxima for the single slit, although there is clearly a maximum right in the center of the slit; hence we have explicitly specified that $m = 0$ is *not* a minimum. The first minima mark the width of this central maximum. We also see from the diffraction equation why we need $a \sim \lambda$ to see any diffraction effects at all. If $a < \lambda$, then the equation (3.17) has no solution for θ: there are no minima, and the central maximum fills the entire screen. On the other hand, if $a \gg \lambda$, the diffraction minima are so closely spaced that they blur together.

3.2.3 Optical Path Length

Consider the following situation: a monochromatic beam of light of wavelength λ in a vacuum passes through a medium of index of refraction n, while a nearby beam misses the medium and continues traveling through the vacuum. When the two beams are recombined, will there be interference? At first glance, it seems like there won't be any interference as long as both beams travel the same distance, since there is no path difference like the $k\Delta x$ of the double-slit setup.

But take a look back at equation (3.9): in the medium with index n, the wavelength of light is reduced by a factor of n while its frequency remains the same. If the wave in vacuum travels a distance d, the total phase (i.e. the argument of the sine or cosine) of the wave increases by

$$\delta_1 = kd = \frac{2\pi d}{\lambda},$$

but the total phase of the wave in the medium increases by

$$\delta_2 = \frac{2\pi d}{\lambda/n} = \frac{2\pi nd}{\lambda}.$$

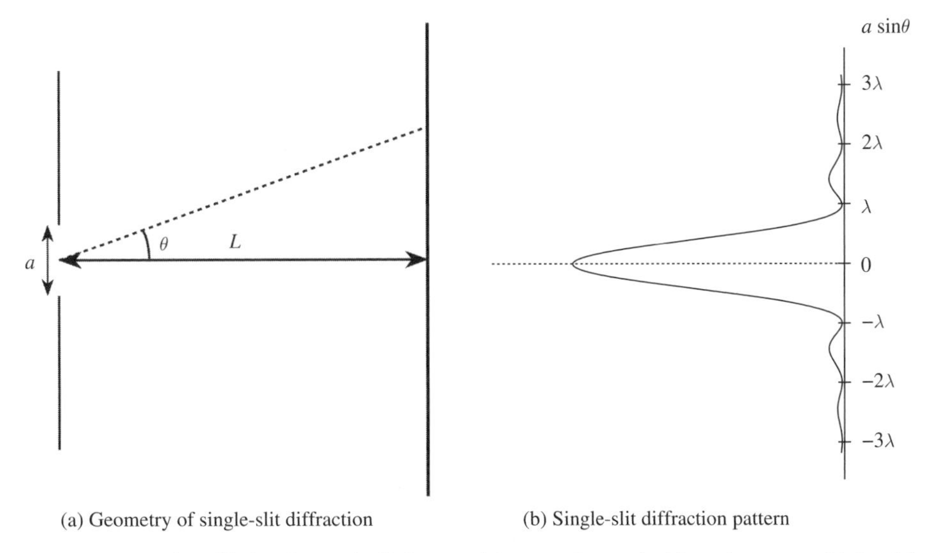

(a) Geometry of single-slit diffraction (b) Single-slit diffraction pattern

Figure 3.5 The diffraction pattern produced behind a single slit has a minimum when $a \sin \theta$ is an integer multiple of the wavelength λ.

Notice that $\delta_1 \neq \delta_2$! This means that, in general, there *will* be interference when the beams are recombined. In fact, the interference pattern will behave precisely as if the wave in the medium has traveled a distance nd while the wave in vacuum traveled a distance d. This suggests that we define the *optical path length* by

$$\Delta x = nd \qquad \text{(optical path length)}, \tag{3.18}$$

where d is the actual distance traveled in the medium. Now that we have this definition, you can feel free to forget all the explanation that came before it: just remember that when you see a situation where a wave passes through a distance d of a medium of index of refraction n, assign it a path length of nd and *not* just d. An easy way of remembering whether the factor of n goes in the numerator or denominator is to consider the limiting case of a medium with $n \to \infty$: there, the wave will be slowed down so much that, as it passes through the medium, it undergoes infinitely many cycles, sending the phase difference to infinity.

3.2.4 Thin Films and Phase Shifts

An additional source of phase shifts is reflection off a boundary between two media. If light is traveling from a medium with index of refraction n_1 toward a medium with index of refraction n_2 and reflects back off the boundary, Maxwell's equations imply

$$n_2 > n_1 : \text{phase shift of } \pi, \tag{3.19}$$

$$n_2 < n_1 : \textit{no} \text{ phase shift.} \tag{3.20}$$

To actually derive these from Maxwell's equations takes an unwieldy amount of algebra, so just memorize them. In a typical thin-film setup, one considers light traveling through air ($n \approx 1$) and striking the boundary of a thin film of soap, oil, or some other material with $n > 1$ and thickness d. The film may be surrounded by air on both sides, or may be placed on another surface with yet another index of refraction. At normal incidence (that is, when the light is shining perpendicular to the surface), light that reflects off the front boundary of the film can interfere with light that passes through the film and reflects off the back boundary. Figure 3.6 shows an example with a film of index of refraction n_2 surrounded by air ($n = 1$) on both sides. Note that the incident wave is supposed to be at normal incidence, and is only shown angled for clarity.

There are now two sources of phase shift: a geometric one, due to the difference in optical path length $2dn_2$ from traversing the thickness of the film twice (the dashed segment in the diagram), and a possible additional phase shift depending on the arrangement of indices of refraction according to rules (3.19) and (3.20). When there are additional phase shifts like

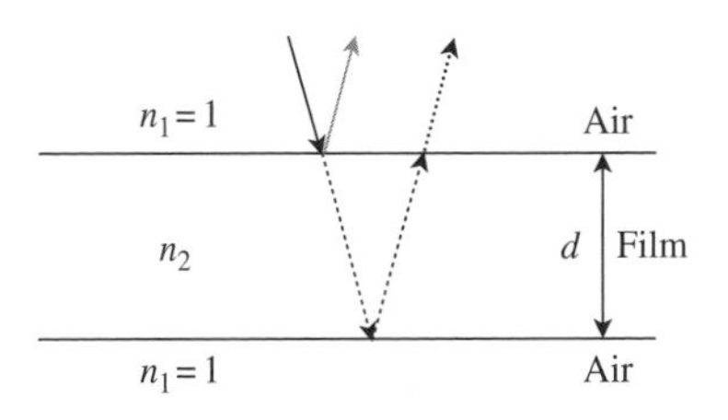

Figure 3.6 Example of reflection off a thin film. If $n_2 > 1$, the light reflecting off the first interface will experience a phase shift of π, but light reflecting off the second interface will have no phase shift. The light reflecting off each interface can interfere either constructively or destructively depending on the thickness d of the film.

this, you can compute the conditions for interference using $\delta = k\Delta x$, tack on an additional phase shift of π, and set this equal to an odd or even multiple of π as appropriate. Or, you can remember that the additional phase shift of π essentially reverses the conditions for constructive and destructive interference. Use whichever method you find conceptually and computationally simplest; the end-of-chapter problems explore the various standard cases.

3.2.5 Miscellaneous Diffraction

A few other types of diffraction scenarios appear on the GRE, mostly because they involve the application of simple formulas. Unfortunately these formulas are rather difficult to derive, and remembering how to do so is frankly a waste of time for the GRE: just memorize them, carefully noting their similarities and differences to the usual diffraction formulas.

- **Circular aperture**. A circular hole will also diffract light, but the angular positions of the diffraction minima now require solving some complicated differential equation. The answer is given numerically, as the Rayleigh criterion:

$$\text{First circular diffraction minimum: } D\sin\theta = 1.22\lambda, \tag{3.21}$$

  where D is the diameter of the hole, the circular analogue to the slit width a in single-slit diffraction. The Rayleigh criterion tells us that the angular separation of two point sources observed through a circular aperture (for example, a telescope) must be greater than $\sin^{-1}(1.22\lambda/D)$ for the two sources to be resolved as two different objects, so that the diffraction maxima of the two sources don't overlap and blur together as one big blob. Equation (3.21) is the limiting case where the first minimum of one source lies exactly at the central maximum of the other source.
- **Bragg diffraction**. This technically belongs to Specialized Topics but is conceptually closely related to the other classic diffraction problems. When x-rays[4] shine on a crystal lattice, they bounce off the atoms forming the lattice and interfere by path-length difference just as in double-slit interference. By modeling the crystal as a set of parallel planes a distance d apart, we get

$$\text{Maxima: } d\sin\theta = n\lambda/2, \tag{3.22}$$

  where θ is the angle of the incident x-rays with respect to the plane of the crystal. The only new things to remember in this formula are the factor of 2, which has the same origin as the path-length difference $2dn_2$ in thin-film interference because the light must traverse the distance between adjacent layers twice, and the fact that θ has a different interpretation than in double-slit interference.

[4] Why x-rays, as opposed to some other kind of light? Simply because the plane spacing d for most crystals of interest is comparable to the wavelength of x-rays.

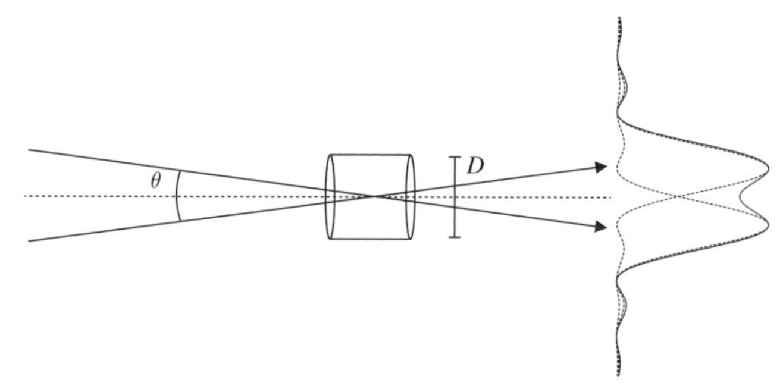

Figure 3.7 Light diffracting through a circular aperture of diameter D will exhibit an interference minimum at $D\sin\theta = 1.22\lambda$. If two objects are separated by an angle θ, this provides a condition, called the Rayleigh criterion, for whether the objects can be resolved through the aperture. At the Rayleigh criterion, the diffraction minimum of the first object coincides with the central maximum for the second object, as shown by the dashed lines on the right. The observed pattern is the sum of the two diffraction patterns, shown by the solid line on the right, which shows that two objects are just barely distinguishable.

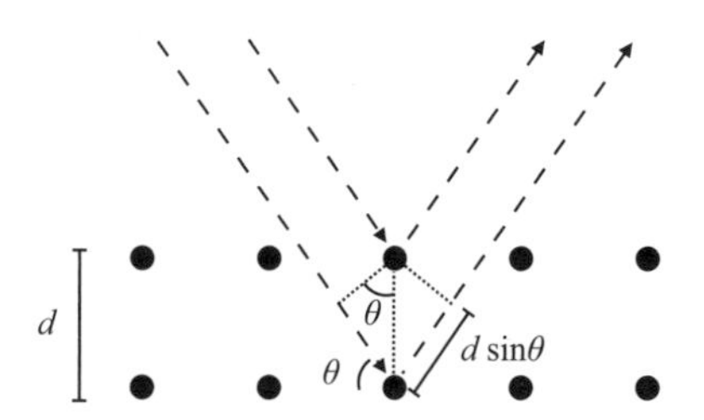

Figure 3.8 X-rays incident on a crystal with interatomic spacing d, at an angle θ relative to the horizontal will experience constructive and destructive interference, known as Bragg diffraction. The x-rays that scatter off the second layer of atoms will travel a total distance $2d\sin\theta$ longer than x-rays that scatter off the first layer. This path difference produces interference in the outgoing x-rays.

3.3 Geometric Optics

Geometric optics is a long-distance approximation to everything discussed in Section 3.2. If the dimensions of the objects in the problem are orders of magnitude larger than the wavelength of light involved, all interference and diffraction effects disappear, and light can be treated as if it travels in straight lines, just like a beam of particles.

3.3.1 Reflection and Refraction

The basic instruments of geometric optics are lenses, which bend (refract) light, and mirrors, which reflect it. The wave

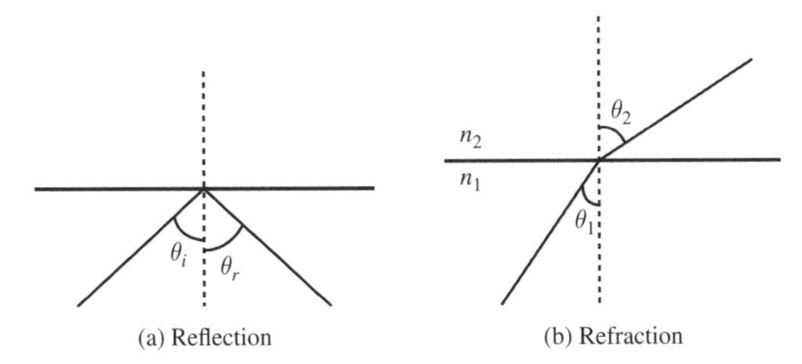

Figure 3.9 Definition of angles used in reflection and refraction problems.

properties of light imply the following laws, which you probably know already:

$$\text{Reflection: } \theta_i = \theta_r \quad \text{(angle of incidence equals angle of reflection)}, \tag{3.23}$$

$$\text{Refraction: } n_1 \sin\theta_1 = n_2 \sin\theta_2 \quad \text{(Snell's law)}. \tag{3.24}$$

As with thin-film phase shifts, the derivations of these equations are long and complicated so the equations must be memorized. Two points are worth mentioning here:

- These laws apply to *any* wave phenomena, not just light: sound, water waves, and so on. In Snell's law, n_1 and n_2 are the indices of refraction of the two media through which the light passes. But by remembering the interpretation of n as the factor by which the speed of light is reduced in the medium, you can easily translate to the relevant equation in terms of the speed of a general wave:

$$c_2 \sin\theta_1 = c_1 \sin\theta_2. \tag{3.25}$$

- Because the equations of electromagnetism are invariant under time reversal, so are geometric optics diagrams: *light rays are reversible*. For a real-world example, if you can see someone's eyes in a mirror, they can see your eyes as well.

3.3.2 Lenses and Mirrors

Suitably interpreted, there is only one equation you need to remember to solve all geometric optics problems:

$$\frac{1}{s} + \frac{1}{s'} = \frac{1}{f}. \tag{3.26}$$

The trick lies in remembering the meanings and sign conventions of the various terms in this equation. Starting from the right-hand side, f represents a *focal length*, the distance from the surface of the optical instrument to a special point called the focus (denoted by F in Fig. 3.10). In geometric optics, all mirrors and lenses are idealized so that all incident light rays parallel to the axis (dotted lines in Fig. 3.10) are reflected or refracted through the focus, respectively. Conversely, by reversibility all light rays passing through the focus are either reflected or refracted so that they come out parallel to the axis. The approximation that rays passing through the center of the lens (dashed lines) travel straight through without refraction, combined with some geometry involving similar triangles, leads to (3.26) for lenses.

For the idealized spherical mirrors used in geometric optics,

$$f = R/2, \tag{3.27}$$

where R is the radius of curvature of the mirror. By convention, f is positive if the center of curvature is on the same side of the mirror as the incoming light, which occurs if the mirror is concave. Convex mirrors have the center of curvature on the opposite side, and f is negative. This time, (3.26) comes from the law of reflection, since rays passing through the exact center of the mirror are reflected back at the same angle.

For lenses, f is positive for converging lenses, which have two convex surfaces, and negative for diverging lenses, with two concave surfaces. The lensmaker's equation gives the focal length of a thin lens in terms of the radii of curvature of the two surfaces:

$$\frac{1}{f} = (n-1)\left(\frac{1}{R_1} - \frac{1}{R_2}\right). \tag{3.28}$$

The sign conventions are very confusing here because a convex surface viewed from the left becomes a concave surface viewed from the right, so the best way to remember the signs is to imagine the simple case of a converging lens. There, the focal length is positive no matter what the radii of curvature, so R_1 is positive and R_2 is negative.

Moving on to the left-hand side of (3.26), s represents the position of an object, and s' represents the position of the image of that object formed from the single lens or mirror. Typically (3.26) is used to solve for the image position s' in terms of s and f. The sign conventions for s and s' are as follows:

$$\text{Positive distances} \iff \text{same side as light rays (incoming for } s, \text{ outgoing for } s'), \tag{3.29}$$

$$\text{Negative distances} \iff \text{opposite side as light rays (incoming for } s, \text{ outgoing for } s'). \tag{3.30}$$

Both s and s' can be positive or negative depending on the situation. The image distance s' is positive if the image is *real* (formed by the intersection of actual light rays), but negative if the image is *virtual* (formed on the opposite side of the optical

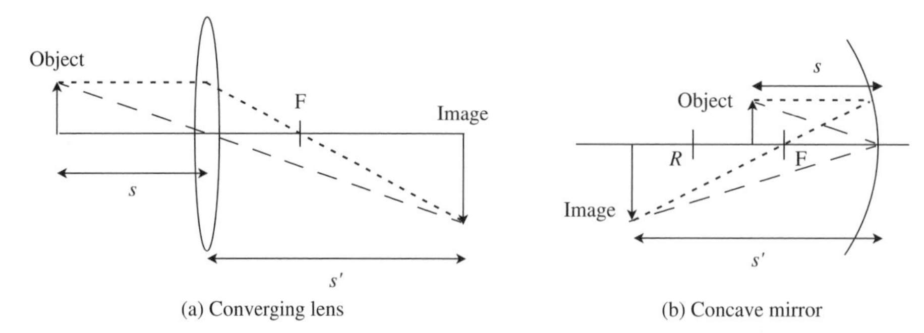

Figure 3.10 Geometry of typical lens and mirror problems, showing raytracing and definitions of common variables. The location of the image of a converging lens in (a) is given by finding the intersection of two lines: (1) draw a line from the top of the object through the center of the lens, and (2) draw a line intersecting the focal point and the point on the lens directly in front the object. The point where these lines intersects is the top of the image. The image of a spherical concave mirror in (b) is given by an analogous procedure, except that the rays reflect off the mirror rather than pass through as they did in the case of the lens. For a spherical concave mirror, the focal length is given by $f = R/2$.

instrument from the outgoing light rays). Typically s is positive, but it can be negative in a configuration of multiple lenses if an image formed by one lens acts as an object for a second lens. If the first image is on the opposite side of the second lens from the *incoming* light rays, its object distance is negative.

Once we have solved for s', we can obtain the magnification $m = -s'/s$. The absolute value $|m|$ is the ratio of the size of the image to the size of the object, and the sign determines the orientation of the image: positive if the image is upright, and negative if inverted. To solve problems involving several lenses and/or mirrors, simply solve (3.26) successively for each instrument. The image formed by the first instrument becomes the object for the second instrument, and so on, with magnifications multiplying at each step. A couple of practice problems are provided at the end of this chapter, but for tons of standard practice problems, just refer to any freshman physics textbook.

3.4 Assorted Extra Topics

3.4.1 Rayleigh Scattering

Why is the sky blue? To answer this, we have to look at how light scatters off small particles. For a given wavelength of light λ there are three regimes depending on the size of the particle a: $\lambda \ll a$, $\lambda \sim a$, or $\lambda \gg a$. The first regime is covered by geometric optics, the second by a theory called Mie scattering, and the last by Rayleigh scattering. We've already discussed geometric optics at length, and Mie scattering is quite complicated because of the coincidence of scales between the wavelength and the scatterer. For the purposes of the GRE you should be familiar with Rayleigh scattering, mostly because the formula takes a rather simple form:

$$I \propto I_0 \lambda^{-4} a^6. \tag{3.31}$$

As with most of the equations in this section, the derivation of this formula is complicated, the prefactors are irrelevant, and all that matters is the λ^{-4} dependence. (You may be asked about the a^6 dependence of the particle size, but such a question hasn't shown up on any of the recent tests.)

The most common physical application of this formula is the scattering of solar light by air and water molecules in the upper atmosphere. The light hitting the atmosphere approaches from only one direction, but is scattered in all directions, and what we observe on Earth is the scattered light. The λ^{-4} dependence means that shorter wavelengths are scattered much more strongly than longer wavelengths, and hence we observe a blue color (the reason we don't see purple has to do with the photoreceptors in our eyes and not with any underlying physics per se). On the other hand, at sunset when we look directly at the Sun, we are receiving light that has *not* been scattered away, and so we see what's left of the spectrum, which has a red color.

If you look back at Section 2.6.2, you'll notice the λ^{-4} dependence in Rayleigh scattering is the same as the ω^4 dependence from dipole radiation: in fact these are equivalent descriptions of the same process, by which light from the Sun causes the molecules to radiate as dipoles. So we can deduce that the blue light from the Sun is *polarized*, accounting for the effectiveness of Polaroid sunglasses in yet another context.

3.4.2 Doppler Effect

To conclude our review of optics and waves, we shift gears a bit from light waves to sound waves. You're undoubtedly familiar with the fact that an ambulance driving by at high speed with its siren blaring has a characteristic sound, where the frequency appears to drop as the ambulance drives past. This is an example of the Doppler effect, which arises any time a source of waves (sound waves, in our example) is moving relative to the observer. If the source is moving towards the observer, the wave crests get squished together, resulting in an apparent increase in the frequency; if the source is moving away, the wave crests get spread out, resulting in a decrease. The precise relationship between the emitted frequency f_0 and the received frequency f is

$$f = \left(\frac{v + v_r}{v - v_s}\right) f_0, \tag{3.32}$$

where v is the velocity of the waves in the medium, v_r is the velocity of the receiver relative to the medium in which the waves propagate, and v_s is the velocity of the source with respect to the medium. The sign conventions for the velocities have v_r and v_s positive if the source and receiver are approaching each other, and negative if they are moving away from each other. This equation is rather tricky to derive, and should be memorized. One shortcut to remember the signs: as v_s approaches v, the observed frequency f should blow up to infinity, accounting for the minus sign in the denominator. Also note that f goes negative if $v_s > v$: this equation simply does not apply when the source or receiver is moving faster than the wave velocity (instead, we need a theory of shock waves and sonic booms). To add a further caveat, (3.32) only applies when all velocities are small compared to the speed of light. There is a corresponding formula for the relativistic Doppler shift of light due to the motion of the source; see Section 6.4.1.

Note that (3.32) *only* applies if the receiver and source are moving *directly* toward or away from each other; otherwise, there is an additional factor of $\cos\theta$, where θ is the angle between the source velocity and the receiver's position. In fact, note that for v_r and v_s constant, f is *constant* in (3.32). The falling-frequency sound characteristic of an approaching and receding ambulance is due entirely to the varying factor of $\cos\theta$ as the ambulance drives by, since the ambulance does not directly approach the receiver but rather has some nonzero distance of closest approach.

3.4.3 Standing Sound Waves

Like any other kind of wave, sound waves involve oscillations. Unlike light waves, which involve oscillating electric and magnetic fields, sound waves involve *pressure* oscillations in a gaseous medium (usually air). Additionally, they are *longitudinal* waves, which means that the pressure oscillations take place along the same direction that the wave is traveling. Thus there is no concept of polarization for ordinary sound waves. However, all the other considerations of Sections 3.1.1 and 3.1.2 apply. One of the most common situations involves standing sound waves in an open or half-open pipe. A pipe of length L will support standing sound waves: an example is an empty bottle, which when you blow across the opening creates a definite pitch. But the longest-wavelength oscillation is not simply of length L. Indeed, for an idealized pipe, the open end must always be a pressure *node*, where the difference in average pressure compared to atmospheric pressure is *zero* because the air in the pipe can equilibrate with the air outside.[5] Similarly, a closed end is a pressure *antinode*, where the pressure difference is maximal because air is pushing against the fixed endcap of the pipe. These physical considerations serve as boundary conditions for figuring out the possible wavelengths of fully open and half-open pipes. When solving these problems, it's useful to depict the pressure waves by drawing parts of sine and cosine curves inside the pipe, as in the cartoons in Fig. 3.11, which show the longest allowed wavelengths.

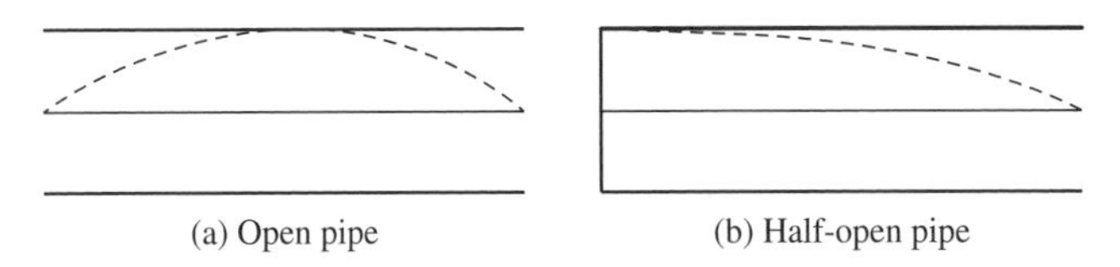

(a) Open pipe (b) Half-open pipe

Figure 3.11 Snapshot at a single point in time of a *pressure* standing wave in a pipe with two ends open (a) or one end open (b). The point at the open end always must have a fixed pressure because it is open to the outside, which has constant pressure. Note that, confusingly, some references will draw similar figures to represent a snapshot of *particle displacement* in the wave. In that case, closed ends look like nodes, because the end of the pipe is fixed, and open ends look like antinodes. In either case, however, the wavelengths of the fundamental and its harmonics are the same.

The vertical axis in these cartoons represents the deviation from average pressure, so both ends of the open pipe are pressure nodes, while the left (closed) end of the half-open pipe is an antinode. We see that the open pipe contains half a wavelength, so the longest-wavelength standing wave has wavelength $2L$, while the half-open pipe contains a quarter wavelength, giving wavelength $4L$ for the lowest mode. The same method will allow you to figure out the other allowed wavelengths, a favorite GRE question.

[5] You can also think of sound waves as displacement waves, where actual chunks of air are moving back and forth, in which case the open end would be a displacement antinode rather than a node since the air is free to slosh back and forth. But in this situation it's often simpler to think in terms of pressure.

3.5 Problems: Optics and Waves

1. Sound waves in air can be described by the equation

$$\frac{\partial^2 \rho}{\partial x^2} = \kappa^2 \frac{\partial^2 \rho}{\partial t^2},$$

where ρ is the deviation from average pressure, and κ is a constant. The speed of sound is

(A) $1/\kappa^2$
(B) $1/\kappa$
(C) $\sqrt{\kappa}$
(D) κ
(E) κ^2

2. Polarized light with polarization vector $\mathbf{n} = 2\hat{\mathbf{x}} + 3\hat{\mathbf{y}}$ is incident on a polarizer oriented at $\mathbf{v} = \hat{\mathbf{x}} + 2\hat{\mathbf{y}}$. The ratio of the intensity of the transmitted light to the initial intensity is

(A) $1/\sqrt{8}$
(B) $\sqrt{8/65}$
(C) $\sqrt{64/65}$
(D) 64/65
(E) 1

3. Let $f(x, t)$ and $g(x, t)$ be two traveling wave solutions to the homogeneous wave equation

$$\frac{\partial^2 f}{\partial t^2} = v^2 \frac{\partial^2 f}{\partial x^2}.$$

Which of the following statements are true?

I. $f + g$ solves the wave equation.
II. fg solves the wave equation.
III. $2f - 3g$ solves the wave equation.

(A) I only
(B) II only
(C) III only
(D) I and III
(E) II and III

4. What is the correct relationship between phase velocity, v_{phase}, and group velocity, v_{group}, for a quantum-mechanical wave packet?

(A) $v_{\text{phase}} = v_{\text{group}}$
(B) $v_{\text{phase}} = 2v_{\text{group}}$
(C) $v_{\text{phase}} = \frac{1}{2}v_{\text{group}}$
(D) $v_{\text{phase}}v_{\text{group}} = c^2$
(E) none of these

5. What is the absolute value of the relative phase between two waves described by $\sin(x - vt + \pi/6)$ and $\cos(x - vt)$?

(A) 0
(B) $\pi/6$
(C) $\pi/3$
(D) $2\pi/3$
(E) π

6. A person standing in the middle of a long straight road sees a truck with its headlights on approaching in the distance. The truck's headlights are 3 m apart. Assuming the headlights are point sources emitting yellow light of wavelength 600 nm and the diameter of the human pupil is 5 mm, approximately how far is the truck from the person when he can first resolve the two headlights as separate sources?

(A) 2 m
(B) 20 m
(C) 200 m
(D) 2 km
(E) 20 km

7. Blue light of wavelength 400 nm and green light of wavelength 500 nm are incident on a slit of width 20 μm, and the light passing through the slit hits a screen 2 m away from the slit. What is the distance on the screen between the first diffraction minimum for blue light and the first minimum for green light?

(A) 1 mm
(B) 5 mm
(C) 1 cm
(D) 4 cm
(E) 5 cm

8. Monochromatic light of wavelength λ is directed at a double-slit arrangement with slit separation d. If the same light is directed at a different double-slit arrangement with slit separation d', the position of the third interference minimum corresponds to the position of the old second interference maximum after the central maximum. What is d' in terms of d and λ?

(A) $4d/5$
(B) $4\lambda^2/5d$
(C) d
(D) $5d/4$
(E) $5d^2/4\lambda$

9. A soap bubble is formed by a thin film of soap (index of refraction 1.5) surrounded on both sides by air. For soap of thickness 1 μm, which of the following wavelengths of light will exhibit constructive interference when reflecting off the bubble at normal incidence?

(A) 400 nm
(B) 500 nm
(C) 600 nm
(D) 800 nm
(E) 900 nm

10. A glass window (index of refraction 1.3) is coated with an antireflective film of thickness 2 μm. Which of the following indices of refraction of the film would cause the intensity of reflected light of wavelength 800 nm from a normally incident beam to be suppressed?

    I. 1.1
    II. 1.4
    III. 1.7

    (A) I only
    (B) II only
    (C) III only
    (D) I and II
    (E) I, II, and III

11. A person at the bottom of a swimming pool looking up at the sky observes the Sun at an angle θ from the horizon. What is the true angle of the Sun, in terms of the index of refraction n of the water? (You may assume the index of refraction of air is 1.)

    (A) $\cos^{-1}(n\cos\theta)$
    (B) $\sin^{-1}(n\cos\theta)$
    (C) $\sin^{-1}(n\sin\theta)$
    (D) $\cos^{-1}(\cos\theta/n)$
    (E) $\sin^{-1}(\sin\theta/n)$

12. A sound wave propagates through a region filled with an ideal gas at constant temperature T. It approaches an acoustically permeable but thermally insulating membrane such that the angle between the wave and the plane of the membrane is 30 degrees. On the other side of the membrane is the same gas at a different temperature T'. What is the minimum value of T'/T such that no sound passes across the barrier? (You may find it useful to know that the speed of sound in an ideal gas is proportional to $\sqrt{T}$.)

    (A) 1/2
    (B) 3/4
    (C) 1
    (D) 4/3
    (E) 2

13. Which of the following MUST be true of the image of an object formed by a general configuration of ideal lenses and mirrors, where m denotes the absolute value of the magnification of the object?

    (A) A real image must have $m > 1$.
    (B) A virtual image must have $m > 1$.
    (C) A real image must be inverted.
    (D) A virtual image must be inverted.
    (E) None of the above.

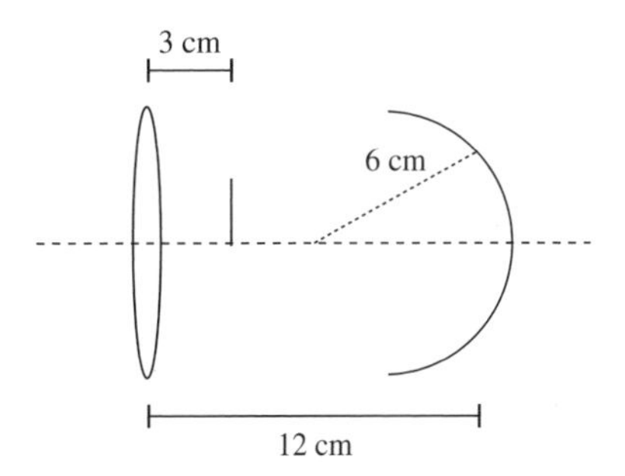

14. A converging lens of focal length 6 cm is placed 12 cm to the left of a concave spherical mirror of radius of curvature 6 cm, as shown in the diagram. An object is placed 3 cm to the right of the lens, in between the mirror and the lens. Which of the following describes the image(s) of the object formed by the lens in this configuration?

    (A) One real image
    (B) One virtual image
    (C) Two real images
    (D) Two virtual images
    (E) One real image and one virtual image

15. A trumpeter on a horse, riding directly towards you, plays a note at 200 Hz. You, standing still, hear a note at 210 Hz. Assuming the speed of sound is 350 m/s, how fast is the horse traveling?

    (A) 8.5 m/s
    (B) 16.7 m/s
    (C) 17.5 m/s
    (D) 333.3 m/s
    (E) 367.5 m/s

3.6 Solutions: Optics and Waves

1. B – This is pure dimensional analysis. κ^2 has units of time^2/length^2, and so $1/\kappa$ has units of velocity.
2. D – Recall that given two vectors **v** and **w**, the cosine of the angle between them is given by

$$\cos\theta = \frac{\mathbf{v}\cdot\mathbf{w}}{|\mathbf{v}||\mathbf{w}|},$$

so using Malus's law,

$$\frac{I}{I_0} = \cos^2\theta = \frac{(\mathbf{n}\cdot\mathbf{v})^2}{(\mathbf{n})^2\,(\mathbf{v})^2} = \frac{(2\cdot 1 + 3\cdot 2)^2}{(2^2+3^2)(1^2+2^2)} = \frac{64}{65},$$

choice D.

3. D – I and III are true by the principle of superposition, but II is not: any linear combinations of f and g are also solutions, but products are not linear combinations.
4. C – You need the Einstein relation $E = \hbar\omega$ and the de Broglie relation $p = \hbar k$, which we discuss further in

Chapter 5. Combined with the classical relation for a particle of mass m, $E = p^2/2m$, this gives the dispersion relation $\omega = \frac{\hbar k^2}{2m}$. The phase velocity is $v_{\text{phase}} = \frac{\omega}{k} = \frac{\hbar k}{2m}$, while the group velocity is $v_{\text{group}} = \frac{d\omega}{dk} = \frac{\hbar k}{m}$, so we see that $v_{\text{phase}} = \frac{1}{2}v_{\text{group}}$, choice C.

5. C – Note that the answer is *not* choice B because there is an inherent phase difference of $\pi/2$ between sine and cosine. To figure out whether the answer should be $\pi/2 + \pi/6$ or $\pi/2 - \pi/6$, ignore the phase shift of $\pi/6$ for now and remember that cosine is shifted to the *left* from sine by $\pi/2$. So $\cos x = \sin(x + \pi/2)$, and the phase difference is $\pi/2 - \pi/6 = \pi/3$, choice C.
6. E – Although probably a little too involved for a real GRE problem, this is a classic example of the Rayleigh criterion. Let d be the distance of the truck from the person in meters. The angular separation θ of the two sources is given by $\tan(\theta/2) = 1.5/d$; since θ is likely small, we can approximate $\tan\theta/2 \approx \theta/2$ to get $\theta \approx 3/d$. Now, the distance D appearing in the Rayleigh criterion is the aperture diameter of 5 mm, so we get (again using the small-angle approximation)

$$(5\text{ mm})(3/d) = 1.22(600\text{ nm}) \implies d = \frac{15\text{ mm}}{1.22(600\times10^{-6}\text{ mm})}\text{ m} \approx 2\times10^4\text{ m} = 20\text{ km}.$$

This is choice E, and also seems physically reasonable – on a clear night and a straight road, we ordinarily have no trouble distinguishing the headlights of an approaching vehicle.
7. C – From the single-slit formula, $a\sin\theta = \lambda$ for the first minimum, so $\sin\theta_{\text{blue}} = 2\times10^{-2}$ and $\sin\theta_{\text{green}} = 2.5\times10^{-2}$. These are small enough that we are justified in using the small-angle approximation, $\sin\theta \approx \tan\theta \approx \theta$. Since $\theta = x/L$, where x is the distance on the screen and L is the distance to the screen (2 m in this case), $x_{\text{blue}} = 4$ cm and $x_{\text{green}} = 5$ cm, and the distance between them is 1 cm, which is C.
8. D – The position of the old second maximum (after the central maximum) is given by $d\sin\theta = 2\lambda$, and the new third minimum is $d'\sin\theta = 5\lambda/2$. Setting the two expressions for $\sin\theta$ equal, we get

$$\frac{2\lambda}{d} = \frac{5\lambda}{2d'} \implies d' = \frac{5}{4}d.$$

Note that this is independent of λ!
9. A – Let $n = 1.5$ be the index of refraction of the soap. At the front boundary, we have $n > n_{\text{air}}$ so there is a phase shift of π. At the back boundary, $n_{\text{air}} < n$ so there is no phase shift. The optical path length is $2nd$, where

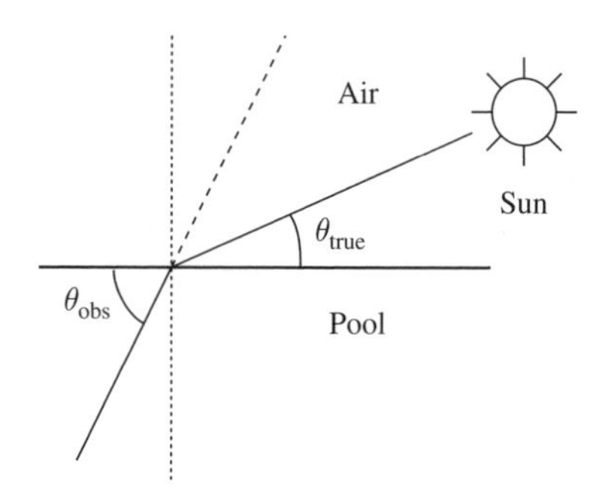

Figure 3.12 Solution for problem 11.

$d = 1$ μm. Thus the total phase shift is $2ndk + \pi$, where k is the wavenumber, and the condition for constructive interference is that the total phase be a multiple of 2π:

$$2dn(2\pi/\lambda) + \pi = 2m\pi \implies \lambda = \frac{4dn}{2m-1}.$$

For $d = 1$ μm, we can take $m = 8$ to get 400 nm, which is A. None of the other choices correspond to possible values of m.
10. D – Now we are looking for destructive interference. There is already a phase shift of π at the boundary between the air and the coating because all the given indices of refraction are greater than 1, so the only question is whether there is an additional phase shift of π at the coating–glass boundary. If the coating has index of refraction $n < 1.3$, there is an additional phase shift, so the condition for destructive interference is

$$2dn = (m - 1/2)\lambda \implies \lambda = \frac{4dn}{2m-1}.$$

If $n > 1.3$, there is no additional phase shift, so we get instead

$$2dn = m\lambda \implies \lambda = \frac{2dn}{m}.$$

Choice I must satisfy the first condition, which it does when $m = 6$. Choice II must satisfy the second condition, which it does when $m = 7$. Choice III fails the second condition (it satisfies the first condition, but that does not apply since $n > 1.3$), so the correct options are I and II, choice D.
11. A – Referring to Fig. 3.12, let $\theta = \theta_{\text{obs}}$, the observed angle of the Sun, and θ_{true} be the true angle. Since we're given angles with respect to the horizon, *not* the normal, we have to be careful applying Snell's law. If α is the angle to the horizontal, and $\beta = \pi/2 - \alpha$ is the angle to the normal, then $\sin\beta = \cos\alpha$, so we can forget about the

normal and just use cosines rather than sines. Since the index of refraction of air is 1, we have

$$\cos\theta_{\text{true}} = n\cos\theta_{\text{obs}} \implies \theta_{\text{true}} = \cos^{-1}(n\cos\theta_{\text{obs}}).$$

This matches choice A. For a quick limiting-cases analysis, notice that if $n = 1$ we should have $\theta_{\text{true}} = \theta_{\text{obs}}$, which gets rid of choice B.

12. D – This is the phenomenon of *total internal reflection*, applied to the unfamiliar context of sound waves. Total internal reflection occurs when Snell's law has no solution for θ_2, so that there is no refracted wave at all: this occurs when $\sin\theta_2 > 1$, or $n_1 \sin\theta_1/n_2 > 1$. Here, θ_1 (which, remember, is the angle to the *normal*) is 60°, so $\sin\theta_1 = \sqrt{3}/2$. We'll see in the following chapter on thermodynamics that the speed of sound in an ideal gas is proportional to $\sqrt{T}$, and the "index of refraction" is proportional to the reciprocal of the wave speed, so we have the condition

$$\frac{\sqrt{T'}(\sqrt{3}/2)}{\sqrt{T}} > 1 \implies \frac{T'}{T} > \frac{4}{3},$$

choice D. Choice E is the classic mistake of taking θ_1 as the angle to the boundary, rather than the angle to the normal, while choice B results from forgetting that n_1 is the reciprocal of the speed.

13. E – This is a little tricky. If there is only *one* lens or mirror, then by $m = -s'/s$ and the sign conventions for s and s', a real image always has $m < 0$ and hence is inverted, leading us to suspect choice C. However, in a more general configuration a real image can serve as the object for another lens or mirror, and a second successive real image causes another inversion and the resulting object is upright. Counterexamples for the remaining choices are easy. For A, place the object beyond the center of curvature of a single concave mirror; for B and D, place the object anywhere outside a single convex mirror.

14. E – There are two possibilities for light rays coming from the object: they can either go to the left and pass straight through the lens, or they can go to the right, hit the mirror first, then pass through the lens. Each of these paths can potentially give rise to an image. For the first possibility, note that the object is inside the focal length of the lens, so we will get a virtual image; you can also see this from $1/s + 1/s' = 1/f$, which shows s' must be negative. This eliminates A and C. As for the second possibility, the mirror has focal length $R/2 = 3$ cm. The object is 9 cm to the left of the mirror, so the lens equation gives $s' = 4.5$ cm, a real image. But this is now outside the focal length of the lens, which will give a real image somewhere far to the left of the lens. Thus choice E is correct.

15. B – Straightforward application of the Doppler effect. Here $v_r = 0$, and we are solving for v_s:

$$\frac{f}{f_0} = \frac{210}{200} = \frac{350}{350 - v_s}$$
$$\implies v_s = 16.7 \text{ m/s}.$$

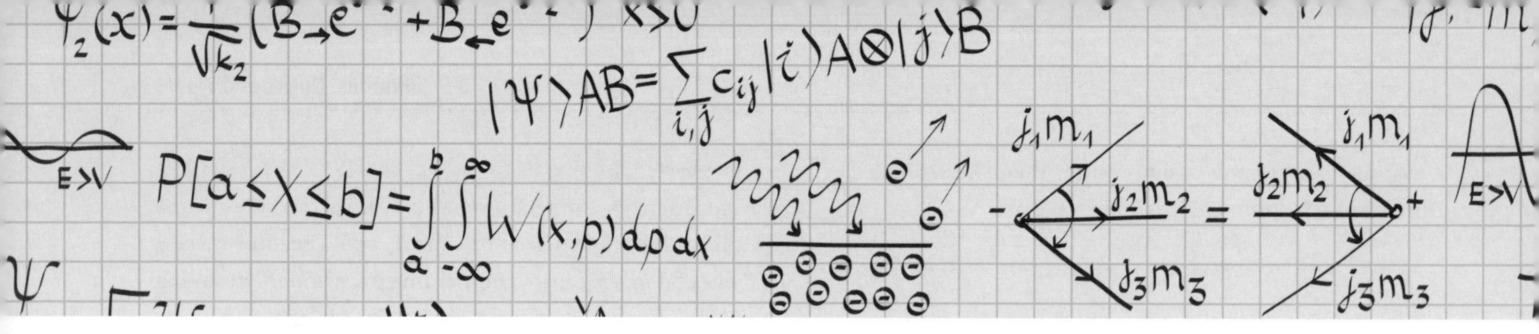

4 Thermodynamics and Statistical Mechanics

Statistical mechanics and thermodynamics form a small but important section of the exam. On the one hand, only 10% of the exam covers these topics. On the other, the questions tend to be at an easy level and students tend to perform poorly because statistical mechanics and thermodynamics are often poorly covered in undergraduate courses. A relatively small amount of knowledge translates into a large gain in performance on these questions.

In this section we will give a sketch of the basic structures of statistical mechanics and thermodynamics, and then we will consider their application to some model systems. Most problems will involve just a few of these model systems, so it is worth understanding them well.

4.1 Basic Statistical Mechanics

4.1.1 Ensembles and the Partition Function

State variables such as temperature and pressure characterize the *macrostate* of a system, and completely specify the macroscopic behavior of the system. On the other hand, many microscopic configurations or *microstates* correspond to the same macrostate. We can imagine preparing several copies of a system with slightly different initial conditions; if, at the time of measurement, the copies have the same state variables, the systems will be macroscopically identical. Statistical mechanics is the business of calculating properties of macrostates without knowing the exact microstate of a system.

To carry out the calculation, we need to identify what *ensemble* applies to our system. An ensemble is the collection of all possible microscopic configurations of the system subject to some constraints, and formalizes the idea of different microstates corresponding to a single macrostate. The most common ensemble, by far, that appears on the GRE is the *canonical ensemble*. The canonical ensemble consists of all possible states of a system with

- fixed particle number (N)
- fixed volume (V)
- fixed temperature (T): the system is allowed to exchange energy with a large heat bath whose heat capacity is assumed to be so large that its temperature stays fixed.

So, a system described by the canonical ensemble can have energy fluctuations, but the state variables N, V, and T must remain fixed. A somewhat less common ensemble is the *microcanonical ensemble*, which has fixed energy E rather than fixed temperature T, and whose temperature can fluctuate accordingly. There are other ensembles associated with other state variables, but these do not ever seem to be tested by the GRE. Unless otherwise specified, we will work with the canonical ensemble in the remainder of this chapter.

With this picture, how do we actually calculate anything? Consider a system with discrete energy states $\{E_i\}$. From a few very basic assumptions one can show that the probability of the system being in a state i is given by

$$p_i = \frac{e^{-\beta E_i}}{\sum_j e^{-\beta E_j}}, \tag{4.1}$$

where

$$\beta = \frac{1}{k_B T}. \tag{4.2}$$

Here k_B is Boltzmann's constant, and probabilities satisfying (4.1) are known as *Boltzmann statistics*; we will often use β and T interchangeably depending on which one is more convenient.[1] Equation (4.1) allows us to calculate extensive quantities, such as average energy E or its variance σ_E^2, by using the p_i as a probability distribution. For example, suppose the system is a magnet, and the energy states E_i have magnetizations M_i. The average magnetization is given by

$$\langle M \rangle = \sum_i p_i M_i.$$

In fact, whenever you encounter any system that has a discrete set of states in any parameter $\mathcal{O}$ (energy, spin, magnetization, etc.), and you need to compute the average (or *expectation value*) value of this parameter, just use the weighted sum of the different states:

$$\langle \mathcal{O} \rangle = \sum_i p_i \mathcal{O}_i, \tag{4.3}$$

which should be familiar from basic probability.

It is equally instructive to rewrite equation (4.1) as

$$p_i = \frac{e^{-\beta E_i}}{Z}, \tag{4.4}$$

where Z is known as the *partition function*,

$$Z = \sum_j e^{-\beta E_j}. \tag{4.5}$$

The advantage of the partition function formalism is that the same quantities we computed above using probabilities can be computed directly from Z. This is important enough to repeat: *if you know the partition function, you can compute all the state variables.* The expectation value of energy, for example, is

$$\langle E \rangle = \sum_i p_i E_i = \frac{\sum_i E_i e^{-\beta E_i}}{Z} = -\frac{\partial}{\partial \beta} \ln Z. \tag{4.6}$$

This last equality is an extremely useful trick because it allows us to compute the average energy directly from the partition function without even thinking about the probabilities or doing complicated sums: just find Z and differentiate its logarithm! On the other hand, for very simple systems, it may be faster to compute the weighted sums by hand.

This n-state system arises very frequently on the GRE. In general, you might be given a system with n discrete states and corresponding energies E_i; the quantum harmonic oscillator is a favorite example, but sometimes the GRE just gives you a list of states and their associated energies. Problems may ask you to compute the probability of the system being in certain states, ratios of probabilities of states, expectation values of energy, expectation values of other quantities, and the heat capacity, to be discussed in Section 4.2.5 below. The formulas in this section should be enough to get you through any of these kinds of problems.

[1] The GRE writes k instead of k_B in the Table of Information at the beginning of the exam, and we will use this notation in the sample exams at the end of this book. To avoid any confusion with the many other uses of the letter k, we will stick to k_B in this chapter.

4.1.2 Entropy

Another quantity of fundamental importance that constrains the behavior of systems in thermodynamics and statistical mechanics is the entropy. Entropy can be conceptually tricky, but for the purposes of the GRE, you will only be expected to know its mathematical definitions, and how to apply it in well-defined thermodynamic scenarios. Here we give two definitions and a useful formula, but more discussion will follow in Section 4.2.3.

- The most transparent and elegant definition is the one given by Boltzmann:

$$S = k_B \ln \Omega, \tag{4.7}$$

  where Ω is the number of microstates corresponding to the system's macrostate. For example, if a two-electron system is in a magnetic field with Hamiltonian $H \propto \mathbf{S} \cdot \mathbf{B}$, the states of the system are the spin singlet with $\mathbf{S} = 0$ and the spin triplet with $\mathbf{S} = 1$. The zero-energy state is the one with $S_z = 0$, for which there are two corresponding microstates: the spin singlet, and the $S_z = 0$ component of the spin triplet.[2] Thus the zero-energy state has entropy $k_B \ln 2$. The formula (4.7) is useful if you ever need to calculate the entropy of a system that either has a small number of states or a simple analytic expression for the number of states.
- Another closely related expression for entropy is

$$S = -k_B \sum_i p_i \ln p_i = \frac{\partial}{\partial T}\left(k_B T \ln Z\right), \tag{4.8}$$

  where p_i is the probability of the system being found in the ith microstate. (Exercise: derive the last equality from the definitions (4.4) and (4.5).) This definition has the advantage of being directly connected to the partition function, so it is useful if you happen to know the exact form of the partition function.

It turns out to be equivalent to (4.7) if we assume something called the "fundamental assumption of statistical thermodynamics" or sometimes the "postulate of equal *a priori* probability." The justification for this is very deep and subtle, but for the purposes of the exam you can assume that these forms of the entropy are equivalent.

[2] If this discussion is unfamiliar to you, see Section 5.5.3.

- For a *monoatomic ideal gas*, the expression for entropy is

$$S = Nk_B\left(\ln\frac{V}{N} + \frac{3}{2}\ln T + \frac{5}{2} + \frac{3}{2}\ln\frac{2\pi mk_B}{h^2}\right), \quad (4.9)$$

where V is the volume, N is the number of particles, T is the temperature, m is the mass of the gas particles, and h is Planck's constant. The important point here is not the constant numerical factors, but the scaling of S with state variables. It is most useful to remember

$$S = Nk_B \ln\frac{VT^{3/2}}{N} + \text{constants}. \quad (4.10)$$

We'll discuss ideal gases more in Section 4.2, but we mention them here just to show what an explicit formula for entropy looks like.

The interpretation of entropy from these definitions is as a measure of the number of possible states that a system could have in a particular macroscopic state. Gibbs is rumored to have called it, more-or-less accurately, the amount of "mixed-up-ness" of a system. If a system could be in many possible states, then it has high entropy. Alternatively, we can think of entropy as a measure of our uncertainty about the underlying state of the system. If a system has low entropy, then it can only be in a small number of possible states, and we can be relatively certain what underlying state it is actually in.

4.1.3 Classical Limit

The partition function formalism is very simple to apply to systems with a small number of states. But classical systems do not generally have a discrete set of states. Because position and momentum are continuous variables, there are formally an infinite number of microstates for a box of gas, for example. Nevertheless, we can extend the formalism above to the continuum limit. The sum in the partition function becomes an integral and we obtain the partition function for N (identical) classical particles:

$$Z_N = \frac{1}{N!h^{3N}}\int e^{-\beta H(\mathbf{p}_1,\ldots\mathbf{p}_n;\mathbf{x}_1,\ldots\mathbf{x}_n)}d^3\mathbf{p}_1\ldots d^3\mathbf{p}_n d^3\mathbf{x}_1\ldots d^3\mathbf{x}_n, \quad (4.11)$$

where $\mathbf{x}_i$ and $\mathbf{p}_i$ are three-dimensional vectors, H is the classical Hamiltonian, h is Planck's constant, and the normalization factor $N!$ accounts for identical particles. The effects of identical particles can be rather tricky and irritating, but it is not too difficult to understand. When particles are identical, states formed by interchanging particles are not counted twice, and dividing by $N!$ accounts for this. That said, N is a fixed quantity in the canonical ensemble, so the factors multiplying the integral in Z almost always disappear when you take the log and differentiate to find the state variables.

4.1.4 Equipartition Theorem

The equipartition theorem is a simple application of equation (4.11), which gives a quick rule for determining the internal energy of a system. This is extremely useful for computing heat capacities, as discussed further in Section 4.2.5. The theorem states that

> *Each quadratic term (degree of freedom) in the Hamiltonian for a particle contributes (1/2)k_BT to the internal energy of the particle.*

For example, a particle in an ideal gas has a Hamiltonian $H = \mathbf{p}^2/2m = p_x^2/2m + p_y^2/2m + p_z^2/2m$, so the internal energy of such a particle is $U = (3/2)k_BT$. The proof of this is an exercise in manipulating Gaussian integrals, noting that we can integrate each quadratic term in the exponential of equation (4.11) one by one, but the details are not relevant for the GRE. See Example 4.1.

Often we can compute the internal energy from the equipartition theorem without ever writing down an explicit Hamiltonian, just by counting quadratic degrees of freedom.

4.1.5 Some Combinatorial Facts

A few basic combinatorial facts occasionally come in handy. The first is the *binomial coefficient*. Suppose that we have N distinguishable marbles and we want to know the number of ways of choosing a group of M of these marbles from a hat. The result is denoted

$$\binom{N}{M} = \frac{N!}{(N-M)!M!}, \quad (4.12)$$

which is read "N choose M."

Another useful identity known as *Stirling's formula* is

$$\ln(n!) \approx n\ln n - n \quad (4.13)$$

for large values of n.

4.2 Thermodynamics

Statistical mechanics reproduces macroscopic physics by analyzing microscopic physics. Thermodynamics ignores the microscopic foundations and sets down rules for how macroscopic systems should behave. Even without the microscopic foundations, thermodynamics gives a complete description of thermal systems. Though thermodynamics is usually taught first, it can be derived from statistical mechanics, so hopefully the laws and formulas will seem straightforward in light of the more formal tone of the previous section.

4.2.1 Three Laws

Thermodynamics can be summarized succinctly in three laws (plus a zeroth law) that completely determine the behavior of a system. The laws of thermodynamics are

1. *Energy cannot be created or destroyed.* This is just the conservation of energy from the statistical mechanical point of view, and it can be stated mathematically as

$$\Delta U = Q - W, \tag{4.14}$$

where ΔU is the change in the internal energy of a system, Q is the heat added *to* the system, and W is the work done *by* the system. Take care that the signs of your terms agree with your definitions of the quantities.

2. *There is no process in which the sole effect is to transfer heat from a body at a lower temperature to a body at a higher temperature.* There are several equivalent formulations of

EXAMPLE 4.1

What is the specific heat of a diatomic gas as a function of temperature?

There are three translational degrees of freedom corresponding to the kinetic energy of the center of mass in the three spatial dimensions. There are two rotational degrees of freedom, corresponding to rotations around the axes perpendicular to the axis connecting the two atoms (since the atoms are assumed to be point-like, there is no energy associated with rotation along the axis connecting the atoms). And there are two vibrational degrees of freedom corresponding to the kinetic and potential energy of the vibrating molecules. Without writing anything, we see that there are seven quadratic degrees of freedom and the internal energy is $(7/2)k_BT$. The explicit Hamiltonian for a diatomic gas composed of two atoms of mass m is

$$H = \frac{p_x^2}{2m} + \frac{p_y^2}{2m} + \frac{p_z^2}{2m} + \frac{L_1^2}{2I_1} + \frac{L_2^2}{2I_2} + \frac{p_s^2}{m} + \frac{1}{2}ks^2,$$

where s is the separation between the atoms, k is the vibrational spring constant, $I_{1,2}$ are the two moments of inertia about the two rotational axes, and $L_{1,2}$ and p_s are the appropriate conjugate momenta (note that there is no factor of $1/2$ in the second-to-last term because vibrations are about the center of mass, and the reduced mass is $m/2$). But clearly, we didn't need to write down the Hamiltonian to get the answer. Other examples include the extreme relativistic gas and two-dimensional gases, whose internal energies you will calculate in the end-of-chapter problems.

Remember that the equipartition theorem is a *classical* statement, and it breaks down when the spacing between energy levels becomes large compared with k_BT. In this regime, the classical assumption that the energy levels form a continuum is no longer valid, and the equipartition theorem will tend to overestimate the internal energy of a system. This usually happens at low temperatures, when degrees of freedom are "frozen out." In the case of a diatomic ideal gas, for example, we saw that the internal energy at high temperatures was $(7/2)k_BT$. We know from quantum mechanics that the low-energy states of the harmonic oscillator are discrete, and it is straightforward to show that the low-energy states of the rigid rotor in three dimensions are also discrete. Using typical values for the harmonic oscillator angular frequency ω and the moment of inertia I, we can estimate the temperature at which quantum mechanics becomes important:

$$k_BT \sim \hbar\omega \implies T \sim 1000\text{ K} \qquad \text{(vibrational)},$$

$$k_BT \sim \frac{\hbar^2}{2I} \implies T \sim 1\text{ K} \qquad \text{(rotational)}.$$

So vibrational degrees of freedom freeze out first; at room temperature, the equipartition theorem still applies to translational and rotational degrees of freedom, and the internal energy is $(5/2)k_BT$. At very low temperatures, if the substance still exists as a gas once rotational degrees of freedom have frozen out, all that is left are the energy states associated with the translational part of the Hamiltonian, or the free particle states. From quantum mechanics we know that such states form a continuum down to low energies. So, at low temperatures, the equipartition theorem will continue to apply to the three translational degrees of freedom and the diatomic gas will have an internal energy approximately equal to $(3/2)k_BT$, the same as for a monoatomic ideal gas. This situation is conveniently summarized in Fig. 4.1, where the temperature axis is logarithmic.

EXAMPLE 4.1 (Cont.)

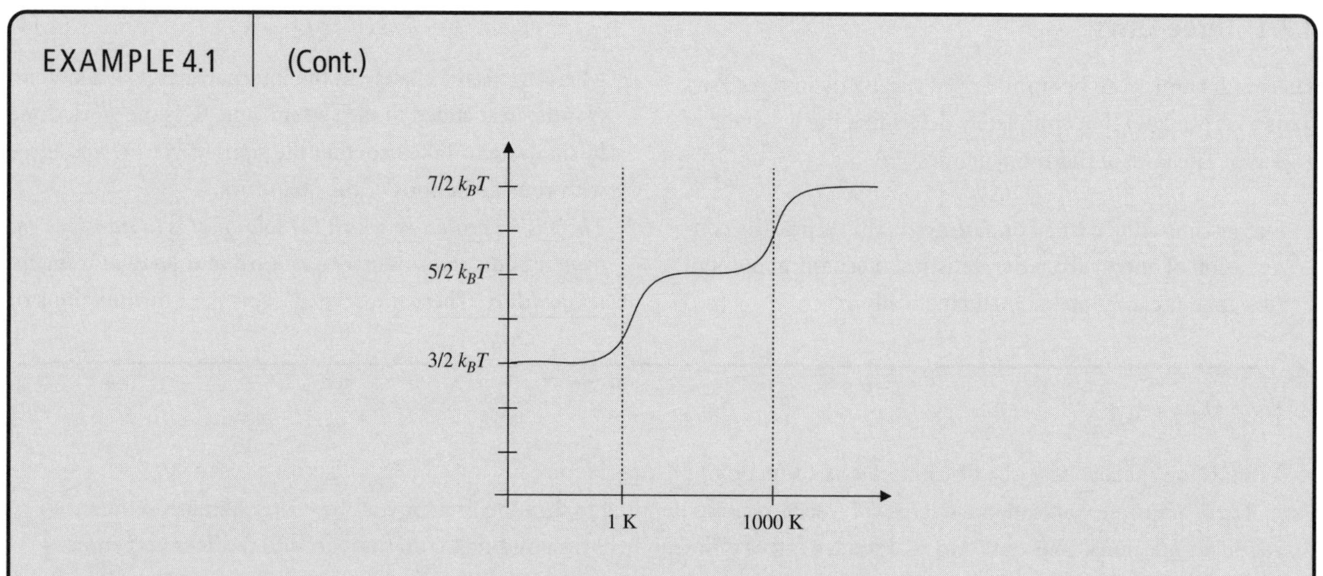

Figure 4.1 A sketch of specific heat C_V of a diatomic gas as a function of temperature. Note the plateaus at $(5/2)k_BT$ and $(7/2)k_BT$ corresponding to the "unfreezing" of rotational and vibrational degrees of freedom, respectively.

the Second Law of Thermodynamics: the previous sentence is the so-called "refrigerator" statement, while the "engine" version is *no system can undergo a cyclic process where heat absorbed from a reservoir at a single temperature is completely converted into mechanical work.* The refrigerator statement says that refrigeration cannot happen spontaneously, and the engine statement says that heat engines must always waste some heat. All equivalent statements of the Second Law imply the mathematical relationship

$$\Delta S \geq \int \frac{\delta Q}{T}, \tag{4.15}$$

which describes the well-known result that the entropy of a thermally isolated system (where $\delta Q = 0$) cannot decrease. Past GREs have had questions involving conceptual applications of the Second Law, so it's useful to know both the words and the mathematical relationship.

3. *Entropy is zero at absolute zero temperature.* This statement follows from the definition of entropy (4.7). At absolute zero, there is only one microstate for a system and so the entropy is zero.

There is an additional law, called the *Zeroth Law*, which states that if two systems are in equilibrium with a third, then they are in equilibrium with each other, easily remembered by the succinct statement "thermometers exist." This is necessary in order for the notion of thermodynamic equilibrium to be well defined.

4.2.2 Gases and Equations of State

Thermodynamic systems are specified by an equation of state that constrains the values that the state variables may assume. The most famous example of this is the equation of state for an ideal gas:

$$PV = Nk_BT, \tag{4.16}$$

where P is the pressure, V is the volume, N is the number of particles, and T is the temperature. You probably know this as the *ideal gas law*, describing how the state variables P, V, and T are related. The only other equation of state you may see on the exam is the equation of state for the van der Waals gas. The van der Waals gas has particles with nonzero size and a pairwise attractive potential, and the equation of state is

$$\left(P + \frac{N^2 a}{V^2}\right)(V - Nb) = Nk_BT,$$

where a measures the attraction between particles, and b measures the size of the particles. Clearly $a = b = 0$ reduces to the ideal gas law. You should definitely not memorize this equation, as it will be given to you if needed, but you should be familiar with its interpretation.

4.2.3 Types of Processes

Thermodynamic processes can be classified into several types that have important properties, mostly relating to holding certain variables constant. The words used to describe many of these processes are often used interchangeably, but this is sloppy: there are very precise meanings, and it is crucial to keep definitions straight to know which formulas apply. Here is a list of the most important processes you will encounter on the GRE.

- **Reversible process**. A process that proceeds in infinitesimal steps such that the system is in equilibrium at each step, *and* each infinitesimal step can be reversed by somehow changing the state of the system. A common example is slowly heating a gas in a container of flexible size. As an infinitesimal amount of heat is added, the gas expands, and comes to equilibrium after enough time has elapsed; this can be reversed simply by cooling the system by removing an equal amount of heat.

  Reversible processes imply several important conditions. If the process involves a gas, the work done in each infinitesimal step is given by

  $$\delta W = P\,dV \quad \text{(reversible)}. \tag{4.17}$$

  Note that this is *not* true if the process is irreversible: see "free expansion" below. In addition, the *total* entropy change of the system and its surroundings must be zero. The entropy of the system *by itself* can change, however, and is given by

  $$\delta Q = T\,dS \quad \text{(reversible)}, \tag{4.18}$$

  where T is the temperature and δQ is the heat input. The integral form of this equation is

  $$\Delta S = \int \frac{\delta Q}{T}, \tag{4.19}$$

  which is just the lower bound of the Second Law expression (4.15): a reversible process has the minimum possible entropy increase for a given δQ. Notice that if $\delta Q < 0$, ΔS is negative! There is no contradiction here: cooling a system reduces its entropy, but increases the entropy of its surroundings by an equal amount because of the heat deposited, so the total entropy change is zero. Finally, notice that, if the initial and final states of the system are the same, the limits of integration degenerate onto each other, and $\Delta S = 0$: *the entropy change of a system undergoing a reversible cycle is zero.*
- **Quasistatic process**. A process that happens infinitely slowly, so that at each instant the system is in thermodynamic equilibrium. Reversible processes are always quasistatic, but a quasistatic process need not be reversible in general.
- **Adiabatic process**. A process for which $\delta Q = 0$: no heat is exchanged between a system and its surroundings.
- **Isentropic process/Reversible adiabatic process**. A process that is both adiabatic and reversible. One can prove that a process is reversible and adiabatic if and only if it has zero entropy change (an isentropic process). Here we mean that the entropy change of the system itself is zero, and consequently the entropy change of the surroundings is zero as well. Compressing a gas in a cylinder by pressing down a piston is the prototypical example. By applying a very small force, the piston is displaced a small amount; each compression step can be reversed by removing the force, which causes the gas to expand again and the system to return to its initial state. Since no heat is exchanged during this process, adiabatic compression is isentropic.

  One major reason that this type of process arises in problems is that for an ideal gas undergoing an isentropic process,

  $$PV^\gamma = \text{constant}, \tag{4.20}$$

  where $\gamma = C_P/C_V$ is the ratio of the heat capacity at constant pressure P to heat capacity at constant volume V. There is a particularly nice formula for γ for ideal gases, which we discuss further in Section 4.2.5.

  Warning! Past GREs have been known to use "adiabatic" to mean "reversible adiabatic." Keep the more restrictive definitions in mind, but be prepared to be flexible based on context because the GRE often does not define its terms in the test questions.
- **Iso-something process**. A process for which some state variable is held constant. "Isothermal" means constant temperature, "isobaric" means constant pressure, and "isochoric" means constant volume, though memorizing these terms is completely unnecessary. The key to figuring out the work done by a gas in such a process is to use the definition of work (4.17), supplanted by the ideal gas law equation of state (4.16), to solve for P in terms of the constant variables. For example, if P is constant, it can be pulled outside the integral and $W = P\Delta V$. If T is constant, then solve for P in terms of T:

  $$P = \frac{Nk_BT}{V} \implies W = Nk_BT \int \frac{dV}{V}.$$

  If V is constant, then no work is done since $dV = 0$. The isothermal process is perhaps the most common, so get familiar with the form of the integral and the appearance of $\ln V$ from integrating dV/V.
- **Free expansion**. This process occurs when a gas suddenly expands from a smaller region to a larger region: think of

a balloon popping, where the gas initially inside the balloon rushes out to occupy the whole volume of the room. The temperature of an ideal gas does not change during free expansion, so we have

$$PV = P'V' \qquad \text{(free expansion)}.$$

Note that free expansion is adiabatic, but *not* reversible: one cannot force all the gas molecules back into the smaller volume once they have expanded, simply by changing the state variables of the system. Free expansion is the prototypical irreversible process, just as irreversible as an egg breaking. The fact that this process is irreversible means that there *is* an entropy change, despite the fact that $\delta Q = 0$; equation (4.19) simply does not apply to irreversible processes. We will see in Example 4.2 how to calculate the entropy change. Similarly, note that, despite the fact that the volume changes, *the gas does no work.* This is consistent with the First Law: $\Delta U = 0$ at constant temperature, $Q = 0$ since the process is adiabatic, and $W = 0$, so we trivially get $0 = 0 - 0$.

4.2.4 Relations Between Thermodynamic Variables

Equations of state tell us what states are accessible to a particular system in terms of the system's state variables. What if we are interested in other variables that are not the state variables of a system? How would we calculate the entropy of an ideal gas, for example? There are a large number of thermodynamic identities that relate, under varying circumstances, all of the variables that we have been discussing.

The first identity is so useful that it is often called the *fundamental thermodynamic identity*. It relates the differentials of state variables U, S, and V:

$$dU = TdS - PdV. \tag{4.21}$$

This equation is simply the infinitesimal version of the First Law (4.14), $dU = \delta Q - \delta W$, supplanted by the definitions (4.18) and (4.17) for δQ and δW respectively. Notice that we write δQ and δW, not dQ and dW. This is because Q and W are not state variables, but refer to small quantities of heat and work added to or done by the system, respectively. In contrast, U is a state function, a mathematical representation of the internal energy of a system, which can thus be sensibly differentiated. Finally, note that (4.21) applies to *all* infinitesimal changes of state, not just reversible ones: this needs a tricky bit of reasoning, so this fact is best simply memorized.

The fundamental thermodynamic identity also implies a definition of temperature and pressure:

$$T = \left(\frac{\partial U}{\partial S}\right)\bigg|_V, \tag{4.22}$$

$$P = -\left(\frac{\partial U}{\partial V}\right)\bigg|_S. \tag{4.23}$$

The vertical bars refer to holding the subscript variable constant, so (4.22) holds at constant V, and (4.23) holds at constant S.

Another important class of thermodynamic relations are the Maxwell relations, which relate partial derivatives of thermodynamic variables. From the expressions for T and P above, we can equate mixed partial derivatives and determine

$$\left(\frac{\partial P}{\partial S}\right)\bigg|_V = -\left(\frac{\partial T}{\partial V}\right)\bigg|_S. \tag{4.24}$$

There are three other thermodynamic potentials, which we give here for completeness, though you almost certainly won't need them for the GRE:

$$dH = TdS + VdP, \tag{4.25}$$

$$dA = -SdT - PdV, \tag{4.26}$$

$$dG = -SdT + VdP. \tag{4.27}$$

The associated Maxwell relations are

$$\left(\frac{\partial T}{\partial P}\right)\bigg|_S = \left(\frac{\partial V}{\partial S}\right)\bigg|_P, \tag{4.28}$$

$$\left(\frac{\partial S}{\partial V}\right)\bigg|_T = \left(\frac{\partial P}{\partial T}\right)\bigg|_V, \tag{4.29}$$

$$-\left(\frac{\partial S}{\partial P}\right)\bigg|_T = \left(\frac{\partial V}{\partial T}\right)\bigg|_P. \tag{4.30}$$

4.2.5 Heat Capacity

The heat capacity of an object is the amount of heat it takes to change the temperature of that object. More precisely, we can define the heat capacity at constant volume and pressure as

$$\left(\frac{\partial Q}{\partial T}\right)_V = C_V, \tag{4.31}$$

$$\left(\frac{\partial Q}{\partial T}\right)_P = C_P, \tag{4.32}$$

respectively.[3] The more common one is C_V, because referring back to the fundamental equation (4.21), constant volume is

[3] Given our admonition to be careful about writing δQ rather than dQ, the abuse of notation ∂Q may seem strange, but it is a common one. Just make sure ∂Q only shows up when taking a derivative with respect to something, and not as a total differential as in (4.21).

EXAMPLE 4.2

Let's calculate the entropy change for an ideal gas in a particular process, the adiabatic free expansion of the gas from volume V_1 to volume V_2. The internal energy depends only on T ($\frac{3}{2}Nk_BT$ for a monoatomic gas, or $\frac{5}{2}Nk_BT$ or $\frac{7}{2}Nk_BT$ for a diatomic gas depending on the temperature), but since temperature is constant in free expansion, $dU = 0$. Thus $TdS = PdV$, and

$$\Delta S = \int_{V_1}^{V_2} P\frac{dV}{T} = \int_{V_1}^{V_2} \frac{Nk_BT}{V}\frac{dV}{T} = Nk_B \ln\left(\frac{V_2}{V_1}\right).$$

Of course, this matches the result you would obtain from using the full formula for the entropy (4.9). It's much more important to remember the steps that go into this derivation than memorize the equation for entropy change itself, since this same kind of reasoning is used often in thermodynamics problems.

equivalent to $dV = 0$, so we can differentiate both sides with respect to T and get

$$\left(\frac{\partial Q}{\partial T}\right)_V = \frac{\partial U}{\partial T}. \tag{4.33}$$

Hence, differentiating the energy you obtain from the equipartition theorem gives you C_V.

In fact, for an ideal gas there is a very simple (but *very* tricky to derive) relation between C_P and C_V:

$$C_P - C_V = Nk_B, \tag{4.34}$$

where N is the number of particles. Note that this is true for *any* ideal gas: monoatomic, diatomic, or something more complicated. This in turn immediately gives you the ratio γ you need for an adiabatic process if you know the internal structure of the gas. For example, a monoatomic gas has internal energy $\frac{3}{2}k_BT$ per particle from the equipartition theorem, so we have

$$C_V = \frac{3}{2}Nk_B \implies C_P = \frac{5}{2}Nk_B \implies \gamma = \frac{C_P}{C_V} = \frac{5}{3}.$$

As defined, C_P and C_V are *extensive* variables, because they depend on the quantity of the substance being heated. If we normalize to the mass of the material, we get an *intensive* quantity c called the *specific heat* (or *specific heat capacity*), typically in units of $\mathrm{J\,K^{-1}\,g^{-1}}$. From dimensional analysis, we can remember the formula for the amount of energy Q required to heat a mass m of specific heat c by ΔT degrees as

$$Q = mc\Delta T. \tag{4.35}$$

The specific heat capacity of water is famously equal to $4.18\,\mathrm{J\,K^{-1}\,g^{-1}}$ at standard temperature and pressure. Because of this, specific heats are sometimes quoted in units of *calories* rather than joules. One calorie is set to be 4.18 J, so that the specific heat capacity of water is conveniently $1\,\mathrm{cal\,K^{-1}\,g^{-1}}$.

We should note that the terms "heat capacity" and "specific heat" are often used interchangeably, and if a distinction is necessary the units and context should tell you which one is meant.

4.2.6 Model Systems

Two thermodynamic systems are almost guaranteed to appear on the GRE: heat engines and ideal gases. Gases also provide a nice system for studying the propagation of sound waves.

- **Heat engines, P–V, and T–S diagrams.** A heat engine is a process that converts heat into work. Schematically, the heat engine absorbs some heat Q_H from a hot reservoir at T_H, expels Q_C of this to a cold reservoir at T_C, and converts the remainder in work $W = Q_H - Q_C$. Since energy Q_C is not converted into work, the efficiency of the heat engine is

$$e = 1 - \left|\frac{Q_C}{Q_H}\right|. \tag{4.36}$$

(Note that there are various sign conventions for Q_C and Q_H, but the absolute value signs make sure $e < 1$ always.) The *maximum* theoretical efficiency for a heat engine is

$$e = 1 - \frac{T_C}{T_H}, \tag{4.37}$$

which is in fact the efficiency of the idealized Carnot cycle. The Carnot cycle consists of four steps. First, the gas undergoes reversible isothermal expansion at the hot temperature T_H. Entropy increases from S_1 to S_2 during this process. Next, the gas expands adiabatically at constant entropy until it has temperature T_C. The gas then is compressed at constant temperature T_C, and entropy decreases from S_2 back to S_1. Finally, the gas is compressed adiabatically,

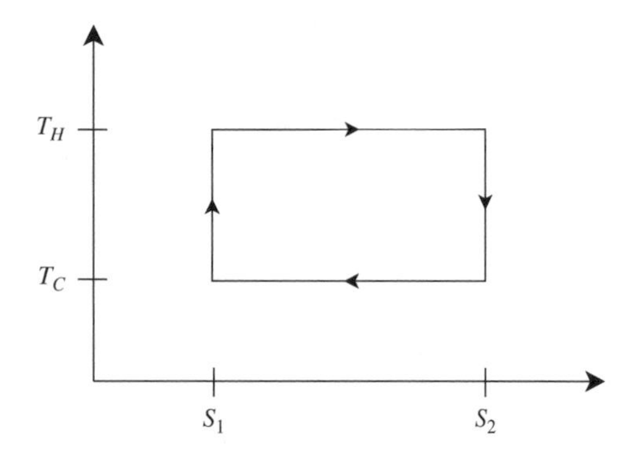

Figure 4.2 An example of a Carnot cycle in the T–S-plane.

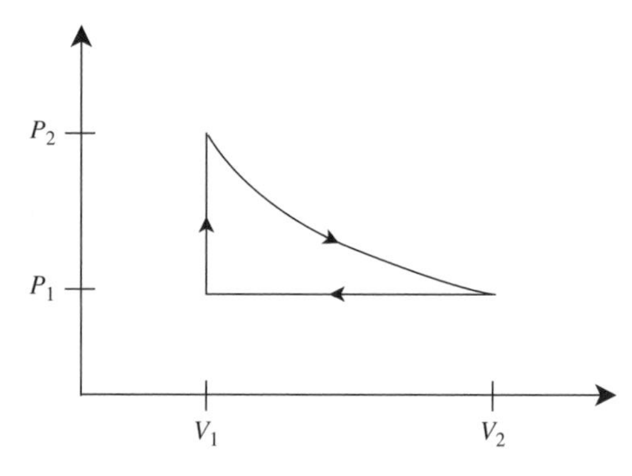

Figure 4.3 An example of a cycle in the P–V-plane. The work done is the signed area enclosed by the curve.

increasing the temperature from T_C to T_H. The entire process is thus a rectangle in the T–S-plane, illustrated in Fig. 4.2.

Since the system returns to its initial state, its change in internal energy over the cycle is zero: $\Delta U = 0$. The First Law then tells us that $Q = W$. The heat input in reversible processes is $Q = \int TdS$, so we can compute the work by

$$W = Q = \int TdS = (T_H - T_C)(S_2 - S_1).$$

In other words, the work is the area in the T–S-plane bounded by the Carnot cycle. Problems on the GRE often involve analyzing cycles of systems in pressure–volume or entropy–temperature diagrams. The Carnot cycle is just one example of a problem that can be analyzed with one of these diagrams.

A similar example with a P–V diagram is shown in Fig. 4.3. This diagram could represent an ideal gas that undergoes compression at constant pressure, an increase in pressure at constant volume via heating, and then isothermal expansion. To find the work done during the process, just find the *area* in the P–V-plane, since that's what the integral $\int PdV$ instructs you to do.

It is important to note that the areas are *signed* quantities: if you reverse the direction of the closed path bounding the area in question, you flip the sign of the area, and hence the sign of the work. This is often a crucial distinction, since it tells you whether the system is doing work on its surroundings, or vice versa. You can determine this sign with a right-hand type rule, but it's just as easy to use physical reasoning: the gas does positive work as it expands, and for a given volume change ΔV, it does more work at higher pressure. So, imagining a rectangle in the P–V-plane, the gas does positive net work when the upper horizontal edge is traversed left-to-right, and the bottom edge is traversed right-to-left. A similar analysis holds for T–S diagrams, so, to summarize,

> *Clockwise paths in the P–V- and T–S-planes do positive work.*

Just remember which axis goes where!

- **Monoatomic ideal gases.** Since ideal gases are so common, at a minimum you should know the internal energy, the entropy, and the root-mean-square (rms) velocity. Sometimes it's handy to know the partition function, because it allows you to calculate so many observables, but it is not essential knowledge.

The Hamiltonian for a particle in an ideal gas is just the Hamiltonian for a free particle in three dimensions. So, using equation (4.11) with $H = \mathbf{p}^2/2m$, we have the partition function for a single particle:

$$Z_1 = \frac{1}{h^3} \int e^{-\beta \mathbf{p}^2/2m} d^3p d^3x = \frac{V}{h^3} \left(2\pi m k_B T\right)^{3/2}.$$

For a gas of N identical particles, the partition function is just the product of all single-particle partition functions times the normalization factor $1/N!$ for identical particles:

$$Z_N = \frac{V^N}{N! h^{3N}} \left(2\pi m k_B T\right)^{3N/2}.$$

From this, we can immediately derive the internal energy of a classical ideal gas using (4.6):

$$U = \frac{3}{2} N k_B T, \tag{4.38}$$

which implies $C_V = (3/2) N k_B$ from the discussion in Section 4.2.5. You can turn this around and use the equipartition theorem to remind yourself about the factor of 3/2 in the exponent of the partition function: there are three quadratic degrees of freedom in the Hamiltonian (one for each spatial dimension), giving the factor of 3/2 in the internal energy.

Related to the average internal energy of a gas is the rms velocity of a gas molecule. This is given by

$$v_{\rm rms} = \sqrt{\frac{3k_BT}{m}}. \quad (4.39)$$

The rotational and vibrational motion of a gas molecule does not contribute to its velocity: a molecule can rotate however it wants, but if it has no translational energy, its rms velocity will be zero. The result of this is that even though monoatomic and diatomic gases have different internal energies, the expressions for their rms velocities are equivalent. This makes the formula very easy to remember: the factor of 3 comes from three dimensions, and the rest of the factors come from dimensional analysis.

We can also calculate the entropy of an ideal gas just by differentiating the partition function, using (4.8), and derive equation (4.9), which we saw earlier, repeated here for convenience:

$$S = Nk_B\left(\ln\frac{V}{N} + \frac{3}{2}\ln T + \frac{5}{2} + \frac{3}{2}\ln\frac{2\pi m k_B}{h^2}\right). \text{ (p.80)} \quad (4.9)$$

This expression is useful for demonstrating the *entropy of mixing*, as shown in Example 4.3.

- **Sound waves in gases.** Sound waves need a medium in which to propagate, and in most everyday situations, this medium is a gas. As discussed in Chapter 3, the speed of sound in a material is given by

$$c = \sqrt{\frac{K}{\rho}},$$

where K is a measure of stiffness called the *bulk modulus* and ρ is the density. The "stiffness" of an ideal gas is tricky to define, but the general result is that

$$K = \gamma P,$$

where $\gamma = C_P/C_V$ and P is the pressure as usual. From this, we have the general result that the speed of sound in an ideal gas is just

$$c = \sqrt{\gamma\frac{P}{\rho}}. \quad (4.40)$$

This formula is also simple to memorize: higher pressure means faster speed of sound, higher density means lower speed of sound, and the square root comes from dimensional analysis. Notice, though, that this formula contains an implicit dependence on the temperature: since $\rho = Nm/V$, the ideal gas law gives

$$\frac{P}{\rho} = \frac{P}{Nm/V} = \frac{PV}{Nm} = \frac{Nk_BT}{Nm},$$

so the speed of sound can be written

$$c = \sqrt{\gamma\frac{k_BT}{m}}. \quad (4.41)$$

4.3 Quantum Statistical Mechanics

Though we discussed the role of identical particles when calculating the entropy, we have not considered the effect of distinguishability in a more general quantum context. Identical particles behave very differently in quantum mechanics than might be expected from classical mechanics; the best-known example is the Pauli exclusion principle (to be discussed in much more detail in the following chapter), which states that two identical fermions can't occupy the same quantum state. When we wish to calculate the energy of some large ensemble of particles, knowing how many particles are allowed to occupy each state with energy ϵ is obviously essential. More precisely, it affects the average occupation number of a particular energy state.

EXAMPLE 4.3

Consider a box that is divided into two equal halves of volume V by an impermeable partition, with the same number N of gas particles in each half. If the gases are identical, then removing the partition does not change the entropy of the system. On the other hand, if the gases are different, then removing the partition produces a change in entropy:

$$\Delta S_{\rm mix} = 2Nk_B\left(\ln\frac{2V}{N} - \ln\frac{V}{N}\right) = 2Nk_B\ln 2.$$

This example demonstrates the importance of whether particles in a system are identical or distinguishable for determining the entropy. Once again, remembering the constant terms is probably not worth your time, since they tend to cancel when calculating entropy differences anyway, as they did in this example.

Deriving the *average occupation number* for bosons and fermions is straightforward, though a little lengthy, so we will not repeat it here and simply state the results. By average occupation number, we mean the average number of particles occupying a single-particle state at some energy ϵ_i. For *identical* fermions (particles with half-integer spin), the occupation number as a function of energy level ϵ_i is described by the *Fermi–Dirac distribution*:

$$F_{\rm FD}(\epsilon_i) = \frac{1}{e^{(\epsilon_i-\mu)/k_BT}+1}. \tag{4.42}$$

Notice that $F_{\rm FD}$ can never exceed 1: this is the Pauli exclusion principle at work, since there can be no more than one fermion per energy level. For identical bosons (integer spin), we have the *Bose–Einstein distribution* instead:

$$F_{\rm BE}(\epsilon_i) = \frac{1}{e^{(\epsilon_i-\mu)/k_BT}-1}. \tag{4.43}$$

Now, the change of sign in the denominator means that, depending on the temperature, arbitrarily many bosons can occupy the same state. At zero temperature, this leads to the familiar concept of *Bose condensation*: all the particles want to sit in the state with lowest energy. Both of these distributions limit to the usual exponential Boltzmann statistics (4.4) when the system can be treated "classically" (large interparticle distance and high temperature), which forces the exponential factor in the denominator to be much greater than 1.

To actually calculate anything with these distributions, we need a few extra ingredients. The new quantity appearing in all the above formulas is μ, the *chemical potential*, which roughly speaking is the energy associated with adding or removing a particle from the system. In the *grand canonical ensemble*, a generalization of the canonical ensemble, the number of particles is allowed to vary but the chemical potential is held fixed. So just as in the canonical ensemble, where we can fix T and ask about average E, in the grand canonical ensemble we can fix μ in the above formulas and ask about average particle number N. But first, we must take into account the fact that each energy level ϵ_i may have a *degeneracy* $g(\epsilon_i)$: for example, a free spin-1/2 fermion has two possible spin states with the same energy (see Section 5.5), so in that case $g(\epsilon_i) = 2$, independent of the energy. To get the average particle number, we just sum the distribution functions over all energy states weighted by the degeneracy:

$$\langle N \rangle = \sum_i g(\epsilon_i)F(\epsilon_i), \tag{4.44}$$

where F is either $F_{\rm FD}$ or $F_{\rm BE}$ as appropriate for fermions or bosons, respectively. If the energy levels are spaced closely together enough, we can approximate the sum by an integral, and instead of a degeneracy factor we use the *density of states* $\rho(\epsilon)$, which counts the number of available states between energies ϵ and $\epsilon + d\epsilon$. We then have

$$\langle N \rangle = \int \rho(\epsilon)F(\epsilon)\, d\epsilon, \tag{4.45}$$

It is unlikely that you would ever have to evaluate integrals like this on the GRE, but knowing the physical meaning of the distribution functions and density of states comes in handy.

4.4 Problems: Thermodynamics and Statistical Mechanics

1. What is the partition function of a one-dimensional quantum harmonic oscillator?
   (A) $\exp\left(\frac{-\hbar\omega}{k_BT}\right)$
   (B) $1-\exp\left(\frac{-\hbar\omega}{k_BT}\right)$
   (C) $\left(1-\exp\left(\frac{-\hbar\omega}{k_BT}\right)\right)^{-1}$
   (D) $\left(2\cosh\frac{\hbar\omega}{2k_BT}\right)^{-1}$
   (E) $\left(2\sinh\frac{\hbar\omega}{2k_BT}\right)^{-1}$
2. At low temperature, a gas undergoing isentropic expansion from pressure P_1 and volume V_1 to pressure P_2 and volume V_2 is seen to satisfy $P_1V_1^{5/3} = P_2V_2^{5/3}$. At higher temperatures, the gas undergoing the same process could satisfy which of the following relations?
   I. $P_1V_1^{5/3} = P_2V_2^{5/3}$
   II. $P_1V_1^{7/5} = P_2V_2^{7/5}$
   III. $P_1V_1^{9/7} = P_2V_2^{9/7}$
   (A) II only
   (B) III only
   (C) I and II only
   (D) II and III only
   (E) I, II, and III
3. A system has two states of energies $-\epsilon$ and 2ϵ. What is the probability of observing it in the higher energy state at temperature T?
   (A) 0
   (B) $\left(1+\exp\left(\frac{3\epsilon}{k_BT}\right)\right)^{-1}$
   (C) $\left(1-\exp\left(\frac{3\epsilon}{k_BT}\right)\right)^{-1}$
   (D) $\left(\exp\left(\frac{\epsilon}{k_BT}\right)+\exp\left(\frac{-2\epsilon}{k_BT}\right)\right)^{-1}$
   (E) 1
4. A diatomic ideal gas of N particles is trapped on a layer of material, such that the gas molecules are free to move only in two dimensions. What is the heat capacity at constant volume of this gas at room temperature? You may

assume that $k_B T \ll \hbar\sqrt{k/m}$, where m is the mass of the molecule and k is the Hooke's law constant associated with vibrational motion.

(A) $(1/2)Nk_B$
(B) Nk_B
(C) $(3/2)Nk_B$
(D) $2Nk_B$
(E) $(5/2)Nk_B$

5. The heat capacity per particle at constant volume of a *relativistic* ideal gas is

(A) $k_B/2$
(B) k_B
(C) $3k_B/2$
(D) $2k_B$
(E) $3k_B$

6. The three-dimensional quantum harmonic oscillator transitions from the $n = 1$ state to the $n = 2$ state. What is the change in entropy?

(A) 0
(B) $\hbar\omega$
(C) k_B
(D) $k_B \ln 2$
(E) $k_B \ln 3$

7. Six cups numbered 1 through 6 can each hold one marble. There are three red marbles and one blue marble. How many different ways are there to fill the cups with all four marbles?

(A) 6
(B) 12
(C) 24
(D) 30
(E) 60

8. Which of the following MUST be true of a closed system undergoing an adiabatic process?

(A) No heat is exchanged with the environment.
(B) Entropy is constant.
(C) The system does no work on its environment.
(D) Entropy increases.
(E) The system is at constant pressure.

9. A box is partitioned into equal volumes, each containing a different ideal monoatomic gas. When the partition is removed, which of the following statements must be true?

(A) The entropy decreases.
(B) The entropy increases.
(C) The entropy does not change.
(D) The change in entropy depends on the mass of particles.
(E) The change in entropy depends on the temperature of the gas.

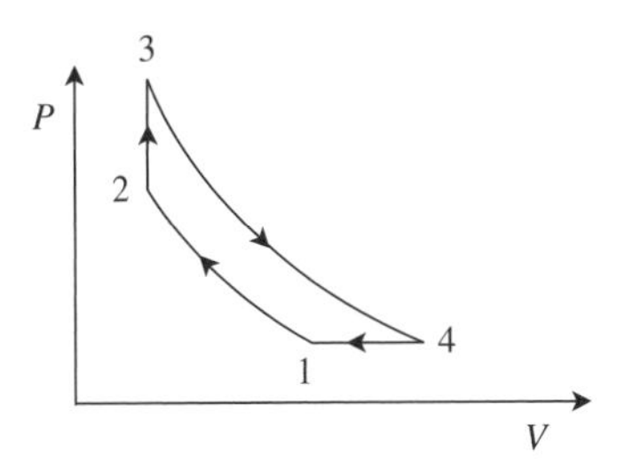

10. The P–V diagram above illustrates the Sargent cycle for an ideal gas. Paths 1–2 and 3–4 are reversible adiabatic, path 2–3 is at constant volume, and path 4–1 is at constant pressure. If T_i denotes the temperature at point i, which temperature is hottest?

(A) T_1
(B) T_2
(C) T_3
(D) T_4
(E) it is impossible to tell from the information given

11. A Carnot engine absorbs 10 J of heat from a hot bath in one cycle and is 90% efficient. How much heat is dissipated to a cold bath?

(A) 10 J
(B) 9 J
(C) 2 J
(D) 1 J
(E) 0.9 J

12. The effects of intermolecular attractions in a van der Waals gas can be modeled by adding a constant term $-a$ to the single-particle Hamiltonian. Which of the following thermodynamic quantities depend(s) on the parameter a?

I. The partition function
II. The internal energy
III. The specific heat

(A) I only
(B) II only
(C) I and II
(D) I and III
(E) I, II, and III

13. At zero energy, the Bose–Einstein distribution for a collection of bosons approaches infinity as the chemical potential approaches zero from below. What is the correct interpretation of this phenomenon?

(A) The average occupation number of the ground state is infinite.

(B) There is no ground state for the ideal Bose–Einstein gas.
(C) The Bose–Einstein distribution is the incorrect distribution function for the ground state, and one should use the Fermi–Dirac distribution instead.
(D) A macroscopic number of particles can occupy the ground state and must be treated separately from the occupation numbers of higher states.
(E) None of the above.

14. A grand canonical ensemble of fermions at chemical potential μ has a density of states given by $\rho(\epsilon) = Ae^{-\kappa\epsilon}$. At zero temperature, what is the average number of particles in the system?
(A) 1
(B) $\frac{A}{\kappa}e^{-\kappa\mu}$
(C) $\frac{A}{\kappa}\left(1 - e^{-\kappa\mu}\right)$
(D) $\frac{A}{\kappa} - e^{-\kappa\mu}$
(E) 0

4.5 Solutions: Thermodynamics and Statistical Mechanics

1. E – The energies of the harmonic oscillator are just $E_n = \hbar\omega(n + 1/2)$, so the partition function is $Z = \sum_{n=0}^{\infty} e^{-\hbar\omega(n+1/2)/(k_BT)}$. Pulling out a factor of $e^{-\hbar\omega/(2k_BT)}$ gives us a geometric series, which we can sum as

$$\sum_{n=0}^{\infty} e^{-\hbar\omega n/(k_BT)} = \frac{1}{1 - e^{-\hbar\omega/(k_BT)}}.$$

Putting back in the factor we pulled out and doing a little manipulation with hyperbolic trig functions gives

$$e^{-\hbar\omega/(2k_BT)}\frac{1}{1 - e^{-\hbar\omega/(k_BT)}} = \left(2\sinh\frac{\hbar\omega}{2k_BT}\right)^{-1}.$$

2. E – In an isentropic process, pressure and volume are related by PV^γ = constant. The problem statement tells us that $\gamma = 5/3$ at low temperatures, which is true for any gas because all degrees of freedom are frozen out except for translational modes. At higher temperatures, if the gas is monoatomic, γ will remain 5/3, but if the gas is diatomic, unfreezing of rotational or vibrational modes can result in $\gamma = 7/5$ or $\gamma = 9/7$, respectively. Thus, all of the listed relations are possible, since the information at low temperatures is not sufficient to decide if the gas is monoatomic or diatomic.

3. B – We can always shift the energies such that one of them is 0. This trick usually avoids unnecessary confusion with the notation. So we solve the problem for a system of energies 0 and 3ϵ. Using the expression for the probability of finding the system in a particular state, we have

$$p_i = \frac{e^{-E_i/(k_BT)}}{Z} = \frac{e^{-3\epsilon/(k_BT)}}{1 + e^{-3\epsilon/(k_BT)}} = \frac{1}{1 + e^{3\epsilon/(k_BT)}}.$$

4. C – The last sentence of the problem statement tells us we may assume that vibrational motion is frozen out. In this case, a diatomic gas in two dimensions has only three quadratic degrees of freedom: two translational for the center of mass, and one rotational (the only possible axis of rotation is perpendicular to the plane on which the gas is trapped). By the equipartition theorem, each contributes $(1/2)k_BT$ to the internal energy U, and $C_V = \partial U/\partial T$, so choice C is correct.

5. E – Unlike the previous problem, we can't just apply the equipartition theorem since the energy–momentum relation for a relativistic particle is *linear* rather than quadratic.[4] On the other hand, the heat capacity of a relativistic gas shows up so often that it may be useful simply to memorize the result. The derivation is rather straightforward, though. In the canonical ensemble, we calculate the partition function in a box of volume V. There is no potential energy, so the energy of a single particle is pure kinetic, $E = |\mathbf{p}|c$. Thus the partition function is

$$Z = \frac{V}{h^3}\int d^3p\, e^{-\beta|\mathbf{p}|c},$$

where $\beta = 1/k_BT$ as usual. Going to spherical coordinates in momentum space and using the fact that the integrand is spherically symmetric,

$$Z = \frac{4\pi V}{h^3}\int_0^\infty p^2 e^{-\beta pc}dp = \frac{8\pi V}{(h\beta c)^3}.$$

(The integral can be done simply by repeated application of integration by parts.) Since we are calculating the heat capacity, we only care about the part of $\ln Z$ that depends on β:

$$\ln Z = -3\ln\beta + \text{const.}$$

Continuing, $E = -\frac{\partial \ln Z}{\partial\beta} = \frac{3}{\beta} = 3k_BT$, so $C_V = \frac{dE}{dT} = 3k_B$, choice E.

6. D – The energy of the three-dimensional harmonic oscillator is $E_n = \hbar\omega(n_x + n_y + n_z + 3/2)$, where $n = n_x{+}n_y{+}n_z$. There are three degenerate states corresponding to the $n = 1$ level. There are six degenerate states

[4] See Chapter 6 for a review of relativity.

corresponding to the $n = 2$ level. From the Boltzmann expression for entropy, the initial entropy is $S_1 = k_B \ln 3$. The final entropy is $S_2 = k_B \ln 6$. The change in entropy is therefore $\Delta S = k_B \ln 2$.

7. E – There are six ways to place the blue marble. For each placement of the blue marble, there are $\binom{5}{3}$ distinct ways to place the remaining three red marbles. So there are

$$6 \times \frac{5!}{3! \times 2!} = 60$$

distinct ways of placing the marbles into the cups. Equivalently, we can choose the four cups out of the six to place the marbles in, then choose the one cup out of the four to place the blue marble in, giving

$$\binom{6}{4} \times 4 = 60,$$

the same answer.

8. A – By definition, an adiabatic process is a process for which there is no heat transfer to the environment. Entropy may be constant if the process is reversible, but can increase if the process is irreversible, as in adiabatic free expansion (incidentally, the pressure also changes in this situation, eliminating choice E). The system can do work on its environment, as it does in the expansion phases of the Carnot cycle.
9. B – The entropy of mixing causes the total entropy to increase. This can be seen immediately from equation (4.9).
10. C – An ideal gas undergoing a reversible adiabatic process has a constant value of PV^γ. This means that $TV^{\gamma-1}$ is constant. For any ideal gas, $C_P > C_V$, so $\gamma - 1 > 0$ and a decrease in volume leads to an increase in temperature. This implies that $T_3 > T_4$ and $T_2 > T_1$. From the ideal gas law, we know that $PV = Nk_BT$, so $T_3 > T_2$ and $T_4 > T_1$. Collecting inequalities, we have $T_1 < T_2 < T_3$ and $T_1 < T_4 < T_3$, so T_3 is hottest.
11. D – From the expression for the efficiency of a Carnot engine, we have $Q_C = Q_H(1 - e)$. Plugging in numbers we find that 1 J of heat is dissipated to the cold bath.
12. C – The effect of the constant term is to change the Hamiltonian to $H = \mathbf{p}^2/2m - a$. This shows up explicitly in the single-particle partition function $Z_1 \propto \int e^{-\beta H}$, but since it is constant, it can be pulled outside the integral:

$$Z_1 = e^{\beta a} Z_{1,\text{ ideal gas}},$$
$$Z_N = e^{N\beta a} Z_{N,\text{ ideal gas}}.$$

This just adds a constant term to the log of the partition function: $\ln Z_N = N\beta a + \cdots$. The internal energy still depends on a: $U = -\partial \ln Z_N/\partial\beta = -Na + \cdots$. However, the a-dependent term no longer depends on T, so taking the derivative to get $C_V = \partial U/\partial T$, this term disappears. Thus C_V is independent of a, so only I and II depend on a.

13. D – This is the statement of Bose condensation. Since the occupation number cannot be infinite for any finite sample of the gas, one must treat the two components of the gas separately: the macroscopic condensate occupying the ground state, and the rest of the particles occupying the excited states.
14. C – As temperature goes to zero, the Fermi–Dirac distribution limits to a step function: 1 for energies less than the chemical potential $\epsilon < \mu$ and 0 for energies above the chemical potential $\epsilon > \mu$. This is a useful fact to remember. Unlike the bosons of the previous problem, the chemical potential of fermions is not bounded by the lowest energy. In this case, we can just integrate the density of states from 0 to μ, and the number of particles is just

$$\langle N \rangle = \int_0^\mu A e^{-\kappa\epsilon}\, d\epsilon = \frac{A}{\kappa}\left(1 - e^{-\kappa\mu}\right).$$

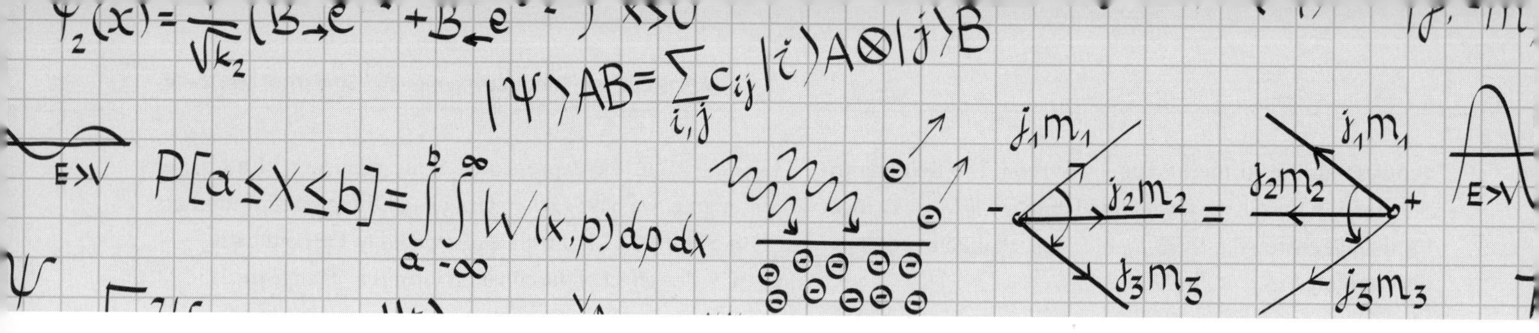

5 Quantum Mechanics and Atomic Physics

Quantum mechanics no doubt seemed somewhat bizarre when you were first introduced to it. Nice classical particles which followed deterministic trajectories were replaced by mysterious wavefunctions and particles that no longer had properties until you measured them – and let's not even mention the long philosophical discourses accompanying explanations of the Copenhagen Interpretation, nonlocal hidden variables, and the like. The good news is that *none of this is relevant* for the GRE, so we won't waste your time with it. This will be a lightning review of how to *compute* things in quantum mechanics: you can leave the deep thought to a situation where you have more than two minutes per question.

While quantum mechanics itself is rather under-represented on the GRE (only 12% of questions, according to ETS), the atomic physics section (10%) is really quantum mechanics in disguise, which is why we include it in the same chapter as quantum mechanics. Throw in a couple questions from the Miscellaneous and Optics and Waves categories, and quantum phenomena really make up about a full 25% of the test, so it pays to know this material in detail.

5.1 Formalism (How To Calculate)

5.1.1 Wavefunctions and Operators

The state of a quantum system, whether a single particle or a collection of 10^{23} particles, is described by a single *complex-valued* function of position and time called the *wavefunction*, usually denoted $\Psi(x,t)$. If there are multiple particles in the system, Ψ is a function of all the coordinates $x_1, x_2, \ldots$ of the various particles as well as time. It's likely that the only situation where you'll be concerned with multiple-particle wavefunctions on the GRE is when dealing with Bose or Fermi statistics, which we'll get to later (see Section 5.5.4), so for now we assume that our quantum system is just a single particle. Given the wavefunction, the rules for calculating quantities of interest in one-dimensional quantum mechanics are the following:

1. The probability that the particle with wavefunction $\Psi(x,t)$ will be found between positions x and $x+dx$ is $|\Psi(x,t)|^2\,dx$.
2. Observables A are represented by Hermitian operators $\hat{A}$ which act on Ψ. The expectation value of an observable A in the state Ψ is
$$\langle A \rangle = \int_{-\infty}^{\infty} \Psi^* \hat{A} \Psi \, dx. \tag{5.1}$$
3. A measurement of an observable A at a time t_0 on a state Ψ will always return one of the eigenvalues of $\hat{A}$. To find the probability that a particular eigenvalue λ_n is observed, expand $\Psi(x,t_0) = \sum_k c_k f_k(x)$ in a basis of orthonormal eigenfunctions $f_k(x)$ for $\hat{A}$ with eigenvalues λ_k. The desired probability is $|c_n|^2$. After measurement, the wavefunction of the particle at time t_0 is now $f_n(x)$.

Let's now examine each of these rules in detail.

1. **Position measurements.** Notice that the quantity that appears is the *complex modulus* $|\Psi(x,t)|^2$, not $\Psi(x,t)$ itself. This is comforting because probabilities must be non-negative real numbers. In fact, the probability of finding the particle *somewhere* had better be exactly 1:
$$\int_{-\infty}^{\infty} |\Psi(x,t)|^2 \, dx = 1. \tag{5.2}$$

This extremely important consistency condition means that the wavefunction must be *normalized.* The fact that we can actually perform this normalization means that Ψ can't do anything funny at $x = \pm\infty$: it has to vanish fast enough for the integral to converge. Always keep normalization in the back of your mind when calculating probabilities on the GRE – if the problem doesn't explicitly state that the wavefunction is normalized, you should normalize it *before* computing anything else. By the way, for a particle living in more than one dimension, this rule generalizes appropriately: the probability of finding a particle in a small box with opposite corners (x, y, z) and $(x + dx, y + dy, z + dz)$ is $|\Psi(x, y, z, t)|^2 \, dx\, dy\, dz$, and the probability of finding a particle in a spherical shell at radius r with thickness dr is $|\Psi(r, \theta, \phi, t)|^2 (4\pi r^2 \, dr)$.[1] Dimensional analysis can also help you with normalization: since $|\Psi|^2$ must be integrated over space to give a probability, which is dimensionless, the wavefunction has dimensions of $(\text{length})^{-d/2}$ in d spatial dimensions.

An aside about normalization: note that we could multiply Ψ by some phase factor, $e^{i\theta}$, which would not change the normalization, since $e^{i\theta}$ and $e^{-i\theta}$ cancel out in $|\Psi|^2$. The interpretation is that Ψ and $e^{i\theta}\Psi$ represent the *same state*. However, *relative* phases between two wavefunctions are physical: if we have a wavefunction such as $\Psi = \psi_1 + e^{i\alpha}\psi_2$, we can multiply Ψ by a phase (which preserves the relative phase $e^{i\alpha}$ between ψ_1 and ψ_2) but we are *not* allowed to multiply ψ_1 and ψ_2 individually by their own phase factors. Only the total wavefunction is defined up to a phase.

2. **Observables.** First, some definitions. An *operator* is just a rule instructing us to *do* something to a function $f(x)$. For example, the operator x says "multiply a function by x," and the operator $5\, d/dx$ says "differentiate a function with respect to x and multiply by 5." The rule is that operators are read from right to left, such that the piece closest to the function acts first. This is important when derivatives are involved: the operator $x\, d/dx$ means "differentiate with respect to x, *then* multiply by x," not the other way around! To distinguish an operator from an ordinary function, we put a hat on it, like $\hat{O}$.

Sometimes, acting on a function $f(x)$ with an operator $\hat{O}$ may return the *same* function, multiplied by a constant c. Here are some examples:

- $3 \cdot f(x) = 3f(x)$ for any function $f(x)$. In other words, the operator "3" acting on any $f(x)$ just returns $f(x)$ multiplied by the constant 3.
- $\dfrac{d}{dx} e^{\lambda x} = \lambda e^{\lambda x}$. The operator d/dx, acting on a certain function $f(x) = e^{\lambda x}$, returns $\lambda f(x)$.

In these cases, we say that $f(x)$ is an *eigenfunction* of the operator $\hat{O}$ with *eigenvalue* c. So in the examples above, all functions whatsoever are eigenfunctions of the operator $\hat{O} = 3$ with eigenvalue 3, but only functions of the form $e^{\lambda x}$ are eigenfunctions of $\hat{O} = d/dx$ with eigenvalue λ. Note that this is *not* always the case: $\dfrac{d}{dx} \sin x = \cos x$, so $\sin x$ is *not* an eigenfunction of d/dx. Indeed, eigenfunctions are special (and extremely useful) because they reduce the potentially complicated action of an operator to something very simple, namely multiplication by a constant.

A *Hermitian* operator $\hat{A}$ is one such that

$$\int_{-\infty}^{\infty} f(x)^* (\hat{A} g(x))\, dx = \int_{-\infty}^{\infty} (\hat{A} f(x))^* g(x)\, dx \tag{5.3}$$

for any $f(x)$ and $g(x)$. Here are some useful facts (definitely worth memorizing!) about Hermitian operators:

- All their eigenvalues are real.
- Eigenfunctions corresponding to different eigenvalues are orthogonal: if $f(x)$ and $g(x)$ are eigenfunctions with different eigenvalues, then $\int_{-\infty}^{\infty} f(x)^* g(x)\, dx = 0$.

Now, in classical mechanics, *all* observables can be built out of the two quantities x and p, position and momentum. For example, kinetic energy is $E = p^2/2m$, the potential energy of a harmonic oscillator is $\frac{1}{2}kx^2$, angular momentum in three dimensions is $\mathbf{r} \times \mathbf{p}$, and so on. In quantum mechanics, these observables are represented by the operators

$$\hat{x} = x, \quad \hat{p} = -i\hbar \frac{\partial}{\partial x}. \tag{5.4}$$

It's easy to see that $\hat{x}$, so defined, is Hermitian according to (5.3): you should check for yourself that $\hat{p}$ is Hermitian, using integration by parts (notice the very important factor of $-i$ which makes this work). Any operators that are the quantum analogues of classical observables (angular momentum, potential energy, and so on) can be built out of these two operators. For the purposes of the GRE, the *only* operators that can't be built out of x and p are the spin operators – we'll treat those in great detail in Section 5.5. We will often refer to "observables" and "operators" interchangeably, sometimes sloppily with the same notation for both (omitting the hat), but remember that "observable" refers to a physical quantity to be measured,

[1] Note, however, that it is conventional to normalize the radial and angular parts of a three-dimensional wavefunction *separately*: see Section 5.4.1.

while "operator" refers to its mathematical representation in the formalism of quantum mechanics.

3. **Measurements.** The radical departure of quantum mechanics from classical mechanics arises from the first sentence of rule 3 – a measurement of A will *always* return an eigenvalue λ_n, and if the eigenvalues are discrete rather than continuous, the measurement of A is *quantized.* To go further, we use the previously mentioned facts about Hermitian operators: since the eigenfunctions are orthogonal, we can build an orthonormal *basis* from them, and express any normalized wavefunction $\Psi(x,t)$ as a linear combination of these basis functions $f_n(x)$. The coefficient c_n is then given by

$$c_n = \int_{-\infty}^{\infty} f_n(x)^* \, \Psi(x,t)\, dx. \tag{5.5}$$

Taking the complex modulus squared of this number then gives the probability of getting eigenvalue λ_n when measuring A on the state Ψ at time t. Once this happens, the state of the system is *no longer* Ψ: it is simply $f_n(x)$. Thus, a subsequent measurement of A immediately following the first measurement is *guaranteed* to return the value λ_n, and no other.

Expectation values are very easy to compute once you have the decomposition of Ψ in an orthonormal basis: if $\Psi(x,t_0) = \sum_k c_k f_k(x)$, then

$$\langle A \rangle = \sum_k \lambda_k |c_k|^2. \tag{5.6}$$

When using this formula, make sure your eigenfunction expansion matches the observable whose expectation value you are trying to compute! Notice that expectation values are *averages* and are not required to be equal to one of the eigenvalues, just as the average of a set of integers need not be an integer.

5.1.2 Dirac Notation

All this talk about eigenvalues and eigenvectors, basis decompositions and normalization should remind you of linear algebra. This is no accident – the formalism of quantum mechanics is best expressed in this language, and Dirac notation provides an extremely convenient and intuitive way to do this. You may find it unfamiliar to think of functions (the wavefunction in particular) as elements of a vector space. This is actually not so bad, and we'll give the "dictionary" now.

Dirac notation represents a vector as a *ket*, like this: $|a\rangle$. Here, a is just a label – we could have written $|1\rangle$, or $|\text{Bob}\rangle$. To each ket $|a\rangle$ is associated another object, the *bra* $\langle a|$, which allows us to take inner products:

$$\text{Inner product of } |a\rangle \text{ and } |b\rangle \equiv \langle b|a\rangle. \tag{5.7}$$

The inner product is a complex number, which in this notation is also called the *bracket* of $|a\rangle$ and $|b\rangle$, hence the names *bra(c)ket* for these objects. The vector space that the kets live in is called *Hilbert space*, which is just a fancy name for a vector space where we are allowed to take the inner products of vectors. We will always work with complex vector spaces, so we define the inner product to behave as follows under complex conjugation:

$$\langle a|b\rangle := \langle b|a\rangle^*. \tag{5.8}$$

The reason for this is that $\langle a|a\rangle = \langle a|a\rangle^*$, so the norm of a vector is a real number.

To show the action of an operator $\hat{A}$ on a vector $|b\rangle$, we make the following convenient definition:

$$\hat{A}|b\rangle \equiv |\hat{A}b\rangle.$$

Note that this is *just notation*, since, as we have previously stated, the text that goes inside the ket is just a label. The power of this notation comes when we now take the bracket with another vector:

$$\langle a|\hat{A}b\rangle := \langle \hat{A}^\dagger a|b\rangle. \tag{5.9}$$

This *defines* $\hat{A}^\dagger$, the Hermitian conjugate of $\hat{A}$. Most importantly, this means that in a bracket $\langle a|\hat{A}b\rangle$, we can let an observable act *either* "on the left" on ket $|b\rangle$, or let its Hermitian conjugate $\hat{A}^\dagger$ act "on the right" on bra $\langle a|$. We have defined $\hat{A}^\dagger$ so we get the same answer either way.

If $\hat{A}$ is Hermitian – in other words, if $\hat{A}^\dagger = \hat{A}$ – then both sides of (5.9) contain the same operator $\hat{A}$, so we might as well define another convenient notation:

$$\hat{A} \text{ Hermitian} \implies \langle a|\hat{A}b\rangle = \langle \hat{A}a|b\rangle \equiv \langle a|\hat{A}|b\rangle.$$

Note the similarity with (5.3). These two conditions are in fact *identical* provided we make the following definitions:

$$\langle x|f\rangle := f(x), \tag{5.10}$$

$$\langle f|g\rangle := \int_{-\infty}^{\infty} f(x)^* g(x)\, dx. \tag{5.11}$$

The second of these just says that the inner product on *function space* is given by f^*g. The first is a little more subtle: it says that a function f should really be thought of not as a vector itself, but as a collection of coefficients $f(x)$, one for each point x. In other words, $|f\rangle$ is the abstract vector, and $\langle x|f\rangle$ represents the decomposition of f along the basis vectors $|x\rangle$.

If the last two sentences didn't make total sense to you, don't worry. This part of Dirac notation is not really relevant for the GRE: we only include it so that if you see a statement like "Let $|1\rangle$ be the ground state of the harmonic oscillator," you'll understand that $|1\rangle$ plays the role of $|f\rangle$, and you won't find yourself wondering what happened to the wavefunction $f(x)$. Indeed, we'll see below that many quantum mechanics problems can be solved totally within the confines of Dirac notation, without ever having to resort to the wavefunction.

One final comment about Dirac notation: if we are working in a finite-dimensional vector space, for example when talking about spin-1/2, then kets are just column vectors, operators are just matrices, and we have the following simple dictionary:

$$\langle b| := (b^T)^*,$$
$$A^\dagger := (A^T)^*,$$

where the superscript T denotes the transpose (of either a vector or a matrix).

5.1.3 Schrödinger Equation

In all of the previous description, we assumed we were *given* the wavefunction. But how do we find it in the first place? The answer is given by the Schrödinger equation:

$$i\hbar\frac{\partial}{\partial t}\Psi(x,t) = \hat{H}\Psi(x,t). \tag{5.12}$$

Here, we have introduced time dependence in the wavefunction, and the operator $\hat{H}$ appearing on the right-hand side is the Hamiltonian operator, which represents the total energy. *Almost always*,

$$\hat{H} = \frac{\hat{p}^2}{2m} + \hat{V}(x) = -\frac{\hbar^2}{2m}\frac{\partial^2}{\partial x^2} + \hat{V}(x), \tag{5.13}$$

where $V(x)$ is the potential energy. The two exceptions are the presence of an external magnetic field, which modifies the first (kinetic) term, and when the potential V also depends on time. Both of these cases require a different kind of analysis, which is more advanced than what you'll need on the GRE.

There is a useful way to read (5.12). We could view the left-hand side as an operator in its own right, and define $\hat{E} := i\hbar\,\partial/\partial t$ to be the total energy operator. Then, by rule 3 of Section 5.1.1, a measurement of the energy will always return an eigenvalue of $\hat{E}$, which by the Schrödinger equation is also an eigenvalue of $\hat{H}$. Therefore, *to find the possible energies of the system, we must find all the eigenvalues of $\hat{H}$.* Suppose we have done this, and we have a system in the state $\psi_n(x)$ with eigenvalue E_n. Then the Schrödinger equation reads

$$i\hbar\frac{\partial}{\partial t}\Psi(x,t) = E_n\Psi(x,t), \tag{5.14}$$

where E_n on the right-hand side is *just a number*, not an operator. Now we can solve this equation:

$$\Psi(x,t) = e^{-iE_nt/\hbar}\,\psi_n(x).$$

The most general wavefunction $\Psi(x,t)$ will just be a linear combination of all the eigenfunctions of $\hat{H}$, with appropriate time dependence $e^{-iE_nt/\hbar}$ tacked on.

This line of reasoning leads to a recipe for finding the time evolution of a quantum-mechanical system with Hamiltonian $\hat{H}$:

1. Solve the eigenvalue equation $\hat{H}\psi(x) = E\psi(x)$ to find a set of *time-independent* eigenfunctions $\psi_n(x)$ (also called *stationary states*) with eigenvalues E_n.
2. Given the wavefunction at time $t = 0$, $\Psi(x,0)$, decompose it along the basis of eigenfunctions $\psi_n(x)$: $\Psi(x,0) = \sum_n c_n\psi_n(x)$.
3. The full time-dependent wavefunction is $\Psi(x,t) = \sum_n c_n e^{-iE_nt/\hbar}\,\psi_n(x)$.

On the GRE, you will *never* have to complete all these steps from scratch. Almost always, you will be given a well-known Hamiltonian, for which the eigenfunctions are either given to you or which you are supposed to remember yourself. You will then be asked about time dependence, or to compute expectation values of various observables in these states. You may also be asked conceptual questions about this procedure: for example, you should check that if $\Psi(x,0) = \psi_n(x)$ (that is, at $t = 0$ the system is in a stationary state of energy E_n), then the probability of getting energy E_n at some other time t is always exactly 1.

Here is a useful list (worth memorizing) of general properties of the time-independent energy eigenfunctions $\psi_n(x)$, valid for any Hamiltonian you will encounter on the GRE:

- ψ_n for different values of n are orthogonal, since they correspond to different energy eigenvalues.
- ψ is always continuous. Its derivative $d\psi/dx$ is also always continuous, *except* at a boundary where the potential $V(x)$ is infinite. This exception will be treated in various contexts in Section 5.3.

- ψ can be taken to be purely real, without loss of generality.[2] Note this is *not* true for the full time-dependent wavefunction, since we must attach the complex exponential factors. However, it does lead to an extremely convenient computational shortcut: if a particle is in a stationary state ψ_n, which is taken to be real, the expectation value of its momentum *must vanish*. The proof is as follows:

$$\begin{aligned}\langle p \rangle &= \int \psi(x)e^{+iE_n t/\hbar}\left(-i\hbar\frac{\partial}{\partial x}\right)\psi(x)e^{-iE_n t/\hbar}dx \\ &= -i\hbar \times (\text{something real}),\end{aligned}$$

  because the exponential factors $e^{\pm iE_n t/\hbar}$ cancel with each other. But expectation values must be real for Hermitian operators, hence $\langle p \rangle = 0$. Caution: this does *not* apply to a superposition of stationary states, for example $\Psi(x,t) = \psi_1 e^{-iE_1 t/\hbar} + \psi_2 e^{-iE_2 t/\hbar}$, because the exponential factors will not cancel completely and $\Psi^*\Psi$ will contain a real term $\cos((E_2 - E_1)t/\hbar)$.
- The ground state ψ_0, corresponding to the lowest energy E_0, has no *nodes*: a node is a point at which the wavefunction vanishes (excluding the case where the wavefunction vanishes at a boundary, as in the infinite square well). Recalling the probabilistic interpretation, this means that there are no points where the particle is guaranteed not to be found. Each successive energy eigenstate has one more node than the previous one: ψ_1 has one node, ψ_2 has two nodes, and so on. So even if you know nothing about a given Hamiltonian, you can say something about its energy eigenfunctions just by looking at their graphs. Indeed, a classic GRE problem gives you sketches of possible wavefunctions for an unspecified Hamiltonian and asks you questions about them.
- If the potential $V(x)$ is *even* (that is, if $V(x) = V(-x)$), then $\psi(x)$ can be taken to have definite parity. This means that $\psi(x)$ is either even, $\psi(x) = \psi(-x)$, or odd, $\psi(x) = -\psi(-x)$. Furthermore, the parity of ψ_n alternates as we change n: The ground state ψ_0 is even, the first excited state ψ_1 is odd, and so on.
- For ψ to be normalizable, we must have $E > V_{\text{min}}$, where V_{min} is the global minimum of $V(x)$. The intuition, borrowed from classical mechanics, is that if the particle has less energy than the minimum of V, its kinetic energy must be negative, which is impossible. As we have emphasized, this classical reasoning does not hold strictly true in quantum mechanics, but it is a good mnemonic.

5.1.4 Commutators and the Uncertainty Principle

If you remember only one thing about operators in quantum mechanics, remember this:

Operators don't commute (in general).

That is, applying $\hat{A}$, followed by $\hat{B}$, is in general not the same thing as applying $\hat{B}$ followed by $\hat{A}$. If you know a little linear algebra, this follows in the finite-dimensional case from the fact that matrices don't commute in general. It's even true in the infinite-dimensional case, though. Example 5.1 shows how this works.

EXAMPLE 5.1

Consider the two operators $\hat{x}$ and $\hat{p}$, defined in (5.4). Let them act on a test function $f(x)$:

$$\begin{aligned}(\hat{x}\circ\hat{p})f(x) &= x\left(-i\hbar\frac{d}{dx}f(x)\right) = -i\hbar(xf'(x)), \\ (\hat{p}\circ\hat{x})f(x) &= -i\hbar\frac{d}{dx}(xf(x)) = -i\hbar(f(x) + xf'(x)) \\ &\implies (\hat{x}\circ\hat{p} - \hat{p}\circ\hat{x})f(x) = i\hbar f(x).\end{aligned}$$

Since the last line is true regardless of the function $f(x)$, we can drop f and write a relation involving only the operators:

$$[\hat{x}, \hat{p}] = i\hbar. \tag{5.15}$$

[2] This is not to say that ψ *must* be real, only that we *can* choose a real basis of energy eigenfunctions. For example, $e^{\pm ipx/\hbar}$ are eigenfunctions of the free particle Hamiltonian, but so are the real linear combinations $\sin(px/\hbar)$ and $\cos(px/\hbar)$. Of course, these latter two are not momentum eigenstates, but that's an added requirement we're not concerned with here.

Equation (5.15) in Example 5.1 is perhaps the most important equation in quantum mechanics. The symbol [,] stands for the *commutator* of two operators: compose them in one order, then subtract the result of composing them in the other order. It can be quite easy to get confused when computing commutators of operators like this, since just writing down $\hat{p} \circ \hat{x} = -i\hbar \frac{d}{dx} x$ might lead us to assume that only x is supposed to be differentiated, and we would lose the second term that we got from the product rule by acting on a test function $f(x)$ above. So, *when computing commutators where the operators involve derivatives, always act on a test function.* One other important thing to note is that the commutator, in general, is *itself an operator*: in this case it's a particularly simple operator, given by multiplication by the constant $i\hbar$. Equation (5.15) is known as the *canonical commutation relation.*

Here are two useful identities for computing commutators of products of operators:

- $[AB, C] = A[B, C] + [A, C]B$
- $[A, BC] = [A, B]C + B[A, C]$

Both are fairly easy to remember, since they resemble product rules for derivatives. But to rederive them in a pinch, just write out the commutators as if they were matrix multiplication: $[A, B] = AB - BA$ and so forth. You can be even more economical and note that one rule follows from the other by changing the order of the commutator using $[A, B] = -[B, A]$, and relabeling.

Commutators are intimately tied up with the uncertainty principle, for the following reason. The commutator measures the difference between the results of applying two operators in different orders, which, according to the rules given above, represents the difference between outcomes of measurements of two observables applied in different orders. If the commutator vanishes, it means that we can measure the two observables in either order, and we're guaranteed to get the same answer. A nonzero commutator $[\hat{A}, \hat{B}]$, however, means that in general, if we measure B and put the system in an eigenstate of $\hat{B}$, a follow-up measurement of observable A will destroy this state and put the system back in a linear combination of eigenstates of $\hat{B}$. In other words, there is a fundamental uncertainty in measurements of A versus measurements of B.

This is made precise by the following statement:

$$\sigma_A^2 \sigma_B^2 \geq \left(\frac{1}{2i} \langle [\hat{A}, \hat{B}] \rangle \right)^2 . \tag{5.16}$$

Here, σ_A^2 is the statistical *variance* of a measurement of A, defined as

$$\sigma_A^2 := \langle A^2 \rangle - \langle A \rangle^2, \tag{5.17}$$

and similarly for σ_B^2. Note that what appears on the right-hand side of (5.16) is the *expectation value* of the commutator $[\hat{A}, \hat{B}]$: this means that the uncertainty *bound* (the right-hand side) depends in general on what state the system is in. Of course, the actual uncertainty (the left-hand side) also depends on the state of the system. In particular, if we can find a state such that the inequality becomes an equality, we call such a state a *minimal-uncertainty state* for the two observables A and B. And once again, if the commutator vanishes identically, so does the uncertainty bound: we can find states for which both uncertainties are zero.

The case you are undoubtedly familiar with is $A = x$, $B = p$, in which case the right-hand side of (5.16) becomes $(i\hbar/2i)^2 = \hbar^2/4$. Taking the square root of both sides gives the familiar relation

$$\sigma_x \sigma_p \geq \frac{\hbar}{2}. \tag{5.18}$$

Notice that, because the commutator $[\hat{x}, \hat{p}]$ is just a number, its expectation value is independent of the state, and the minimum uncertainty is always the same, $\hbar/2$.

It is a very important fact (derived in all basic quantum mechanics books) that *the position-space wavefunction of a minimum-uncertainty state is a Gaussian.* Indeed, because it's a minimum-uncertainty state, its momentum–space wavefunction is *also* a Gaussian. Even if we're not dealing with a minimum-uncertainty state, most systems do not conspire to exceed the uncertainty bound by huge amounts, so the following "folklore" statement,

$$\Delta x \Delta p \approx \hbar, \tag{5.19}$$

holds quite generally. Note the missing factor of 2, and the replacement of the precisely defined σ_x and σ_p by the rather vague $\Delta x \Delta p$; this is because this statement is *only intended to give an order-of-magnitude estimate.* Nevertheless, it is quite useful, as you will see in the problems.

A similar "folklore" statement holds for energy and time,

$$\Delta E \Delta t \approx \hbar. \tag{5.20}$$

The standard application of (5.20) is to decay processes: Δt represents the lifetime (mean lifetime, or half-life, or whatever, since this is just an order-of-magnitude estimate) of the unstable state, and ΔE represents the uncertainty in energy of the decay process. For example, when an unstable particle with an extremely short lifetime decays at rest, its decay products can have widely varying total energy because of the large value of ΔE. Equally well, when a short-lived atomic excited state decays by emitting a photon, the energy of the photon is not precisely determined, but has a spread ΔE. In both of these contexts, ΔE is known as the *width* (either of the excited state, or of the emission line).

Finally, one very important remark regarding energy and commutators:

If an operator $\hat{O}$ commutes with the Hamiltonian, the corresponding observable is conserved.

In other words, we can simultaneously diagonalize $\hat{O}$ and the Hamiltonian, and label states of the system by eigenvalues of $\hat{O}$ and energies at a given time. The above statement guarantees that, at any subsequent time, these labels don't change.

5.1.5 Problems: Formalism

1. A particle has the wavefunction $\Psi(x) = A(1-x^2)$ for $|x| \leq 1$, and $\Psi(x) = 0$ elsewhere. What is the probability the particle will be found in the region $x < 0$?

   (A) 0
   (B) 1/4
   (C) 1/2
   (D) 3/4
   (E) 1

2. Let ψ_1 and ψ_2 be energy eigenstates of a time-independent Hamiltonian with energies E_1 and E_2. At time $t = 0$, a system is in state $\frac{1}{\sqrt{2}}(\psi_1 - \psi_2)$. At time t, what is the probability that a measurement of the energy of the system will return E_1?

   (A) 0
   (B) $1/\sqrt{2}$
   (C) 1/2
   (D) $\cos[(E_2 - E_1)/\hbar]$
   (E) $\cos[(E_2 + E_1)/\hbar]$

3. Let $|a\rangle$ and $|b\rangle$ denote momentum eigenstates with eigenvalues a and b respectively, where $a \neq b$. What is $\langle a|\hat{p}|b\rangle$?

   (A) a
   (B) b
   (C) $|ab|$
   (D) $|ab|^{1/2}$
   (E) 0

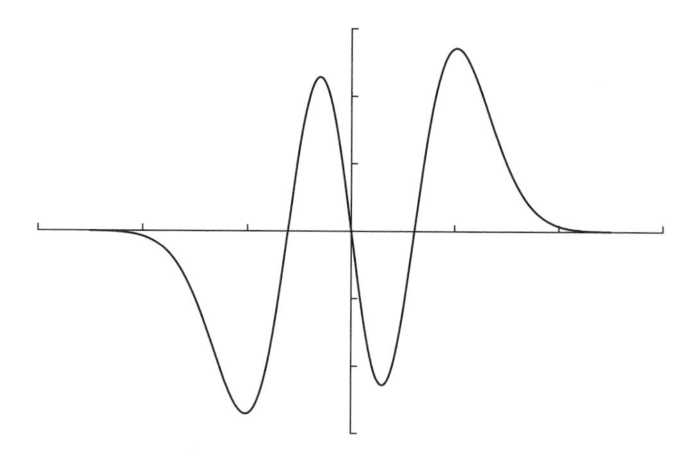

4. The wavefunction shown in the diagram above represents one of the excited states of the harmonic oscillator. What is the energy of the state?

   (A) $\hbar\omega/2$
   (B) $3\hbar\omega/2$
   (C) $5\hbar\omega/2$
   (D) $7\hbar\omega/2$
   (E) $9\hbar\omega/2$

5. Which of the following is a Hermitian operator?

   I. $\begin{pmatrix} 1 & 0 \\ -1 & 0 \end{pmatrix}$

   II. $\begin{pmatrix} 0 & -i \\ i & 0 \end{pmatrix}$

   III. $\begin{pmatrix} 1 & 2 \\ 2 & 1 \end{pmatrix}$

   (A) I only
   (B) II only
   (C) III only
   (D) I and III
   (E) II and III

6. Let $|s\rangle$ and $|t\rangle$ denote orthonormal states. Let $|\Psi_1\rangle = |s\rangle + 2i|t\rangle$ and $|\Psi_2\rangle = 2|s\rangle + x|t\rangle$. What must the value of x be so that $|\Psi_1\rangle$ and $|\Psi_2\rangle$ are orthogonal?

   (A) i
   (B) $-i$
   (C) 1
   (D) -1
   (E) $i/\sqrt{5}$

7. An unstable particle with a lifetime of 1.0×10^{-23} s and a mass of 500 MeV/c^2 is measured in a new experiment to have a mass of 450 MeV/c^2. The mass resolution

of the experiment is 10 MeV/c^2. The difference between the observed mass and the expected mass is most likely due to

(A) violation of conservation of energy
(B) the uncertainty principle
(C) experimental error
(D) time dilation
(E) the Mossbauer effect

8. The nitrogen molecule consists of two nitrogen atoms joined by a covalent bond with length approximately 100 pm. What is the approximate kinetic energy of one of the covalently bonded electrons?

(A) 4 meV
(B) 4 eV
(C) 4 keV
(D) 4 MeV
(E) 4 GeV

5.2 Harmonic Oscillator

5.2.1 One Dimension

The Hamiltonian for the quantum harmonic oscillator in one dimension is

$$H = \frac{\hat{p}^2}{2m} + \frac{1}{2}m\omega^2\hat{x}^2 \qquad \text{(harmonic oscillator).} \tag{5.21}$$

In this form, it's totally useless – we only include it to remind you that if a problem says "A system has Hamiltonian given by equation (5.21)," it's just talking about a harmonic oscillator. It may be written in terms of the spring constant $k = m\omega^2$ instead of the angular frequency ω, but no matter: if the potential is quadratic, you have a harmonic oscillator on your hands. Just remember to match the form given above by relating ω to the coefficient of the quadratic.

A clever change of variables brings it to the following, much more useful, form:

$$H = \hbar\omega\left(a^\dagger a + \frac{1}{2}\right). \tag{5.22}$$

How we arrived at the form (5.22) is irrelevant for GRE purposes – what's important is that there exists an operator a (called a *lowering operator* or *annihilation operator*), and its Hermitian conjugate $a^\dagger$ (called a *raising operator* or *creation operator*), which are linear combinations of $\hat{x}$ and $\hat{p}$ such that H can be transformed as above. It's worth memorizing the commutation relation of a and $a^\dagger$, since it's very simple:

$$[a, a^\dagger] = 1. \tag{5.23}$$

The reason (5.22) is useful is that we can read off the eigenstates of H right away. Suppose there exists a state called[3] $|0\rangle$ which is killed by a: $a|0\rangle = 0$. Then

$$H|0\rangle = \frac{\hbar\omega}{2}|0\rangle,$$

and $|0\rangle$ is an eigenstate of H with eigenvalue $\hbar\omega/2$. Indeed, one can prove that this is the lowest-energy eigenstate of H, so another fact worth memorizing is *the ground state of the harmonic oscillator has energy* $\hbar\omega/2$. Using the commutation relation (5.23) and the Hamiltonian (5.22), one can also prove that acting with $a^\dagger$ on $|0\rangle$ produces yet another energy eigenstate, with energy $3\hbar\omega/2$. (Exercise: check this yourself.) We can continue this process indefinitely, so we have derived the spectrum of the harmonic oscillator:

$$H|n\rangle = \hbar\omega\left(n + \frac{1}{2}\right)|n\rangle, \qquad n = 0, 1, 2, \ldots \tag{5.24}$$

As usual, the states $|n\rangle$ are orthogonal, because they are eigenvectors of H with different eigenvalues. They are also assumed to be normalized. However, $a^\dagger|n\rangle$ is *not* automatically normalized to $|n+1\rangle$ – the normalization factor will be provided to you on the test if you need it, but it's important to keep in mind that when calculating expectation values, $|n\rangle$ is normalized while $a^\dagger|n\rangle$ and $a|n\rangle$ are not. For convenience, we'll give you the normalization factors,

$$a^\dagger|n\rangle = \sqrt{n+1}|n+1\rangle; \quad a|n\rangle = \sqrt{n}|n-1\rangle,$$

but remember, you need *not* memorize these.

A standard question about the harmonic oscillator asks you to calculate the expectation value of some observable written in terms of a and $a^\dagger$. We'll walk you through this calculation once in Example 5.2 because it is both an excellent example of the use of Dirac notation, but also an illustration of how to calculate expectation values by using only orthonormality and commutation relations.

We'll finish this lightning review of the harmonic oscillator with some bits of trivia.

- The ground state of the harmonic oscillator happens to be a minimum-uncertainty state, so its position-space wavefunction is a Gaussian. You probably won't need its full expression, but you can get an estimate of its width from dimensional analysis: the parameters of the harmonic oscillator Hamiltonian are $\hbar$, m, and ω, and the only combination of these with the dimensions of length is $\sqrt{\hbar/m\omega}$.
- The position-space wavefunctions are related to so-called *Hermite polynomials.*

[3] Remember, this is *not* the zero vector! The 0 inside the ket is just a label. However, the action of a on $|0\rangle$ *does* give the zero vector.

EXAMPLE 5.2

Let's find the expectation value of $(a + a^\dagger)^2$ in the state $|3\rangle$. We want to calculate

$$\langle 3|(a + a^\dagger)^2|3\rangle = \langle 3|a^2 + aa^\dagger + a^\dagger a + (a^\dagger)^2|3\rangle.$$

Notice that because a and $a^\dagger$ don't commute, the order matters when expanding out the square, so we can't just combine the two middle terms. Now comes the clever part. Examining the first term, a^2 will act on the $|3\rangle$ on the right to give something proportional to $|1\rangle$; but $|3\rangle$ and $|1\rangle$ are orthogonal, so this term vanishes. Equally well, a^2 can act on the $\langle 3|$ on the left to give something proportional to $\langle 5|$ (remembering (5.9), a acts like its Hermitian conjugate $a^\dagger$ when acting on a bra). But $|5\rangle$ and $|3\rangle$ are also orthogonal, so this term still vanishes. An identical argument holds for the $(a^\dagger)^2$ term. Thus the only two terms that contribute are the two middle terms, which each raise once and lower once, bringing us back to state $|3\rangle$ which has nonzero inner product with itself. Using the normalization relations,

$$aa^\dagger|3\rangle = \sqrt{4}(a|4\rangle) = \sqrt{4}\sqrt{4}|3\rangle = 4|3\rangle,$$

and similarly,

$$a^\dagger a|3\rangle = \sqrt{3}(a^\dagger|2\rangle) = \sqrt{3}\sqrt{3}|3\rangle = 3|3\rangle.$$

Let's check that this makes sense using the commutation relation:

$$[a, a^\dagger] = 1 \implies aa^\dagger - a^\dagger a = 1,$$

and indeed, subtracting $a^\dagger a|3\rangle$ from $aa^\dagger|3\rangle$ gives just $|3\rangle$. So instead of calculating both terms separately, we could have combined them using the commutation relation – either method is fine. Going back to our original expectation value,

$$\langle 3|a^2 + aa^\dagger + a^\dagger a + (a^\dagger)^2|3\rangle = (4 + 3)\langle 3|3\rangle = 7,$$

since by assumption $|3\rangle$ is normalized. That's all there is to it.

- All energy eigenstates of the harmonic oscillator obey the virial theorem, which states for the harmonic oscillator

$$\langle T\rangle = \langle V\rangle = \frac{E_n}{2}. \quad (5.25)$$

In general, this theorem does *not* apply to superpositions of energy eigenstates, but in certain particular cases it does – see the problems for an example.

5.2.2 Three Dimensions

The generalization of the harmonic oscillator Hamiltonian to three dimensions is simple: it's just three identical copies of the one-dimensional version. The quadratic potential is the reason this works: $r^2 = x^2 + y^2 + z^2$, so a potential which is quadratic in r, the three-dimensional distance to the origin, is the sum of quadratic potentials in the three rectangular coordinates x, y, and z. This is *very particular* to the harmonic oscillator, but also very convenient. It means that the energy eigenfunctions are products of the energy eigenfunctions for the coordinates x, y, and z, and the energies are sums of the individual energies:

$$\psi_N(x, y, z) = \psi_{n_1}(x)\psi_{n_2}(y)\psi_{n_3}(z);$$

$$E_N = \left(N + \frac{3}{2}\right)\hbar\omega \text{ with } N = n_1 + n_2 + n_3. \quad (5.26)$$

In particular, this means that while the ground state is nondegenerate (all the n_i must be 0 for N to be 0), the first excited state is three-fold degenerate, because the three permutations

$$(n_1, n_2, n_3) = (1, 0, 0),\ (0, 1, 0),\ (0, 0, 1)$$

all give the same energy. (Of course, the same general arguments would hold if we had a system confined to two dimensions, though the details change – be careful!)

5.2.3 Problems: Harmonic Oscillator

1. A particle of mass m in a harmonic oscillator potential with angular frequency ω is in the state $\frac{1}{\sqrt{2}}(|1\rangle + |4\rangle)$. What is $\langle p^2 \rangle$ for this particle?
   (A) $3\hbar\omega/2$
   (B) $9\hbar\omega/2$
   (C) $6\sqrt{2}m\hbar\omega/2$
   (D) $3m\hbar\omega$
   (E) $6m\hbar\omega$

2. Which of the following is NOT true of the spectrum of the one-dimensional quantum harmonic oscillator?
   (A) The ground state energy is equal to the classical ground state energy.
   (B) There are an infinite number of bound states.
   (C) The energy levels are equally spaced relative to the ground state.
   (D) The ground state saturates the uncertainty principle bound.
   (E) The spectrum is nondegenerate.

3. A charged particle confined to two dimensions and subject to an external magnetic field can be modeled by a two-dimensional harmonic oscillator potential, $V(x, y) = \frac{1}{2}m\omega^2(x^2 + y^2)$. What is the degeneracy of the state with energy $3\hbar\omega$?
   (A) 1
   (B) 2
   (C) 3
   (D) 4
   (E) There is no state with this energy.

5.3 Other Standard Hamiltonians

There are four other classic one-dimensional quantum-mechanical Hamiltonians that it pays to be familiar with. Most of the technical information in this section (energies and eigenfunctions) will likely be given to you on the test, so you need not memorize it, but being intimately familiar with it means much less time spent puzzling over a complicated-looking formula. What *is* important to memorize is the methodology for approaching each particular Hamiltonian, as this can be very difficult to derive from scratch and very easily lead to lots of wasted time. These four Hamiltonians all admit *bound states*, which we'll study first; the last two also admit *scattering states*, whose analysis is a little different, so we treat it separately.

Keep in mind as we proceed that the essential difference between bound and scattering states is that bound states have discrete energy eigenvalues, whose values are determined by enforcing boundary conditions on the wavefunction. In the case where $V(x)$ goes to zero as x goes to $\pm\infty$, bound states are the ones with $E < 0$, and scattering states have $E > 0$. Along the same lines, if the potential goes to infinity at $x = \pm\infty$, as in the infinite square well or the harmonic oscillator, *every* state is bound.

As a guide to your studying, the following four Hamiltonians are listed in order of decreasing priority. Only the square well is listed explicitly on the official ETS list of topics, and the free particle is important in its own right, as a basis for many other solutions of the Schrödinger equation. However, you may not see the delta-function well or the finite square well on your exam, so don't work too hard on them. Scattering is a bit of a wild card: you will probably see *something* related to scattering, but it will likely be a conceptual rather than a computational question.

5.3.1 Infinite Square Well

This Hamiltonian is particularly simple:

$$H = -\frac{\hbar^2}{2m}\frac{d^2}{dx^2} + V(x), \quad V(x) = \begin{cases} 0, & 0 \le x \le a, \\ \infty, & \text{otherwise.} \end{cases}$$

The eigenfunctions are found by requiring the wavefunction to vanish at $x = 0$ and $x = a$, the endpoints of the well. An important subtlety arises here: usually we require the wavefunction *and* its derivatives to be continuous, but for the infinite square well, this is impossible. In general, when the potential is *infinite* at a boundary, the derivative of the wavefunction will *not* be continuous there – we'll see another example of this below, with the delta-function potential. The best we can do is to make the wavefunction continuous by vanishing at the endpoints, and let the derivative be what it is. Solving the differential equation $H\psi = E\psi$ gives the normalized wavefunctions and energy eigenvalues:

$$\psi_n = \sqrt{\frac{2}{a}}\sin\left(\frac{n\pi x}{a}\right), \quad E_n = \frac{n^2\pi^2\hbar^2}{2ma^2}.$$

Unlike the harmonic oscillator, we start counting from $n = 1$, since $n = 0$ would give a wavefunction that is identically zero, hence not normalizable. So once again, the ground state has nonzero energy $E_1 = \pi^2\hbar^2/2ma^2$. We can almost derive the formula for the energies just by pure dimensional analysis: the parameters of the Hamiltonian are $\hbar$, m, and a, and the only combination with the units of energy is $\hbar^2/ma^2$. We can't get the factors of π or 2 correct from this argument, but it does tell us that, if we double the mass, we halve the ground state energy, and if the well expands by a factor of 2, then each energy changes by a factor of 1/4. This kind of reasoning *is* important on the GRE, so you should get familiar with it.

By the way, the infinite square well in three dimensions, defined by

$$V(x,y,z) = \begin{cases} 0, & 0 \le x,y,z \le a, \\ \infty, & \text{otherwise,} \end{cases}$$

behaves the same way as the three-dimensional harmonic oscillator: the wavefunctions are just products of the one-dimensional versions, and the energies just add. Note that this is *not* the same as the infinite *spherical* well, which would have $V(x,y,z) = 0$ for $r < a$, where $r = \sqrt{x^2+y^2+z^2}$. This is an entirely different beast, which we will cover in Section 5.4 on three-dimensional quantum mechanics.

5.3.2 Free Particle

By definition, a free particle isn't acted upon by any forces, so there is no potential and the Hamiltonian is simply

$$H = -\frac{\hbar^2}{2m}\frac{d^2}{dx^2}.$$

Solving this system reduces to finding the eigenfunctions of d^2/dx^2, which are easily checked to be exponentials, $e^{\pm ikx}$ and $e^{\pm\kappa x}$. If we adopt the convention that k and κ must both be real, then only the oscillating exponentials $e^{\pm ikx}$ are eigenfunctions with positive energy:

$$\psi(x) = e^{\pm ikx}, \quad E = \frac{\hbar^2k^2}{2m}. \tag{5.27}$$

We gave this equation a number because it *is* worth memorizing. The energy shouldn't take too much work to memorize: remembering the de Broglie formula,

$$p = \hbar k, \tag{5.28}$$

(5.27) just says the energy is equal to $p^2/2m$, just as for a classical particle. But there are no boundary conditions anywhere to be found, so nothing restricts the value of k: the free particle can have *any momentum at all*. The form of the eigenfunctions is no more difficult: they represent *waves* (hence the name wavefunction) with constant modulus throughout all of space. This last statement implies that the energy eigenfunctions for the free particle are *not normalizable*.

From here, most books launch into a long story about Fourier transforms, wave packets, and such, but we just list a few salient points:

- We can construct a normalizable wavefunction by forming a continuous superposition of wavefunctions $\psi(x)$ with different values of k. The more values of k we throw in, the more possible values of momentum we could measure for the particle. It's not important for the GRE to know how to do this superposition, just to know it can be done in principle. The resulting wavefunction is called a *wave packet*. In fact, by a clever choice of coefficients, we can construct a *minimum-uncertainty wave packet*, which will of course be a Gaussian in x.
- The non-normalizability of the energy eigenstates (which are, incidentally, momentum eigenstates as well) just says there is no such thing as a particle with a perfectly defined value of momentum. This makes sense in the context of the uncertainty principle: the uncertainty in position would have to be infinite.
- The same story will hold for positive-energy solutions of an arbitrary Hamiltonian, wherever the potential is zero: these are called *scattering states*. The eigenfunctions will be oscillating exponentials, and while we could form wave packets to make the whole thing normalizable, this is rarely necessary in practice. Scattering problems require their own tricks of the trade and will be treated in Section 5.3.5 below.
- The formula for the energy can be read as a *dispersion relation* for a free quantum particle. Einstein's relation

$$E = \hbar\omega \tag{5.29}$$

implies that $\omega(k) = \hbar k^2/2m$, so $\omega(k)$ is quadratic in k. This should be contrasted with the case of a classical wave, which has $\omega = ck$, where c is the wave velocity.

5.3.3 Delta Function

Recall that a delta function $\delta(x)$ is zero everywhere except at $x = 0$, where the delta function is infinite. So if we let $V(x) = -A\delta(x)$ (a delta-function potential well), the Hamiltonian we end up with,

$$H = -\frac{\hbar^2}{2m}\frac{d^2}{dx^2} - A\delta(x),$$

is the same as the free-particle Hamiltonian, *except* at the single point $x = 0$. The fact that the potential is infinite there shouldn't scare you: we've already dealt with an infinite potential in the infinite square well above. It just means we have to be careful about the boundary conditions for $d\psi/dx$.

Exploiting the similarity with the free particle, the wavefunction will be an exponential to the left and to the right. Whether we get an oscillatory exponential $e^{\pm ikx}$ or a growing/decaying exponential $e^{\pm\kappa x}$ depends on whether we want to consider positive or negative energy solutions; in other words, scattering states or bound states. Here we consider bound states. The wavefunctions ψ_- on the left of the delta function and ψ_+ on the right must decay at $x = \pm\infty$, so we must have $\psi_- \propto e^{\kappa x}$ and $\psi_+ \propto e^{-\kappa x}$. (It's the same value of κ for both because both pieces must be energy eigenfunctions with the same eigenvalue.) Now the crucial part: *κ is determined by the boundary conditions enforced by the delta function.* To find these boundary conditions, we use the trick of integrating the Schrödinger equation $H\psi = E\psi$ on an infinitesimal interval $(-\epsilon, \epsilon)$ about $x = 0$:

$$\int_{-\epsilon}^{\epsilon} \left(-\frac{\hbar^2}{2m} \frac{d^2}{dx^2} \psi(x) \right) dx - A \int_{-\epsilon}^{\epsilon} \delta(x)\psi(x)\, dx = E \int_{-\epsilon}^{\epsilon} \psi(x)\, dx.$$

Now take $\epsilon \to 0$. The term on the right-hand side vanishes, because ψ is continuous and we're integrating it over an interval whose size goes to zero. The left-hand side is more interesting:

$$\int_{-\epsilon}^{\epsilon} \left(-\frac{\hbar^2}{2m} \frac{d^2}{dx^2} \psi(x) \right) dx - A \int_{-\epsilon}^{\epsilon} \delta(x)\psi(x)\, dx = -\frac{\hbar^2}{2m} \left(\frac{d\psi}{dx} \right) \Bigg|_{-\epsilon}^{\epsilon} - A\psi(0),$$

where we have used the fact that the delta function integrates to 1 over any interval containing zero. The first term measures the discontinuity in $d\psi/dx$ about $x = 0$, so we can solve for κ in terms of A and $\psi(0)$. Rather than do this, though, we will once again point out that κ is essentially determined by dimensional analysis. This time our dimensional parameters are $\hbar$ and m, as usual, and the constant A, which has *dimensions of energy × length*. This follows from the fact that $\delta(x)$ has dimensions of 1/length, since it integrates to a pure number. The only combination of these units with dimensions of 1/length is $mA/\hbar^2$, and indeed this is correct even up to numerical factors:

$$\psi(x) = \frac{\sqrt{mA}}{\hbar} e^{-mA|x|/\hbar^2}, \quad E = -\frac{mA^2}{2\hbar^2}.$$

We could also get E (up to the factor of 1/2) from an identical dimensional analysis argument. As usual, it's useless to memorize the wavefunction and the energy; what matters is the method. An upshot of this analysis is *the delta-function potential admits only one bound state*, with energy and wavefunction given above.

5.3.4 Finite Square Well

The finite square well is similar to the infinite square well, except the potential well has finite depth. For what follows

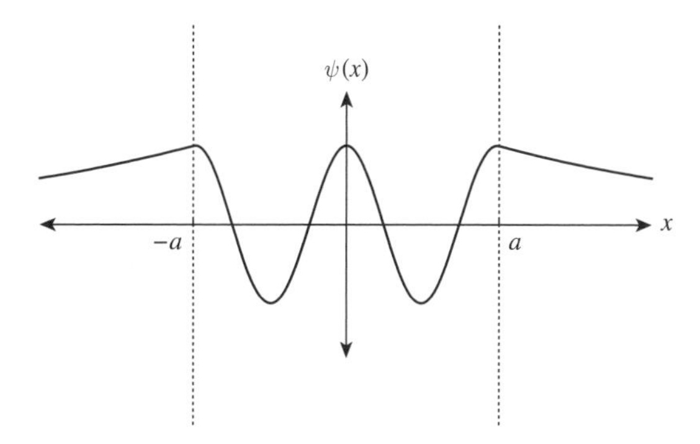

Figure 5.1 Sketch of the wavefunction for a bound state of the finite square well.

it will be convenient to center the well at $x = 0$, so the Hamiltonian is as follows:

$$H = -\frac{\hbar^2}{2m}\frac{d^2}{dx^2} + V(x), \quad V(x) = \begin{cases} -V_0, & -a \le x \le a, \\ 0, & \text{otherwise.} \end{cases}$$

We'll consider bound states, which have $E < 0$. We will exploit the fact that $V(x)$ is even, so ψ can be chosen to have definite parity; this means we only have to find the wavefunction for $x < 0$, and the rest of it will be determined by $\psi(x) = \psi(-x)$ or $\psi(x) = -\psi(-x)$. Outside the well, where $V = 0$, the solutions are as for the free particle: the normalizable one is $e^{\kappa x}$. Inside the well, we are solving the differential equation $-\frac{\hbar^2}{2m}\frac{d^2}{dx^2}\psi - V_0\psi = E\psi$, and by moving V_0 to the right-hand side we get the same equation as for the free particle. However, since the minimum of the potential is $-V_0$, we must have $E > -V_0$, and $E + V_0 > 0$. So instead of being decaying exponentials, the solutions should be oscillating sines and cosines. The wavefunction then looks as shown in Fig. 5.1.

This is essentially all you need to know about the finite square well. The constants k and κ, and from them the energies, are determined by solving a transcendental equation arising from the boundary conditions, and this is way beyond the level of stuff you're expected to do in two minutes on the GRE. One final piece of trivia: since the potential is even, the ground state is even. As the well gets shallower and shallower, the excited states disappear one by one, until all that is left is a single bound state, which is even.

5.3.5 Scattering States: Reflection and Transmission

The delta-function and the finite square well potentials share the feature that they are both *localized*: the potential is zero for $x \ll 0$, becomes nonzero in some small region, and is then zero again for $x \gg 0$. This is the setup for a scattering problem, where something (the *incident wave*) comes in from the left, interacts with the potential, and then separates into two parts: the *reflected wave*, which travels back to the left, and the *transmitted wave*, which goes through the potential and travels to the right. Following the convention that e^{ikx} with $k > 0$ represents a plane wave traveling to the right, the wavefunction for a particle incident from the left on a generic potential localized to $|x| < a$ can be written as:

$$\psi(x) = \begin{cases} Ae^{ikx} + Be^{-ikx}, & x \leq -a, \\ \text{something}, & -a \leq x \leq a, \\ Ce^{ikx}, & x \geq a. \end{cases}$$

Here, A is the amplitude of the incident wave, B is the amplitude of the reflected wave, and C is the amplitude of the transmitted wave. Determining the ratios of these amplitudes as a function of k is well beyond what you're expected to know for the GRE, but the following qualitative knowledge is important:

- The behavior of the wavefunction in the region $|x| < a$ depends on the energy and the height of the potential. For example, in a square *well*, the wavefunction is sinusoidal, but in a square *barrier* of height $+V_0$, if $0 < E < V_0$, the wavefunction will decay exponentially inside the barrier. This is the phenomenon of *tunneling*: a classical particle wouldn't have enough energy to get over the barrier, but a quantum particle can.
- The delta-function potential has the curious property that the reflection and transmission probabilities are the *same* for a delta-function well ($V(x) = -A\delta(x)$) and a delta-function barrier ($V(x) = A\delta(x)$).
- The probabilities of reflection and transmission go as the square of the wavefunction:

$$R = \frac{|B|^2}{|A|^2}, \quad T = \frac{|C|^2}{|A|^2}.$$

Conservation of probability requires $R + T = 1$.

One simple scattering problem whose exact solution *did* appear on a recent test is that of a step potential:

Rather than derive the solution here, we'll illustrate how this setup might appear on the test in Problem 3 below. Note that this potential is not localized, so there are subtle changes in calculating the transmission coefficient; see Griffiths for details.

5.3.6 Problems: Other Standard Hamiltonians

1. Which of the following is true of the wavefunction of a particle in an energy eigenstate of the infinite square well?

   I. It vanishes at the boundaries of the well.
   II. It is discontinuous at the boundaries of the well.
   III. Its derivative is discontinuous at the boundaries of the well.

   (A) I only
   (B) II only
   (C) I and II
   (D) I and III
   (E) II and III

2. A free particle has the wavefunction $\sin(kx)$. The particle has

   I. a definite value of position
   II. a definite value of momentum
   III. a definite value of energy

   (A) I only
   (B) II only
   (C) III only
   (D) I and II
   (E) II and III

3. A particle of mass m and energy E is incident from the left on a step potential,

$$V(x) = \begin{cases} 0, & x \leq 0, \\ V_0, & x \geq 0, \end{cases}$$

   where $E > V_0$. The wavefunction in the region $x \leq 0$ is $Ae^{ik_L x} + Be^{-ik_L x}$, and the wavefunction in the region $x > 0$ is $Ce^{ik_R x}$. Which of the following gives the transmission probability for the particle to be found in the region $x > 0$?

   (A) 0
   (B) 1
   (C) $\frac{k_L}{k_R}$
   (D) $\left(\frac{k_L - k_R}{k_L + k_R}\right)^2$
   (E) $\frac{4k_L k_R}{(k_L + k_R)^2}$

5.4 Quantum Mechanics in Three Dimensions

The generalization from one dimension to three dimensions is really quite simple. In three dimensions, the momentum operator is a vector: $\hat{\mathbf{p}} = -i\hbar\nabla$. In Cartesian components, $\hat{p}_x = -i\hbar\partial/\partial x$, $\hat{p}_y = -i\hbar\partial/\partial y$, $\hat{p}_z = -i\hbar\partial/\partial z$. The position operator generalizes similarly: $\hat{x} = x$, $\hat{y} = y$, and $\hat{z} = z$. So a general Hamiltonian in three dimensions is

$$H = -\frac{\hbar^2}{2m}\nabla^2 + V(\mathbf{r}) \qquad \text{(three dimensions)}, \tag{5.30}$$

where ∇^2 is the three-dimensional Laplacian, $\nabla^2 = \partial^2/\partial x^2 + \partial^2/\partial y^2 + \partial^2/\partial z^2$. The final ingredient is the extension of the canonical commutation relations, and all there is to remember is that *different coordinates commute with each other.* So we have

$$[\hat{x}, \hat{p}_x] = i\hbar, \quad [\hat{y}, \hat{p}_y] = i\hbar, \quad [\hat{z}, \hat{p}_z] = i\hbar, \tag{5.31}$$

but

$$[\hat{x}, \hat{y}] = 0, \quad [\hat{x}, \hat{p}_y] = 0, \quad [\hat{x}, \hat{p}_z] = 0, \ldots \tag{5.32}$$

So we can feel free to form observables such as $\hat{x}\hat{y}^2\hat{p}_z$ without worrying about ambiguities coming from nonzero commutators.

5.4.1 Radial Equation and Effective Potential

Suppose the potential depends only on the distance from the origin: $V = V(r)$. Then, just as in classical mechanics, we can separate out the angular and radial parts of the problem (Section 1.6). In classical mechanics, we used conservation of angular momentum to reduce the dynamics to a one-dimensional problem involving an effective potential. A very similar thing happens in quantum mechanics, but the technical details are different: we will use the technique of *separation of variables* to obtain a radial equation involving an effective potential, and an angular equation whose solutions will be treated in Section 5.4.2.

For a central potential $V = V(r)$, it is most convenient to work in spherical coordinates. From the general equation (5.30), this will involve the Laplacian in spherical coordinates. No doubt you've seen the exact form of this object before: it's quite complicated and totally unnecessary to memorize, but what's important is that it can be separated into a radial piece and an angular piece,

$$\nabla^2 = \nabla_r^2 + \frac{1}{r^2}\nabla_{\theta,\phi}^2.$$

If we then assume that the wavefunction is a product of a radial part and an angular part, $\psi(\mathbf{r}) = R(r)Y(\theta, \phi)$, we can use the usual technique of separation of variables to get separate equations for R and Y. For completeness, the radial equation is

$$-\frac{\hbar^2}{2m}\frac{d^2u}{dr^2} + \left[V + \frac{\hbar^2}{2m}\frac{l(l+1)}{r^2}\right]u = Eu,$$

where $u(r) = rR(r)$ and $l(l+1)$ is a separation constant related to the eigenvalue of Y under the operator $\nabla_{\theta,\phi}^2$. So just as in classical mechanics, we get an effective potential, equal to the original potential plus a piece that depends on angular momentum (as will be seen below). Note that because of this effective potential term, the infinite spherical well has rather different eigenfunctions for $l \neq 0$ than the infinite square well in three dimensions.

Because we've separated the wavefunction into a radial and an angular piece, it is conventional to normalize each piece separately:

$$\int_0^\infty |R(r)|^2 r^2\, dr = 1,$$
$$\int_0^{2\pi}\int_0^{\pi} |Y(\theta, \phi)|^2 \sin\theta\, d\theta\, d\phi = 1. \tag{5.33}$$

Note how the spherical volume element $dV = r^2 \sin\theta\, dr\, d\theta\, d\phi$ gets split up between the two normalization integrals.

5.4.2 Angular Momentum and Spherical Harmonics

The story of how angular momentum relates to symmetries in quantum mechanics is a subtle and beautiful one, which for better or for worse has no place on the GRE. This section will simply be a review of the relevant rules for dealing with systems involving angular momentum.

Just as angular momentum was defined in classical mechanics as $\mathbf{L} = \mathbf{r} \times \mathbf{p}$, the quantum operator for orbital angular momentum $\hat{\mathbf{L}}$ is defined similarly: $\hat{\mathbf{L}} = \hat{\mathbf{r}} \times \hat{\mathbf{p}}$. In components:

$$\hat{L}_x = \hat{y}\hat{p}_z - \hat{z}\hat{p}_y, \tag{5.34}$$
$$\hat{L}_y = \hat{z}\hat{p}_x - \hat{x}\hat{p}_z, \tag{5.35}$$
$$\hat{L}_z = \hat{x}\hat{p}_y - \hat{y}\hat{p}_x. \tag{5.36}$$

By using the commutation relations for $\hat{\mathbf{r}}$ and $\hat{\mathbf{p}}$, we get the following commutation relations for $\hat{\mathbf{L}}$:

$$[\hat{L}_x, \hat{L}_y] = i\hbar\hat{L}_z, \text{ and cyclic permutations of } x, y, z. \tag{5.37}$$

You *may* be given these relations on the GRE (one recent test gave them, another did not), so it's a good idea to be safe and memorize them anyway.

Another useful operator is the dot product of $\hat{\mathbf{L}}$ with itself, known as the *total (orbital) angular momentum* $\hat{L}^2$:

$$\hat{L}^2 := \hat{L}_x^2 + \hat{L}_y^2 + \hat{L}_z^2. \tag{5.38}$$

Most importantly, $\hat{L}^2$ *commutes with all the other* $\hat{L}_i$ (exercise: check this!). So we can find simultaneous eigenfunctions of $\hat{L}^2$ and *one* of the $\hat{L}_i$, conventionally chosen to be $\hat{L}_z$. In spherical coordinates, these operators take a familiar form:

$$\hat{L}^2 = -\hbar^2 \nabla^2_{\theta,\phi},$$
$$\hat{L}_z = -i\hbar \frac{\partial}{\partial \phi}.$$

where $\nabla^2_{\theta,\phi}$ is the angular part of the Laplacian introduced in Section 5.4.1.

The simultaneous eigenfunctions of these two operators are the *spherical harmonics*, $Y_l^m(\theta, \phi)$, where θ and ϕ are the usual angles in a spherical coordinate system. Despite the complicated notation, they are just certain functions of θ and ϕ, whose explicit forms you will probably never need for the GRE. Their eigenvalues, labeled by the orbital quantum number l and the azimuthal quantum number m, *are* important and are given as follows:

$$\hat{L}_z Y_l^m = m\hbar\, Y_l^m, \tag{5.39}$$
$$\hat{L}^2 Y_l^m = l(l+1)\hbar^2\, Y_l^m. \tag{5.40}$$

Here, m and l are integers, with $l \geq 0$. However, l and m are not totally independent: given a value of l, the corresponding allowed values of m are

$$m = l, l-1, l-2, \ldots, -l. \tag{5.41}$$

These restrictions should make intuitive sense: l is the quantum-mechanical analogue of the length of the angular momentum vector, which must be non-negative, and m is one component of that vector, which can't exceed the total length of the vector. Note that these intuitive reasons are *not* strictly true, since quantum operators behave nothing like classical vectors, but they are useful mnemonics for remembering the correct relations.

Some important facts about (normalized) spherical harmonics:

- They are orthonormal:

$$\int_0^{2\pi} \int_0^{\pi} (Y_l^m(\theta, \phi))^* Y_{l'}^{m'}(\theta, \phi) \sin\theta \, d\theta \, d\phi = \delta_{ll'}\delta_{mm'}, \tag{5.42}$$

  which follows from the fact that they are eigenfunctions for the Hermitian operators $\hat{L}_z$ and $\hat{L}^2$ with different eigenvalues.
- The ϕ dependence of Y_l^m is always of the form $e^{im\phi}$. The θ dependence is more complicated, being contained in the so-called Legendre polynomials, which will always be given to you on the test should you need them.
- Y_0^0 is simply a constant, with no θ or ϕ dependence – again there's no need to memorize its value (although it's easy to work out), since you will always be given the functional form on the test should you need it.
- If the angular part of a particle's wavefunction is exactly equal to one of the spherical harmonics Y_l^m, the particle has a *definite value* of total angular momentum l and z-component of angular momentum m. More generally, the spherical harmonics are a *complete* set of functions, which means that the angular part of any wavefunction on the sphere can be expressed as a linear combination of them. So to answer the question "what is the probability a system will be found with total angular momentum l and z-component of angular momentum m," we must decompose the wavefunction into spherical harmonics and compute the probability from the coefficients as discussed in Section 5.1. This just requires a lot of ugly integrals, so on the GRE you will most likely be given the decomposition into spherical harmonics, or will be asked a question about a system with definite values of l and m.

5.4.3 The Hydrogen Atom

The hydrogen atom receives so much attention in quantum mechanics texts because it is one of the very few exactly solvable problems that accurately represents a realistic system. Some of this attention is also for historical reasons: the birth of quantum mechanics began with the Bohr formula for the hydrogen energy levels, which, despite being derived using a lucky combination of classical mechanics and blind intuition, turned out to be correct. Since it is treated in such great detail, it's also a favorite on the GRE, and straddles the two categories of quantum mechanics and atomic physics.

The potential energy for the hydrogen atom is the Coulomb potential for an electron of charge $-e$ and a proton of charge $+e$, so the Hamiltonian is

$$H = -\frac{\hbar^2}{2\mu}\nabla^2 - \frac{e^2}{4\pi\epsilon_0}\frac{1}{r} \quad \text{(hydrogen atom).} \tag{5.43}$$

Note the μ in the denominator of the kinetic term, rather than m_e: this is because, just as in classical mechanics, we must use the *reduced mass* $\mu = m_1 m_2/(m_1 + m_2)$ to reduce the two-body problem of electron and proton to an equivalent one-body problem. In the case of hydrogen, μ is very close to m_e because the proton is so much heavier than the electron. Nevertheless, the reduced mass is what shows up in all the formulas for the energies and the wavefunctions, and by writing μ we can use (5.43) as a model for all hydrogen-*like* systems (for example, positronium, with a positron replacing the proton), except with different values of the reduced mass.[4] This applies to all the formulas in this section.

From here, it is simply several pages of algebra to derive the energies and the wavefunctions from the radial equation. This is done in all quantum mechanics texts, so instead we will get the answer by dimensional analysis. The dimensionful parameters in this problem are μ, $\hbar$, and the combination $e^2/4\pi\epsilon_0$, which has units of energy $\times$ length. As usual, there is only one way we can combine these to get a length, which is so special that it gets its own name, the *Bohr radius*:

$$a = \frac{4\pi\epsilon_0\hbar^2}{\mu e^2}. \qquad (5.44)$$

(Sometimes this is denoted a_0 rather than a; we use the two interchangeably in this book.) Because of the factor of μ, there is a different "Bohr radius" for each hydrogen-like atom. A favorite GRE question asks how the Bohr radius of positronium differs from that of hydrogen: because $\mu = m_e/2$ for positronium and $\mu \approx m_e$ for hydrogen, the Bohr radius of positronium is double that of hydrogen. You could play the same game with muonium, where a muon (essentially a heavy electron) replaces the electron, or with any number of other fundamental particles.

Because of the electrostatic attraction between the electron and the proton, intuition suggests that the wavefunction of the electron should be largest near the origin. Since the only length in the problem is a, a reasonable guess for the (un-normalized) ground state wavefunction is

$$\psi_1(r) \propto e^{-r/a}, \qquad (5.45)$$

and indeed, this is correct. As for the ground state energy, it has the same form as for the infinite square well of width a (the other standard Hamiltonian with a given length scale), but the numerical factors are different:

$$-E_1 = \frac{\hbar^2}{2\mu a^2} = \frac{\mu e^4}{2(4\pi\epsilon_0)^2\hbar^2} = 13.6 \text{ eV for hydrogen.} \qquad (5.46)$$

(The minus sign is there because all energies are measured relative to $V(\infty) = 0$.)

The importance of memorizing this single number, 13.6 eV, cannot be overemphasized. Not only does it tell you that the energy scale of atomic processes is of the order of a few eV, it can be used to derive approximate ionization energies for hydrogenic atoms, and its numerical value is often a GRE question in and of itself. In fact, both expressions for E_1 are worth memorizing; luckily, this is probably the most complicated formula you'll have to memorize for the test. Again, the numerical factors are not important but the dependence on μ is: the second formula, containing all explicit factors of μ, shows that the ground state energy is proportional to the reduced mass. So, in positronium, halving the reduced mass compared to hydrogen gives a binding energy that is half that of hydrogen.

Another trap to beware of is the e^4 in the second formula for E_1: this does *not* mean that a helium ion (an electron orbiting a nucleus of charge $2e$) has ground state energy $2^4 \times E_1$. Rather, $e^4 = (e^2)^2$ is the square of the e^2 appearing in the Coulomb potential, which is the product of the electron charge and the nuclear charge. For a nucleus of charge $2e$, the Coulomb potential is stronger by a factor of 2, so E_1 is greater by a factor of only $2^2 = 4$.

The last thing to remember is that the energies decrease as $1/n^2$:

$$-E_n = \frac{\hbar^2}{2\mu a^2}\frac{1}{n^2}, \quad n = 1, 2, 3, \ldots, \qquad (5.47)$$

where n is the *principal quantum number*, which gives the famous Rydberg formula for the frequency of light emitted in hydrogen transitions:

$$f \propto \frac{1}{n_f^2} - \frac{1}{n_i^2}.$$

You might have noticed that the formula for E_n doesn't depend on the angular quantum numbers l and m: this is a special property of the Coulomb potential, and this degeneracy will go away when we include various small effects that we have neglected. As this is more properly an atomic physics topic, we will treat it in Section 5.7 at the end of this chapter.

We finish with some important trivia about the hydrogen atom:

- The energies don't depend on l or m, but of course the wavefunctions do, with the angular dependence carried in the spherical harmonics. The radial part of the wavefunction is determined by so-called *Laguerre polynomials*, which depend on l but not m.
- The wavefunction is *zero* at the origin for states with $l \neq 0$, so the states with zero angular momentum have higher probabilities of being found near the origin. In particular, the state $n = 1$ has the highest probability of all the states of being found near $r = 0$.
- If we include the speed of light c as a dimensionful parameter, we can form a *dimensionless* constant called the *fine-structure constant*:

[4] You may also see the term *hydrogenic*, which usually refers to a bound state of a negatively charged particle orbiting a positively charged particle or nucleus.

$$\alpha = \frac{e^2}{4\pi\epsilon_0\hbar c} \approx 1/137. \tag{5.48}$$

Note that this does not involve the reduced mass μ: it just characterizes the general quantum-mechanical strength of electromagnetic interactions. Perturbations to the hydrogen atom are usually expressed in terms of powers of α, and the reason they can be treated as perturbations is the fortunate situation $\alpha \ll 1$.

5.4.4 Problems: Quantum Mechanics in Three Dimensions

1. A spin-zero particle has angular wavefunction

$$\frac{1}{\sqrt{2}}(Y_3^2(\theta,\phi) + Y_2^1(\theta,\phi)),$$

where $Y_l^m(\theta,\phi)$ are the normalized spherical harmonics. What is the expectation value of the total spin L^2?

(A) $2\hbar^2$
(B) $5\hbar^2/2$
(C) $3\hbar^2$
(D) $9\hbar^2$
(E) $18\hbar^2$

2. The radiation emitted by the first excited state of hydrogen as it drops to the ground state is called the Lyman alpha line, and has wavelength 122 nm. An excited hydrogen atom is observed radiating light at wavelength 488 nm. This most likely results from the transition

(A) $n = 2$ to $n = 1$
(B) $n = 3$ to $n = 1$
(C) $n = 3$ to $n = 2$
(D) $n = 4$ to $n = 1$
(E) $n = 4$ to $n = 2$

3. It is theoretically possible for an electron and a neutron to form a bound state where the binding force is due to *gravity*, rather than electrostatic attraction. Let G_N denote Newton's constant and m_e and m_n denote the mass of the electron and neutron, respectively. Under the approximation $m_e \ll m_n$, which of the following gives the "Bohr radius" of the gravitationally bound electron?

(A) $\dfrac{\hbar^2}{G_N m_e^2 m_n}$
(B) $\dfrac{\hbar^2}{G_N m_e m_n}$
(C) $\dfrac{\hbar^2}{G_N m_e m_n^2}$
(D) $\dfrac{\hbar^2 G_N}{m_e m_n}$
(E) $\dfrac{\hbar^2 G_N m_e}{m_n^2}$

4. A new hydrogen-like atom is discovered where the particle orbiting the proton has mass $2m_e$ and charge $2e$, where m_e and e are the charge and mass of the electron, respectively. What is the binding energy of this atom, in terms of the binding energy E of ordinary hydrogen?

(A) $E/2$
(B) E
(C) $2E$
(D) $4E$
(E) $8E$

5.5 Spin

It is an observational fact that all quantum particles carry an "intrinsic" quantum number known as *spin*, whose operators $\hat{S}_x$, $\hat{S}_y$, and $\hat{S}_z$ obey the same commutation relations as the angular momentum operators. In contrast to total orbital angular momentum, whose value depends on what state the particle is in, a particle's total spin is a characteristic of the particle itself and never changes. Spin is usually denoted s, where s plays the same role as l in the nomenclature of angular momentum: for example, a particle with $s = 1/2$ has $\hat{S}^2$ (the dot product of $\hat{\mathbf{S}}$ with itself) eigenvalue of $s(s+1)\hbar^2 = \frac{3}{4}\hbar^2$ always and forever, and we say the particle has *spin-1/2*.

In contrast to the orbital angular momentum l, which must be an integer, s can be *either an integer or a half-integer*. Other than that, all the same rules still apply: the z-component of spin, denoted m_s, can still range from $-s$ to s in integer steps, and can change depending on what state the particle is in. But since s is always fixed, there are always the same number of possible m_s for a given particle, and the Hilbert space has dimension $2s + 1$. For spin-1/2, we have only two states: spin-up, with $m_s = +1/2$, denoted by $|\uparrow\rangle$, and spin-down, with $m_s = -1/2$, denoted by $|\downarrow\rangle$. There is no such convenient notation for spin $s = 1$ and higher, but for the purposes of the GRE, pretty much all you have to know is spin-1/2.

5.5.1 Spin-1/2

Because the Hilbert space for spin is finite-dimensional, the spin operators, instead of being differential operators, are just matrices. The only ones you have to be familiar with are those for spin-1/2, known as the Pauli matrices: $\hat{S}_x = \frac{\hbar}{2}\sigma_x$, $\hat{S}_y = \frac{\hbar}{2}\sigma_y$, and $\hat{S}_z = \frac{\hbar}{2}\sigma_z$, where

$$\sigma_x = \begin{pmatrix} 0 & 1 \\ 1 & 0 \end{pmatrix}, \quad \sigma_y = \begin{pmatrix} 0 & -i \\ i & 0 \end{pmatrix}, \quad \sigma_z = \begin{pmatrix} 1 & 0 \\ 0 & -1 \end{pmatrix}.$$

Don't bother memorizing these because they will be given to you on the GRE, but do become intimately familiar with their properties so you're not forced to rederive anything during the exam. Notice that the σ's are Hermitian matrices, so the corresponding operators are Hermitian, as observables should be. Since σ_z is diagonal, the vectors corresponding to $|\uparrow\rangle$ and $|\downarrow\rangle$ are particularly simple: $|\uparrow\rangle = \binom{1}{0}$ and $|\downarrow\rangle = \binom{0}{1}$. You can check that these have the correct eigenvalues under $\hat{S}_z$, namely $+\hbar/2$ and $-\hbar/2$, respectively. In the case of spin-1/2, it's particularly simple to verify that $\hat{S}^2$ commutes with all the other spin operators, since $\hat{S}^2$ turns out to be proportional to the identity matrix.

With a finite-dimensional Hilbert space, we have to slightly modify our rules for computing things such as expectation values. Luckily, we have already taken care of most of this in Section 5.1.2. Instead of integrating to compute an inner product, we just take a dot product, and the action of Hermitian observables is by matrix multiplication. For example, the expectation value of $\hat{S}_x$ in the state $|\uparrow\rangle$ is computed as follows:

$$\langle\uparrow|\, \hat{S}_x \,|\uparrow\rangle = \frac{\hbar}{2}\left[(1\ 0)\begin{pmatrix} 0 & 1 \\ 1 & 0 \end{pmatrix}\begin{pmatrix} 1 \\ 0 \end{pmatrix}\right] = \frac{\hbar}{2}\left[(1\ 0)\begin{pmatrix} 0 \\ 1 \end{pmatrix}\right] = 0.$$

Another common question asks you the probability of measuring a particular value of $\hat{S}_z$ for a given state. If the normalized wavefunction is $\Psi = \begin{pmatrix} \alpha \\ \beta \end{pmatrix}$ (in this context, normalized means $|\alpha|^2 + |\beta|^2 = 1$), then the probability of measuring $\hat{S}_z = +\hbar/2$ is $|\alpha|^2$, and the probability of measuring $-\hbar/2$ is $|\beta|^2$. To see why this is true, write the wavefunction in Dirac notation:

$$\Psi = \alpha\begin{pmatrix} 1 \\ 0 \end{pmatrix} + \beta\begin{pmatrix} 0 \\ 1 \end{pmatrix} = \alpha\,|\uparrow\rangle + \beta\,|\downarrow\rangle,$$

which is just the expansion of a vector in its basis components. Now we can just apply the usual rules of quantum mechanics.

To avoid doing repeated calculations, it's extremely useful to remember the eigenvectors of $\hat{S}_x$ and $\hat{S}_y$. Instead of doing the usual steps of finding the characteristic equation of the matrix, finding the eigenvalues, and then solving for the eigenvectors, we can take a convenient shortcut. The possible values of m_s are $\pm\hbar/2$ no matter what basis we happen to be working in, since that's what it means for a particle to have spin-1/2. We chose the $\hat{S}_z$ basis because it was conventional, but it had no physical meaning. Thus, the eigenvalues of $\hat{S}_x$ and $\hat{S}_y$ must *also* be $\pm\hbar/2$, and thus the eigenvalues of the Pauli matrices σ_i are all ± 1. Now, we can figure out the eigenvectors almost by just staring at the matrices:

$$|\uparrow\rangle_x = \frac{1}{\sqrt{2}}\begin{pmatrix} 1 \\ 1 \end{pmatrix}, \qquad |\downarrow\rangle_x = \frac{1}{\sqrt{2}}\begin{pmatrix} 1 \\ -1 \end{pmatrix}, \tag{5.49}$$

$$|\uparrow\rangle_y = \frac{1}{\sqrt{2}}\begin{pmatrix} 1 \\ i \end{pmatrix}, \qquad |\downarrow\rangle_y = \frac{1}{\sqrt{2}}\begin{pmatrix} 1 \\ -i \end{pmatrix}. \tag{5.50}$$

In the ket notation, we would say $|\uparrow\rangle_y = \frac{1}{\sqrt{2}}\,(|\uparrow\rangle + i\,|\downarrow\rangle)$, and so on. These are simple enough to memorize (just remember an extra factor of i for the y vectors), but by remembering what the eigenvalues are, it's at most a one-minute derivation if you forget. In any problem where you'll need this information, finding the eigenvectors is 90% of the work, so this would be a minute well spent.

One final detail, which you will likely not have to know for the GRE but will be important for what follows here: there is a spin analogue to the raising and lowering operators of the harmonic oscillator. The linear combinations

$$\hat{S}_+ := \hat{S}_x + i\hat{S}_y, \quad \hat{S}_- := \hat{S}_x - i\hat{S}_y \tag{5.51}$$

are called the spin raising and lowering operators. You can show they deserve their names by acting on $|\uparrow\rangle$ and $|\downarrow\rangle$ with them. You should find

$$\hat{S}_+\,|\uparrow\rangle = 0, \quad \hat{S}_-\,|\uparrow\rangle = \hbar\,|\downarrow\rangle, \tag{5.52}$$

and

$$\hat{S}_+\,|\downarrow\rangle = \hbar\,|\uparrow\rangle, \quad \hat{S}_-\,|\downarrow\rangle = 0. \tag{5.53}$$

The spin raising operator $\hat{S}_+$ turns a down spin into an up spin, but kills the up spin vector. Likewise, the lowering operator $\hat{S}_-$ turns an up spin into a down spin, and kills the down spin.

These relations generalize to higher spin: what the lowering operator does, in general, is preserve the total spin s but reduce the value of m_s by one unit of $\hbar$. When acting on the lowest possible m_s, it annihilates that state. The raising operator does just the opposite. Beware, though: just like their harmonic oscillator counterparts, the states created by raising and lowering operators are not automatically normalized!

5.5.2 Spin and the Wavefunction

The next step is to detail how spin fits into the more general picture of a particle's wavefunction. The simplest way to think about it is that the total wavefunction is always (a linear combination of) a *product* of a spatial wavefunction and a spin wavefunction. See Example 5.3 for an illustration. For systems of identical particles, the symmetry properties of both pieces of the wavefunction can have important effects, as we will see in much more detail in Section 5.5.4 below, so keeping track of both pieces is crucial.

EXAMPLE 5.3

A spin-1/2 particle in the ground state of the infinite square well and $m_s = +1/2$ would have the total wavefunction

$$\Psi(x,t) = \left[\sqrt{\frac{2}{a}}\sin\left(\frac{\pi x}{a}\right) \otimes \begin{pmatrix}1\\0\end{pmatrix}\right] e^{-iE_1 t/\hbar}.$$

The funny symbol $\otimes$ stands for a tensor product, and is there to remind you that the spin part and spatial part are separate entities which never get mixed. The usual position and momentum operators act on the spatial part, but don't touch the spin part; likewise, the spin operators act on the spin part but leave the spatial part alone. In practice, this means that *spin operators always commute with spatial operators*: in particular, $[\hat{\mathbf{L}}, \hat{\mathbf{S}}] = 0$. Acting with some operators for practice:

$$\hat{p}\,\Psi(x,t) = \left[-i\hbar\frac{\pi}{a}\sqrt{\frac{2}{a}}\cos\left(\frac{\pi x}{a}\right) \otimes \begin{pmatrix}1\\0\end{pmatrix}\right] e^{-iE_1 t/\hbar},$$

$$\hat{S}_y\Psi(x,t) = \left[\frac{\hbar}{2}\sqrt{\frac{2}{a}}\sin\left(\frac{\pi x}{a}\right) \otimes \begin{pmatrix}0\\i\end{pmatrix}\right] e^{-iE_1 t/\hbar}.$$

Notice that we can pull the constants that multiply either piece out to the front of the whole wavefunction: that's all the tensor product means. It's just a way of writing a product of two completely independent pieces where constants can be absorbed into either one. This means that we are free to normalize each piece separately, which is always done in practice. Finally, note that if a particle has spin-0, the spin part of its wavefunction is just 1: the system is completely described by just a spatial wavefunction.

It's important to remember that, in general, the Hamiltonian *can* contain both spatial operators and spin operators. Indeed, this is exactly what happens when a quantum system is exposed to a magnetic field: a particle's spin makes it act like a magnetic dipole, and there is a corresponding term in the Hamiltonian representing the energy of a magnetic dipole in an external field. While it's true that the spatial and spin parts never get mixed up with each other, keep in mind that the Hamiltonian can act on both pieces, and thus both pieces can contribute to the total energy of the system.

5.5.3 Adding Spins

Suppose we have a particle of spin s and a particle of spin s'. The system of the two particles taken together also has a total spin, but it's *not* simply $s + s'$. In fact, that's just one of many possibilities. The possible spin states of the system are

$$\text{Spin } s \text{ and spin } s': s_{\text{tot}} = s+s',\ s+s'-1,\ s+s'-2, \ldots, |s-s'|. \tag{5.54}$$

In other words, we can get any spin between $|s - s'|$ and $s + s'$ in integer steps. Thankfully, m_s values *do* add,

$$m_{\text{tot}} = m_s + m'_{s'}, \tag{5.55}$$

which restricts the possibilities for s_{tot}. For instance, if you have a spin-1 particle with $m_s = +1$ and a spin-1/2 particle with $m_s = +1/2$, then $m_{\text{tot}} = +3/2$, and the system must be in the state with $s_{\text{tot}} = 3/2$. On the other hand, if the spin-1/2 particle had $m_s = -1/2$, then $m_{\text{tot}} = +1/2$ and *either* $s_{\text{tot}} = 3/2$ or $s_{\text{tot}} = 1/2$ are allowed. In fact, the system can be described as a linear combination of states with different values of s_{tot}.

The more common situation, though, is when we know the total spin and m_s value of the system, but *not* of the individual particles. The all-important application is to two spin-1/2 particles, so we'll walk through that case in detail in Example 5.4.

Note that the way to add more than two spins is to add them in groups of two. For example, consider a system of three spin-1/2 particles. Adding the first two spins gives possible values of $s = 1$ and $s = 0$, and now adding the third spin gives possible values $s = 3/2, 1/2, 1/2$. Note the duplication of $s = 1/2$: one copy came from adding spin-1/2 to spin-1, and the second copy came from adding spin-1/2 to spin-0. Finally, this formalism applies not only to spins, but to any operators obeying the angular momentum commutation relations, most notably orbital angular momentum, as discussed in Section 5.4.2. Perhaps the most important example of addition of

EXAMPLE 5.4

Our goal is to classify the states of two spin-1/2 particles in terms of total spin and m_s. The possible total spins are $1/2 + 1/2 = 1$, or $1/2 - 1/2 = 0$. As described above, if both of the particles have their maximum value of m_s, then our job is easy: the system is in the state $s = 1$, $m_s = 1$. We denote this as follows:

$$s = 1, m_s = 1 : |\uparrow\rangle\, |\uparrow\rangle .$$

Now, the *total* spin operator is $\hat{\mathbf{S}} = \hat{\mathbf{S}}^{(1)} + \hat{\mathbf{S}}^{(2)}$, where the superscripts 1 and 2 refer to particles 1 and 2. As with spin-1/2, we can form the total raising and lowering operators by just adding together the contributions for particles 1 and 2, with the understanding that the 1 operators act only on the first spin, and the 2 operators act only on the second. Using the relations (5.53), we can act on the state $|\uparrow\rangle\,|\uparrow\rangle$ with the lowering operator:

$$\hat{S}_- = (\hat{S}_-^{(1)} + \hat{S}_-^{(2)})\, |\uparrow\rangle\, |\uparrow\rangle = \hbar(|\downarrow\rangle\, |\uparrow\rangle + |\uparrow\rangle\, |\downarrow\rangle).$$

By the rules for the action of the lowering operator, this must (after normalization) represent a state with $s = 1$ but one less unit of m_s:

$$s = 1, m_s = 0 : \frac{1}{\sqrt{2}} (|\downarrow\rangle\, |\uparrow\rangle + |\uparrow\rangle\, |\downarrow\rangle) .$$

Finally, acting once more with the lowering operator, we obtain the state with $m = -1$:

$$s = 1, m_s = -1 : |\downarrow\rangle\, |\downarrow\rangle .$$

This is precisely as expected, since we could have written down this state from the beginning: it has the minimum possible values of m_s for both particles, so it must have the largest of the available s values.

There is one state remaining, since we started with a Hilbert space of dimension 4 (2 states for particle 1 times 2 states for particle 2) and we only have three states so far. We know this state must have $s = 0$, and hence $m = 0$. Since m values add, it must be a linear combination of $|\uparrow\rangle\,|\downarrow\rangle$ and $|\downarrow\rangle\,|\uparrow\rangle$, and since it has a different eigenvalue for $\hat{S}^2$ from the $s = 1, m = 0$ state, it must be orthogonal to that state. This uniquely fixes the remaining state to be

$$s = 0, m_s = 0 : \frac{1}{\sqrt{2}} (|\uparrow\rangle\, |\downarrow\rangle - |\downarrow\rangle\, |\uparrow\rangle) . \tag{5.56}$$

This state is given its own name, the *singlet*, and it shows up everywhere. It's the unique combination of two spin-1/2 states with total spin zero, and it's also the only combination antisymmetric under the interchange of particles 1 and 2. The other three states, with $s = 1$, are called the *triplet* states, and they are symmetric under the interchange of 1 and 2.

angular momenta is in forming the total angular momentum of a system, $\hat{\mathbf{J}} = \hat{\mathbf{L}} + \hat{\mathbf{S}}$, where as usual $\hat{\mathbf{L}}$ is the orbital angular momentum operator and $\hat{\mathbf{S}}$ is the spin operator. For a given spin s and orbital angular momentum l, the possible eigenvalues for $\hat{J}^2$ and $\hat{J}_z$ are then given by (5.54) and (5.55) with $s' = l$. You'll see several examples of this in the problems, both in this section and in Section 5.7.

5.5.4 Bosons and Fermions

Particles with integer versus half-integer spin behave so radically differently that they are given their own names: integer spin particles are *bosons*, and half-integer spin particles are *fermions*. The reason for their difference in behavior only arises when we consider collections of these particles:

> *Under the interchange of two identical particles, boson wavefunctions are symmetric, and fermion wavefunctions are antisymmetric.*

Note the caveat "identical!" This only applies to systems of three electrons, or five photons, not an electron and a proton: while the electron and proton are indeed both fermions, they are not identical, so exchanging these particles need not transform the wavefunction in this manner. This means, for example, that if we have a collection of three identical bosons with coordinates x_1, x_2, and x_3, the wavefunction of the system $\Psi(x_1, x_2, x_3)$ will obey

$$\Psi(x_1, x_2, x_3) = \Psi(x_1, x_3, x_2) = \Psi(x_2, x_1, x_3) = \cdots$$

In other words, the wavefunction remains the same under any permutation of particles 1, 2, and 3: it is *totally symmetric.*

Fermions, on the other hand, pick up a minus sign under the interchange of particles: the wavefunction is *totally antisymmetric.* For a system of two identical fermions at x_1 and x_2, the wavefunction $\Phi(x_1, x_2)$ satisfies

$$\Phi(x_1, x_2) = -\Phi(x_2, x_1).$$

For more than two particles, two consecutive swaps result in an overall plus sign, so one has to keep track of the so-called sign of the permutation, but this is more than you'll need for the exam. Remember that overall phase factors don't change the quantum state, so the interchange of fermions doesn't change the state of the system: this is a reflection of the fact that identical quantum particles are indistinguishable even in principle. However, the minus sign can result in interesting interference effects. More importantly, it restricts the possible states of multi-fermion systems, which is encapsulated in the Pauli exclusion principle:

> *No two identical fermions can occupy the same quantum state.*

This follows from the antisymmetry of the wavefunction: if two fermions were in the same state $\psi(x)$, then the only way to make an antisymmetric combination out of $\psi(x_1)\psi(x_2)$ would be $\psi(x_1)\psi(x_2) - \psi(x_2)\psi(x_1) = 0$, and the whole wavefunction just vanishes.

The whole story becomes a little trickier when we include spin. Remember, the spin and spatial parts of the wavefunctions are independent, so for the *total* wavefunction to have a certain symmetry property, the symmetries of the spin part and the spatial part must combine in the correct way. Usually, this means both the spin and spatial parts have definite symmetry or antisymmetry.[5] For a concrete example, consider the electrons in a helium atom. The two-electron wavefunction $\Psi(\mathbf{r}_1, \mathbf{r}_2) = \psi(\mathbf{r}_1, \mathbf{r}_2)\chi(\mathbf{s}_1, \mathbf{s}_2)$ must be antisymmetric overall, so we can have either a symmetric ψ and an antisymmetric χ, or an antisymmetric ψ and a symmetric χ. You'll finish this analysis in the problems below.

[5] These are the only possibilities for a two-particle state. For three or more particles, there can be states of mixed symmetry, which can combine to give an overall symmetric or antisymmetric total wavefunction, but you won't have to worry about these on the GRE.

To conclude, we list two important facts relating spin and symmetry.

- When adding n identical spin-1/2 particles, the states with the highest value of s (namely $s = n/2$) are *always* symmetric. We saw this above for the case of two spin-1/2 particles: the triplet states were all symmetric. Continuing on, adding three spin-1/2 particles gives possible values $s = 3/2$ and $s = 1/2$, and all four states with $s = 3/2$ are symmetric under interchange of spins. On the other hand, the $s = 1/2$ states have mixed symmetry, neither symmetric nor antisymmetric.
- All of the symmetry arguments apply to *subsystems* of identical particles. For example, there are six fermions in the ^{4}He atom: two protons, two neutrons, and two electrons. We are free to consider the symmetry or antisymmetry of just the identical electrons, as we did above, while ignoring the fermions in the nucleus. However, the total spin of the system will of course have to include contributions from all the fermions.

5.5.5 Problems: Spin

1. A spin-1/2 particle is initially measured to have $S_z = \hbar/2$. A subsequent measurement of S_x returns $-\hbar/2$. If a third measurement is made, this time of S_z again, what is the probability the measurement returns $\hbar/2$?

   (A) 0
   (B) 1/4
   (C) 1/2
   (D) 3/4
   (E) 1

2. A deuterium atom, consisting of a proton and neutron in the nucleus with a single orbital electron, is measured to have total angular momentum $j = 3/2$ and $m_j = 1/2$ in the ground state. Let $|\uparrow\rangle_p$ and $|\uparrow\rangle_n$ denote spin-up protons and neutrons, respectively, and $|\downarrow\rangle_p$ and $|\downarrow\rangle_n$ denote spin-down protons and neutrons. Assuming the nucleus has no orbital angular momentum, the spin state of the nucleus could be

   I. $\frac{1}{\sqrt{2}}(|\uparrow\rangle_p|\downarrow\rangle_n - |\downarrow\rangle_p|\uparrow\rangle_n)$
   II. $|\uparrow\rangle_p|\uparrow\rangle_n$
   III. $\frac{1}{\sqrt{2}}(|\uparrow\rangle_p|\downarrow\rangle_n + |\downarrow\rangle_p|\uparrow\rangle_n)$
   IV. $|\downarrow\rangle_p|\downarrow\rangle_n$

   (A) I only
   (B) I and III

(C) II and IV
(D) II and III
(E) III and IV

3. In the ground state of helium, which of the following gives the total spin quantum numbers of the two electrons?

   (A) $s = 0, m_s = 0$
   (B) $s = 1, m_s = 1$
   (C) $s = 1, m_s = 0$
   (D) $s = 1, m_s = -1$
   (E) $s = 1/2, m_s = 1/2$

4. Two spin-1/2 electrons are placed in a one-dimensional harmonic oscillator potential of angular frequency ω. If a measurement of S_z of the system returns $\hbar$, which of the following is the smallest possible energy of the system?

   (A) $\hbar\omega/2$
   (B) $\hbar\omega$
   (C) $3\hbar\omega/2$
   (D) $2\hbar\omega$
   (E) $5\hbar\omega/2$

5.6 Approximation Methods

Exactly solvable quantum mechanics problems are few and far between. The hydrogen atom is pretty much the only example of a realistic system with a closed-form solution, and as soon as we include relativistic and spin-related corrections, we lose the exact solution. To this end, numerous approximation schemes have been developed. By *far* the most important, both on the GRE and in the everyday life of a physicist, is perturbation theory, discussed below. The variational principle and the adiabatic theorem are both important, but the variational principle is too involved to actually apply on a GRE-type question, and the adiabatic theorem is more of a heuristic that applies to only certain types of problems. So we'll discuss perturbation theory in great detail, but spend only a short paragraph on the other two methods.

5.6.1 Time-Independent Perturbation Theory: First and Second Order

Suppose the Hamiltonian H of a quantum system can be written as

$$H = H_0 + \lambda H',$$

where λ is a dimensionless number that is small compared to 1. Suppose further that we *know* the exact energies E_n^0 and corresponding eigenfunctions ψ_n^0 of H_0, and that H' is independent of time. Then perturbation theory gives us a recipe for computing corrections to these energies and eigenfunctions, as a power series in the strength of the perturbation λ. To start, the first-order energy shift to the nth level is

$$E_n = E_n^0 + \lambda \left\langle \psi_n^0 | H' | \psi_n^0 \right\rangle. \tag{5.57}$$

In other words, the first-order correction to the nth energy E_n^0 is proportional to λ and the *expectation value of H' in the unperturbed state.* So we don't need to solve the Schrödinger equation anew in order to find these first-order shifts: all we have to do is compute an expectation value for some given observable H', which is usually pretty easy.

If the first-order shift happens to vanish, then we must go to second order in λ to find a nonzero correction. The formula is considerably more complicated, but we present it here for completeness:

$$E_n = E_n^0 + \lambda^2 \sum_{m \neq n} \frac{|\langle \psi_m^0 | H' | \psi_n^0 \rangle|^2}{E_n^0 - E_m^0}. \tag{5.58}$$

Rather than memorize this formula, just remember its important features:

- It still only involves brackets of H' with the unperturbed states, except instead of being expectation values, they are *off-diagonal matrix elements* $H'_{mn} = \langle \psi_m^0 | H' | \psi_n^0 \rangle$.
- The numerator is always non-negative, so the sign of the denominator is important: it tends to push neighboring energies away from each other. In other words, if E_m^0 is only a little smaller than E_n^0, the denominator will be small and positive, so E_n will be pushed even further away from E_m^0. This is known as the *no level-crossing* phenomenon. In fact, remembering that levels don't cross is an excellent way to remember the sign of the denominator in the first place.
- If there is degeneracy in H_0, that is if $E_n^0 = E_m^0$ for some m, the expression is undefined because the denominator goes to zero.

This last observation is the basis of degenerate perturbation theory. Before discussing this, we should mention that there is a formula similar to (5.58) for the first-order correction to the wavefunctions, but it's never been asked for on the GRE and so is probably not worth learning.

If H_0 is degenerate, then there are multiple linearly independent eigenfunctions corresponding to the same energy eigenvalue. We are free to apply the first-order formula as long as we choose the basis of eigenfunctions appropriately. The solution, which you should be able to quote in your sleep, is

Diagonalize the perturbation in the subspace of degenerate states.

This sounds scary but it's really not. For example, say some state of H_0 was twofold degenerate, with eigenstates ψ_a and ψ_b. Then in the degenerate subspace, the perturbation is a 2 × 2 matrix:

$$H'_{\text{degen}} = \begin{pmatrix} H'_{aa} & H'_{ab} \\ H'_{ba} & H'_{bb} \end{pmatrix} \equiv \begin{pmatrix} \langle\psi_a|H'|\psi_a\rangle & \langle\psi_a|H'|\psi_b\rangle \\ \langle\psi_b|H'|\psi_a\rangle & \langle\psi_b|H'|\psi_b\rangle \end{pmatrix}.$$

Now diagonalize this matrix, and the eigenvalues are your first-order shifts. Clean and simple. In fact, you're not even required to find the eigenvectors first: the eigenvalues give you all the information about the energy that you need. Also, if it happens that a state has some large degeneracy, but most of the matrix elements vanish, you can simply exclude those states with vanishing matrix elements from consideration until you get down to a matrix that doesn't have any rows or columns that are identically zero. We'll see some examples of this in Section 5.7.4.

Incidentally, the problem of the denominator vanishing is taken care of by excluding the degenerate states from the sum (5.58). But again, this is too advanced for the GRE: all you really need to know about second-order perturbation theory is that it only matters if the first-order shift vanishes. Because it's so important, we'll repeat the mantra of degenerate perturbation theory again: *diagonalize the perturbation*. This will be extremely useful when we have all sorts of competing perturbations H', of different strengths, as we'll see Section 5.7.4.

5.6.2 Variational Principle

The variational principle is a method for approximating the ground state energy of a system for which the Hamiltonian is known. It is based on the simple observation that *for any normalized wavefunction, the expectation value of the energy is greater than the ground state energy E_0.*[6] Roughly, this is because a normalized wavefunction that is not the ground state wavefunction will contain admixtures of other excited state wavefunctions, which push the expectation value of the energy above E_0. This leads to a method of calculating the ground state energy by picking a trial wavefunction with an adjustable parameter (say, a Gaussian of variable width), then calculating $\langle E\rangle$ and minimizing it with respect to this parameter. You won't have to do this on the GRE, but it's worth knowing how the procedure works since you may be asked questions about it.

[6] Of course, it's equal to E_0 if you happened to pick the ground state wavefunction itself.

5.6.3 Adiabatic Theorem

Suppose a particle is in the nth eigenstate of a Hamiltonian H. The adiabatic theorem states that if we slowly change H to H', then at the end of the process the particle will end up in the corresponding eigenstate of H'. For the purposes of the GRE, that's all you need to know: whenever you see "slowly" on a quantum mechanics problem, you should immediately think of the adiabatic theorem. The standard applications take a one-dimensional Hamiltonian, such as the infinite square well or the harmonic oscillator, and "slowly" change one of the parameters: for example, letting the well expand slowly to some other width, or slowly changing the spring constant of the harmonic oscillator. Then the final energy is determined by the corresponding eigenstate of the Hamiltonian with the new parameters. This is really simple in practice, and is best illustrated through an example problem – see below.

5.6.4 Problems: Approximation Methods

1. A harmonic oscillator Hamiltonian of angular frequency ω is perturbed by a potential of the form $V = K(a - a^\dagger)^2$, where K is a constant and the operators a and $a^\dagger$ satisfy $a^\dagger|n\rangle = \sqrt{n+1}|n+1\rangle$ and $a|n\rangle = \sqrt{n}|n-1\rangle$. What is the energy shift of the state with unperturbed energy $3\hbar\omega/2$, to first order in perturbation theory?

   (A) $-3K$
   (B) $-K$
   (C) 0
   (D) K
   (E) $2K$

2. A particle in the ground state of an infinite square well between $x = 0$ and $x = a$ is subject to a perturbation $\Delta H = Vx$, where V is a constant. What is the first-order shift in the energy?

   (A) $\dfrac{V\sqrt{2a^3}}{\pi}$
   (B) $-Va/2$
   (C) 0
   (D) $Va/2$
   (E) $2Va/\pi$

3. A particle of mass m is subject to an infinite square well potential of size a, and is found to have energy E. The well is now expanded *slowly* to size $2a$. What is E'/E, where E' is the expectation value of the energy of the particle after the expansion has finished?

   (A) 0
   (B) 1

(C) $1/\sqrt{2}$
(D) 1/2
(E) 1/4

5.7 Atomic Physics Topics

5.7.1 Bohr Model

Why you would need to remember an obviously incorrect formulation of quantum mechanics is beyond us, but questions on the Bohr model of the hydrogen atom have appeared on recent GRE exams, so here is a brief review.

- The electron moves in *classical* circular orbits around the nucleus, called energy shells or energy levels, with *quantized* values of angular momentum: $L = n\hbar$ with $n = 1, 2, \ldots$
- Electrons in a given shell do not radiate as they move around the nucleus – this accounts for the stability of the atom. This was a big problem in early twentieth-century physics, since classically, the electron would radiate at all times during its orbit (since it's moving in a circle, it's accelerating) and spiral into the proton, collapsing the atom.
- The energy of transitions between energy shells can be shown to match the Rydberg formula given in Section 5.4.3.

If asked about the Bohr model, you will likely get either a simple calculation question about angular momentum quantization, or a conceptual question about why the Bohr model rescued the stability of the atom.

5.7.2 Perturbations to Hydrogen Atoms

The hydrogen atom as described in Section 5.4.3 is a great approximation, but it's not the whole story. There are several important corrections to the spectrum of hydrogen, all of which give interesting observable effects. In the discussion below (and in Section 5.7.4), keep in mind that, because L^2, S^2, and $J^2 = (\mathbf{L} + \mathbf{S})^2$ commute with everything, states of hydrogen can always be labeled by any combination of s, l, and j. However, in the presence of some of these perturbations, the corresponding m-values may *not* be conserved.

- **Fine structure.** These corrections are order α^2 smaller than the Bohr energies, and arise from two very different physical phenomena: replacing the electron kinetic energy term in the Hamiltonian with the full relativistic form, and including spin–orbit coupling between the electron's orbital angular momentum and its spin, of the form $\mathbf{L} \cdot \mathbf{S}$. Note that this is *not* the interaction of the electron's orbit with the *nuclear* spin: nuclear spin only gets into the game when we discuss hyperfine structure below. Actually calculating the energy shifts requires way more math than you'll have time for on the GRE, but the upshot is that the hydrogen energy levels acquire a dependence on the *total* spin j: since $\mathbf{J}$ commutes with $\mathbf{L} \cdot \mathbf{S}$ (exercise: check this), m_j is also conserved. The energies are still degenerate in m_j, but the degeneracy in l is broken: states with different values of l can have different energies.
- **Lamb shift.** This effect is smaller than fine structure by an *additional* factor of α, and splits the $2s$ and the $2p$ levels with $j = 1/2$, which are degenerate even including fine structure since they both have the same j. The physical mechanism comes from quantum electrodynamics, far beyond the scope of the GRE, but the Lamb shift could appear as a trivia-type question.
- **Hyperfine structure.** This correction is a magnetic dipole–dipole interaction between the spins of the electron and the proton, known as spin–spin coupling. The energy corrections are of the same order in α as fine structure, but are further suppressed by the ratio m_e/m_p because the gyromagnetic ratio of point particles depends on their masses. Because of the smallness of the effect, this perturbation is known as hyperfine structure. The result is that the ground state of hydrogen is split depending on whether the two spins are in the singlet or triplet state; the triplet has the higher energy (roughly because the spins are aligned, and magnetic dipoles want to be anti-aligned), and the wavelength of the emitted photon in a transition between these two states is about 21 cm, with energy about 5×10^{-6} eV.

It's extremely unlikely that you'll have to calculate anything relating to these perturbations, but their relative strengths and the physical mechanisms responsible for them is certainly fair game for the GRE. In passing, we will mention one calculational trick that is useful when dealing with fine structure:

$$\begin{aligned} J^2 &= (\mathbf{L} + \mathbf{S})^2 \\ &= L^2 + 2\mathbf{L} \cdot \mathbf{S} + S^2 \\ \implies \mathbf{L} \cdot \mathbf{S} &= \frac{1}{2}\left(J^2 - L^2 - S^2\right). \end{aligned} \tag{5.59}$$

For a system with definite values of j, l, and s, this lets you calculate the eigenvalues of the spin–orbit operator $\mathbf{L} \cdot \mathbf{S}$ immediately.

5.7.3 Shell Model and Electronic Notation

Let's go back to high school chemistry. There, you learned that an atom has various energy shells, numbered 1, 2, ... You also learned that each electron could occupy one of many orbitals, labeled $s, p, d, f, \ldots$ This is just saying that atomic electrons lie in hydrogenic states ψ_{nlm}: the energy shells are labeled by n, and the orbitals are labeled by l (s is $l = 0$, p is $l = 1$, d is $l = 2$, and f is $l = 3$). This naming scheme is stupid but unfortunately must be memorized. For a given n, there are n^2 possible combinations of l and m, but since an electron has spin-1/2, it can be either spin-up or spin-down in any of these states. This gives a total of $2n^2$ possible orbitals in each energy shell. Since m can take any value from $-l$ to l, there are $2(2l + 1)$ possible states in each orbital. The counting for the first few goes as follows:

$1s$ 2 states

$2s$ 2 states

$2p$ 6 states

$3s$ 2 states

$3p$ 6 states

$3d$ 10 states

Electronic notation represents the electron configuration of an atom by putting the number of electrons in each orbital as an exponent next to the orbital name. For example, hydrogen is $1s^1$, helium is $1s^2$, and the noble gas neon is $1s^2 2s^2 2p^6$. The detailed description of how the shells get filled is more chemistry than physics, but a few points to remember:

- The shells fill in order, preferring smaller values of l, up through argon with atomic number 18, where the rules break down due to complicated interactions among the electrons and the nucleus. So, continuing after helium, lithium has $1s^2 2s^1$, beryllium has $1s^2 2s^2$, boron has $1s^2 2s^2 2p^1$, and so on.
- Noble gases are chemically inert because they have totally filled energy shells. This means that there are the maximum possible number of electrons in each orbital.
- Alkali metals have one "extra" electron compared to the noble gases: for example, sodium has $3s^1$ in addition to the $1s^2 2s^2 2p^6$ of neon. Their chemical tendency is to shed this extra electron to form ions. Similarly, halogens have one electron fewer, like fluorine ($1s^2 2s^2 2p^5$), and want to gain an electron. This accounts for salts such as NaF, where sodium transfers an electron to fluorine in an ionic bond.

5.7.4 Stark and Zeeman Effects

Placing an atom in external electric or magnetic fields tends to split degenerate energy levels into closely spaced multiplets. The splitting by an electric field is called the Stark effect, and the splitting by a magnetic field is called the Zeeman effect. Calculating the energy shifts usually involves the whole machinery of degenerate perturbation theory, so you're unlikely to have to do a full-blown calculation on the GRE, but there are some important facts and conceptual statements about each that you should be aware of.

Stark Effect

- The change in the Hamiltonian from the potential energy of a charge $-e$ in a uniform electric field $\mathbf{E}$ is
$$\Delta H = e\mathbf{E} \cdot \mathbf{r}. \tag{5.60}$$
If $|\mathbf{E}|$ is small, this can be treated as a perturbation, at least for small r.
- There is *no change* in the ground state energy of hydrogen or any hydrogenic atom, to first order in $|\mathbf{E}|$.
- The lowest-energy states to show a first-order shift are the $n = 2$ states. The states with $m = \pm 1$ are unperturbed, but the $2s$ state and the $2p$ state with $m = 0$ are split. The magnitude of the splitting is likely unimportant for the GRE, but we can get pretty close just by dimensional analysis. Dimensional analysis tells us the energy splitting must be of the form $\Delta E = ke|\mathbf{E}|d$, where k is a constant and d is some length. The only length scale of the hydrogen atom is the Bohr radius, so we can take $d = a_0$ and all that is left undetermined is some number k.

Zeeman Effect

- The Hamiltonian responsible for this effect is the interaction of *both* the electron's orbital angular momentum and spin with a magnetic field $\mathbf{B}$:
$$\Delta H = \frac{e}{2m}(\mathbf{L} + 2\mathbf{S}) \cdot \mathbf{B}. \tag{5.61}$$
The exact numerical factors are probably not important to memorize, but they have names: $e/2m$ is the electron's classical *gyromagnetic ratio*, and the extra factor of 2 in front of the spin operator is because the quantum gyromagnetic ratio happens to be twice the classical value. The magnetic field picks out a preferred vector, so it is conventional to measure spins with respect to the direction of $\mathbf{B}$, which is equivalent to taking $\mathbf{B}$ to point in the $\hat{\mathbf{z}}$ direction.
- The main pedagogical purpose of the Zeeman effect is to illustrate the concept of "good quantum numbers." In play

are three operators, **L**, **S**, and **J** $=$ **L** $+$ **S**. Depending on the magnitude of **B**, different combinations of these operators may be conserved (or approximately so), and hence different combinations of eigenvalues may label the energies.

- If $|\mathbf{B}|$ is small, the Zeeman Hamiltonian is a perturbation on top of fine structure, for which we have already seen that the energies are labeled by j, l, and m_j. The weak-field Zeeman effect splits the j states according to m_j, with the lowest energy for the most negative m_j: physically, the electron spin wants to be anti-aligned with the magnetic field, since this is energetically favorable. This splitting of energy levels according to spin can be seen in the famous *Stern–Gerlach experiment*, where an inhomogeneous magnetic field splits a beam of atoms into two, effectively performing a measurement of m_j. If spin were a classical phenomenon, one would have expected the beam to smear out continuously depending on the projection of the spin vector onto the direction of **B**; the splitting into two sharp components was a striking demonstration that spin was quantized.
- If $|\mathbf{B}|$ is large, we treat fine structure as a perturbation on top of the Zeeman Hamiltonian, the reverse of the weak-field case. Since we have taken **B** to point in the $\hat{\mathbf{z}}$-direction, L_z and S_z *both* commute with the Zeeman Hamiltonian. This means that l, m_l, and m_s are now conserved before fine structure comes into the picture. The total spin j and m_j are *not* conserved, because the magnetic field provides an external torque. The energy of the Zeeman states depends on m_l and m_s in the same way as for the weak-field effect, and fine structure causes these states to develop a dependence on l as well.

5.7.5 Selection Rules

A rough characterization of atomic physics is the study of what happens when you poke atoms with light. That being said, the emission and absorption of electromagnetic radiation by atoms is really a topic for quantum field theory. Thankfully, it can be well approximated by time-dependent perturbation theory in ordinary quantum mechanics, and many results summarized in a list of rules known as *selection rules*, which are actually quite easy to remember. One caveat: the following rules apply only to the *electric dipole approximation*, which assumes that the wavelength of the electromagnetic radiation is large compared to the size of the atom, so that the spatial variation of the field is negligible and the atom feels a *homogeneous* electric and magnetic field which oscillates sinusoidally in time. When we say that certain things can or can't happen, we are working only in the context of this approximation. We emphatically do *not* mean that these processes can't *ever* happen: in fact, any quantum process that is not forbidden by conservation laws (energy, momentum, charge, and so on) must occur with some nonzero probability.

The following selection rules refer to transitions between hydrogenic orbitals ψ_{nlm} and $\psi_{n'l'm'}$. To a decent approximation, all atomic electrons are in one of these spatial wavefunctions, so we are considering excitation or de-excitation of a single atomic electron between two of these states. Of course, since we are dealing with quantum mechanics, the energy of the emitted or absorbed photon can only take certain discrete values, corresponding to the energy difference between the final and initial states.

- *No transitions occur unless* $\Delta m = \pm 1$ *or* 0. You can remember this by conservation of the z-component of angular momentum: the photon has spin 1, and hence must carry off $-\hbar$, 0, or $\hbar$ units of angular momentum in the z-direction. If the incident electric field is oriented in the z-direction, we can be more precise: no transitions occur unless $\Delta m = 0$.
- *No transitions occur unless* $\Delta l = \pm 1$. Again, this sort of follows by the rules of addition of angular momentum, but the case $l = l'$ is conspicuously absent.

Note that the $2s \to 1s$ transition in hydrogen is technically forbidden by these rules, since this transition has $\Delta l = 0$. This transition does happen, but it gets around the rules because it occurs by *two*-photon emission. This also implies that if you shine light on the ground state of hydrogen, the first excited states you will populate are the $2p$ states, with $l = 1$.

5.7.6 Blackbody Radiation

One of the triumphs of early quantum mechanics was the derivation of the radiation spectrum of a *blackbody*. This is an idealized object which absorbs all the radiation that hits it, while reflecting none of it, so that any radiation emitted from the blackbody is a result of its overall temperature. Unlike in atomic transitions, this radiation is not all at one frequency: there is a whole spectrum. The word "blackbody" is actually a bit misleading, because if this spectrum peaks in the visible region, the body will appear to be the color that corresponds to the peak frequency. This is in fact the case for the Sun, which to a good approximation is a blackbody whose spectrum peaks in the yellow part of the visible region. By using the dependence of the blackbody spectrum on temperature, one can measure the temperature of an object by observing its color.

In equations, the *power spectrum*, which is the radiated power per unit area of the blackbody per unit solid angle per unit frequency (what a mouthful!) as a function of frequency, is

$$I(\omega) \propto \frac{\hbar\omega^3}{c^2} \frac{1}{e^{\hbar\omega/k_B T} - 1}. \tag{5.62}$$

You probably don't need to memorize this equation, but it is important for historical reasons: the classical formula contains only the first factor proportional to ω^3, which would cause the power to grow without bound as the frequency of the radiation increased. This "ultraviolet catastrophe" was averted by Planck, who supplied the second factor and ushered in the era of quantum mechanics: you should recognize it as the Bose–Einstein factor (4.43) for identical photons, which are bosons. (The chemical potential for photons is $\mu = 0$, since there is no conservation law for photon number: they can be created or destroyed in physical processes.)

The two facts you *do* need to remember about blackbody radiation are:

- Integrating (5.62) over everything but the area of the blackbody, we obtain from the power spectrum the *Stefan–Boltzmann law*:

$$\frac{dP}{dA} \propto T^4, \tag{5.63}$$

  where T is the temperature of the blackbody. The exact prefactor isn't worth memorizing (for what it's worth, it's some dimensionful constant called the Stefan–Boltzmann constant), but the T^4 dependence is crucial.
- The location of the peak of the spectrum (5.62) is given by *Wien's displacement law*:

$$\lambda_{\text{max}} = (2.9 \times 10^{-3}\,\text{K} \cdot \text{m})\, T^{-1}. \tag{5.64}$$

  The numerical prefactor *is* important this time: it's one of those constants that really should appear in the currently useless Table of Information at the top of your test, but it's not there. We strongly recommend that you memorize this number (approximate 2.9 to 3 if you like), since although the constant has shown up explicitly on recent tests, its appearance or nonappearance on past GRE tests is inconsistent. Notice that this formula is in terms of wavelength, not frequency: this is conventional. If you know the temperature of a blackbody, you can use this formula to find the wavelength at which the power spectrum is at a maximum, or vice versa. (Try it yourself: plug in the wavelength of yellow light, and find the temperature of the Sun.) Also note that the maximum wavelength is inversely proportional to temperature: as the body gets hotter, λ_{max} gets smaller, shifting toward the ultraviolet.

5.7.7 Problems: Atomic Physics Topics

1. In the Bohr model of the hydrogen atom, let r_1 and r_2 be the radii of the $n = 1$ and $n = 2$ orbital shells, respectively. What is r_2/r_1?

   (A) 1/2
   (B) $1/\sqrt{2}$
   (C) 1
   (D) 2
   (E) 4

2. The Bohr model is inconsistent with the modern picture of quantum mechanics because it predicts which of the following?

   (A) The electron will not lose energy as it orbits the nucleus.
   (B) The electron is confined to distinct energy shells.
   (C) Angular momentum of the atom is quantized.
   (D) The ground state has nonzero orbital angular momentum.
   (E) The energy levels go as $1/n^2$, where n is the principal quantum number.

3. An atom with the electron configuration $1s^2 2s^3$ is forbidden by which of the following?

   (A) Conservation of angular momentum
   (B) Hund's rule
   (C) The Pauli exclusion principle
   (D) The uncertainty principle
   (E) None of the above

4. Which of the following is the most likely decay chain of the $3s$ state of hydrogen? Ignore all relativistic and fine-structure effects.

   (A) $3s \to 1s$
   (B) $3s \to 2s \to 1s$
   (C) $3s \to 2p \to 1s$
   (D) $3s \to 2p \to 2s \to 1s$
   (E) The $3s$ state is stable

5. Observations of the early universe show that it behaves as an almost perfect blackbody, with associated radiation known as the cosmic microwave background (CMB). Given that the spectrum of the CMB peaks in the microwave region at a wavelength of 1.06 mm, which of the following is closest to the temperature of the CMB?

(A) 0.03 K
(B) 0.3 K
(C) 3 K
(D) 30 K
(E) 300 K

6. A space heater whose heating elements are maintained at a constant temperature can heat a cold room from 15 °C to 25 °C in time t. If the temperature of the heating elements is doubled, how much time will it take to heat the room?

(A) $t/16$
(B) $t/8$
(C) $t/4$
(D) $t/2$
(E) t

5.8 Solutions: Quantum Mechanics and Atomic Physics

Formalism

1. C – We could calculate the normalization constant, and then do the integral $\int_{-\infty}^{0} |\Psi(x)|^2\, dx$ to find the probability. But it's much simpler to stare at the wavefunction and just write down the correct answer, 1/2. The reason is that $\Psi(x)$ is symmetric with respect to $x = 0$, so the area under the curve of $|\Psi(x)|^2$ is equal for x positive and negative, independent of what A is. This is typical of the kind of shortcuts you might see on a GRE problem.
2. C – Let $\Psi(0) = \frac{1}{\sqrt{2}}(\psi_1 - \psi_2)$. The full time-dependent wavefunction is given by $\Psi(t) = \frac{1}{\sqrt{2}}(\psi_1 e^{-iE_1 t/\hbar} - \psi_2 e^{-iE_2 t/\hbar})$. However, note that the complex moduli of the coefficients multiplying ψ_1 and ψ_2 are the *same* as they were at $t = 0$, since the exponential factors have modulus 1 at any time t. So the probability of measuring E_1 is the same at time t as at time zero: namely, $(1/\sqrt{2})^2 = 1/2$. This is a general feature of time-independent potentials: even in superpositions of energy eigenstates, the energy eigenvalues and relative probabilities of energy measurements are *constant* in time. This is no longer true for time-dependent potentials, but luckily you won't have to worry about those on the GRE.
3. E – Since $|b\rangle$ is an eigenstate of $\hat{p}$, we have $\hat{p}|b\rangle = b|b\rangle$. But since $\hat{p}$ is a Hermitian operator, eigenfunctions corresponding to different eigenvalues are orthogonal, and

$$\langle a|\hat{p}|b\rangle = b\langle a|b\rangle = 0.$$

We also could have proceeded by process of elimination. Since $\hat{p}$ is Hermitian, we have $\langle a|\hat{p}|b\rangle = \langle b|\hat{p}|a\rangle$, so the answer must be symmetric in a and b, which eliminates choices A and B. The answer must also have dimensions of momentum (the same units as a and b), which leaves D and E. It's hard to see how a square root might come out of a calculation like this, so E seems the most reasonable choice.
4. D – Recall that the nth state of the harmonic oscillator has energy $(n+1/2)\hbar\omega$. If you don't remember this, make sure you study Section 5.2! This wavefunction has three nodes, so it must represent the third excited state; since the harmonic oscillator starts counting with $n = 0$, this means it's the $n = 3$ state, with energy $7\hbar\omega/2$.
5. E – For a matrix to be a Hermitian operator, it must be equal to its Hermitian conjugate; in other words, transposing the matrix and complex-conjugating the entries must give back the original matrix. For matrices that consist of purely real entries, this just means the matrix must be symmetric: I fails but III passes. It turns out that II is also Hermitian, since transposing switches the i and $-i$, but conjugating switches them back. Incidentally, II is one of the Pauli matrices, and if you recognized this you could immediately determine that it is Hermitian – see Section 5.5.
6. B – We want the inner product $\langle \Psi_1|\Psi_2\rangle$ to vanish. In taking the inner product, remember that we have to *complex-conjugate* the coefficients of $|\Psi_1\rangle$:

$$(1)(2) + (-2i)(x) = 0 \implies x = -i.$$

If we had forgotten to conjugate, we would have ended up with the trap answer A.
7. B – This is an application of the energy–time uncertainty principle, combined with a little special relativity which tells us a particle of mass m has rest energy mc^2. From $\Delta E\,\Delta t \approx \hbar$, we plug in the lifetime of the particle for Δt and use the numerical value of $\hbar$ (which would be given in the Table of Information at the top of the test, here we want units of eV · s) to find $\Delta E \approx 66$ MeV. In other words, a mass difference of greater than 66 MeV/c^2 from the central value of 500 MeV/c^2 represents one standard deviation, and is expected to happen about 32% of the time. Here the mass difference is only 50 MeV/c^2, so this may be expected to happen about 50% of the time. On the other hand, the mass difference is five times the expected experimental error, and a 5-sigma event happens much more rarely. So the difference is most likely due to the uncertainty principle, choice B.

8. B – This is a classic application of the "folklore" uncertainty principle (5.19):

$$(100 \text{ pm})(\Delta p) \approx \hbar \implies \Delta p \approx 2 \text{ keV}/c.$$

To do the arithmetic fast, we could use the fact that $hc = 1240$ eV · nm (also given in the Table of Information on the most recently released test) to get the momentum in natural units of eV/c. Now, find the energy by

$$E = \frac{p^2}{2m_e} \approx \frac{4 \text{ keV}^2/c^2}{1 \text{ MeV}/c^2} = 4 \text{ eV}.$$

This makes complete sense, since we know that chemical processes take place on the scale of eV, rather than MeV (the scale of nuclear physics) or GeV (the scale of particle accelerator physics). Using just this knowledge, we probably could have zeroed in on choice B right away, but it's nice to see it come out of the uncertainty principle.

Harmonic Oscillator

1. D – The key here is the virial theorem, supplemented by a bit of clever reasoning. Since $\langle T \rangle = E_n/2$ for any energy eigenstate, and $T = p^2/2m$, we have $\langle p^2 \rangle = mE_n$ for any energy eigenstate $|n\rangle$. Now, the virial theorem does not generally apply to superpositions of energy eigenstates, but in this particular case it does. The operator $\hat{p}$ is a linear combination of a and $a^\dagger$, so $\hat{p}^2$ contains products of at most two of these operators, such as $(a^\dagger)^2$ or $a^\dagger a$. Thus $\langle 1|\hat{p}^2|4\rangle = 0$, because applying a raising or lowering operator twice can never give a product of nonorthogonal states; we always get orthogonal combinations such as $\langle 2|3\rangle$ or $\langle 1|2\rangle$. So *there are no cross terms*, and the virial theorem still applies to each diagonal term in the expectation value. Denoting our state by $|\psi\rangle = \frac{1}{\sqrt{2}}(|1\rangle + |4\rangle)$, we have

$$\langle \psi|\hat{p}^2|\psi\rangle = \frac{1}{2}(\langle 1|\hat{p}^2|1\rangle + \langle 4|\hat{p}^2|4\rangle).$$

Using the virial theorem gives

$$\langle p^2 \rangle = \frac{1}{2}m(E_1 + E_4) = \frac{m}{2}(3\hbar\omega/2 + 9\hbar\omega/2) = 3m\hbar\omega,$$

choice D.

2. A – The classical ground state of the harmonic oscillator has zero energy, corresponding to a particle sitting at $x = 0$, where the potential vanishes, and having no kinetic energy. But the ground state of the quantum oscillator has energy $\hbar\omega/2$.
3. C – Let (n_x, n_y) label the levels of the independent x- and y-coordinate oscillators. The total zero-point energy is $\hbar\omega/2 + \hbar\omega/2 = \hbar\omega$ in two dimensions, so the state with energy $3\hbar\omega$ has $n_x + n_y = 2$, which can occur in three ways: (2,0), (1,1), and (0,2). Thus this state has a degeneracy of 3.

Other Standard Hamiltonians

1. D – This is easiest to see from the explicit form of the wavefunction, $\sin(n\pi x/L)$. Even without the functional form, we can still eliminate B, C, and E because II is clearly false: the wavefunction is *always* continuous.
2. C – Clearly I is false because a definite value of position would mean the wavefunction must be a delta function $\delta(x - x_0)$. In fact, this wavefunction represents a superposition of two momentum eigenstates, e^{+ikx} and e^{-ikx}, with equal but opposite values of momentum $p = \hbar k$. So II is false. However, the energy $\hbar^2 k^2/2m$ doesn't care about the sign of k, so the particle does have a definite value of energy. Only III is true, so C is correct.
3. E – The incident wave has amplitude $|A|$, the reflected wave has amplitude $|B|$, and the transmitted wave has amplitude $|C|$, so the transmission probability is $|C|^2/|A|^2$ multiplied by an extra factor which takes into account the different speeds of the waves on the left and the right. Rather than solving for this quantity using conservation of probability, continuity of the wavefunction and its derivatives and so on, we can just use logic. Choices A and B are clearly false, for any finite nonzero value of V_0. Furthermore, we should have the transmission probability going to 1 when $k_L = k_R$, since that corresponds to the step disappearing. This eliminates D. Furthermore, since probabilities are proportional to *squared* amplitudes, the only way a ratio like k_L/k_R would show up would be if solving the continuity conditions gave a $\sqrt{k}$ factor: since derivatives of the wavefunction will only give k and not $\sqrt{k}$, this seems rather unlikely. So E seems best, and indeed it is the correct answer. Notice that this kind of careful reasoning, while it may seem involved, took *much* less time than the corresponding calculations would have. In our opinion this is the best way to approach these kinds of quantum mechanics problems on the GRE.

Quantum Mechanics in Three Dimensions

1. D – Since the spherical harmonics are eigenfunctions of $\mathbf{L}^2$, Y_3^2 has eigenvalue $3(3+1)\hbar^2 = 12\hbar^2$ and Y_2^1 has eigenvalue $2(2+1)\hbar^2 = 6\hbar^2$. The squares of the coefficients in the given wavefunction are 1/2 for both spherical harmonics, so the expectation value is $(1/2)6\hbar^2 + (1/2)12\hbar^2 = 9\hbar^2$, choice D.

2. E – The given wavelength is exactly 4 times the wavelength of the Lyman alpha line, which comes from the $n = 2$ to $n = 1$ transition. Letting n_i and n_f be the initial and final states of the 488 nm transition, the Rydberg formula (inverted because we're dealing with wavelength rather than frequency) gives

$$4 = \frac{\frac{1}{1^2} - \frac{1}{2^2}}{\frac{1}{n_f^2} - \frac{1}{n_i^2}},$$

and this is solved just by doubling the n values for Lyman radiation: $n_i = 4$ and $n_f = 2$. This is choice E.

3. A – The gravitational potential is $G_N m_e m_n / r$, and the corresponding electric potential in the hydrogen atom is $e^2/4\pi\epsilon_0 r$, so we simply replace $e^2/4\pi\epsilon_0$ by $G_N m_e m_n$ in the formula for the Bohr radius. Since we're approximating $m_e \ll m_n$, we can use m_e instead of the reduced mass. This gives

$$a = \frac{\hbar^2}{m_e} \frac{1}{G_N m_e m_n} = \frac{\hbar^2}{G_N m_e^2 m_n},$$

which is A. We could also have made progress by dimensional analysis, since only A and C have the correct units. By the way, the reason we never see this bound state is that, plugging in the numbers, $a = 1.2 \times 10^{29}$ m; for comparison, the radius of the observable universe is 4.7×10^{26} m!

4. E – From (5.46), the ground state energy is proportional to the reduced mass and the square of the "electron" charge. This gives a total factor of $2(2^2) = 8$, choice E.

Spin

1. C – The first measurement is irrelevant. After the second measurement the particle is in the $-\hbar/2$ eigenstate of $\hat{S}_x$, which is $\frac{1}{\sqrt{2}}\begin{pmatrix} 1 \\ -1 \end{pmatrix}$. The components of $|\uparrow\rangle$ and $|\downarrow\rangle$ in this state have equal magnitude, so the probability of measuring either one is 1/2, choice C.

2. D – The ground state has zero orbital angular momentum, so combined with the given information about the nucleus having no orbital angular momentum, we only have to worry about spins. To get $j = 3/2$ the nucleus must be in the triplet state, which rules out I. The z-component of spin of the electron can be either $1/2$ or $-1/2$, and so the nucleus could either have $m_j = 0$ (with $m_s = 1/2$ for the electron), or $m_j = 1$ (with $m_s = -1/2$ for the electron). So both II and III are viable options, choice D.

3. A – This is a classic problem and well worth remembering, so we'll reason through it carefully. Because the interaction of electron spins with the nucleus is an order of magnitude smaller than the Coulomb interaction between the electrons and the protons (see Section 5.7), the ground state is determined by the spatial wavefunction, and prefers both electrons to be in the same lowest-energy orbital, the 1s orbital. Since both electrons have the same spatial wavefunction, the spatial piece is symmetric, so the spin configuration that comes along for the ride must be antisymmetric. The possible spin configurations of two electrons are the singlet and the triplet, as discussed in Section 5.5.3. But only the singlet (choice A) is antisymmetric, so this must be the correct spin state. Helium is such a GRE favorite that it's best just to memorize the conclusion of this argument, rather than rederiving it each time: *the ground state of helium is the singlet state.*[7] But really, you already knew this from high school chemistry, when you wrote the electron configuration of hydrogen as $\uparrow$, the configuration of helium as $\uparrow\downarrow$, and so on. While technically an imprecise shorthand, it's a good mnemonic nonetheless.

4. D – Total $S_z = \hbar$ means the electrons must be in the triplet state, which is symmetric. For a totally antisymmetric wavefunction, the spatial wavefunction must be antisymmetric. This knocks out the ground state, where both electrons are in the $n = 0$ state of the harmonic oscillator, since after antisymmetrization this vanishes identically. So the next available state is an antisymmetrized version having $n = 0$ and $n = 1$:

$$\psi_{\text{spatial}} = \frac{1}{\sqrt{2}}(|0\rangle_1 |1\rangle_2 - |1\rangle_1 |0\rangle_2).$$

This is an energy eigenstate with energy $\hbar\omega/2 + 3\hbar\omega/2$, choice D.

Approximation Methods

1. A – The state with energy $3\hbar\omega/2$ is $|1\rangle$, so the shift we're looking for is $\langle 1|V|1\rangle$. Expanding out V isn't really that helpful. Instead, note that the only terms that survive in the expectation value contain exactly one a and exactly one $a^\dagger$, since otherwise we would have something of the form $\langle m|n\rangle$ which vanishes by orthogonality. There are two such terms: $-aa^\dagger$ and $-a^\dagger a$. (Remember to keep track of the order, since these operators don't commute!) We have

$$\langle 1|(-aa^\dagger)|1\rangle = -(\sqrt{2})^2\langle 2|2\rangle = -2,$$

[7] Note that there is an important subtlety in this discussion: the electron–electron repulsion in helium means that the ground states are *not* really the pure hydrogen orbitals. However, this line of reasoning does give the correct experimentally observed answer, which is all that matters. So feel free to regard this argument as a mnemonic for remembering the correct answer.

$$\langle 1|(-a^\dagger a)|1\rangle = -(\sqrt{1})^2\langle 0|0\rangle = -1,$$

where in the first line we acted on $|1\rangle$ with $a^\dagger$ and on $\langle 1|$ with a on the right, and similarly in the second line with the roles of a and $a^\dagger$ reversed. Adding these up gives $\Delta E = K(-2-1) = -3K$, choice A.

2. D – The ground state is $\psi(x) = \sqrt{2/a}\sin(\pi x/a)$, and the first-order shift is given by

$$\Delta E = \int \psi^*(x)\Delta H\psi(x)\,dx = V\frac{2}{a}\int_0^a x\sin^2(\pi x/a)\,dx.$$

This is likely more difficult than any integral you'll actually see on the GRE, but just in case, let's briefly review how to do it: use the half-angle identity to write

$$\sin^2(\pi x/a) = \frac{1}{2}\left(1 - \cos(2\pi x/a)\right).$$

Integrating the first term against x gives

$$V\frac{2}{a}\frac{1}{2}\int_0^a x\,dx = \frac{V}{2a}a^2 = \frac{Va}{2}.$$

Do the second integral $\int x\cos(2\pi x/a)\,dx$ by parts; the boundary term vanishes because $\sin(2\pi x/a)$ vanishes at both the endpoints, so the remaining integral is

$$-V\frac{2}{a}\frac{1}{2}\frac{a}{2\pi}\int_0^a \sin(2\pi x/a)\,dx = 0,$$

because $\cos(2\pi x/a)$ is equal to 1 at both endpoints. So the energy shift is simply $Va/2$, choice D.

3. E – After an energy measurement the particle is in some eigenstate of the infinite square well, say the nth state. Under a slow expansion of the well, the adiabatic theorem tells us the particle remains in the nth state of the infinite square well. Since the energies of the infinite square well are $\frac{n^2\pi^2\hbar^2}{2ma^2}$, the energy of the final configuration is $1/4$ of the original energy, independent of n.

Atomic Physics Topics

1. E – We can eliminate A, B, and C right away since the $n = 2$ shell has more energy than $n = 1$, and hence must be further away from the nucleus. The angular momentum for a circular orbit is $L = mvr$, and the orbital velocity can be determined from classical mechanics by setting the centripetal force equal to the Coulomb force:

$$\frac{1}{4\pi\epsilon_0}\frac{e^2}{r^2} = \frac{mv^2}{r} \implies v \sim r^{-1/2}.$$

Thus $L \sim r^{1/2}$, and from $L_1 = \hbar$ and $L_2 = 2\hbar$, we get $r_2/r_1 = L_2^2/L_1^2 = 4$.

2. D – In the Bohr model, $L = n\hbar$, where $n > 0$, since otherwise the electron would be sitting right on top of the nucleus (its classical orbital radius would be zero). However, in the modern picture the ground state is ψ_{100}, which has orbital quantum number $l = 0$.

3. C – The 2s orbital has $l = 0$, and hence $m = 0$, so electrons in this orbital are distinguished only by the direction of their spin. There are only two independent spin states, so a third electron would have to be in the same state as one of the other two, violating the Pauli exclusion principle for fermions.

4. C – By the selection rules, 3s must decay to a state with $l = 1$, so the only option is the 2p state. From there, we again must have $\Delta l = \pm 1$, and so 1s is the only remaining possibility.

5. C – This is a straightforward application of Wien's law. Solving for T, we get $T = (2.9\times 10^{-3}\text{K}\cdot\text{m})/\lambda_{\text{max}}$, and approximating Wien's constant by 3 and the wavelength by 1 mm, we get $T \approx 3$ K. (Even without any knowledge of Wien's constant, we could have eliminated choice E since the universe certainly isn't at room temperature.) By the way, questions on the CMB are fairly likely to appear on the GRE in some form, so it's worth remembering both the wavelength and the temperature if you can.

6. A – Treating the space heater as a blackbody, the power (energy per unit time) is proportional to the fourth power of temperature. So doubling the temperature increases the power by a factor of 16, and for a constant desired energy, this means the time taken to deliver this energy will decrease by a factor of 16.

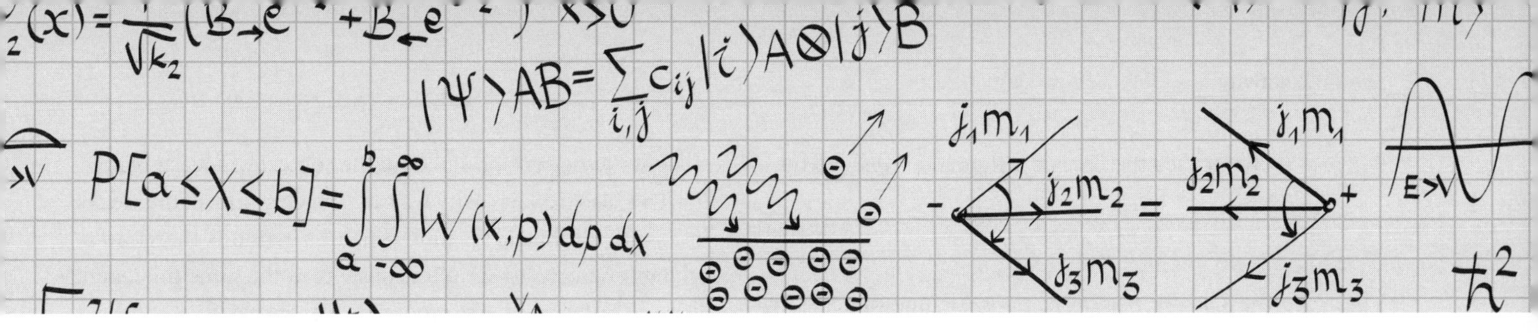

6 Special Relativity

Problems on relativity in the GRE often look simple, but can be tricky as they tend to deal more with conceptual issues than with detailed calculations. Throughout this chapter, keep in mind how important it is to define things precisely: concepts such as time, length, and simultaneity, which seem so obvious in classical mechanics, can actually be quite subtle. Our discussion will necessarily be brief, so by all means consult our recommended references if you want more details.

6.1 Relativity Basics

There are two simple postulates from which all of special relativity follows:

1. The speed of light in vacuum is a constant (denoted c) in all inertial reference frames.
2. The laws of physics are identical in all inertial reference frames.

The crucial idea in both postulates is the *inertial reference frame,* which for the purposes of the GRE just means an observer traveling at constant velocity – that is, in a straight line with constant speed. So another way of stating the second postulate is that the apparent laws of physics do not change as long as the observer is in a frame that is moving at a constant velocity. We should add that only inertial frames with velocity $v < c$ are allowed: this is the familiar statement that no signals propagate faster than light.

These postulates lead directly to the famous Lorentz transformations, which relate the coordinates used to describe two inertial frames. Since this can get a little confusing, let's first consider a simpler example, where the velocities are small compared to c and we can use ordinary coordinate transformations familiar from classical mechanics. Let system S be a person standing on the side of the highway, and system S' be a car traveling on the highway at constant speed v. We'll let (t, x) represent the time and distance along the highway, respectively, for the person in S, and (t', x') represent the analogous quantities for the car in S', where $x' = 0$ corresponds to the front bumper of the car. The first thing to do is synchronize the origin of coordinates in both systems: let $t = t' = 0$ be the instant where the front bumper of the car passes the person on the side of the highway, and let this occur at $x = x' = 0$. The car in S' is moving with constant speed, so its position at time t is $x = vt$. In the frame of the car, then, an object at position x' with respect to the front bumper has traveled an additional distance vt as seen by the person in S: $x = x' + vt$, or $x' = x - vt$. Finally, in classical mechanics, clocks tick at the same rate everywhere, so $t = t'$. Thus we have the *Galilean transformations* relating the coordinates (t, x) to (t', x'):

$$t' = t,$$
$$x' = x - vt.$$

We can check that these make sense: at $t' = t = 0$, we have $x' = x$, corresponding to the synchronizing of clocks described earlier.

However, in special relativity, this simple behavior goes out the window. At velocities near the speed of light, *time and space mix with each other.* The analogous transformations between S and S' are

$$t' = \gamma\left(t - \frac{v}{c^2}x\right), \tag{6.1}$$

$$x' = \gamma\left(x - vt\right), \tag{6.2}$$

where γ is often called the *Lorentz factor*, and is defined by

$$\gamma = \frac{1}{\sqrt{1 - v^2/c^2}}. \tag{6.3}$$

These are the *Lorentz transformations* for one-dimensional motion along the x-axis. They tell you, given coordinates (t, x) in S, what the analogous coordinates (t', x') are in S'. Notice how the t' equation involves *both* x and t, which is very much unlike the simple statement that $t' = t$ in classical mechanics. The inverse transformations, which express S coordinates in terms of S' coordinates, are

$$t = \gamma\left(t' + \frac{v}{c^2}x'\right), \tag{6.4}$$

$$x = \gamma\left(x' + vt'\right). \tag{6.5}$$

These are extremely easy to remember: just flip the sign of v! This is the concept of "relativity" in action: if S' is moving to the right with respect to S, then from the point of view of an observer in S', it's S that is moving to the left. (Just picture how the scenery seems to fly by backwards as you're driving in a car.)

We're now going to look in detail at the consequences of these transformations. In what follows, think of S as a collection of clocks, one at each point x, all of which are synchronized with each other: a clock at $x = x_1$ and another at $x = x_2$ will all read the same time t. Similarly, S' is a different collection of clocks, all traveling together at velocity v with respect to S, and all synchronized with each other. However, they are *not* all synchronized with the clocks in S! The best we can do is synchronize the origin of coordinates, as we did in our simple example above: we define $t = t' = 0$ to be the instant that the $x' = 0$ clock in S' passes the $x = 0$ clock in S. This is consistent with the Lorentz transformations we wrote down: at $(t, x) = (0, 0)$, we also have $(t', x') = (0, 0)$.

6.1.1 Simultaneity

Suppose two events are *simultaneous* in system S: that is, event A occurs at position x_A, while event B occurs at position x_B, and they both happen at time t. Looking at the t' equation, we find

$$t'_A = \gamma\left(t - \frac{v}{c^2}x_A\right), \quad t'_B = \gamma\left(t - \frac{v}{c^2}x_B\right).$$

These are not the same! In fact, unless $x_A = x_B$ (in which case A and B might as well have been the same event), the times measured in S' are different, and so A and B are *not* simultaneous in S'. Since simultaneity is a concept that is so ingrained in our intuition, and we use it to build other concepts such as length and causality, this is a red flag which we will have to be very careful of in the context of special relativity. This also explains why we can't synchronize all the clocks in S' with all the clocks in S: synchronization is the same as asking for clocks to simultaneously read the same time, and, as we've just seen, that's impossible for two clocks at different positions.

6.1.2 Time Dilation

Suppose we're sitting in S, and we wait for a time interval Δt. How much time elapses in frame S'? Here we have to carefully define what we mean by "time elapsed": since we're thinking of S' as a set of clocks, one at each position, we want to follow a *single* clock, at a *fixed* position x', for a time Δt. So the appropriate equation is (6.4), relating Δt to $\Delta t'$ at fixed x':

$$\Delta t = \gamma \Delta t' \quad (\text{fixed } x'). \tag{6.6}$$

Notice that $\gamma \geq 1$, since $v < c$, so for a given interval of time that elapses in the moving frame S', *more* time elapses in the stationary frame, by a factor of γ. This is known as

Time dilation: moving clocks run slower by a factor of γ.

Here is where things start to get confusing. An easy mistake to make is to use fixed x instead of fixed x': plugging this into the equation for t', we would find $\Delta t' = \gamma \Delta t$, which is precisely the *opposite* of the correct result! The problem is that fixed x corresponds to looking at a whole sequence of clocks in S' as they fly by, one by one, which does *not* correspond to measuring any kind of time elapsed in S' because of the issue of synchronization mentioned earlier. The moral of the story is that, while "time dilation" sounds nice and simple, it's very easy to get mixed up by exactly whose time is being dilated. Luckily, the instances you'll see on the GRE are all pretty standard, and you'll see many examples in the practice problems.

6.1.3 Lorentz Contraction

Suppose we are sitting in S, and want to measure the length of an object in S'. For simplicity, let's put the back end of the object at $x' = 0$ and the front end at $x' = L'$. Now, to measure the length of the object as seen from S, we need to note the positions of the two ends at the *same* S-time t. This brings in simultaneity, but that's OK since the clocks in S are all synchronized with each other, so we can talk about events that are simultaneous in S. Applying the x' equation (6.2) at $x' = 0$

and $x' = L'$, we have

$$0 = \gamma(x_1 - vt),$$
$$L' = \gamma(x_2 - vt).$$

Subtracting the first from the second, and defining $L = x_2 - x_1$ as the length in S, we obtain

$$L' = \gamma L \quad (\text{fixed } t). \tag{6.7}$$

Again, the two definitions of length differ by the same factor γ. But note the direction! Since L, the length as measured in S, is smaller than L', we remember this as

Length contraction: moving objects are shortened by a factor of γ.

Note that we don't say "moving objects *appear* shortened." As far as we can define the concept of length, they *are* shortened.

6.1.4 Velocity Addition

Here's a familiar situation from classical mechanics: you're riding in a car at 100 km/h, and you throw a ball forwards out the window at 5 km/h. Ignoring air resistance, from the point of view of someone standing on the side of the highway, the ball travels at $100 + 5 = 105$ km/h. In our language of inertial reference frames, we would say that the stationary observer defines a reference frame S, the car defines a frame S', and the ball defines a third reference frame S'': we just calculated the velocity of S'' with respect to S, given its velocity with respect to S'.

With this formalism, it's easy to re-evaluate this situation in the context of special relativity, and as expected it differs from the classical result. Suppose S' travels at velocity v, and the ball travels at velocity u with respect to S'. Then a bit of algebra with the Lorentz transformation equations gives the *Einstein velocity addition rule* for the velocity w as seen in frame S:

$$w = \frac{v + u}{1 + vu/c^2}. \tag{6.8}$$

This formula is useful to memorize since it is tricky to derive, but simple to remember. Here are a couple sanity checks that will make memorizing this formula easy. The factor of $1/c^2$ in the denominator (required for dimensional consistency) means that at velocities very small compared to the speed of light, $u, v \ll c$, the formula reduces to the usual addition of velocities, $w = u + v$. That gives you the numerator. On the other hand, plugging in $u = c$ (corresponding to shooting a laser beam out of the car, instead of throwing a ball), we find $w = c$. This recovers the first postulate of relativity, that the velocity of light is the same in all inertial frames: here, we've shown it to be true for a frame moving at velocity v. Finally, note that the signs must always match: if we throw the ball backwards rather than forwards, we should change the sign of u in both the numerator and the denominator. By the way, this velocity addition formula only applies to one-dimensional motion, but that's the only case you'll see on the GRE: the general formula is considerably more complicated and isn't worth memorizing.

6.2 4-Vectors

We're now going to introduce some notation that will make the previous results easy to remember, and generalizes easily to other useful physics situations. Define

$$x^0 = ct, \qquad x^1 = x, \qquad x^2 = y, \qquad x^3 = z, \tag{6.9}$$

which emphasizes the fact that space and time are treated on a similar footing in relativity: they're both just coordinates. Note that the superscripts are labels, *not* exponents! This notation is totally standard, so, as confusing as it is, we'll stick with it because it will match what you'll see on the exam. We can collect these coordinates into a single object called a *4-vector*:

$$x^\mu = (x^0, x^1, x^2, x^3) = (ct, x, y, z). \tag{6.10}$$

In this notation, the superscript μ is again a label that takes the values 0, 1, 2, or 3.

6.2.1 Lorentz Transformation Matrices

Now, we can write the Lorentz transformations as a matrix equation involving the vector x^μ. Defining

$$\beta = v/c \tag{6.11}$$

as the velocity of S' in units of c, we have

$$\begin{pmatrix} x^{0'} \\ x^{1'} \\ x^{2'} \\ x^{3'} \end{pmatrix} = \begin{pmatrix} \gamma & -\gamma\beta & 0 & 0 \\ -\gamma\beta & \gamma & 0 & 0 \\ 0 & 0 & 1 & 0 \\ 0 & 0 & 0 & 1 \end{pmatrix} \begin{pmatrix} x^0 \\ x^1 \\ x^2 \\ x^3 \end{pmatrix}. \tag{6.12}$$

The top 2×2 block of the matrix reproduces (6.1) and (6.2), but is much easier to remember because all the annoying factors of c have been absorbed into the various symbols γ, β, x^0, and x^1. The rest of the matrix comes from the fact that a Lorentz transformation along the x-axis does not touch the y or z coordinates. By the way, a Lorentz transformation is often called a *boost*. It is unlikely that you will see a problem on the GRE that will ask you to simply plug in numbers to the Lorentz transformation, but we have encountered GRE questions that required identifying the form of the Lorentz transformation. The generalization to boosts along the other coordinate axes is straightforward, and you'll see an example in the problems at the end of this chapter.

In fact, there are several other quantities whose Lorentz transformation properties use that exact same matrix. In addition to position $x^\mu = (ct, x, y, z)$, which we have already discussed, the other useful 4-vectors are

$$\text{Energy–momentum: } p^\mu = (E/c, \mathbf{p}), \tag{6.13}$$

$$\text{Current density: } j^\mu = (c\rho, \mathbf{J}), \tag{6.14}$$

$$\text{Wavevector: } k^\mu = (\omega/c, \mathbf{k}). \tag{6.15}$$

The first one needs some clarification: when we write $\mathbf{p}$, we mean the *relativistic* momentum, which differs from the usual definition of momentum by a crucial factor of γ:

$$\mathbf{p} = \gamma m \mathbf{v} = \frac{m\mathbf{v}}{\sqrt{1 - |\mathbf{v}|^2/c^2}}. \tag{6.16}$$

With this caveat, all of the 4-vectors listed above satisfy the matrix equation (6.12) for a boost along the x-axis, with the components x^0, x^1, etc. replaced by the appropriate components of the 4-vector. Note that not every random collection of four objects satisfies this property, just like not every collection of three objects transform correctly under rotations, as would be true for an ordinary vector. But for the purposes of the GRE, you don't need to know where these 4-vectors come from: in fact, it's probably sufficient just to memorize the energy–momentum 4-vector, as it is by far the most common.

Speaking of the energy–momentum 4-vector, we should mention a couple of key properties you're probably already familiar with, but are very important for the GRE. Consider a particle of mass m. In its rest frame, its velocity is zero, so $p^\mu = (E_0/c, 0, 0, 0)$. It turns out that the *rest energy* E_0 is

$$E_0 = mc^2, \tag{6.17}$$

an equation so famous it barely even needs explaining. In another inertial frame, the zeroth component of p^μ will still contain a contribution from the rest energy, and we define the remainder as the kinetic energy:

$$T = E - mc^2. \tag{6.18}$$

In fact, plugging p^μ into (6.12) we see that the energy in a frame other than the rest frame is given by

$$E = \gamma mc^2, \tag{6.19}$$

so

$$T = (\gamma - 1)mc^2. \tag{6.20}$$

It's an excellent exercise to Taylor expand this last equation for $v \ll c$ and see that we recover the correct nonrelativistic expression for the kinetic energy T.

6.2.2 Relativistic Dot Product

With this covariant notation it is easy to write down an extremely important quantity, the 4-vector product, or *relativistic dot product*. The dot product of two 4-vectors is defined to be[1]

$$a \cdot b \equiv a^0b^0 - a^1b^1 - a^2b^2 - a^3b^3. \tag{6.21}$$

At this point many relativity texts go into enormous detail about covariant versus contravariant indices, the metric tensor, Einstein summation convention, and so forth. Forget about all that: (6.21) is all you ever have to remember for the GRE about the relativistic dot product. Its key property is that it is *invariant under Lorentz transformations*, in exactly the same way that the ordinary 3-vector dot product is invariant under rotations of the coordinate axes. Practically speaking, this means that you can evaluate the dot product in *any inertial frame* you want: you'll get the same answer no matter which frame you use. The individual components of the 4-vectors will change, but the combination $a \cdot b$ remains the same. This often allows us to work in the reference frame with the simplest physics.

There are two important special cases of this formula, both involving dotting a 4-vector with itself. The first will give us a classification of spacetime events based on the sign of the dot product, and the second is a useful formula relating energy and momentum. The power of both of these results is that, because they use the invariant dot product, they are independent of the reference frame, and hold regardless of which coordinate system we choose for the 4-vectors themselves.

- **Invariant interval.** Given two position 4-vectors x^μ_A and x^μ_B, we can define the *displacement 4-vector*

$$\Delta x^\mu \equiv x^\mu_B - x^\mu_A$$

that represents the spacetime vector between two events, A and B, occurring at x_A and x_B respectively. The reason it's easier to work with displacement rather than position is that relative positions, rather than absolute positions, are

[1] **Warning!** Some texts define the dot product with an extra overall minus sign, so be careful! This convention is fairly standard, but be prepared to be flexible about sign conventions depending on where you've learned relativity previously.

actually meaningful. Now, dotting Δx^μ with itself gives the spacetime "distance" between the two events, also known as the *invariant interval*, $(\Delta x)^2$. (The notation p^2 for the relativistic dot product of a 4-vector p^μ with itself is standard, but don't confuse it with the square of a scalar, or the ordinary dot product!) Crucially, this quantity can be positive, negative, or zero. Each of these cases has a name and a corresponding physical interpretation:

$$\text{Timelike: } (\Delta x)^2 > 0, \tag{6.22}$$

$$\text{Spacelike: } (\Delta x)^2 < 0, \tag{6.23}$$

$$\text{Lightlike or null: } (\Delta x)^2 = 0. \tag{6.24}$$

Two events that are timelike-separated are in causal contact: there exists an inertial frame where both events occur at the same *place*. It's useful to imagine this frame as a spaceship that travels between events A and B: in the frame of the spaceship, both events occur at the same spatial point (namely, the origin of the spaceship coordinate system), but at different times (corresponding to how long it takes the spaceship to travel between them).

For events that are spacelike-separated, there exists an inertial frame where both events occur at the same *time*. This gives a precise condition for simultaneity: simultaneous events must be spacelike-separated (though not all spacelike-separated events are simultaneous, since that depends on the frame). Incidentally, these events are *not* in causal contact: if there is a frame such that event A occurs before event B, there exists another frame such that the order is reversed, and event B occurs before event A! Thus the whole notion of causality doesn't make sense for spacelike-separated events. The intuition is that these kinds of events are so "far away" from each other that there is no inertial frame traveling slower than light that can go between them.

Finally, lightlike-separated events correspond to paths of light rays: A and B are lightlike-separated if and only if they lie on a trajectory traveling at the speed of light. The signs and names here are quite tricky, but the best way to remember them is to consider the simplest of all possible displacement 4-vectors: $\Delta x^\mu = (c\Delta t, 0, 0, 0)$, corresponding to sitting in the same place for a time Δt. This 4-vector clearly has $(\Delta x)^2 > 0$, and it only has a time component (hence "timelike"), and furthermore it represents the displacement in the frame where the two events occur at the same place.

- **Energy–momentum formula.** The second important application of the relativistic dot product is to the energy–momentum 4-vector. Consider a particle of mass m. As we noted above, in its rest frame the energy–momentum 4-vector is $p^\mu = (mc, 0, 0, 0)$, which satisfies $p \cdot p = m^2c^2$. On the other hand, plugging its components into a general frame from (6.13) into (6.21), we find $p \cdot p = E^2/c^2 - \mathbf{p}^2$. Setting these expressions equal and rearranging, we find the very useful formula

$$E^2 = \mathbf{p}^2c^2 + m^2c^4. \tag{6.25}$$

This lets us determine a particle's energy given its momentum without ever having to deal with its velocity, which can save a lot of time in calculations.

6.3 Relativistic Kinematics

One of the main applications of relativity is to kinematics problems: systems of particles that decay and collide, but whose speeds are large and so must be treated relativistically. It's important to remember that whenever relativity is involved, we *must* use the relativistic energy $E = \gamma mc^2$ and relativistic momentum $\mathbf{p} = \gamma m\mathbf{v}$: forgetting the factors of γ will likely lead to trap answers. But apart from that, the setup of the problems should be very familiar from classical mechanics. Despite the importance of these kinds of problems in the physics curriculum, they are fairly underrepresented on the GRE, at least based on our experience and the exams that have been released so far. A careful treatment with many examples can be found in both Griffiths's book on electrodynamics as well as his text on elementary particles.

6.3.1 Conserved vs. Invariant

The reason for introducing all the previous definitions is that the quantities we've defined satisfy certain properties that make calculations easier. As you undoubtedly remember from classical mechanics, momentum is conserved in the absence of external forces. For any relativity question you'll see on the GRE, we can drop that caveat about external forces: *relativistic energy–momentum is conserved.* Note that because the Lorentz transformations mix up space and time components, asking for the relativistic momentum to be conserved *implies* that the whole energy–momentum 4-vector is conserved, which includes the relativistic energy as its first component. We can write the conservation as a 4-vector equation:

$$\sum_i p_i^\mu = \sum_f p_f^\mu, \tag{6.26}$$

where p_i are the incoming 4-vectors and p_f are the outgoing 4-vectors. As with ordinary vector equations, this means that each component of the total 4-momentum must match before and after the collision.

We've also introduced the relativistic dot product, which is invariant under Lorentz transformations. The GRE loves to test the subtleties of these two definitions, so let's be totally clear:

Conserved = same before and after. **Invariant** = same in every reference frame.

A very common question will give a list of quantities and ask whether they are conserved, invariant, both, or neither. For example, the total momentum of a system is conserved, but is not invariant, because it can be transformed to zero by going to the center-of-momentum frame. The kinetic energy of a system is neither conserved (since it's not the whole relativistic energy, but only a part of it) nor invariant (because it can be changed by transforming to another frame). An additional example can be found in the problems at the end of the chapter.

6.3.2 Exploiting the Invariant Dot Product

A standard trick in kinematics problems is to exploit two key properties of the relativistic dot product:

- It takes the same value in any reference frame.
- The square of a particle's energy–momentum 4-vector is equal to its mass squared with a factor of c^2: $p^2 = m^2c^2$. (You can put the c's in the right place by remembering that the whole energy–momentum 4-vector has units of momentum.)

This trick, suitably applied, will (almost) always let you calculate energies and momenta without ever having to compute a Lorentz factor γ or a velocity β. The idea is to choose a

EXAMPLE 6.1

Suppose we have a particle of mass M at rest, decaying to two particles of masses m_2 and m_3. What is the energy of m_2?

Let p_1 be the 4-vector of M, and p_2, p_3 be the 4-vectors of m_2 and m_3. By conservation of momentum, $p_1 = p_2 + p_3$, but for reasons we'll see in a moment, we actually want to write this as

$$p_1 - p_2 = p_3.$$

Now we square both sides using the relativistic dot product:

$$p_1^2 + p_2^2 - 2p_1 \cdot p_2 = p_3^2.$$

Note that the usual algebraic rules for squaring a sum apply to the relativistic dot product as well. Now, since M is at rest, $p_1 = (Mc, 0, 0, 0)$. We don't know p_2 yet, but we can always write it as $(E_2/c, \mathbf{p}_2)$. Note that $p_1 \cdot p_2$ exactly isolates E_2:

$$p_1 \cdot p_2 = (Mc)(E_2/c) - 0 = ME_2,$$

which is what we're looking for! We don't know p_3 either, but by the second property of the dot product, $p_3^2 = m_3^2c^2$. Making these replacements, we have

$$M^2c^2 + m_2^2c^2 - 2ME_2 = m_3^2c^2,$$

and solving for E_2 gives

$$E_2 = \frac{(M^2 + m_2^2 - m_3^2)c^2}{2M}.$$

The key step here was to move p_2 to the other side so as to isolate E_2 in the dot product: if we wanted the energy of m_3, we would have done the same with p_3. Just squaring the conservation equation $p_1 = p_2 + p_3$ directly would not have helped, since it would have involved the dot product $p_2 \cdot p_3$ of two 4-vectors we know nothing about. At most, you'll see one of these types of problems on the GRE, but it's still an important trick which can save you precious minutes compared to calculating Lorentz factors directly.

EXAMPLE 6.2

Suppose that a galaxy is moving toward us at some substantial fraction of the speed of light, and emitting red light: what wavelength of light do we receive? The motion towards us means that the light will be blueshifted, so the wavelength will decrease, and we should flip the sign of β so that the numerator is smaller than the denominator:

$$\lambda_{\text{rec}} = \sqrt{\frac{1-\beta}{1+\beta}}\lambda_{\text{emit}}.$$

Similarly, from $\lambda f = c$, we can find the change in f by taking the reciprocal:

$$f_{\text{rec}} = \sqrt{\frac{1+\beta}{1-\beta}}f_{\text{emit}}.$$

frame, and a combination of 4-vectors, such that the square has as many zeros as possible and so is easier to calculate. Example 6.1 shows how this technique is used.

6.4 Miscellaneous Relativity Topics

Here are a couple of odd topics not covered by the previous discussion, but which have appeared frequently on the GRE.

6.4.1 Relativistic Doppler Shift

We've already covered the Doppler shift in Chapter 3, but the relevant formulas change slightly when we include the effects of relativity. Recall that the formula for the Doppler shift depended on both the velocity of the source and the velocity of the emitter. But according to special relativity, if we're asking about the Doppler shift of *light*, there are no privileged reference frames, this distinction is meaningless, and the shift can only depend on the *relative* velocity between the source and the observer. If this relative velocity is $v = \beta c$, then the change in wavelength of the emitted light is

$$\frac{\lambda'}{\lambda} = \sqrt{\frac{1+\beta}{1-\beta}}. \tag{6.27}$$

This simple-looking equation is actually quite tricky to derive, so we recommend simply memorizing it. The signs, as well as the corresponding formula for the frequency shift, can be deduced from context and some physical intuition. See Example 6.2.

As with velocity addition, the Doppler shift formula only applies to collinear motion, where the source and emitter move along the same line. There also exists a transverse Doppler effect, but we are not aware of it ever having shown up on the GRE thus far.

6.4.2 Pythagorean Triples

You'll likely be expected to do some number-crunching in the relativity questions, either by calculating length contractions, Doppler shifts as above, or energies of particles in collisions. These all involve the ubiquitous factor $\gamma = 1/\sqrt{1-\beta^2}$, and taking square roots by hand is annoying. Luckily, since the GRE is made to minimize calculations, the presence of the square root *tells* you exactly which numbers to expect: Pythagorean triples! Only certain values of β make the square root easy to compute, and they're the ones for which 1 and β form two parts of a Pythagorean triple where 1 is the hypotenuse. An extremely common example is $\beta = 0.6$, which belongs to the triple $(0.6, 0.8, 1)$, better known as $(3, 4, 5)$. Since things will usually be given in terms of decimals rather than fractions, we're looking for triples with nice denominators, so $(0.28, 0.96, 1)$ (derived from $(7, 24, 25)$) is probably more common than one derived from $(5, 12, 13)$. In any case, it may be helpful to spend just a few minutes reminding yourself of the small Pythagorean triples, just to save a few minutes on arithmetic. For the most common triple, $(0.6, 0.8, 1)$, we have

$$\beta = 0.6 \implies \gamma = 1.25, \tag{6.28}$$

$$\beta = 0.8 \implies \gamma = 5/3. \tag{6.29}$$

6.5 Relativity: What to Memorize

There was a good deal of information presented in this chapter, but since relativity only makes up 6% of the exam, it's important not to go overboard memorizing equations. We recommend memorizing *only* the following, which are simple to state but too time-consuming to derive on the spot in the exam:

- Definitions:

$$\beta = v/c \tag{6.11}$$

$$\gamma = \frac{1}{\sqrt{1 - v^2/c^2}} = \frac{1}{\sqrt{1-\beta^2}} \tag{6.3}$$

$$x^\mu = (ct, x, y, z) \tag{6.10}$$

$$p^\mu = (E/c, \mathbf{p}) \tag{6.13}$$

$$\mathbf{p} = \gamma m \mathbf{v} \tag{6.16}$$

- Lorentz transformation matrix for boost along the x-axis:

$$\begin{pmatrix} x^{0'} \\ x^{1'} \\ x^{2'} \\ x^{3'} \end{pmatrix} = \begin{pmatrix} \gamma & -\gamma\beta & 0 & 0 \\ -\gamma\beta & \gamma & 0 & 0 \\ 0 & 0 & 1 & 0 \\ 0 & 0 & 0 & 1 \end{pmatrix} \begin{pmatrix} x^0 \\ x^1 \\ x^2 \\ x^3 \end{pmatrix} \tag{6.12}$$

- Addition of velocities:

$$w = \frac{v+u}{1+vu/c^2} \tag{6.8}$$

- Relativistic Doppler shift:

$$\frac{\lambda'}{\lambda} = \sqrt{\frac{1+\beta}{1-\beta}} \tag{6.27}$$

- 4-vector dot product:

$$a \cdot b \equiv a^0 b^0 - a^1 b^1 - a^2 b^2 - a^3 b^3 \tag{6.21}$$

- Rest energy of a particle of mass m:

$$E_0 = mc^2 \tag{6.17}$$

Everything else can be derived very quickly from these. In particular, time dilation and length contraction can be derived from (6.12) after specializing to the position 4-vector, the invariant interval and the energy–momentum relationship $E^2 = \mathbf{p}^2c^2 + m^2c^4$ can be derived from (6.21), and so forth. If you feel comfortable with it, an excellent additional simplification is just to set $c = 1$ in all the formulas in this chapter. These are the units typical for particle physics, and you can always restore the factors of c by dimensional analysis.

6.6 Problems: Special Relativity

1. System $\overline{S}$ travels with constant velocity $v \neq 0$ in the $\hat{\mathbf{x}}$-direction with respect to system S. If two events, separated by a distance $x \neq 0$, occur simultaneously at time t in S, do they occur simultaneously in $\overline{S}$?

   (A) Yes, always
   (B) No, never
   (C) Only if $x < vt$
   (D) Only if $x > vt$
   (E) Only if $x < ct$

2. System B travels with respect to system A at constant velocity $\mathbf{v} = \beta c\hat{\mathbf{z}}$. Assuming the origins of both coordinate systems coincide, which of the following represents the Lorentz transformation matrix from the coordinates (ct', x', y', z') of system B to the coordinates (ct, x, y, z) of system A? ($\gamma = 1/\sqrt{1-\beta^2}$.)

   (A) $\begin{pmatrix} \gamma & -\gamma\beta & 0 & 0 \\ -\gamma\beta & \gamma & 0 & 0 \\ 0 & 0 & 1 & 0 \\ 0 & 0 & 0 & 1 \end{pmatrix}$

   (B) $\begin{pmatrix} \gamma & \gamma\beta & 0 & 0 \\ \gamma\beta & \gamma & 0 & 0 \\ 0 & 0 & 1 & 0 \\ 0 & 0 & 0 & 1 \end{pmatrix}$

   (C) $\begin{pmatrix} \gamma & 0 & 0 & -\gamma\beta \\ 0 & 1 & 0 & 0 \\ 0 & 0 & 1 & 0 \\ -\gamma\beta & 0 & 0 & \gamma \end{pmatrix}$

   (D) $\begin{pmatrix} \gamma & 0 & 0 & \gamma\beta \\ 0 & 1 & 0 & 0 \\ 0 & 0 & 1 & 0 \\ \gamma\beta & 0 & 0 & \gamma \end{pmatrix}$

   (E) $\begin{pmatrix} 1 & 0 & 0 & 0 \\ 0 & \gamma & 0 & \gamma\beta \\ 0 & 0 & 1 & 0 \\ 0 & \gamma\beta & 0 & \gamma \end{pmatrix}$

3. A particle of mass M and energy E decays into three identical particles of equal energy. What is the magnitude of the momentum of one of the decay products of mass m?

   (A) $\dfrac{E}{3c}$
   (B) $\dfrac{1}{3}Mc$
   (C) $\sqrt{\dfrac{E^2}{9c^2} + m^2c^2}$
   (D) $\sqrt{\dfrac{E^2}{9c^2} + M^2c^2}$
   (E) $\sqrt{\dfrac{E^2}{9c^2} - m^2c^2}$

4. An explosion occurs at the spacetime point $(ct, \mathbf{x})$ in one frame, and at $(ct', \mathbf{x}')$ in another frame related by a Lorentz transformation. If $(ct)^2 > |\mathbf{x}|^2$, we can conclude:

   (A) There exists a frame where $t' = 0$.
   (B) There exists a frame where $\mathbf{x}' = 0$.
   (C) There exists a frame where $(ct')^2 = |\mathbf{x}'|^2$.

(D) There exists a frame where $(ct')^2 < |\mathbf{x}'|^2$.
(E) None of the above.

5. The classical cyclotron frequency of an electron in a uniform magnetic field is ω_0. What is the cyclotron frequency of an electron of velocity v, as measured by a stationary observer?

(A) ω_0
(B) $\omega_0\sqrt{1 - v^2/c^2}$
(C) $\omega_0(1 - v^2/c^2)$
(D) $\dfrac{\omega_0}{\sqrt{1 - v^2/c^2}}$
(E) $\dfrac{\omega_0}{1 - v^2/c^2}$

6. A massive particle has energy E and relativistic momentum $\mathbf{p}$. Which of the following is true of the quantity $E^2 - \mathbf{p}^2c^2$?

I. It is conserved in elastic collisions.
II. It is invariant under Lorentz transformations.
III. It is equal to zero.

(A) I only
(B) II only
(C) I and II
(D) II and III
(E) I, II, and III

7. The USS *Enterprise*, moving at speed $0.5c$ with respect to a nearby planet, fires a photon torpedo of speed c at a Romulan warship, initially 6000 km away, which is retreating away from the *Enterprise* at constant velocity. According to the *Enterprise's* clock, the torpedo made contact with the warship 0.1 seconds after firing. How fast was the warship traveling, in the frame of the planet?

(A) $\frac{13}{28}c$
(B) $\frac{13}{16}c$
(C) $\frac{13}{14}c$
(D) c
(E) $\frac{13}{10}c$

8. A space-car speeds towards an intergalactic traffic light. The traffic light is red, emitting light of wavelength 750 nm, but the driver sees it as green, at wavelength 500 nm. How fast was the car traveling?

(A) $\dfrac{1}{5}c$
(B) $\dfrac{2}{7}c$
(C) $\dfrac{5}{13}c$
(D) c
(E) The given wavelengths are not consistent with any speed.

9. Spaceship 1, carrying a meter stick, flies past Spaceship 2, carrying a 1 liter container. The occupants of Spaceship 2 measure the meter stick on Spaceship 1 to be 0.5 m long. What volume do the occupants of Spaceship 1 measure for the container on Spaceship 2? Both spaceships travel along parallel trajectories and all dimensions should be measured parallel to the axis of their trajectories.

(A) 0.125 L
(B) 0.25 L
(C) 0.5 L
(D) 1 L
(E) 2 L

10. An 8 kg mass is traveling at 30 m/s. What is the approximate difference between its classical kinetic energy and its relativistic kinetic energy?

(A) 27 pJ
(B) 27 nJ
(C) 27 μJ
(D) 27 mJ
(E) 27 J

6.7 Solutions: Special Relativity

1. B – Following the same method as in the discussion of simultaneity, we arrive at the same equation using the Lorentz transformations:
$$t'_A = \gamma\left(t - \frac{x_A v}{c^2}\right), \quad t'_B = \gamma\left(t - \frac{x_B v}{c^2}\right).$$
These are only equal if $v = 0$ or if $x_B = x_A$, both of which are excluded by the problem statement.
2. D – Since the boost is along the z-axis, we want components of the matrix that mix up the x^0 and x^3 components, and those are the corners of the matrix, as in choices C and D. We're asked for the transformation *from B to A*, so we want the *inverse* Lorentz transformations, which don't have the minus signs. Choice D has the correct signs for the inverse transformations.
3. E – Each final-state particle has the same energy, $E/3$. We now apply the energy–momentum relation (6.25):
$$(E/3)^2 = \mathbf{p}^2c^2 + m^2c^4$$
$$\implies |\mathbf{p}| = \sqrt{\frac{E^2}{9c^2} - m^2c^2},$$
which is choice E. We could also have arrived at this answer purely by logical reasoning: for a given energy E,

as the mass m of the decay products increases, eventually there will not be enough available energy to produce them. Choice E is the only one that displays this behavior: indeed, $|\mathbf{p}|$ goes imaginary when $E < 3mc^2$.

4. B – The given information is equivalent to saying that the displacement vector between $(ct, \mathbf{x})$ and the origin $(0, \mathbf{0})$ is timelike. Thus, by our discussion of the invariant interval, there is a frame where the two events occur at the same place. Since the origins coincide for systems related by Lorentz transformations, this place is $\mathbf{x} = 0$. A is characteristic of a spacelike event, and C and D contradict the given information because the invariant interval never changes sign. Notice how the phrasing of this question doesn't commit itself to a particular choice of sign convention for the invariant interval: this is typical of GRE questions on this topic.
5. B – There are at least two valid solution methods. The first is to apply time dilation: in the electron frame, an interval of time $\Delta t'$ is related to frequency by $\omega_o \propto 1/\Delta t'$. The electron's clock runs slow when it is moving, so the interval Δt measured by a stationary observer is longer by a factor of γ. Again, using the inverse relation of frequency and time, we have

$$\Delta t = \gamma \Delta t',$$
$$\frac{1}{\Delta t} = \frac{1}{\gamma}\frac{1}{\Delta t'}$$
$$\implies \omega = \sqrt{1 - v^2/c^2}\,\omega_0.$$

Stated more simply, the stationary observer's time is dilated by γ, so the frequency observed is reduced by γ. Another method is to recall that the formula for cyclotron motion, $p = qBR$, holds relativistically as long as p is interpreted as the relativistic momentum γmv. The classical cyclotron frequency is $\omega_0 = qB/m$, and doing the algebra shows that the factor of γ ends up in the same place, $\omega = qB/(m\gamma)$.
6. C – By the energy–momentum relation (6.25), the given quantity is equal to m^2c^4, which is conserved in elastic collisions where the outgoing particles are the same as the ingoing particles (since the particle's mass doesn't change). It is also invariant, either because the mass of a particle doesn't depend on its reference frame, or because it is equal to p^2, the square of the energy–momentum 4-vector. It is never identically zero unless the particle is massless, but this case is excluded by the problem statement.
7. C – This problem involves the addition of velocities formula with a small twist. For our setting, the addition of velocities formula is

$$s = \frac{u + v}{1 + \frac{uv}{c^2}},$$

where u is the speed of the warship in the *Enterprise* frame, v is the speed of the *Enterprise* in the planet frame, and s is the speed of the warship in the planet frame. We are given v in the problem, and we are solving for s. To determine u, we divide the distance Δx traveled by the warship in the *Enterprise* frame while the photon torpedo is in transit by the time Δt taken for the photon torpedo to contact the warship in the *Enterprise* frame. This gives

$$u = \frac{\Delta x}{\Delta t}.$$

On the other hand, we know that, since the torpedo travels at c, we must have

$$\Delta t = \frac{\Delta x + x_0}{c},$$

where x_0 is the distance between the *Enterprise* and the warship when the torpedo is fired. This implies that

$$\Delta x = c\Delta t - x_0,$$

and therefore that

$$u = \frac{c\Delta t - x_0}{\Delta t} = c - \frac{6 \times 10^6 \text{ m}}{0.1 \text{ s}} = 2.4 \times 10^8 \text{ m/s} = 0.8c,$$

with $c = 3 \times 10^8$ m/s. Plugging this result into our expression for s above, we find

$$s = \frac{0.5c + 0.8c}{1 + (0.8c)(0.5c)/c^2} = \frac{1.3}{1.4}c = \frac{13}{14}c,$$

which is C.
8. C – A straightforward application of the relativistic Doppler shift formula gives

$$\frac{750 \text{ nm}}{500 \text{ nm}} = \sqrt{\frac{1+\beta}{1-\beta}}$$
$$\implies \beta = \frac{5}{13},$$

so the car's speed is $\frac{5}{13}c$. Note that the wavelength decreases (the light is blueshifted) because you are traveling towards the source, but the problem was kind enough to give you this fact for free.
9. C – The Lorentz contraction factor between Spaceship 2 and Spaceship 1 is $\gamma = 2$, and so each ship will measure the other's length *in the direction of motion* by a factor of $1/\gamma = 0.5$. But as we can see from the Lorentz transformation equations, there is no change to the coordinates in the perpendicular directions, so volumes are only contracted by a factor γ, from the single length contraction in the direction of motion.

10. A – Recall that the total relativistic energy of a particle is given by $E = \gamma mc^2$, so the relativistic kinetic part is $T = \gamma mc^2 - mc^2 = (\gamma - 1)mc^2$. If we Taylor expand the $(\gamma - 1)$ factor, the leading term is the classical kinetic energy, and the subsequent terms are the higher relativistic corrections:

$$
\begin{aligned}
T &= \left(\left(1 - \frac{v^2}{c^2}\right)^{-1/2} - 1\right) mc^2 \\
&= \left(1 + \frac{v^2}{2c^2} + \frac{(-1/2)(-3/2)}{2!}\frac{v^4}{c^4} + \cdots - 1\right) mc^2 \\
&= \frac{1}{2}mv^2 + \frac{3}{8}\frac{mv^4}{c^2} + \cdots .
\end{aligned}
$$

Plugging in the numbers,

$$\frac{3}{8}\frac{mv^4}{c^2} = 27 \text{ pJ},$$

which is choice A. Even without doing an exact Taylor expansion, we could have reasoned as follows: since the velocity is small, the difference is likely to be *extremely* small, which means it is suppressed by powers of c. The only quantities with units of energy are mv^3/c and mv^4/c^2, which correspond roughly to choices C and A, respectively. Odd powers of c are rare in quantities involving energy, so we might make an educated guess towards choice A.

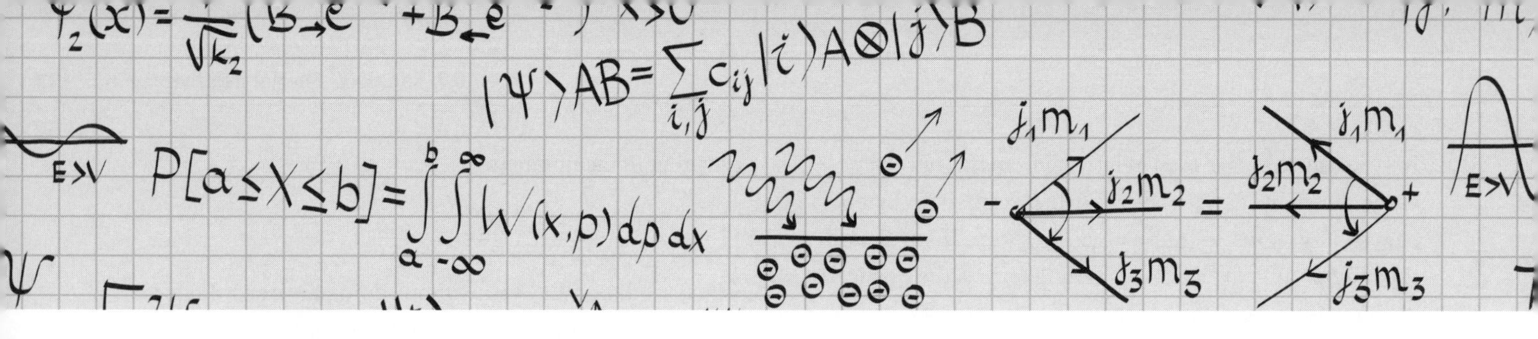

7 Laboratory Methods

The Laboratory Methods section of the GRE is an odd duck. Some questions (such as graph reading or basic statistics) cover things you learned in middle school, while others (such as lasers or radiation detection) deal with things you'll never see until a lab class, or even (if you're a theorist like one of the authors) until your second year of graduate school. The purpose of this chapter is to remedy that problem and briefly review *all* the material you may see on a GRE. Keep in mind that Laboratory Methods questions only make up 6% of the GRE, which means that it's *not* worth memorizing every type of laser medium in detail only to get one question right on your exam. Use this chapter as a reference to shore up any knowledge you may be missing, but by all means don't spend too much time on it.

7.1 Graph Reading

We won't insult your intelligence by telling you how to read a graph. But here are some less common features to watch out for.

7.1.1 Dimensional Analysis

Problem solving with dimensional analysis is mostly discussed in Chapter 9. Here we just mention one rather obvious application to graph reading, because it showed up on a recently released GRE:

- **Read the axis labels.** In particular, note if they carry dimensions. One question on a recent GRE asked for the expression of a slope of a line in terms of some fundamental constants, and the question could be solved entirely by finding the dimensions of the ratio y/x by looking at the respective axes.

7.1.2 Log Plots

Linear plots are useful for displaying data obeying linear relations, but for data obeying a power law or an exponential relation, a log plot is most useful. In this kind of graph, equal intervals on the x- or y-axes correspond to constant *multiples*, rather than constant differences. Or in terms of logarithms, equal intervals correspond to constant differences in $\log_{10} x$ or $\log_{10} y$.[1] For example, equally spaced ticks on the y-axis may represent 1, 10, 100, 1000, etc. Most likely, if you encounter this kind of plot on the GRE, you will see a log–log plot, where *both* axes are divided logarithmically as described above. Occasionally, you may see a log–linear plot, where one axis has a linear scale and the other has a log scale. In either case, examining the scale of the graph will tell you which you're dealing with.

Here are a few facts that come in handy when working with log plots:

- Log–log plots *never* show zero on the axes. This is because if $x = 0$ or $y = 0$, $\log x$ or $\log y$ is $-\infty$. Instead, the graph is simply cut off at some point.
- A straight line on a log–log plot corresponds to a power-law, $y = ax^b$. The slope is b and the constant a can be determined by finding the y value corresponding to $x = 1$.
- A straight line on a log plot, where the y-axis is logarithmic and the x-axis is linear, corresponds to an exponential growth law, $y = C \cdot 10^{bx}$. C is the y-intercept, and the slope

[1] This is one of the very few times in your scientific life you will use base-10 logs rather than base-e.

is b. If the exponent is some other base than 10, for example $y = C \cdot 3^{bx}$, you can take logarithms of both sides to find $\log_{10} y = \log_{10} C + bx \log_{10} 3$, from which we identify the slope as $b \log_{10} 3$ (the y-intercept is still C).

- Similarly, a straight line on a log–linear plot (the opposite of a log plot, logarithmic x-axis and linear y-axis) corresponds to logarithmic growth, $y = C \log_{10}(bx)$. This time the situation is reversed: the x-intercept is b and the slope is C. If the function you're plotting involves the natural logarithm, $y = C \ln(bx)$, you can convert to base-10 with $\ln(bx) = \log_{10}(bx)/\log_{10} e$ to find a slope of $C/\log_{10} e \approx 2.3C$.

7.2 Statistics

You are undoubtedly familiar with the basic statistical concepts of mean, median, and mode; we will not review these here. We will also not review the general theory of statistics, because the only kinds of statistics questions you'll see on the GRE will be applied to particular scenarios. Specifically, you'll see questions dealing with data where measurement error must be quantified, and counting problems where probabilities are given by the Poisson distribution.

7.2.1 Error Analysis

For a *sample* of data points $\{x_1, x_2, \ldots, x_n\}$ taken from a much larger underlying *population*, one can get an estimate of their spread by computing the *sample variance*:

$$\sigma_S^2 = \frac{1}{n-1} \sum_{i=1}^{n} (x_i - \bar{x})^2, \tag{7.1}$$

where $\bar{x}$ is the sample mean, i.e. the average of the sample data points. You may be used to seeing n rather than $n-1$ in the denominator. The difference between the two is clearly only important for small sample sizes, and is meant to correct for the fact that the *true* variance (rather than the sample variance) must be calculated using the *true* mean, instead of the sample mean. This distinction is unlikely to be important on the GRE, but it is easy to remember which to use. If you are estimating the variance of a full population using all members of the population, put n in the denominator. For example, if you want the variance of the height of students in a school, and you have measured the heights of every student, then there is an n in the denominator. If you are using a sample to estimate the variance, then put $n-1$ in the denominator. For example, if you measured the heights of a random sample of 50 out of the 1000 students in the school, then you would use $n-1$.

Measurement errors are typically quoted in terms of the *standard deviation*, which is the square root of the variance. For instance, if a mass measurement is given as 5 ± 2 kg, the mean $\bar{x}$ of the sample is 5 kg, the standard deviation σ is 2 kg, and the variance is 4 kg^2. For a large number of measurements, the distribution of measurement results approaches a Gaussian with mean $\bar{x}$ and standard deviation σ, so the $\pm$ means that the probability of the true value of the measured quantity falling outside the range $\bar{x} \pm \sigma$ is 32%. Turning this around, a measurement error may be computed by some other means, and *used* as if it represents the variance of a Gaussian. There are a couple of standard manipulations you'll be expected to do with measurement error:

- **Propagation of error.** Suppose that you measure a number $X \pm \sigma_X$ and a number $Y \pm \sigma_Y$. What is the uncertainty on, say, the ratio r of X and Y? This is the question answered by *propagation of error*. *Uncorrelated* errors σ add in quadrature. That is, if a measurement is quoted with two separate uncorrelated sources of error, say statistical σ_{stat} and systematic σ_{sys}, the *total* error is

$$\sigma_{\text{tot}} = \sqrt{\sigma_{\text{stat}}^2 + \sigma_{\text{sys}}^2}. \tag{7.2}$$

This simple relation can be generalized quite easily for a variable that is a function of some other variables, $z(x_1, x_2, \ldots, x_n)$. Given errors on the x_i, the error on z is essentially given by the chain rule:

$$\sigma_z^2 = \sum_{i=1}^{n} \left(\frac{\partial z}{\partial x_i} \right)^2 \sigma_{x_i}^2. \tag{7.3}$$

Because the trend of the GRE seems to be to eschew calculus entirely, you probably won't need this formula, but we include it here for completeness. A few specific instances of this formula for combining multiple sources of uncertainty tend to arise frequently and are easy to remember. If $A \pm \sigma_A$ and $B \pm \sigma_B$ are two measurements with *uncorrelated* errors and a is a constant factor with no uncertainty (e.g. the number π), then we have the following combinations and associated uncertainties:

$$\begin{aligned} f &= aA, & \sigma_f &= a\sigma_A, \\ f &= A \pm B, & \sigma_f &= \sqrt{\sigma_A^2 + \sigma_B^2}, \\ f &= AB, & \frac{\sigma_f}{f} &= \sqrt{\left(\frac{\sigma_A}{A}\right)^2 + \left(\frac{\sigma_B}{B}\right)^2}, \\ f &= A/B, & \frac{\sigma_f}{f} &= \sqrt{\left(\frac{\sigma_A}{A}\right)^2 + \left(\frac{\sigma_B}{B}\right)^2}. \end{aligned}$$

Notice the similarity in form of the propagation of error for $f = AB$ and $f = A/B$ to the propagation of error for

$f = A \pm B$; these cases are related by taking logarithms $\ln AB = \ln A + \ln B$ and $\ln A/B = \ln A - \ln B$.

- **Weighted averages.** Suppose the same quantity is measured in two different ways, yielding two different values x and y with two different errors σ_x and σ_y. These measurements can be combined to give a single value X using a weighted average, where the weights are the errors themselves:

$$X = \frac{x/\sigma_x^2 + y/\sigma_y^2}{1/\sigma_x^2 + 1/\sigma_y^2}, \tag{7.4}$$

$$\sigma_{\text{tot}}^2 = \frac{1}{1/\sigma_x^2 + 1/\sigma_y^2}. \tag{7.5}$$

In other words, the data points with smaller errors are weighted more strongly in the weighted average, and the total variance is the harmonic mean of the variances, divided by the sample size. These equations generalize simply to more than two measurements.

- **Uncertainty.** A statement such as "this measurement has an uncertainty of 10%" means that the sample mean $\bar{x}$ and the error σ satisfy $\sigma/\bar{x} = 0.1$.

Finally, we will mention one silly piece of nomenclature that was probably drilled into your head in high school chemistry, but which you've long since forgotten:

- *Precise* measurements have small variance.
- *Accurate* measurements are close to the true value.

You can come up with examples where any combination of these two is true or false: precise but not accurate, accurate (on average) but not precise, and so on.

7.2.2 Poisson Processes

The *Poisson distribution* describes the probability of counting 1, 2, 3, etc. events in a fixed time, when the events occur randomly at a known constant rate. Some classic examples described by the Poisson distribution are the clicks in a Geiger counter measuring radioactive decays, or photons arriving in a telescope. Since the Poisson distribution describes the probability of "counting" a certain number of events, it is often called *counting statistics*. Mathematically,

$$P(n) = \frac{\lambda^n e^{-\lambda}}{n!}. \tag{7.6}$$

Here, λ is the expected (or average) number of counts in a given time interval, and $P(n)$ is the probability of observing exactly n counts in that same time interval. You should probably memorize this formula, but it's really not that hard to remember where all the n's and λ's go by noting that, like any probability distribution, we must have $\sum_{n=0}^{\infty} P(n) = 1$. Indeed, summing over n gives the Taylor series for e^{λ}, which cancels with $e^{-\lambda}$ to give 1 as required. Even if you don't memorize equation (7.6), definitely memorize these important facts:

- **The standard deviation of a Poisson measurement is $\sigma \approx \sqrt{N}$ for large N.** When we observe a Poisson process, we usually are interested in measuring the rate. We might, for example, measure $N = 100$ decays of a radioactive source in 1 second. But then what is the error of the measurement? The Poisson distribution has the useful property that if N events are expected, then the standard deviation of the distribution is $\sigma = \sqrt{N}$ (this approximation is typically safe for $N > 20$). This means that the error on a measurement of N events is just $\sqrt{N}$. In our example above, we would say that the measured rate was 100 ± 10 Hz.
- **$P(0) = e^{-\lambda}$.** This is a measure of how rare the process is: if λ is small, you are relatively likely to observe no events.
- **The time between Poisson events follows an exponential distribution.** More specifically, if one event from a Poisson distribution with mean λ occurs at time $t = 0$, the time t of the next event's arrival is distributed as a function of time according to $P(t) = \lambda e^{-\lambda t}$, the *Poisson waiting time*. Here t is measured in whatever units of time are used to define λ.

7.3 Electronics

The electronics portion of Laboratory Methods takes over where the circuits portion of Electricity and Magnetism left off. Now, instead of just applying a constant voltage to a simple circuit, we are interested in the response of basic circuit elements to a time-varying voltage, the behavior of more advanced circuit elements, and the basics of digital logic.

7.3.1 AC Behavior of Basic Circuit Elements

You're already familiar with the three basic circuit elements (resistors, capacitors, and inductors) from E+M. In the context of electronics, these devices are usually described in terms of their AC behavior; in other words, their response to an alternating current. Roughly, capacitors and inductors can behave as if they carry resistance when hit with an alternating current of various frequencies, and it's convenient to treat all three circuit elements on the same footing. This is done using the concept of *impedance*, Z, a complex number which obeys Ohm's law, $V = IZ$. For an alternating current $V = V_0 e^{i\omega t}$, Z contains information about *both* the magnitude and the phase of the resulting current, allowing considerable calculational simplifications. It's probably useful to remember the following:

$$\text{Capacitor: } Z = \frac{1}{i\omega C}, \tag{7.7}$$

$$\text{Inductor: } Z = i\omega L, \tag{7.8}$$

$$\text{Resistor: } Z = R. \tag{7.9}$$

As always, ω refers to the angular frequency of the supply voltage V.

Looking only at the magnitudes of these quantities, we see that, at high frequencies, capacitors have small impedances – in other words, they tend to cause only small voltage drops, and behave like short circuits. Inductors, on the other hand, have the opposite behavior: at high frequencies, they behave like open circuits, where no current can flow. This makes sense because at high frequencies the capacitor is barely being charged, and easily goes through many tiny charge–discharge cycles without saturating its maximum charge for a given voltage. Inductors are hindered by their self-inductance, which tends to resist large changes in current, so at very high frequencies they don't let any current pass at all. See Example 7.1.

Essentially, impedance is just a clever way for remembering all these arguments and encoding them in a simple mathematical formula so you don't have to reproduce the argument every time. It also tells you how things behave when you add them in series or in parallel: using $V = IZ$, we get

$$\text{Series: } Z_{\text{tot}} = Z_1 + Z_2 + \cdots Z_n, \tag{7.10}$$

$$\text{Parallel: } Z_{\text{tot}}^{-1} = Z_1^{-1} + Z_2^{-1} + \cdots Z_n^{-1}. \tag{7.11}$$

These formulas contain all the usual formulas for resistors, capacitors, and inductors in series, as well as all the information about phase lag in RLC circuits, in one convenient package.

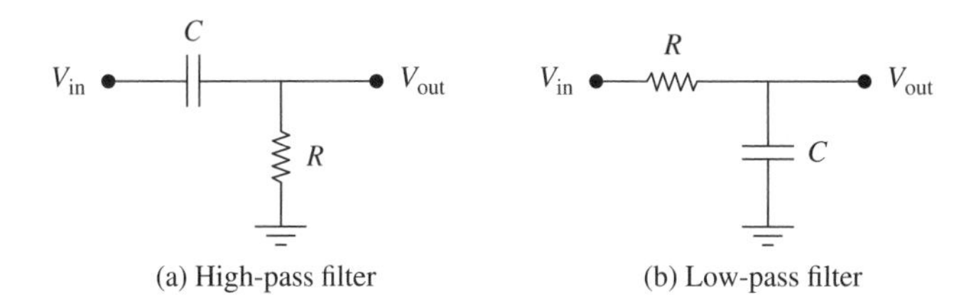

Figure 7.1 Example circuit diagrams for high- and low-pass filters.

For the GRE, the most common application of these impedance formulas will be to high-pass and low-pass filters. Examples of these related circuits are shown in Fig. 7.1.

In both cases, the circuit has distinctive behavior because the capacitor acts like a short circuit at high frequencies. If the resistor is connected to ground as in Fig. 7.1(a), at high frequencies the impedance of the resistor will dominate, and the voltage drop across C will be negligible – in other words, high frequencies will pass but low frequencies are attenuated. On the other hand, if the capacitor is connected to ground, the reverse is true: at high frequencies the capacitor shorts and all the current flows to ground, so V_{out} is near zero. You may find it a helpful mnemonic to remember that in a low-pass filter the capacitor is "low," that is, connected to ground. One could also build RL high-pass and low-pass filters, with the roles of the two circuit elements reversed because of the opposite impedance behavior of capacitors and inductors.

Finally, we should mention how the resonant behavior of LC circuits can be derived using impedance. For a circuit with just one inductor and one capacitor, we have

$$Z_{LC} = \frac{1}{i\omega C} + i\omega L = \frac{-i(1 - \omega^2 LC)}{\omega C}.$$

The numerator vanishes when $\omega = \omega_0 \equiv 1/\sqrt{LC}$, the resonant frequency of an LC circuit. All circuit elements have some small intrinsic resistance, which means that the impedance is never perfectly zero, but frequencies near $1/\sqrt{LC}$ still have small impedance, so this circuit acts as a

EXAMPLE 7.1

One important example of thinking in terms of frequency is when a constant (DC) voltage source V is suddenly switched on at $t = 0$. This situation can be very roughly described by an ultra-high-frequency event ($\omega = \infty$) at $t = 0$, gradually relaxing to low frequencies ($\omega = 0$) at $t = \infty$; this is only a mnemonic, with the precise statement given by taking the Fourier transform, which is unnecessary for the GRE. The above arguments then tell you that the voltage across an uncharged capacitor at $t = 0$ is zero, but increases to V as $t \to \infty$; on the other hand, the voltage across an inductor is V at $t = 0$, but tends to 0 as $t \to \infty$. Similarly, at $t = 0$ there is a large current going through the capacitor, but no current going through the inductor. This is helpful in recognizing the correct graph of V, Q, or I for an RL or RC circuit, a very standard GRE question.

bandpass filter. Indeed, adding a resistor will end up giving a real part to Z, such that Z can never exactly vanish no matter what the frequency: however, resonance is still defined as *the frequency where the imaginary part of the impedance vanishes.*[2] Recall from Section 1.7.2 that a damped oscillator can oscillate at a frequency different from the natural frequency ω_0. The resonant frequency for electrical circuits as defined here is the driving frequency at which the *current* through the circuit will be maximized. This corresponds to the frequency at which a mechanical oscillator will have a kinetic energy resonance; see Chapter 3 of Thornton and Marion for further details.

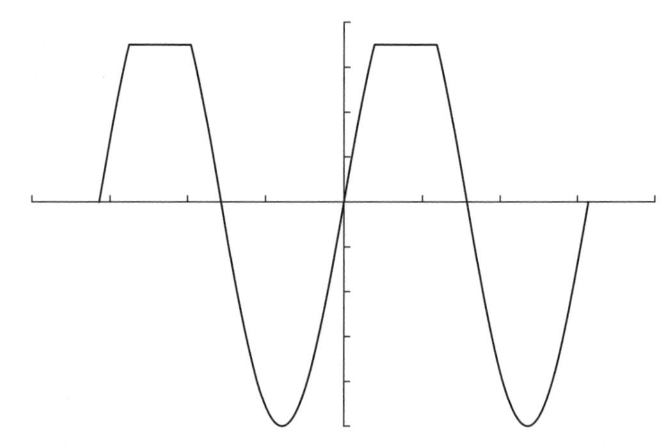

Figure 7.3 Waveform characteristic of clipping.

7.3.2 More Advanced Circuit Elements

The most complicated circuit we can make from just the three basic elements is an RLC circuit, which we already examined in Chapter 2. If we add a few other key circuit elements, whose circuit diagram icons are shown in Fig. 7.2, we get more interesting behavior.

- **Diode.** This device uses properties of semiconductors to ensure that *current can only flow in one direction.* In a circuit diagram, the triangle in Fig. 7.2(a) points in the direction current is allowed to flow. However, no current can flow at all until a minimum *bias voltage* is applied across the diode – typically this is about 0.7 V for a silicon diode. Apart from that bias voltage, the voltage drop across a diode is approximately independent of the current. Uses of diodes include turning an alternating current into a direct current (this is known as a rectifier circuit) and to reroute current away from sensitive electrical components (if the voltage surges, the diode starts conducting, resulting in an almost short circuit if the voltage is high enough).
- **Op-amp.** Short for operational amplifier, this device has two inputs and one output. The output voltage is proportional to the difference between the two input voltages, usually by factors as large as 10,000. However, the op-amp has a maximum possible output voltage, so if the difference between input voltages is too large, the output will *saturate* and distort the signal; that is, there will no longer be a linear relationship between input and output voltage. This is known as *clipping*, and has a very distinctive waveform (as well as a distinctive sound if the signal is audio), shown in Fig. 7.3. Note how the top of the sine wave has been "flattened."

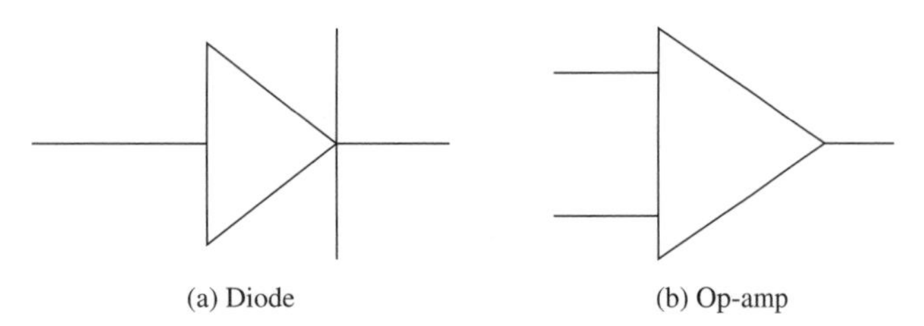

Figure 7.2 Symbols for two common advanced circuit elements: diodes and op-amps.

[2] Depending on whether the circuit elements are in series or parallel, the resonant frequency may actually be a maximum of $|Z|$, rather than a minimum, but the definition of the resonant frequency is still the same.

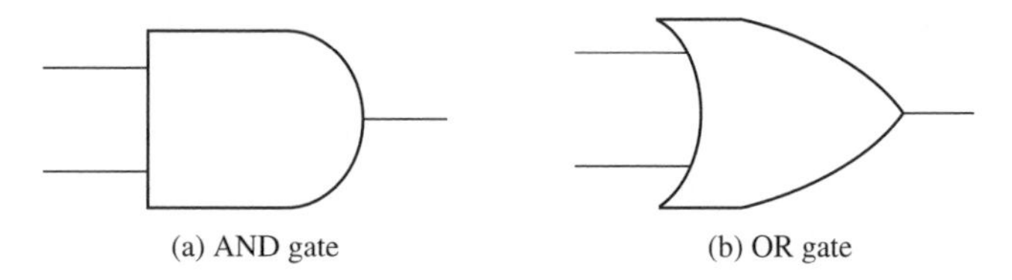

Figure 7.4 Circuit diagram symbols for basic logic gates.

7.3.3 Logic Gates

The basis of modern electronics is digital circuitry, where circuit element output voltages take discrete values rather than continuous ones. A "high" output voltage is interpreted as the digit 1, and a "low" voltage is interpreted as 0, so Boolean logic can be implemented in electronic circuits. The two main logic gates are AND and OR, and their symbols are shown in Fig. 7.4.

For two inputs A and B, AND outputs $A \cdot B$, while OR outputs $A + B$. Here we are using Boolean logic notation (which also shows up on the GRE): note that this is *not* the same as binary arithmetic. Instead, it's easiest to decipher with 0 standing for "false" and 1 standing for "true." So AND returns true only if inputs A *and* B are true, otherwise it returns false. Similarly, OR returns true if inputs A *or* B are true, so only returns false if both A and B are false. These results can be summarized in "truth tables," with the example for the OR gate shown

Table 7.1 Truth table for an OR gate.

A	B	A OR B
0	0	0
0	1	1
1	0	1
1	1	1

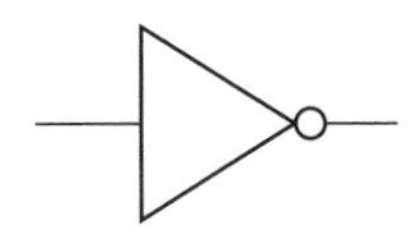

Figure 7.5 Circuit diagram symbol for a NOT gate.

in Table 7.1, but it's often easier to just remember what these gates do by their names.

Both AND and OR gates can be modified by inverting either of the inputs or the output, which is symbolized with a circular "bubble" on the circuit diagram. Alternatively, the circuit element that inverts an input is called a NOT gate, and looks like the symbol for an op-amp but with only one input and a bubble on the output, as shown in Fig. 7.5.

In Boolean logic, inversion is represented with a bar: $\overline{A}$. Inverting the outputs to AND and OR gates gives so-called NAND and NOR: it's a possibly useful fact that *all* basic logic gates can be constructed exclusively from combinations of either NAND and NOR gates.

One final piece of trivia on which you may be tested is De Morgan's laws, which are stated in Boolean algebra as follows:

$$\overline{A \cdot B} = \overline{A} + \overline{B}, \tag{7.12}$$

$$\overline{A + B} = \overline{A} \cdot \overline{B}. \tag{7.13}$$

In other words, a NAND or NOR gate (left-hand side) is just an OR or AND gate with inverted inputs (right-hand side), respectively.

7.4 Radiation Detection and Instrumentation

This section is based largely on the excellent book by Knoll, mentioned in the Resources list at the very beginning of this book. As always, we'll be brief, but feel free to check out that reference if you want more details.

A useful general concept when dealing with subatomic particle interactions is the *cross section* (Fig. 7.6). Imagine that you're shooting a stream of bullets at a bowling ball of radius R. The surface area of the ball is $4\pi R^2$, but the surface that the bullets "see" is the area of the shadow cast by the ball, πR^2. This effective area where collisions can take place is called the cross section, and usually given the symbol σ. Subatomic particles are point particles, so this analogy breaks down in that regime, but we can still associate an effective cross section with a collision event by taking into account the quantum-mechanical probability for the collision to occur. So whenever you see "cross section" in a problem on the GRE, think of "effective collision probability." Occasionally you might be asked to compute a cross section, given other numbers such as luminosity (number of particles per unit time); this will *always* be pure dimensional analysis.

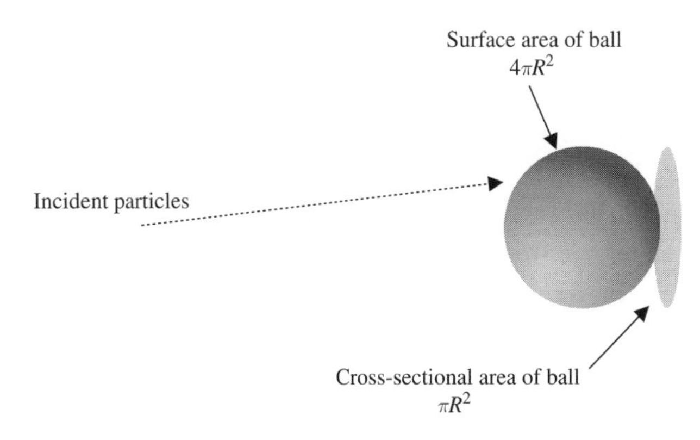

Figure 7.6 The effective scattering cross section of a ball of radius R for incoming particles is the projected area of the ball in the plane perpendicular to the direction of the incoming particles.

7.4.1 Interaction of Charged Particles with Matter

Charged particles come in many different types, but for purposes of the GRE, you really only need to know how electrons and nuclei interact as they pass through bulk matter. Common nuclei could include protons, alpha particles (helium-4 nuclei), or heavier nuclei which are the byproducts of fission reactions. For kinetic energies below the approximate binding energy per nucleon of most elements (a few MeV), both electrons and nuclei overwhelmingly interact with matter by the electromagnetic force. If the interaction occurs with atomic electrons, the electrons can either be excited to higher energy levels (*excitation*) or be stripped from the atom (*ionization*), both of which cause the incident particles to lose energy. The fact that the masses of electrons and nuclei differ by several orders of magnitude results in a variety of differences in the interactions.

- **Range**. Nuclei are stopped faster than electrons: the average path length for an alpha particle is 10^{-5} m, while for a high-energy electron it is 10^{-3} m. Another way of stating this is that dE/dx, the energy loss per unit length, is much higher for nuclei than for electrons (for kinetic energies $\lesssim 1$ GeV).

- **Collision target**. Nuclei interact almost exclusively with atomic electrons; interactions of heavy particles with nuclei are so rare that they can be ignored for most practical purposes, although historically they did play a role in Rutherford's gold foil experiment, which established the existence of the nucleus from scattering by incident alpha particles. Electrons can interact with either atomic electrons or atomic nuclei – the latter interaction is still rare, but can lead to measurable effects in detectors.
- **Path shape**. Nuclei tend to travel in straight lines, because they interact primarily with the much lighter atomic electrons (think of a bowling ball continuously colliding with a sea of ping-pong balls). Electrons tend to bounce around and scatter through wide angles much more often.
- **Energy loss**. Nuclei lose energy exclusively due to collisions, rather than by emitting radiation. Since the mass of an incident nucleus is very different than that of the electrons with which it interacts, nuclei lose only a small amount of energy in each collision. In other words, they are *continuously* losing energy as they interact. On the other hand, when incident electrons undergo collisions with atomic electrons, the target has the same mass as the incident particle, and so elementary kinematics implies that the electron can lose a large fraction of its energy from a single collision. Furthermore, unlike heavy particles, electrons can lose energy through *bremsstrahlung* (literally "braking radiation"), where in the presence of an electric field the electron emits a high-energy photon (usually in the X-ray region), which carries off a large fraction of its energy. This process is rare compared to collisional losses, but occurs more often in materials with high atomic number because the electromagnetic interaction that provokes bremsstrahlung is proportional to nuclear charge. The rates of energy loss for both nuclei and electrons are in general strongly dependent on the initial energy, but at relativistic speeds *all* particles lose a roughly similar amount of energy per unit distance traveled (approximately 1 keV/cm in air). This value corresponds to the minimum of the energy-loss curve for both heavy and light particles, so a relativistic particle is referred to as a *minimum ionizing particle* because it deposits the minimum possible amount of energy per unit distance in the medium.

7.4.2 Photon Interactions

Photons are uncharged, so they don't interact in quite the same way that charged particles do. However, because they mediate the electromagnetic force, they interact in various ways with the other charged particles. Sometimes these underlying interactions produce more charged particles, which then propagate through the detector as described above. Problems involving a qualitative understanding of these interactions appear frequently on the GRE. There are three important types of underlying interaction:

- **Photoabsorption** (or **photoelectric absorption**). The photon is completely absorbed by an atom, and an electron is emitted in its place, with energy $E_\gamma - E_b$, where E_γ is the incident photon energy and E_b is the electron binding energy. This is the dominant process for low-energy photons, up to a few keV. If the photon is absorbed on a sample of bulk material rather than an isolated atom, there are additional surface effects, collectively described by the *work function* of the material, which carries units of energy. The maximum energy that the emitted electrons can have when light is shined on a material is then

$$E_{\text{max}} = E_\gamma - \phi. \tag{7.14}$$

  Problems involving the work functions of various materials occur often enough that this jargon and notation is worth remembering.
- **Compton scattering**. The photon scatters *in*elastically off an atomic electron, and the scattered electron is ejected from the atom. The wider the photon scattering angle, the more energy it loses to the electron. This is the dominant process for medium-energy photons (tens of keV to a few MeV), and sometimes for low-energy photons as well if the absorber has small atomic number. This is probably a good time to bring up the *Compton wavelength* of a particle of mass m,

$$\lambda = \frac{h}{mc}. \tag{7.15}$$

  Unlike the de Broglie wavelength (see Section 5.3.2), the Compton wavelength doesn't depend on the momentum of the particle, but only on its mass. It shows up in the formula for the wavelength shift of light due to Compton scattering:

$$\Delta\lambda = \frac{h}{mc}(1 - \cos\theta). \tag{7.16}$$

  This formula is rather difficult to derive (although it's a good exercise in relativistic kinematics), so should be memorized. You may also be asked for the energy shift of the scattered photon, for which you should use the Einstein relation, $E = hf = hc/\lambda$.

- **Pair production**. If $E_\gamma > 2m_ec^2$, the electric field near a nucleus can induce the photon to produce an electron–positron pair. This is the dominant process for high-energy photons (tens of MeV and above).

Note that photoabsorption is an interaction with the *entire* atom, Compton scattering is an interaction with atomic *electrons*, and pair production is an interaction with the atomic *nucleus*. The probabilities for all three processes are proportional to powers of Z, the atomic number of the absorber, since Z is also the number of atomic electrons available for Compton scattering. More specifically, the probability of pair production is roughly proportional to Z^2, for Compton scattering it is proportional to Z, and for photoabsorption it is roughly proportional to Z^4. The purpose of using high-Z materials such as tungsten is to increase the likelihood these kinds of interactions will occur, and the strong dependence of the photoabsorption probability on Z explains its dominance at low energies.

7.4.3 General Properties of Particle Detectors

By definition, particle detectors are designed to see incoming particles. Once you know a particle is there, the next obvious thing to do is measure its energy; devices that do this are often called *calorimeters*. To measure energy, the detector takes advantage of the natural process of energy loss in the material, and uses the stuff that absorbs the energy (atomic electrons, photoelectrons, produced electron–positron pairs, and so on) to produce a signal. Since for charged particles the number of interactions is usually proportional to the incident particle's energy, simply collecting the produced electrons, counting their charge, and turning that into an electrical current may be enough. Other times, it may be necessary to amplify the signal somehow.

One common case where signal amplification is needed is for photon detection, since only one electron is produced per photon, so we will go through the operation of a generic photon detector in a little more detail because it covers lots of subcomponents that might show up on the GRE. To increase the photon interaction cross section, we want a high-Z material – a common choice is NaI/Tl, sodium iodide doped with thallium, with the iodine providing the high atomic number of $Z = 53$. An incoming photon produces a single electron by one of the three methods mentioned. It happens that NaI/Tl is also a *scintillator*, which means that a passing charged particle will produce visible light, with the intensity of light produced roughly proportional to the electron energy. These visible photons are then directed to a *photomultiplier tube*, which uses a cascade of photoelectric effects to produce a macroscopic current of electrons. These electrons are finally read out by some kind of analyzer, which converts the current to a digital voltage which becomes the raw data. In an ideal world, the photon energy would be directly proportional to the output voltage, and the detector could be calibrated by irradiating it with a photon source of known energy.

7.4.4 Radioactive Decays

A substance that undergoes radioactive decay will have an exponentially decaying number density.

$$N = N_0 e^{-t/\tau}, \tag{7.17}$$

where τ is the mean lifetime. The lifetime is related to the half-life $t_{1/2}$ by

$$t_{1/2} = \tau \ln 2. \tag{7.18}$$

The half-life represents the average amount of time for half the initial particles to remain, while the lifetime represents the average amount of time for a fraction $1/e \approx 0.37$ of the initial particles to remain.

Here is one piece of trivia that has shown up on recent exams. If the substance can decay in several ways (through multiple "decay channels"), then the total lifetime is related to the individual lifetimes τ_1, τ_2, etc. by

$$\frac{1}{\tau} = \frac{1}{\tau_1} + \frac{1}{\tau_2} + \cdots. \tag{7.19}$$

7.5 Lasers and Interferometers

For some reason, questions about names and properties of lasers have become increasingly common on the GRE, despite the fact that the underlying physics of stimulated emission belongs to time-dependent quantum-mechanical perturbation theory and is outside the scope of the test.

7.5.1 Generic Laser Operation

Here's a nontechnical outline of how a generic laser works. Start with a quantum-mechanical system (the *medium*) with at least two energy levels, a ground state and an excited state. The medium could consist of free atoms, organic molecules, or any number of more exotic substances, several of which will be discussed below. Using some external power source

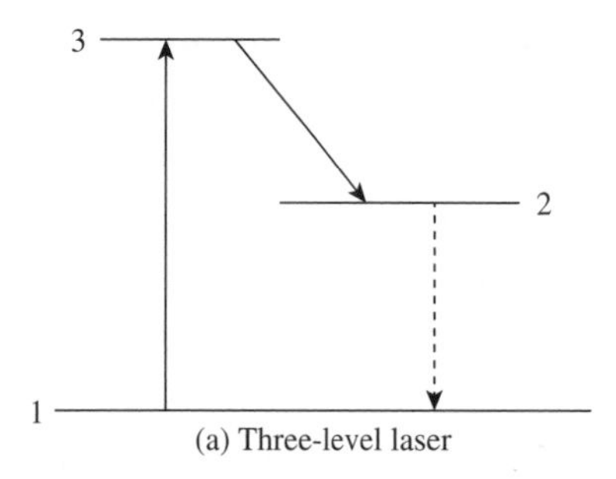

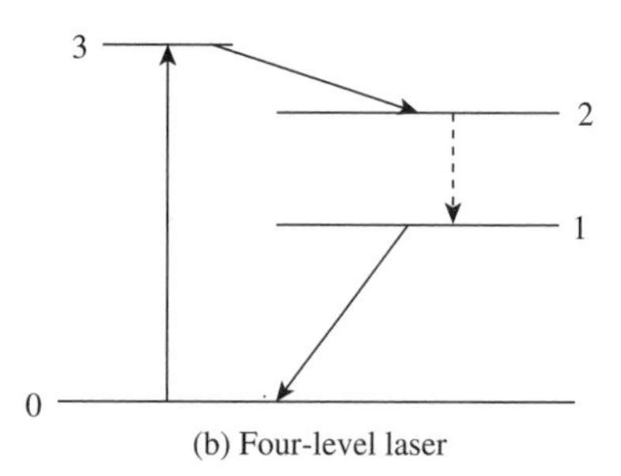

Figure 7.7 Energy levels for three and four-level lasers. The solid arrows represent fast decays, and the dashed arrows represent slow decays which result in laser light.

(an *optical pump*), excite more than half of the medium to the excited state: this can be done with an electrical spark, for example. With a majority of the medium in the excited state, we say that *population inversion* has been achieved. Now, the excited states will tend to decay down to the ground state by *spontaneous emission*, emitting a photon in the process. If this photon is absorbed by a particle in the ground state, it will be excited, and there will be no net change in the system. However, time-dependent perturbation theory shows that the photon can also be absorbed by another excited state, which will be "stimulated" to emit *two* photons and drop to the ground state. This process of photons from decaying excited states being absorbed by other excited states, called *stimulated emission*, starts a chain reaction, the product of which is an exponentially large number of photons, all with exactly the same frequency and phase: this is laser light.

In a real laser, this idealized description must be modified slightly. A careful stat-mech analysis shows that if the system has *only* two levels, it is impossible to achieve population inversion: once the populations of the ground and excited states become equal, the processes of absorption and stimulated emission exactly compensate each other, and there is no chain reaction. Furthermore, the excited state usually decays pretty fast, so we need a *metastable state* between the excited state and the ground state. In Fig. 7.7, level 3 is the excited state reached by pumping, level 2 is the metastable state, and level 1 some other state between level 2 and the ground state. The decay $3 \to 2$ is fast, but the slower decay $2 \to 1$ (dashed arrows) produces the laser light. If level 1 is (or is very close to) the ground state, the system is said to be a *three-level* laser, but if level 1 is significantly above the ground state, we call it a *four-level* laser.

7.5.2 Types of Lasers

Lasers are generally distinguished by their medium and how the transfer between energy levels is achieved. The main examples in the first three categories (indicated in parentheses) do tend to show up as GRE trivia, so you should at least have a passing familiarity with each type.

- **Solid-state lasers (Nd:YAG).** The laser medium is a crystal or glass, and the transitions are between atomic energy levels. In a Nd:YAG laser, the crystal is $Y_2Al_5O_{12}$ (yttrium aluminum garnet, or YAG), with some of the Y ions replaced by Nd. The Nd atomic levels are split by the electric field of the YAG crystal, giving a four-level system.
- **Collisional gas lasers (He–Ne).** The laser medium is a gas or mixture of gases, and the transitions are due to collisions between the atoms: an excited electron from one gas transfers its kinetic energy to excite an electron in another gas molecule. In a He–Ne laser, there are a huge number of possible laser levels, but a specific wavelength can be selected by placing the laser in a resonant cavity, just as one excites certain EM modes using a conducting cavity in ordinary electrodynamics.
- **Molecular gas lasers (CO_2).** The laser medium is again a gas, and the transitions are *vibrational* energy levels. Carbon dioxide is a standard example because it's cheap, widely available, and its triatomic structure gives it a rather rich vibrational spectrum.
- **Dye lasers.** The laser medium is a liquid, usually an organic dye dissolved in water or alcohol. The transitions are related to the electron-transfer properties along chains of carbon atoms which give dyes their characteristic color. Interestingly, the laser does not tend to operate at the wavelength corresponding to the ordinary visible color of the dye, but because the electron transport chain is extremely efficient, laser operation is still possible at other frequencies.
- **Semiconductor or diode lasers.** The laser medium is a semiconductor (discussed in more detail in Section 8.2). Here, the pumping process excites the conduction band, and the transitions are electron–hole annihilation between

electrons in the bottom of the conduction band and holes at the top of the valence band. This gives rise to photons (known as recombination radiation) which form the basis of the laser light.

- **Free electron lasers.** As the name suggests, the laser medium is simply a collection of electrons, not bound to any atom or molecule. When forced to accelerate back and forth in an external electric field, the electrons will emit bremsstrahlung (see Section 7.4) at a frequency depending on their oscillation frequency. There are no discrete energy levels here, so it's a bit of a stretch to call this a laser, although a semiclassical analysis shows that there is amplification.

7.5.3 Interferometers

An interferometer is a device that takes advantage of the wave properties of light to measure distances and velocities very sensitively. Undoubtedly, the most famous interferometer is the Michelson–Morley model, shown in Fig. 7.8, used to disprove the idea of the ether in pre-special relativity days. This is the type that will show up on the GRE if you're asked about interferometers, so we'll confine our attention to this model. Monochromatic light is shined on a half-silvered mirror, which reflects half the light to the mirror marked A, and lets the other half through to a second mirror marked B. The light from both mirrors then bounces back to the half-silvered mirror, which splits the incoming light again. The portion that travels to the detector contains contributions from both paths, which interfere with each other when they reach the detector. If the optical path lengths along the two arms are different, the detector will record a pattern of interference fringes, as discussed in much more detail in Chapter 3. One then *counts* the number of fringes visible on the screen: if this number changes, it means that the optical path length difference between the two arms has changed, either by one of the mirrors moving, a change in the index of refraction along one of the arms, or both. By using the double-slit equation $d \sin\theta = m\lambda$, the number of fringes crossing a certain position on the detector (fixed θ) can be used to measure d given λ, or vice versa.

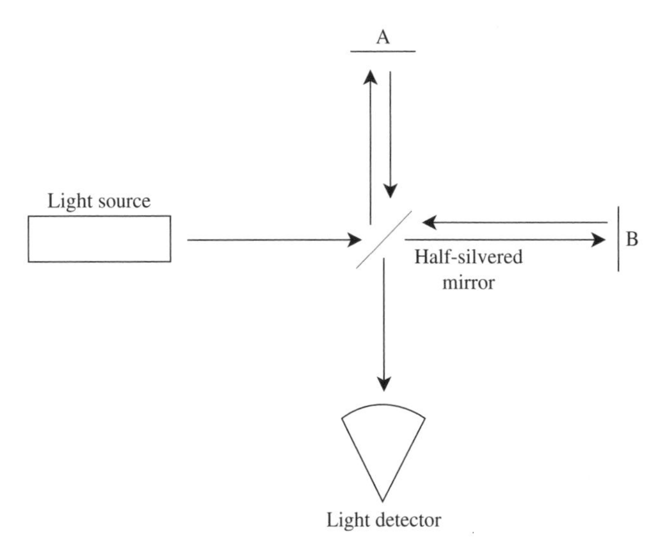

Figure 7.8 Michelson–Morley interferometer.

7.6 Problems: Laboratory Methods

1. In radiation detection, the term "minimum ionizing particle" could refer to

   (A) a photon with energy 10 keV
   (B) a neutron with kinetic energy 1 MeV
   (C) an alpha particle with kinetic energy 5 MeV
   (D) a proton with kinetic energy 10 MeV
   (E) an electron with kinetic energy 50 MeV

2. Event A is drawn from a Gaussian probability distribution with standard deviation σ_A, and event B is drawn from a Gaussian with standard deviation σ_B. If A and B are independent events, the probability distribution for the sum of A and B is a Gaussian with standard deviation

   (A) $\sigma_A + \sigma_B$
   (B) $\sqrt{\sigma_A \sigma_B}$
   (C) $\sqrt{\sigma_A^2 + \sigma_B^2}$
   (D) $\dfrac{1}{1/\sigma_A + 1/\sigma_B}$
   (E) none of these

3. Which of the following probability distributions best describes the probability of obtaining heads 3 times when a fair coin is flipped 10 times?

   (A) Binomial distribution
   (B) Gaussian distribution
   (C) Student's t distribution
   (D) Log-normal distribution
   (E) χ^2 distribution

4. The number N of radioactive atoms of a particular isotope remaining in a sample as a function of time t is found to obey $N(t) = N_0 e^{-\lambda t}$. What is the half-life of the sample in terms of λ?

   (A) $\lambda \ln 2$
   (B) $\dfrac{\lambda}{\ln 2}$

(C) $\frac{\ln 2}{\lambda}$
(D) $\frac{1}{\lambda \ln 2}$
(E) $\lambda^{\ln 2}$

Input 1	Input 2	Result
0	0	1
0	1	1
1	0	1
1	1	0

5. The above "truth table" represents which of the following logic gates?

(A) OR
(B) AND
(C) NOR
(D) NAND
(E) NOT

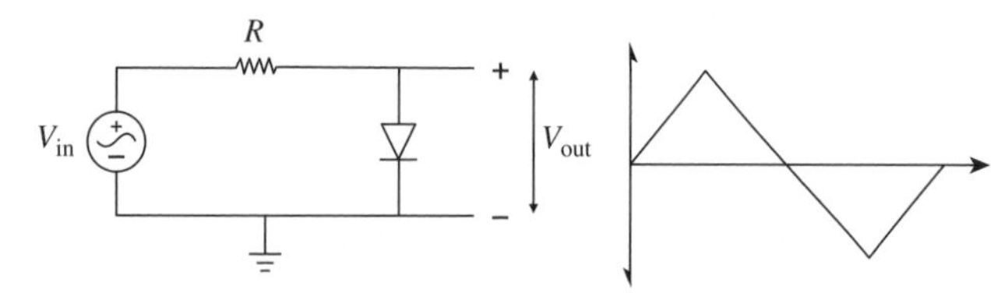

6. The circuit diagram on the left above is driven by an alternating-current generator, whose input voltage V_{in} is shown as a function of time in the plot on the right. Which of the following best represents the shape of the output voltage V_{out}?

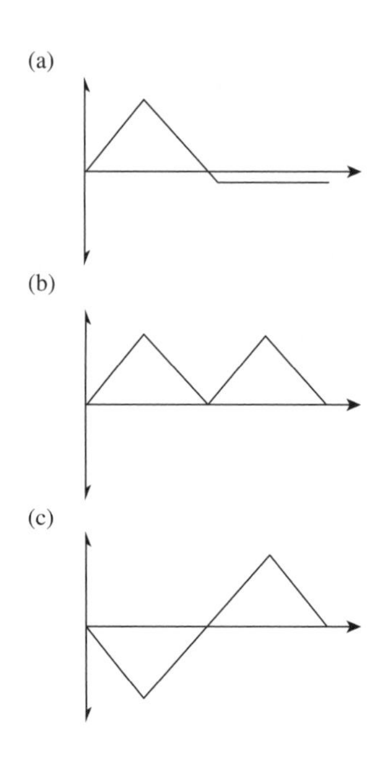

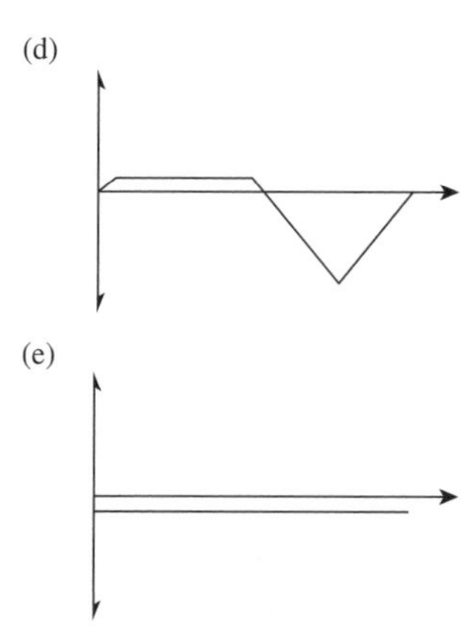

7. A student holding a Geiger counter near a radioactive sample hears five clicks in a 10-second time window. Based on this measurement, what is the probability of hearing exactly one click in a subsequent 10-second time window?

(A) e^{-5}
(B) $5e^{-5}$
(C) $5e^{-2}$
(D) $\frac{5e^{-2}}{2}$
(E) $2^4 e^{-5}$

8. A narrow bandpass filter is centered at 1 MHz. What combination of inductor and capacitor can be used to create such a filter?

(A) 25 nF and 10 nH
(B) 250 nF and 100 nH
(C) 1 μF and 1 μH
(D) 20 μF and 10 μH
(E) 200 μF and 10 μH

9. A Michelson–Morley interferometer can be used to detect gravitational waves, which compress and expand the arms of the interferometer to give an effective path-length difference ΔL. If the interferometer arms have length $L = 5$ km and a laser of wavelength 1000 nm is used, at what value of $\Delta L/L$ will the central interference maximum become a minimum?

(A) 10^{-14}
(B) 10^{-12}
(C) 10^{-10}
(D) 10^{-8}
(E) 10^{-6}

10. A semiconductor laser with a band gap of 4 eV will produce which of the following kinds of light?

(A) Infrared
(B) Red

(C) Green
(D) Ultraviolet
(E) X-ray

7.7 Solutions: Laboratory Methods

1. E – Minimum ionizing particles must be relativistic, and only choice E has energy much greater than its mass. Photons are never minimum ionizing particles because they are neutral and their interactions are qualitatively different from those of charged particles.
2. C – The sum of Gaussian random variables is also a Gaussian random variable. For the same reason that experimental uncertainties add in quadrature, the standard deviation of the probability distribution function for the sum of two Gaussian variables is the quadrature sum of the two distributions.
3. A – The binomial distribution describes any situation where there are a fixed number of trials with binary outcomes (though not necessarily equal odds), and it gives the probability for obtaining $n = 1, 2, 3, \ldots$ successes. The scenario of flipping a coin is completely analogous to this. The Gaussian distribution is a good approximation to the binomial distribution in the limit of large statistics when the probability of success is near 50%. The other three distributions are asymptotic distributions of test statistics commonly used in hypothesis testing, and they have nothing to do with flipping a coin.
4. C – You can obtain the answer quickly from equation (7.18), but in case you can't remember where the ln 2 goes, it's easy to derive the answer from scratch. We want $N(t)$ to drop by half, compared to (say) its value at $t = 0$, so we solve:
$$\frac{N_0}{2} = N_0 e^{-\lambda t_{1/2}}.$$
Taking logs gives $-\ln 2 = -\lambda t_{1/2}$, so $t_{1/2} = \ln 2/\lambda$, choice C.
5. D – It's simplest to recognize that if we switch all the 1's and 0's in the "Result" column, we end up with an AND gate, so the given table must represent a NAND gate.
6. D – When V_{in} is positive, the diode is forward biased, which means that the diode is effectively a wire and $V_{\text{out}} = 0$. (Actually, because of the built-in 0.7 V bias of silicon diodes, V_{in} must be greater than 0.7 V for forward biasing to occur.) When V_{in} is negative, the diode is reverse biased and effectively an open circuit, so $V_{\text{out}} = V_{\text{in}}$. Thus, up to the 0.7 V diode bias effect, V_{out} keeps only the negative portions of V_{in}.
7. B – This is a straightforward application of the Poisson distribution. The average number of events in a 10-second window is $\lambda = 5$, and the number of desired events is $n = 1$, so plugging into (7.6) gives us B.
8. B – A circuit with an inductor and capacitor acts as a bandpass filter since the inductor filters high frequencies and the capacitor filters low frequencies. As an example, the resonant frequency of a circuit containing just an inductor and capacitor in series is given by $\omega = 1/\sqrt{LC}$, or
$$f = \frac{1}{2\pi\sqrt{LC}}.$$
In order to have $f = 1$ MHz, we require
$$LC \sim \frac{1}{3.6 \times 10^{13}} \sim 3 \times 10^{-14}\ \text{s}^2,$$
which is closest to B. We have used here the approximation of $\pi = 3$: a convenient trick for estimating order of magnitudes.
9. C – If the central maximum becomes a minimum, the path length has shifted by half a wavelength, so we solve $\Delta L = \lambda/2$ with $\lambda = 1000$ nm to find $\Delta L = 500$ nm. Then $\Delta L/L = 10^{-10}$, choice C. This is roughly how the LIGO detector works; the sensitivity is increased by reflecting the light many times to increase the effective arm length, and by shifting the baseline path length by $\lambda/2$ so that destructive interference occurs in the absence of gravitational waves, allowing a small amount of light to serve as a signal.
10. D – This is a tricky bit of trivia that may show up on the GRE. Semiconductor lasers create laser light by electron–hole recombination, with the energy of the photon on the order of the band gap. Ultraviolet light corresponds to photon energies of 3–100 eV. You can remember this from the fact that the $n = 2$ to $n = 1$ transition in hydrogen is in the ultraviolet, with an energy of $13.6\ \text{eV} \times \left(1 - \frac{1}{2^2}\right) = 10.2$ eV. This is also related to the 2014 Nobel Prize, see Section 8.4.

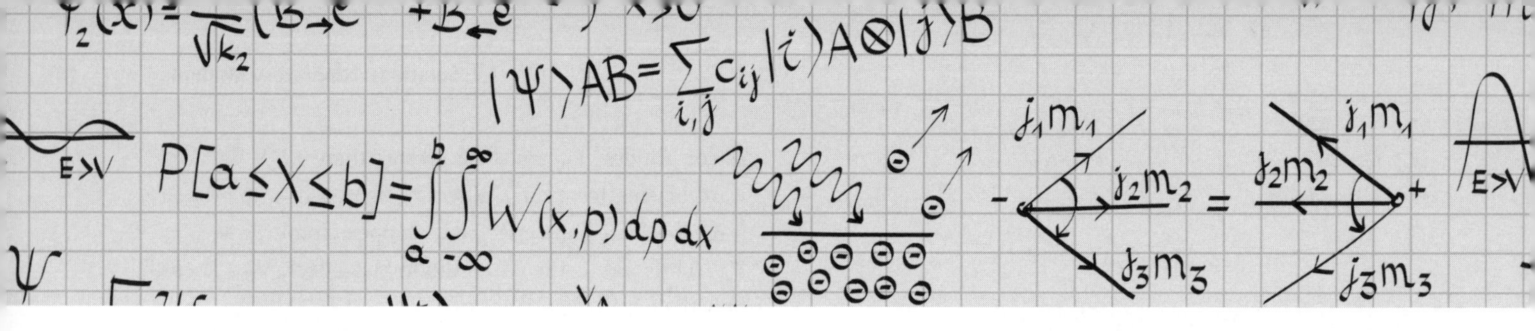

8 Specialized Topics

The Specialized Topics questions on the Physics GRE are probably the most unique aspect of the test. It's hard to think of any other test (other than TV game shows) in which a full 10% is random assorted knowledge. This may seem daunting, but with smart preparation, these questions actually offer a huge advantage.

The special topics questions are almost entirely pure knowledge recall, otherwise known as fact regurgitation. This is the kind of knee-jerk memorization you probably hated in high-school chemistry or biology. When confronted by a special topics question, you'll either know it or you won't. If you know it, that's one question down in under 10 seconds, which gives you a *huge* bonus on time for the more difficult calculational questions. If you don't know it, you probably won't be able to figure out the answer just by reasoning through it, and you may waste 5 or more minutes second-guessing yourself when stuck between two equally appealing answer choices. The optimum strategy, then, is to amass a basic knowledge of as many areas of cutting-edge physics as possible, *just enough* to make the associations between "buzzwords" and concepts that will allow you to recall the required knowledge.

Luckily, this kind of studying is *dead easy.* Every couple days, take a break from your normal Physics GRE practice and just *read.* Pick up a basic textbook in an advanced subject you're unfamiliar with (for example, if you're aiming towards high-energy, choose an introductory solid-state physics or electrical engineering textbook), and don't bother working any problems; just read the book as if it were a novel. You might learn something new and interesting, but that's not really the point: by reading this way, you'll be forming connections and associations in your memory that you might not even be aware of. It's likely you won't be able to remember exactly what you read, but if prompted by a keyword that shows up on the GRE, your memory will spring into action with that feeling of "I've seen this somewhere before." That's really all you need for these kinds of questions.

To give you a head start, we've collected here some of the material that is most likely going to be tested in these kinds of questions. The tone of this chapter will be much more informal than the rest of the book, and some concepts are purposefully not explained in gory detail: as mentioned above, you won't need to know these kinds of details, so consider this leisure reading. When you're sick of doing problems, revisit this chapter and read a few paragraphs.

8.1 Nuclear and Particle Physics

8.1.1 The Standard Model: Particles and Interactions

The modern description of the particles and forces found in nature is a relativistic quantum field theory called the Standard Model. Relativistic quantum field theory is a framework for reconciling quantum mechanics with special relativity. This framework has the surprising result that each charged[1] particle has an *antiparticle*, with identical mass but opposite charges – the first particle such discovered was the anti-electron, more commonly known as the positron. The spin-1 *bosons* of the Standard Model mediate the fundamental forces: photons mediate the electromagnetic force, $W^{\pm}$ and Z bosons

[1] Here "charged" is meant in a general sense and does not refer only to electric charge: for example, the neutron is electrically neutral but carries both weak charge and baryon number, and hence has an antiparticle, the antineutron.

mediate the weak nuclear force, and gluons mediate the strong nuclear force. (There are also hypothetical spin-2 gravitons, which mediate the gravitational force, but because gravity is so weak compared to the other three forces, individual gravitons have not been experimentally observed.) The photon and gluons are massless, whereas the W and Z bosons are extremely heavy, about 90 times the mass of the proton. For group-theoretical reasons, there are eight gluons. The photon and the Z are their own antiparticles, whereas the W^+ is the antiparticle of the W^-. The spin-1/2 *fermions* of the Standard Model, collectively known as *matter*, can be organized in three *generations*. In each generation, there are two *quarks*, one electron-type particle, and one *neutrino*. The first generation contains the up quark (u), down quark (d), electron (e), and electron neutrino (ν_e); the second contains the charm quark (c), strange quark (s), muon (μ), and muon neutrino (ν_μ); and the third contains the top quark (t), bottom quark (b), tau (τ), and tau neutrino (ν_τ). Each generation is successively heavier than the next, with the muon more than 200 times the mass of the electron, and the tau about 20 times the mass of the muon. Because of the large mass hierarchy between generations, third- and second-generation particles will tend to decay to first-generation particles. Indeed, *all* the stable matter in the universe consists of first-generation particles (plus all flavors of neutrinos floating around – more on this later).

The quarks interact via the strong nuclear force, also known as *color*. The mathematical description of the strong force involves assigning one of three "colors" (red, green, blue) to every quark, such that each generation really contains six quarks (up-red, up-blue, up-green, etc.). Quarks also carry electric charge: up-type quarks have charge $+2/3$, and down-type have $-1/3$, in units of the magnitude of the electron charge. In fact, they interact via the weak nuclear force as well: emitting or absorbing a W-boson can change the *flavor* of a quark, from an up-type to a down-type or vice versa, and can also change its generation. The electron-type particles and neutrinos, collectively known as *leptons*, interact via the electromagnetic and weak forces, but *not* the strong force. The electron, muon, and tau all have charge -1, but the neutrinos are electrically neutral. Antiparticles of the quarks and leptons are usually either given the prefix "anti-" or labeled by their opposite charge (μ^+, pronounced "mu-plus," for the antimuon), with the exception of the positron, mentioned above. Similarly, the quarks and neutrinos also have antiparticles, usually denoted by putting a bar over each particle's symbol; for example: $\bar{u}$ for up antiquark, or $\bar{\nu}_e$ for electron antineutrino. Like the leptons, the antiquarks are distinguished from the quarks by having the opposite charge. Since the u has a charge of $+2/3$, for example, the $\bar{u}$ has a charge of $-2/3$.

The strong, weak, and electromagnetic forces vary widely in their strengths and experimental signatures. As the name suggests, the strong force is the strongest, causing strongly interacting particles to decay with lifetimes on the order of 10^{-23} seconds. Next is the electromagnetic force: particles that decay electromagnetically have lifetimes of about 10^{-18}–10^{-16} seconds (note that longer lifetimes mean *weaker* forces), and the telltale signature is the emission of a photon, though this is not required. Finally, particles decaying by the weak force have the longest lifetimes, about 10^{-10}–10^{-8} seconds, and the telltale signature is the emission of a neutrino, though once again this is not always present. The weak decay you're probably most familiar with is beta decay (discussed further below), which provided the first evidence for the neutrino, because the varying energy spectrum of the emitted electron suggested a three-body rather than two-body decay. Questions about which force is responsible for which process are relatively common among special topics questions on the GRE. As we've emphasized, it's impossible to tell which force is responsible just by looking at the decay products, or even the lifetime: there are particles that conspire to have extraordinarily long lifetimes despite decaying strongly, and particles that can decay into the same final state by two different forces. But the combination of these two factors is a useful guide: if you see a particle with a lifetime of 10^{-17} seconds that decays to two photons, you can be pretty sure electromagnetism was responsible.

8.1.2 Nuclear Physics: Bound States

All ordinary matter in the universe is protons, neutrons, and electrons. We've already addressed electrons above: these are elementary particles. However, protons and neutrons (collectively known as *nucleons*) are *composite* – in the framework of quantum field theory, they are teeming seas of quarks and gluons constantly popping in and out of existence. At low energies, where nuclear physics is applicable, we can simplify this description considerably using the *quark model*. Here, the proton is considered a bound state of two up quarks and a down quark, written *uud*, for a total charge of $2(2/3) - 1/3 = +1$, and the neutron is a bound state of two down quarks and an up quark (*udd*), for a total charge of $2(-1/3) + 2/3 = 0$. Due to a property of the strong force called *confinement*, free quarks cannot be seen in nature, and instead they collect themselves into bound states such as protons and neutrons. All of these bound states are color neutral, also referred to as *color singlets*. If we collide strongly interacting particles together at higher and higher energies, we can form all kinds of different bound states, heavier than the proton and neutron. Some may be thought of as excited states of nucleons: for example, the first excited state of the proton is known as the Δ^+, which has the same quark content as the proton but is so much heavier that it can be considered a distinct particle. In general, bound states of quarks fall into two categories: *mesons*, made of a quark and an antiquark, and *baryons*, made of three quarks. (Antibaryons have three antiquarks.) Note that because of the rules for adding spins, mesons may have either spin-1 or spin-0, whereas baryons may have spin-3/2 or spin-1/2. In the 1960s, it was observed that mesons and baryons made out of only the three lightest quarks (up, down, and strange) arranged themselves into interesting patterns, which Gell-Mann called the *Eightfold Way*. Historically, mesons and baryons were the vehicles by which new generations of quarks were discovered: see Griffiths for more information. Color was originally introduced as a description of the strong force in order to explain the quark content of some of these baryons: a baryon made of three identical quarks, for example *sss*, would violate the Pauli exclusion principle since it's supposed to be a fermion but its wavefunction is symmetric. Putting the quarks in an antisymmetric color state, the color singlet mentioned above, fixes this problem. The lightest baryons are the proton and the neutron. The fact that the neutron is slightly heavier than the proton means that the neutron can decay via $n \to p + e^- + \bar{\nu}_e$. Indeed, free neutrons *do* decay, with a lifetime on the order of 15 minutes, but inside a nucleus the constant strong interactions[2] with protons keep them from decaying immediately. In certain nuclei, though, the neutron can decay via quantum tunneling, in a process better known as *beta decay*. As the nuclei get bigger and bigger, the electromagnetic repulsion between protons starts to cancel the attractive effects of the strong force, and whole chunks of the nucleus can break off: this leads to *alpha decay*, emission of bound states of two protons and two neutrons (in other words, helium-4 nuclei). Speaking of nuclear sizes, a good fact to remember is that typical nuclear diameters are femtometers, or 10^{-15} m. The final type of radiation, *gamma* radiation, is the emission of photons from an excited state of a nucleus, which doesn't change the proton/neutron composition of the nucleus.

In addition to processes where nuclei can break apart (*fission*), sufficiently small nuclei can also join together through *fusion*. This requires enormous temperatures and pressures in order to overcome the electromagnetic repulsion between the protons, but can also release enormous amounts of energy; indeed, both the Sun and the hydrogen bomb are powered by fusion reactions. In the Sun, successive protons are fused onto larger and larger nuclei, with some being converted to neutrons along the way, all the way up to ${}^4_2$He. In the standard picture of the genesis of heavy elements in the early universe, light nuclei continued to fuse as a result of supernovae, all the way up to iron (atomic number 26), which is the most stable nucleus. Heavier nuclei become progressively more and more unstable, up to lead (atomic number 82), beyond which all heavier nuclei will eventually decay.

8.1.3 Symmetries and Conservation Laws

The general rule of particle physics is that anything that *can* happen, *will* happen, unless it is forbidden by a symmetry or a conservation law. For example, the electron is the lightest negatively charged particle (excluding quarks, which as we've discussed can't exist free in nature), so it is forbidden from decaying by conservation of charge. But the proton is heavier than the positron – why doesn't it decay? To explain this, and other similar observations, a new law called *conservation of baryon number* was introduced. Baryons get baryon number $+1$, antibaryons get -1, and everything else is assigned zero. Similarly, the fact that an extra neutrino is *always* produced in beta decay suggests *conservation of lepton number*, which is really three separate conservation laws (conservation of electron number, muon number, and tau number), where in each generation the lepton and its associated neutrino are given lepton number $+1$ and corresponding antiparticles get -1. See Example 8.1.

The connection between conservation laws and symmetries is provided by *Noether's theorem*, which states that each conserved quantity is associated with a symmetry transformation that leaves the Lagrangian invariant. In fact, you've already seen this in classical mechanics: if a Lagrangian doesn't depend on a coordinate q, the associated conjugate momentum $\partial L/\partial \dot{q}$ is conserved. Noether's theorem relates this to the fact that the Lagrangian is invariant under changes of the coordinate q. In quantum field theory, the symmetries associated with charge, baryon number, and lepton number all act by multiplication by a phase $e^{i\alpha}$. This transformation acts the same everywhere in space, so it is called a *global* symmetry. It's also a *continuous* symmetry since the parameter α can vary continuously across real numbers, each one corresponding to a different symmetry transformation. A symmetry that acts differently at different points, such as $e^{i\alpha(x)}$ where $\alpha(x)$ is

[2] In particle physics, reactions are sometimes written without the + sign for multiple particles in the initial or final state, so neutron decay may also be written $n \to pe^-\bar{\nu}_e$.

EXAMPLE 8.1

We can apply these rules to find the possible decay modes of the muon μ^-. By conservation of muon number, we must have a muon neutrino among the decay products. By conservation of charge, we must have a lighter negatively charged particle: the only stable option is the electron. But by conservation of electron number, we must have an accompanying neutral particle with electron number -1, namely the anti-electron-neutrino. This fixes one decay mode (in fact, the dominant one) to be

$$\mu^- \to e^- + \nu_\mu + \bar{\nu}_e.$$

Other decays are possible, but they must all contain extra pairs of particles with net charge and lepton number zero: for example, a pair of photons, or an electron–positron pair. In fact, a general rule of thumb says that the dominant decay mode will be the one with the fewest particles in the final state, because each final-state particle comes with a suppression factor of $1/(2\pi)^4$ in the quantum-mechanical amplitude.

a function of position x, is known as a *gauge symmetry*, and underlies the quantum formulation of electromagnetism.

There are also various *discrete* symmetries that are important in nature. The symmetry operation P, or *parity*, reverses the orientation of space, which is another way of saying P takes a configuration of particles to its mirror image. The symmetry operation C, *charge conjugation*, exchanges particles and antiparticles. Finally, T, or *time reversal*, does what it sounds like. A very important theorem in quantum field theory states that all Lorentz-invariant local quantum field theories must be symmetric under the combined action of all these operations, known as *CPT*. However, it is an important and striking fact that the Standard Model does *not* respect each of these symmetries individually. The weak interaction is said to be *maximally parity-violating*: the classic experiment that proved this looked at beta decay of cobalt-60, and found that the decay products were preferentially produced in one direction relative to the spin of the nucleus. In fact, the weak interaction doesn't conserve CP either: the main evidence for this comes from the neutral kaon system (for the record, kaons are the lightest mesons that contain strange quarks). By the CPT theorem, this means that T by itself must also be violated, and there are various experiments involving the precession of muon spins in a magnetic field to demonstrate this.

8.1.4 Recent Developments

For the Standard Model to be mathematically consistent, it must contain at least one additional particle: the *Higgs boson*, whose discovery was at long last announced on July 4, 2012. This particle is responsible for giving mass to all elementary particles, via a mechanism known as *spontaneous symmetry breaking*. For more details on both, see Section 8.4 below – both the particle and the mechanism are important enough to each deserve their own Nobel Prize! For the measured value of the Higgs boson mass, 125 GeV, to be consistent with the principles of quantum field theory, many physicists believe that there must exist a further symmetry of nature known as *supersymmetry*. If this is the case, each elementary particle has a *superpartner* with exactly the same charges, but with spin differing by 1/2. As of this writing, the Large Hadron Collider is hot on the trail of these hypothetical particles, but none have yet been discovered.

On firmer experimental footing is the discovery that *neutrinos have mass*. This was deduced by observing *neutrino oscillations*, a phenomenon by which one flavor of neutrino (say an electron neutrino) is emitted from a source, but later detected as another flavor (say a tau neutrino). Unfortunately, this kind of measurement only permits one to determine mass differences, not absolute masses, but it is known that all neutrinos are extremely light, with masses several orders of magnitude less than the mass of the electron.

8.2 Condensed Matter Physics

8.2.1 Crystal Structure

An ideal crystal is constructed by the infinite repetition of identical structural units in space. In the simplest crystals the structural unit is a single atom, as in copper, silver, gold, iron, aluminum, and the alkali metals. But for other materials the smallest structural unit may contain many atoms or molecules. For simplicity, we'll call this unit an "atom" in

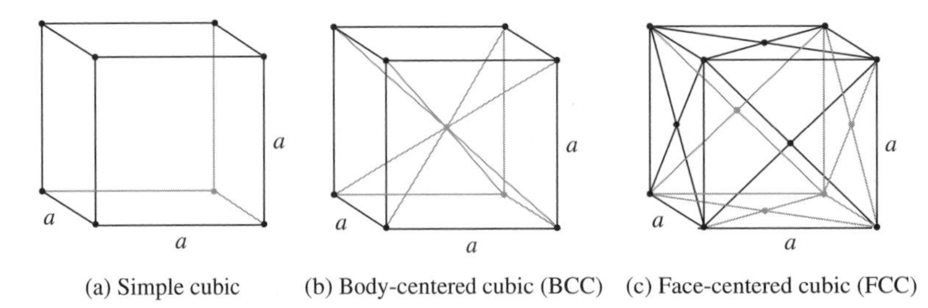

Figure 8.1 Common crystal lattices.

everything that follows, just to keep from using too many words. The examples below illustrate some common crystal lattices.

The key here is *repetition*: the crystal structure is infinite, so we have to define a repeating pattern, known as a *unit cell*. The cubes drawn above are known as *conventional unit cells*, and they make the cubic symmetry apparent by containing a cube with atoms at every vertex, but some add extra points to it: the body-centered cubic (BCC) contains an atom at the center, and the face-centered cubic (FCC) contains atoms at the centers of each of the faces. To construct the rest of the crystal, you can tesselate space in all three dimensions using these unit cells.

Despite making the symmetry manifest, it may be the case that the conventional unit cell is not the smallest repeating pattern, where smallest means "containing the least number of atoms." This smallest pattern is known as the *primitive unit cell*. This is the case for the cubics: the primitive unit cell for the BCC is an octahedron with half the volume of the conventional cell, and the primitive unit cell for the FCC is a parallelepiped with one quarter the volume of the conventional cell. Similarly, the volumes of the conventional unit cells are all equal, but the interatomic distances are all different: for a cube of side a, the simple cubic has distance a, the BCC has distance $a\sqrt{3}/2$, and the FCC has distance $a\sqrt{2}/2$. A favorite GRE question type gives you the volume of a primitive unit cell and asks for the interatomic distance or the volume of the conventional unit cell. By the way, these lattice structures are examples of *Bravais lattices*: there are 14 of them, but the only ones that show up on the GRE with any frequency are the three cubic types given above.

There are two other ideas associated with crystal lattices that may show up on the exam. The *reciprocal*, or *dual*, lattice is the Fourier transform of the original lattice. Just like the Fourier transform of a collection of position vectors $\mathbf{x}$ (position space) is a collection of momentum vectors $\mathbf{p}$ (momentum space), the reciprocal lattice can also be considered a "space" in its own right. The vectors that make up the dual lattice, called the *reciprocal lattice vectors*, are the normal vectors to the planes formed by the original lattice. This is rather complicated to visualize, but the following facts are true:

- The simple cubic is its own dual lattice. For a lattice of side length a, the dual lattice has side length $2\pi/a$ (the 2π comes from the Fourier transform, and the $1/a$ since the lattice vectors are supposed to have units of wavenumber).
- The body-centered cubic and face-centered cubic lattices are dual to one another.
- The dual to a hexagonal lattice is another hexagonal lattice, but rotated through a 30° angle.

The primitive unit cell of the reciprocal lattice is so important that it is given its own name: the *(first) Brillouin zone*.

8.2.2 Electron Theory of Metals

In solid-state physics, the free electron model is a simple model for the behavior of electrons in the crystal structure of a metallic solid: the atomic nuclei and the core electrons are pinned in place, and all the valence atomic electrons are *delocalized*, not belonging to any nucleus and free to roam around the metal. Roughly speaking, this explains why metals conduct electricity. If electrons behaved according to the laws of classical mechanics, at zero temperature they would just sit still with zero energy. But electrons are quantum particles, more specifically *fermions*, and according to the Pauli exclusion principle, no two electrons can occupy the same quantum state. So one electron can have zero momentum with spin up, the next can have zero momentum with spin down, but any additional electrons are going to have nonzero momentum.

Because of the exclusion principle, at zero temperature, the electrons in the metal fill up a sphere in momentum space, the *Fermi sphere*.[3] The electrons which are free to roam around the metal sit on the surface of this sphere (the *Fermi surface*), with wavevectors $|\mathbf{k}| = k_F$, the *Fermi wavevector*. (Everything in this business is named after Fermi.) Recalling

[3] In real systems, the shape of the volume in momentum space is not exactly spherical, so the general term is *Fermi sea*.

the usual relations from quantum mechanics, this means the Fermi momentum is $p_F = \hbar k_F$, and the Fermi energy is $E_F = \hbar^2 k_F^2/2m$. The remaining ingredient is a formula for k_F in terms of the number density n of electrons:

$$k_F = (3\pi^2 n)^{1/3}. \tag{8.1}$$

The details of this derivation involve lots of factors you don't want to worry about. Instead, let's pause for a quick sanity check: n has units $1/(\text{length})^3$, so $n^{1/3}$ has units of $1/\text{length}$, appropriate for a wavevector. The Fermi energy is then

$$E_F = \frac{\hbar^2}{2m}(3\pi^2 n)^{2/3}. \tag{8.2}$$

The *density of states* is the number of possible free electron states at a given energy E, given by

$$\rho(E) = \frac{V\sqrt{2}}{\pi^2 \hbar^3} m^{3/2}\sqrt{E}, \tag{8.3}$$

where we have collected factors to emphasize the scaling which the GRE will care about, $\rho(E) \propto m^{3/2}\sqrt{E}$. Note that integrating the density of states should give the total number of electrons,

$$N = \int_0^{E_F} \rho(E)\, dE. \tag{8.4}$$

Combining the expressions for the Fermi energy and the density of states gives the density of states at the Fermi surface,

$$\rho(E_F) = \frac{3}{2}\frac{N}{E_F}. \tag{8.5}$$

Since most of the electrons are buried in the Fermi sphere, at low temperatures, the number of conduction electrons N_C in the metal is *not* N, but instead is approximately

$$N_C \approx \rho(E_F)(k_B T) \sim N\frac{k_B T}{E_F}, \tag{8.6}$$

which represents the number of electrons close enough to the Fermi surface to be able to escape due to thermal fluctuations. In a typical metal, the Fermi energy is way above room temperature, and often far above the melting point, so this is very often a good approximation. As always, the prefactors of π and 3 and such are not worth remembering (indeed, we dropped the prefactor 3/2 in the heuristic discussion above), but the power-law dependences ($k_F \propto n^{1/3}$, $E_F \propto n^{2/3}$) are very standard GRE questions. As we noted above, a quick way to remember this is to get k_F from dimensional analysis, then E_F from the dispersion relation for a massive particle.

8.2.3 Semiconductors

A semiconductor has electrical conductivity intermediate in magnitude between that of a conductor and an insulator. Semiconductors differ from metals in their characteristic property of *decreasing* electrical resistivity with increasing temperature. This is evidence that the free electron approximation doesn't work in semiconductors, and, indeed, what is happening is that most electrons are trapped in the Fermi sea and are forbidden from being excited until they overcome the *band gap*. At high temperatures, the thermal energy of electrons increases, and more of them can overcome the band gap. Free from the constraints of the exclusion principle, which traps their lower-energy brethren, they can conduct electricity, thus increasing the conductivity (and decreasing the resistivity) at high temperatures.

Current conduction in a semiconductor occurs via these mobile or "free" electrons and holes (the not-so-creative name for the absence of an electron), collectively known as charge carriers. Electrons, of course, have negative charge, so the absence of an electron (a hole) has *positive* charge; in semiconductor physics, one thinks of the hole as an actual positively charged particle in its own right. Adding a small number of impurity atoms to the semiconductor, known as *doping*, greatly increases the number of charge carriers within it. There are two ways to do this:

- **p-type doping.** When a doped semiconductor contains excess holes it is called *p-type* (the "p" stands for "positive"). An example of p-type doping is on silicon: an atom from group 13 of the periodic table, such as boron or aluminum, is substituted into the crystal lattice for silicon.
- **n-type doping.** When a doped semiconductor contains excess free electrons it is known as *n-type* ("n" for "negative"). The most common example for n-type doping is atomic substitution in group 14 solids (silicon, germanium, or tin), which contain four valence electrons, by group 15 elements (phosphorus, arsenic, antimony) which contain five loosely bound valence electrons.

8.2.4 Superconductors

Superconductivity is a phenomenon where certain materials, when cooled below a characteristic critical temperature, have exactly *zero* electrical resistance. Superconductivity is a quantum-mechanical phenomenon, and cannot be understood within the confines of classical electromagnetism. There are two key aspects of superconductivity that are GRE favorites:

- **Meissner effect.** The Meissner effect is an expulsion of a magnetic field from a superconductor during its transition to the superconducting state. In a weak applied field, a superconductor "expels" nearly all magnetic flux, meaning that the magnetic field inside the superconductor is nearly zero (more specifically, it falls off exponentially fast as a function of distance from the surface). It does this by setting up electrical currents near its surface. The magnetic field of these surface currents cancels the applied magnetic field within the bulk of the superconductor. As the field expulsion, or cancellation, does not change with time, the currents producing this effect (called *persistent currents*) do not decay with time. Therefore the conductivity can be thought of as infinite: a superconductor.

- **Cooper pairs.** Cooper pairs are part of the *BCS theory*, which was used to explain superconductivity. A Cooper pair is a specific state of two electrons, weakly bound to each other, such that the pair has total energy lower than the Fermi energy, which would be the energy of the electrons at zero temperature. Thus it is energetically favorable for the electrons to pair up. There are many more details, but just remember that the Cooper pair is "responsible" for superconductivity, and quantum mechanically this pair of electrons behaves collectively like a *boson*, rather than a fermion.

8.3 Astrophysics

The most likely astrophysics topic you'll be tested on is *cosmological redshift*, simply because there are some easy formulas associated with it. Astrophysical evidence clearly indicates that our universe is expanding, but that statement needs to be made more precise for it to have any physics content. What we mean is that the spacetime metric ds^2, which defines what we mean by distance, is *not* constant like we assume in special relativity, but actually changes with time, and does so in such a way that the distance between spacelike-separated points grows with time. In equations,

$$ds^2 = dt^2 - a(t)^2(dx^2 + dy^2 + dz^2),$$

where $a(t)$ is called the *scale factor*. The fact that space is expanding is just the statement that $a(t)$ is an increasing function of t. There's not much we can do with this metric without the tools of general relativity (which you *certainly* don't have to know for the GRE!), but we can write down a few facts without derivation:

- **Photons are redshifted by the scale factor.** If λ_0 is the wavelength of a photon today, and λ_T was its wavelength at a specific time T, then

$$\frac{\lambda_0}{\lambda_T} = \frac{a(\text{today})}{a(T)}. \tag{8.7}$$

Since $a(t)$ is an increasing function of time, this means that the wavelength of a photon is *increased* by a factor of the ratio of distances *now* to distances *then*: this is the origin of the term redshift.

- **Blackbody temperatures are inversely proportional to the ratio of scale factors.** This just follows from (8.7) and Wien's displacement law, $\lambda_{\text{max}} \propto 1/T$; as the universe expands by a factor of 2, blackbodies cool by a factor of 2. The typical application of this result is to the cosmic microwave background temperature, which to an excellent approximation follows a perfect blackbody spectrum. For more about the cosmic microwave background, see the discussion of the 2006 Nobel Prize in Section 8.4.

- **Hubble's law: recession velocity is proportional to distance.** Due to the expansion of space, distant objects appear to be receding from us. If a galaxy appears to be moving away from us with velocity v, then Hubble's law states

$$v = H_0 D, \tag{8.8}$$

where D is the measured distance between the distant galaxy and us, and H_0 is the *Hubble parameter*. H_0 is sometimes called the Hubble "constant," but this is misleading, because H_0 actually changes with time. The details of this are not important for the GRE: just remember that if a galaxy is twice as far away, it will appear to be moving twice as fast.

- **Redshift can be used as a measure of time.** This is just a convention, but a very useful one. Define the *redshift* z by

$$z(T) = \frac{\lambda_0}{\lambda_T} - 1. \tag{8.9}$$

In other words, z measures how much a photon has been redshifted since time T. The -1 is to ensure that $z(\text{today}) = 0$. Thus, positive redshifts correspond to times in the past, and negative redshifts to times in the future. It's certainly a funny way to measure time, but astronomers often refer to events that happened "at redshift 3," which just means the time T such that $\frac{\lambda_0}{\lambda_T} - 1 = 3$; note that this means the photon wavelength has actually been redshifted by a factor of 4, not 3! This is a tricky convention which may well show up on the test.

Finally, a couple of qualitative statements about the expanding universe.

- **Cosmological redshift is *not* a Doppler shift.** It's true that the distances between galaxies are increasing as time goes on, but that does *not* mean they are moving apart from each other. This is a common misconception, and the picture to keep in mind is an inflating balloon: if you glue coins to the surface, as the balloon inflates each coin will move away from each other coin, but the coins are not actually changing their position on the surface. Thus, there is no motion of the source or the receiver to cause a Doppler shift; cosmological redshift is due to *the expansion of space itself.* That being said, there *can* be a component of redshift that comes from the motion of galaxies relative to each other (known as *peculiar velocities*). On recent GREs the wording of the relevant questions did not distinguish between Doppler and cosmological redshift, but the meaning should be clear from context. If you're given or asked for a *speed*, the question is asking about the relativistic Doppler shift; if you're asked about *distances*, the question refers to cosmological redshift.

- **Gravitationally bound systems are exempt.** The metric with the expanding factor $a(t)$ only applies to the universe on large scales, and is *not* the metric for every point in space. Indeed, general relativity tells us that the presence of matter changes the spacetime metric, and in particular systems that contain enough matter to become gravitationally bound (like our solar system) do *not* undergo expansion. In the picture above, the coins that we glued to the surface of the balloon are not themselves expanding. Thus, distances do not seem to expand within our solar system, although we observe distant galaxies receding from each other.

We'll end this section on a topic central to current research: *dark matter*. Since 1933, astronomers have noticed a mismatch between the amount of visible mass in galaxies, and the mass inferred from applying Newtonian gravity and the virial theorem to these galaxies. Specifically, galaxies were rotating too fast for their dynamics to be accounted for solely by the light-emitting matter visible inside. Further astrophysical observations involving the expansion of the universe and the cosmic microwave background radiation, as well as direct observations of colliding galaxy clusters such as the Bullet Cluster, strengthened the case for some kind of gravitating but non-light-emitting matter, dubbed dark matter, whose mass abundance in the universe is more than five times that of ordinary matter. Because it doesn't emit light (photons), dark matter can't be charged (if it is, it must have extraordinarily small electric charge, much smaller than the electron charge), but it may interact with the weak nuclear force. This is the model for the most popular candidate for dark matter, the WIMP or "weakly interacting massive particle," though there are a huge number of other models. Dark matter has not yet been directly detected on Earth, nor has it been unambiguously produced at particle accelerators, but many experiments are underway to try to determine its mass and interactions.

8.4 Recent Nobel Prizes

The Nobel Prizes in Physics provide an excellent source for random trivia about current developments in physics. Anecdotally (though we don't have enough data to back up this claim), the GRE likes to throw in questions dealing with recent Nobel Prizes, so here is a quick summary.

- **2017: Detection of gravitational waves.** A fascinating prediction of general relativity is that accelerating objects should emit gravitational radiation, just as accelerating charges emit electromagnetic radiation. Einstein calculated the emitted power from a test mass shortly after developing general relativity and was firmly convinced it would be too small to ever be detected; in a triumph of physics and engineering, the LIGO (Laser Interferometry Gravitational-Wave Observatory) experiment observed gravitational waves on Earth from the collision and merger of two black holes. The 2017 Nobel Prize was given for the *direct* detection of gravitational waves, but earlier, in 1993, the Nobel Prize was awarded for the *indirect* detection of gravitational radiation. There, the radiating system was a pair of pulsars (neutron stars which emit observable electromagnetic radiation at regular intervals), whose orbits lost a tiny bit of energy as gravitational waves were radiated away over time. The energy loss rate exactly matched the predictions of general relativity. By contrast, the direct detection of such waves exploits the stretching and straining of spacetime when such a wave passes through the Earth. LIGO used a two-armed interferometer (conceptually, just a Michelson-Morley interferometer) to look for a signal where the distance along one interferometer arm contracted slightly while the perpendicular arm expanded. The actual distance perturbations were 10^{-18} m, smaller than nuclear diameters, which (amazingly) could be detected due to an accumulated phase shift when a laser beam traversed the interferometer many times.
- **2016: Topological phases of matter.** This is a fascinating topic and the source of much cutting-edge research in condensed matter physics and theoretical physics in general, but is far too advanced to find a place on the GRE.

- **2015: Neutrino oscillations.** For many years after the introduction of neutrinos to explain the spectrum of beta decay, it was thought that neutrinos were exactly massless. This would have provided a convenient explanation for why neutrinos are always detected with their spin opposite to their direction of motion (all neutrinos are left-handed), since massless spin-1/2 particles carry a fixed helicity, or handed-ness. This turned out not to be the case: neutrinos do have (very tiny) masses, and that causes them to undergo *flavor oscillations*. In any reaction involving the weak force, neutrinos are produced with a definite flavor: electron-type, muon-type, or tau-type. However, the free-particle Hamiltonian does not commute with the weak interaction Hamiltonian, and thus the flavor eigenstates are *not* momentum or mass eigenstates. By the principles of ordinary quantum mechanics, a flavor eigenstate will then evolve into a superposition of mass eigenstates, with probabilities governed by the Schrödinger equation. The detection proceeds by the weak Hamiltonian, so the mass eigenstates get decomposed into flavor eigenstates again, but after propagation a pure electron neutrino has a nonzero probability of being detected as a muon neutrino. This provided a solution to the puzzle of solar neutrinos, where the *pp* chain, which worked so spectacularly well for explaining fusion in the Sun, predicted far more electron neutrinos than were detected on Earth. A similar effect was detected using neutrinos produced from cosmic ray collisions with the atmosphere. As a related bit of trivia, electron neutrinos are typically detected with inverse beta decay, $\nu_e + n \to p + e$.

- **2014: Blue LEDs.** Light-emitting diodes (LEDs) are an excellent practical example of a *p–n* junction. When a bias voltage is applied across the junction, electrons from one side recombine with holes from the other side, and photons are emitted with energy of order the band gap. This causes the LED to shine light at a frequency given by the photon energy. Red and green LEDs were relatively easy to make because the required energies of 1–3 eV correspond to band gaps in readily available doped semiconductors (1.4 eV for GaAs and 2.2 eV for GaP). Finding a material with a so-called *direct band gap* of above 3 eV was difficult – early attempts with materials such as ZnSe and SiC were inefficient since these materials had *indirect band gaps*, meaning the recombination process had to be accompanied by the emission or absorption of a phonon. The prize was given for the fabrication of GaN, with a direct band gap of 3.4 eV, in the ultraviolet.
- **2013: The Higgs mechanism and the Higgs boson.** The role of the Higgs boson has already been mentioned in Section 8.1.4 above, but we can elaborate a little further. A puzzling fact about the forces of the Standard Model is that the W and Z bosons are massive, but all the other force carriers (photons, gluons, and gravitons) are massless. However, putting in a mass "by hand" for these particles results in various inconsistencies, among them the nonsensical prediction that the probability for certain scattering processes is greater than 1! The Higgs mechanism resolves this puzzle by postulating that the W and Z bosons start out massless like the other force carriers, but inherit their mass from the *vacuum expectation value* of the Higgs field, which is constant but nonzero everywhere in space. The same mechanism ends up giving mass to all the other Standard Model matter particles, which also must start out massless for the symmetry properties of the Standard Model to work out correctly. Higgs's insight was that there is an additional particle, the Higgs boson, left over after this mechanism has worked its magic. The Higgs boson couples to matter with strength proportional to the particle's mass; this means it couples most strongly to the heaviest Standard Model particle, the top quark, providing the basis for some of the search strategies that discovered the Higgs boson at the Large Hadron Collider.
- **2012: Measuring and manipulating individual quantum systems.** Quantum systems are hard to study because a measurement of the system typically changes the state of the system. Worse, even if you *try* not to measure a system, the environment that surrounds your system does the measuring for you – this effect is known as *decoherence*, and tends to destroy quantum correlations if the system is at nonzero temperature. The 2012 Nobel Prize was given for two experimental setups to delay this decoherence for as long as possible, and permit certain nondestructive measurements that do not change the state of the system. The former goal is accomplished by cooling systems to very low temperatures and trapping single atoms or photons, and the latter goal uses the clever trick of encoding a measurement of one quantity (for example, the number of photons in a cavity) into the phase of another quantum system (an atom traversing the cavity). A measurement of the phase of the atom will of course destroy the atom's quantum state, but won't touch the photon, allowing it to stay in the cavity and be measured again.
- **2011: Accelerated expansion of the universe.** We've already talked about how the universe is expanding, with a scale factor $a(t)$ that grows with time, $\dot{a}(t) > 0$. The 2011 Nobel Prize was given for the discovery that $\ddot{a}(t) > 0$; in other words, *the expansion of the universe is accelerating*. The main evidence for this striking fact came from

surveys of supernovae that correlated distance with recession velocity, using the more sophisticated time-dependent version of Hubble's law and the relationship between the Hubble parameter and the scale factor. The discovery of accelerated expansion has many important consequences, including the fact that the main component of the energy density of the universe today is not matter, nor dark matter, but whatever substance is responsible for the acceleration, called *dark energy* for lack of a better word. Most evidence suggests that dark energy is actually a *cosmological constant*, a homogeneous energy density throughout all of space with the surprising properties that it exerts negative pressure and does not dilute as the universe expands. But the jury is still out on the nature of dark energy, which is considered one of the most important unsolved problems in fundamental physics today.

- **2010: Isolation of graphene.** Two condensed matter physicists isolated graphene by ripping flakes of it off a lump of graphite using sticky tape. Carbon occurs in many different forms in nature (charcoal, diamond, fullerenes, and so forth), known as *allotropes*. Graphite is one of these, and in fact is made up of a huge number of stacked two-dimensional layers known as graphene. In graphene, carbon atoms are arranged in a hexagonal lattice, which results in highly unusual behavior of the covalently bonded electrons: they behave like *massless particles*, with a linear dispersion relation $\omega \propto k$, rather than massive ones, which have quadratic dispersion $\omega \propto k^2$. This makes graphene an excellent conductor, and the fact that it is one atom thick gives rise to many possible engineering applications.

- **2009: Optical fibers and charge-coupled devices.** As you probably know, optical fibers exploit total internal reflection to transmit light over large distances with very little attenuation. The prize was given for a method of fabricating impurity-free glass fibers: this is probably too engineering-heavy to find a place on the GRE.

- **2008: Spontaneously broken symmetry.** One half of this prize was awarded for "the discovery of the mechanism of spontaneous broken symmetry in subatomic physics." To explain how this works in detail would be far beyond what's needed for the GRE, but a few buzzwords might be useful. A system with a certain underlying symmetry can have a ground state that does not respect that symmetry. The standard example is trying to balance a pencil on its tip: even if there is no preferred direction for the pencil to fall, it will fall eventually, and pick a direction in doing so. A similar situation can happen in particle physics, and any time the quantum-mechanical ground state does not respect a symmetry that was originally present in the theory, a massless spin-0 particle appears called a *Nambu–Goldstone boson*. Nambu's original application was to the BCS theory of superconductivity, where gauge invariance is spontaneously broken, giving rise to a massless phonon. But the most well-known application is to the spontaneous breaking of the $SU(2) \times U(1)$ gauge symmetry of elementary particle physics, for which the Nambu–Goldstone boson is (the massless partner of) the Higgs boson. The second half of the prize was awarded for an interesting technical result which implies that CP violation requires at least three families of quarks.

- **2007: Giant magnetoresistance.** Probably too specialized to find a place on the GRE.

- **2006: CMB anisotropy.** This prize was awarded for discovering that the low-temperature bath of photons pervading the universe, known as the *cosmic microwave background*, has an almost perfect blackbody spectrum, and that the deviations from this spectrum may be hints of structure formation in the early universe. Shortly after the Big Bang, the universe was so hot and dense that photons had an extremely small mean free path. Hence we can't use light to determine anything about the very early universe, because the photons didn't travel in straight lines. After about 380,000 years, though, the universe expanded and cooled enough that the photons hit a *surface of last scattering*, after which they were free to stream through the universe unimpeded. This coincided with the *epoch of recombination*, when electrons and protons combined to form hydrogen atoms. At this time, the temperature of the universe was about 3000 K, but since then the universe has expanded so drastically that these photons are now highly redshifted, as we discussed earlier. Their temperature today is 2.7 K, corresponding to a blackbody peak at a wavelength of 1.9 mm – this is in the microwave region, hence the name. At a level of about one part in 100,000, the spectrum of this radiation bath deviates from the blackbody spectrum, corresponding to density perturbations in the early universe; many of these have been correlated with the positions of galaxies and galaxy clusters today.

8.5 Problems: Specialized Topics

1. Which of the following is a possible decay mode of the π^+? (Note: $\overline{\nu}$ denotes an antineutrino.)

   (A) $e^+\overline{\nu}_e$
   (B) $e^+\nu_e$

(C) $\mu^+\overline{\nu}_\mu$
(D) $\mu^-\nu_e$
(E) $\mu^-\overline{\nu}_\mu$

2. A typical hadron has a lifetime of about 10^{-23} seconds. A new particle is discovered with lifetime of 10^{-10} seconds, and its principal decay mode includes a neutrino. This new particle most likely decays due to the

(A) weak force
(B) strong force
(C) gravitational force
(D) electromagnetic force
(E) Higgs force

3. In the quark model, the Δ^{++}, a spin-3/2 resonance of charge +2, consists of which of the following combinations of quarks?

(A) uuu
(B) uud
(C) udd
(D) ddd
(E) ud

4. The extreme uniformity of the temperature of the cosmic microwave background radiation is evidence for which of the following theories of the early universe?

(A) Quintessence
(B) Big Bang
(C) Inflation
(D) Modified Newtonian dynamics
(E) Scalar-tensor theory

5. The Meissner effect refers to the expulsion of magnetic fields from a superconductor. The exponential decrease in field strength inside the conductor can be modeled by giving the photon an effective

(A) positive electric charge
(B) negative electric charge
(C) mass
(D) spin
(E) color charge

6. The electrical conductivity of a relatively pure semiconductor increases with increasing temperature primarily because:

(A) The scattering of the charge carriers decreases.
(B) The density of the charge carriers increases.
(C) The density of the material decreases due to volume expansion.
(D) The electric field penetrates further into the material.
(E) The lattice vibrations increase in amplitude.

7. The electronic heat capacity of a metal in the Drude model is given by the equipartition theorem $C_{\text{el}} = \frac{3}{2}nk_B$, where n is the number density of electrons. In the Sommerfeld model, $C_{\text{el}} = \frac{1}{2}\pi^2 nk_B(k_BT/E_F)$, which is in better agreement with experiment. The difference between the two formulas can be interpreted as saying:

(A) Only electrons near the Fermi surface participate in heat conduction.
(B) Electron–electron collisions cannot be neglected.
(C) Conduction electrons are confined to the surface of the metal.
(D) Ions also participate in heat conduction.
(E) Electrons in a metal can be treated as classical particles with a modified dispersion relation.

8. Energy levels in a periodic solid are labeled by a discrete set of wavevectors. In a solid with $N \approx 10^{23}$ atoms, energy levels can be labeled by a continuous wavevector in the first Brillouin zone because:

I. Every vector in reciprocal space can be related to a vector in the first Brillouin zone by addition of an integer multiple of a reciprocal lattice vector.
II. The spacing between allowed wavevectors in the first Brillouin zone is proportional to $1/N$.
III. Eigenstates of the solid Hamiltonian are eigenstates of the momentum operator $i\hbar\nabla$.

(A) I only
(B) II only
(C) III only
(D) I and II
(E) I, II, and III

9. Metal A is twice as dense as metal B. Atoms of metal A and metal B have an equal number of valence electrons. If 1 kg of both metals are held at the same temperature, the ratio of the number of charge carriers in metal A to those in metal B is

(A) 1
(B) $2^{1/3}$
(C) $2^{2/3}$
(D) 2
(E) $2^{3/2}$

10. Nuclei are held together by the strong nuclear force. At approximately what distance does the Coulomb repulsion between two protons overtake the attractive strong force?

(A) 2×10^{-6} m
(B) 2×10^{-9} m

(C) 2×10^{-12} m
(D) 2×10^{-15} m
(E) 2×10^{-18} m

11. The top quark is not observed to form bound states because:
    (A) It does not interact with the strong force.
    (B) It is electrically neutral.
    (C) It is too heavy and decays before it can form bound states.
    (D) It has never been produced at particle colliders.
    (E) Its bound states decay only to neutrinos, which are invisible.
12. The Pound–Rebka experiment conclusively demonstrated which of the following properties of light and matter?
    (A) Photons are exactly massless.
    (B) Light is redshifted or blueshifted in a gravitational potential.
    (C) The speed of light in a solid is reduced by its index of refraction.
    (D) Semiconductors preferentially absorb light at discrete frequencies.
    (E) Gamma rays have greater penetration depth than alpha particles.
13. Consider three radioactive isotopes N_1, N_2, and N_3, with atomic number, mass number, and atomic mass in MeV given in the table above. Which of the following decays are possible?
    (A) $N_1 \to N_2$ by beta emission
    (B) $N_3 \to N_2$ by positron emission
    (C) $N_3 \to N_1$ by alpha emission
    (D) $N_3 \to N_2$ by gamma emission
    (E) $N_2 \to N_1$ by electron capture
14. What is the value of the line integral $\oint \frac{1}{z}\,dz$, where z is a complex number and the line integral is taken around a counterclockwise path encircling the origin in the complex plane?
    (A) 0
    (B) 1
    (C) πi
    (D) $2\pi i$
    (E) $-2\pi i$

	Z	A	m (MeV)
N_1	10	20	18.6
N_2	11	20	18.0
N_3	12	20	18.8

8.6 Solutions: Specialized Topics

1. B – The only decay that does not violate conservation of lepton number or charge is B, with one anti-electron and one electron neutrino, and a final charge of $+1$.
2. A – The presence of a neutrino is a dead ringer for the weak force. The longer lifetime is also important, since the weak force is (by its very name) weaker than the strong force, but electromagnetic decays are also slow.
3. A – Recalling that the up quark has charge $+2/3$, the only combination with charge $+2$ is uuu. This is also consistent with spin-3/2, which is one possible state from addition of spins applied to three spin-1/2 quarks.
4. C – Inflation was a period of exponential expansion of the universe, responsible for "smoothing out" all inhomogeneities in the early universe. This manifests itself in the fact that the cosmic microwave background has almost the same temperature at diametrically opposite points on the sky, despite the fact that light had no time to travel between them at the surface of last scattering.
5. C – An exponentially decreasing force law is characteristic of a massive particle. This is the same physics responsible for the exponentially decreasing strong force outside the nucleus due to exchange of massive pions.
6. B – The conductivity for a semiconductor increases as the number of free electrons increases. The density of free electrons in the material is increased by thermal excitation, as the thermal energy is enough to excite some electrons across the gap into the conduction band.
7. A – The Drude model treats the electrons as a classical gas, while the Sommerfeld model incorporates quantum effects, in particular the Fermi–Dirac distribution. The ratio of the specific heats in the two models is approximately $k_B T/E_F$, which is also how the number of conduction electrons scales in a metal. Only electrons near the Fermi surface can be thermally excited and carry current.
8. D – I is true by the definition of the Brillouin zone. II is true by the Born–von Karman boundary conditions, which are almost certainly beyond the scope of the GRE, but the result should be intuitively clear because the number of states in the Brillouin zone must equal the number of primitive unit cells in the crystal, which is proportional

to N. Since the Brillouin zone has fixed volume, the spacing between states must scale as $1/N$, which for large N means the states effectively form a continuum. Finally, III is false because electrons in a periodic lattice satisfy Bloch's theorem, and Bloch functions are not in general eigenstates of the free-space momentum operator.

9. B – Recall that the number of charge carriers N_C in a metal is proportional to Nk_BT/E_F, where N is the total number of valence electrons in the metal. We have $E_F \propto n^{2/3}$, and $N \propto n$, where n is the number density of electrons, so for equal masses of materials with the same number of valence electrons, $N_C \propto n^{1/3}$, implying choice B.
10. D – You should be familiar with the fact that the characteristic range of the nuclear force is 1 fm, or 10^{-15} m.
11. C – The top quark is the heaviest quark and decays on a timescale of 10^{-25} s, much shorter than the typical lifetime 10^{-23} s of hadronic bound states.
12. B – The Pound–Rebka experiment demonstrated the gravitational blueshift of light falling in a gravitational potential, as predicted by general relativity.
13. A – Working by process of elimination, heavier isotopes can only decay to lighter isotopes, so $N_2 \to N_1$ is forbidden, eliminating E. C is forbidden both because the mass difference between N_3 and N_1 is lighter than the mass of an alpha particle (about 4 MeV), and because alpha emission must reduce the mass number of the nucleus by 4. D is forbidden by charge conservation since $N_3 \to N_2$ is effectively a proton converting into a neutron, which must be accompanied by emission of a positively charged particle. This leaves A and B. It turns out that positron emission is forbidden for a somewhat subtle reason: the mass difference of the atoms $m_{N_3} - m_{N_2}$ must be greater than $2m_e \approx 1$ MeV, because after the emission of a positron with mass m_e, N_2 is an ion N_2^- with an extra electron. It must shed this extra electron to transition to the neutral atomic state with mass m_{N_2}, so the transition $N_3 \to N_2$ actually requires the emission of *two* particles of mass m_e (plus neutrinos, which are effectively massless). This leaves only choice A.
14. D – You are unlikely to encounter any complex analysis on the GRE, but because the Specialized Topics description includes "mathematical methods," we included this problem as an example of a basic math fact you might have to know. Cauchy's theorem tells you the value of this integral is $2\pi i$.

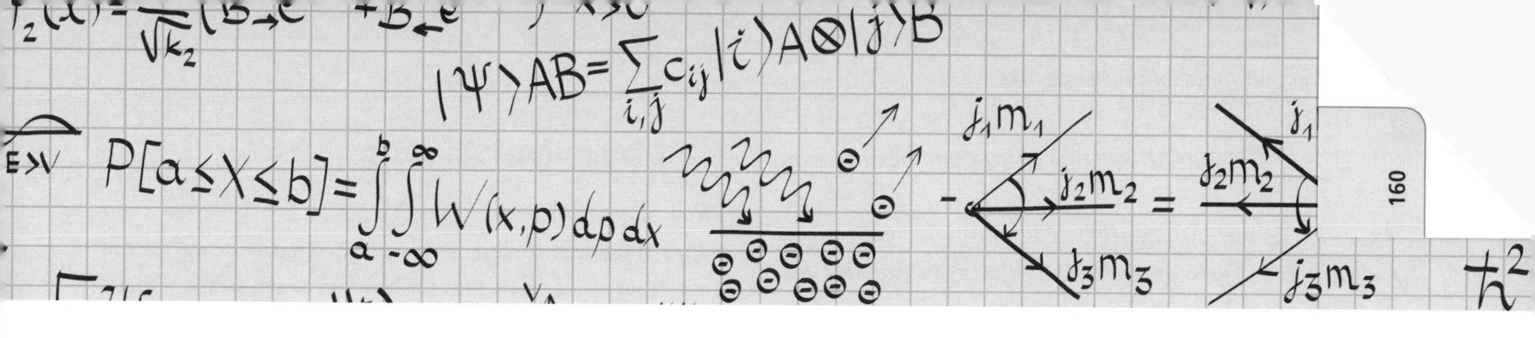

9 Special Tips and Tricks for the Physics GRE

One of the main reasons we wanted to write a Physics GRE review book is that none of the existing review materials address *both* general test-taking strategies and strategies specific to physics problems. We'll make some general suggestions applicable for multiple-choice tests at the end of this chapter, but we'll start with several important and physics-specific tips and tricks.

9.1 Derive, Don't Memorize

If you're just beginning your GRE preparation, and you've started looking through your freshman year textbook, you're probably overwhelmed by the sheer number of formulas. If you're like most physics students, you probably don't even remember learning many of them! But for better or for worse, the Physics GRE is a test of outside knowledge, and you need to know certain formulas to answer many of the questions. And the formula sheet provided at the beginning of the test is worse than useless: numerical values of constants you'll never need, a couple random definitions, and three moments of inertia. Obviously we're going to need an efficient way to remember all the missing formulas.

Richard Feynman (famous twentieth-century physicist and co-inventor of quantum electrodynamics) has a wonderful piece of advice on this sort of thing: "knowledge triangulation." No one can possibly remember all the formulas, but if you can remember a few key facts, you can reconstruct most of the rest of your knowledge, and "triangulate" unknown facts from known ones. The key to this is remembering the basic steps in the important derivations in all the key areas tested on the Physics GRE.

Try this: divide up your formulas into categories based on how involved the derivations are. Class 1 would be the absolute basics, such as $\mathbf{F} = m\mathbf{a}$, expressions for kinetic energy ($\frac{1}{2}mv^2$ for translational, $\frac{1}{2}I\omega^2$ for rotational), the rest energy $E_0 = mc^2$ of a particle, and so on. These are essentially definitions of important physical quantities, rather than actual formulas. Class 2 would be formulas that you could quickly derive in a couple steps from the Class 1 formulas. This might include formulas for recoil velocities in one-dimensional elastic collisions where one mass is at rest (apply conservation of momentum and energy) and the cyclotron frequency of a charged particle in a magnetic field (use the fact that the magnetic field provides the centripetal force required for uniform circular motion). Class 3 would be any formula or equation that you expect will take more than two or three lines of algebra to derive, such as normal mode frequencies for a pair of coupled springs or second-order energy shifts in quantum-mechanical perturbation theory.

Now, focus your attention on memorizing the Class 1 formulas, and the steps in the derivations that lead to the Class 2 formulas. Start a formula sheet containing the Class 3 formulas, adding them as you come across them in your studying, and memorize them as you go. Also, include a sketch of the derivations of the Class 2 formulas, but *don't* include the formula itself. Your notes might look like this:

EM boundary conditions at a conductor: apply Maxwell's equations using infinitesimally thin pillboxes and loops

That way, every time you review your formula sheet, you'll force yourself to rederive these formulas. If you find you can't do this after several tries, promote it to a Class 3 formula and write it down.

Of course, this classification is a very individual process, and will depend strongly on which subjects you consider your strengths or weaknesses. But a good target is to have no more than ten Class 3 formulas for the major subjects (classical mechanics, electricity and magnetism), and no more than five Class 3 formulas for each of the smaller subject areas. Anything else is probably overkill, assuming you're familiar enough with the basics to know the Class 1 formulas by heart. And despite what the GRE formula sheet may suggest, moments of inertia are *not* worth memorizing. We would consider the formula $I = mr^2$ for a point mass a Class 1 formula, and everything else Class 2 (just integrate, or use the parallel axis theorem).

You can go even further and develop mnemonics for memorizing Class 3 formulas by treating them as Class 2 formulas, and doing a quick-and-dirty "derivation." Here are a couple of examples. The formula for the Bohr radius of the hydrogen atom, $a_0 = 4\pi\epsilon_0\hbar^2/m_e e^2$, is both completely ubiquitous in quantum mechanics and a huge mess. But instead of memorizing the expression, you can cheat slightly and derive it using mostly classical mechanics and a little quantum mechanics. Apply the uncertainty principle in the form $\Delta r \Delta p \sim \hbar$ to the Bohr model of the hydrogen atom, where we assume the electron executes uniform circular motion in the Coulomb field of the proton. Putting $\Delta r = r$ and $\Delta p = p$, and turning the $\sim$ sign into an $=$ sign, we obtain precisely the Bohr radius. (Try it yourself!) Strictly speaking, of course, this derivation is completely bogus: the p appearing in the uncertainty relation should really be the *radial* momentum, the right-hand side should be $\hbar/2$, and setting $\Delta r = r$ is dubious at best. However, if you just treat this derivation as a mnemonic, you have a two-line derivation of a Class 3 formula, which takes it off your list of formulas to memorize. A simpler example, but one that may be a little too advanced for the Physics GRE, is the Schwarzschild radius of a black hole. Treat light like a "particle" of mass m and kinetic energy $\frac{1}{2}mv^2$, and find the starting radius R for which the escape velocity from a body of mass M is the speed of light $v = c$. You'll find the mass m cancels out, and that light can only escape to infinity for $R > 2GM/c^2$, the Schwarzschild radius. Again, the right answer for the wrong reasons, but it's quick and it works.

Keep an eye out for mnemonics like this, and you should be able to keep your formula sheet to a manageable size. That way you can devote more of your study time to reviewing and doing practice problems, rather than cramming your brain full of formulas.

9.2 Dimensional Analysis

Physical quantities have units. This may not seem like a profound statement, but it is an extraordinarily powerful tool for getting order-of-magnitude answers to physical questions, without ever doing involved computations. On the GRE, it offers an interesting alternative problem-solving method thanks to the multiple-choice format. The *very first* thing you should do when you see a tough-looking question is to scan the answer choices to see if they all have the same units. If *not*, there's a decent chance that only one of the answer choices has the correct units, and by identifying the units you want for the problem in question, you can get to the correct answer by dimensional analysis alone.

Here are some answer choices similar to those that appeared on a 2008 ETS-released test:

(A) h/f
(B) hf
(C) h/λ
(D) λf
(E) $h\lambda$

Without even knowing the question, only one of these choices can possibly be correct, because they all have different units. A question this easy is relatively rare, but you might expect to see a few problems on each test that can be solved with dimensional analysis.

A somewhat more common example is:

(A) $R\sqrt{l/g}$
(B) $R\sqrt{g/l}$
(C) $R\sqrt{2l/g}$
(D) Rg/l
(E) $R^2l/2g$

Assuming R and l stand for lengths (g always has its usual meaning of gravitational acceleration), a quick scan shows that A and C have the same units, while all the others are different. So once we know which units we're looking for, at best we've solved the problem, and at worst we're down to two choices, A and C.

Because dimensional analysis can at least be used as a check on the answers of many problems on the GRE, it's an excellent fallback tool in case you forget exactly how to approach a problem or draw a complete blank. It pays to get *very* comfortable with computing units for quantities, so here's Example 9.1 to practice with.

Since this kind of dimensional analysis comes up so often, we *strongly* recommend coming up with your own method for solving these dimensional equations. Some combination

EXAMPLE 9.1

Which of the following gives the uncertainty Δx^2 for the ground state of the harmonic oscillator?

(A) $\dfrac{\hbar}{2m\omega}$

(B) $\dfrac{\hbar^2}{m\omega}$

(C) $\dfrac{\hbar\omega}{m}$

(D) $\dfrac{\omega}{2\hbar m}$

(E) $\dfrac{\hbar\omega}{m^2}$

We're looking for a quantity with units of (length)^2. First, let's do the dimensional analysis the straightforward way, listing the dimensions of all the variables as powers of mass M, length L, and time T, the three fundamental units in the SI system:

- $\hbar$: ML^2T^{-1}
- m: M
- ω: T^{-1}

The most general combination we can form is $\hbar^a m^b \omega^c$, and we want this to have units of L^2, so we get a system of linear equations in a, b, and c that we can solve:

$$\begin{aligned} a + b &= 0 \\ 2a &= 2 \\ -a - c &= 0. \end{aligned}$$

It's straightforward to see that $a = 1$, $b = -1$, and $c = -1$; in other words, $\hbar/m\omega$, choice A. We're off by a factor of 2, but who cares: only choice A has the correct units. In fact, writing down the linear equations was probably a waste of time, since we could have just as easily stared at the list of units for $\hbar$, m, and ω and determined that the quantity we were looking for was $\hbar/m\omega$ right away.

For an alternate method, we could have avoided the ugly units of $\hbar$ by remembering that $\hbar\omega$ has nice units of energy. One form of energy is kinetic energy, $\frac{1}{2}mv^2$, so to get units of L^2 we need to divide energy by one power of M and multiply by two powers of T. This gives

$$\hbar\omega \times \frac{1}{m} \times \frac{1}{\omega^2} = \frac{\hbar}{m\omega},$$

as before. Note how much faster this was than actually computing the uncertainty for the harmonic oscillator, either by using operator methods or the position-space wavefunction!

of memorizing the MLT units for common constants, remembering useful combinations of constants with nice units like q^2/ϵ_0, and mnemonic methods would be an excellent start.

9.3 Limiting Cases

A careful analysis of limiting cases is one of the most efficient ways to check your work on physics problems. This is especially true for the GRE, where you'll often be able to hone in on the correct answer choice by considering limiting cases, even when dimensional analysis fails.

What exactly constitutes a "limiting case," of course, depends on the problem. Some of the more common ones include letting a quantity such as a mass, velocity, or energy go to zero or infinity, and seeing if the result makes sense in this limit. Here's a simple example: say you have a block of mass m on an inclined plane at an angle θ from the horizontal,

EXAMPLE 9.2

Consider the classic problem of a wheel of mass M and radius R up against a ledge of height h, shown in Fig. 9.1. What horizontal force F do you have to apply at the axle to roll the wheel up over the ledge? (Try this problem yourself before reading the rest of the discussion.)

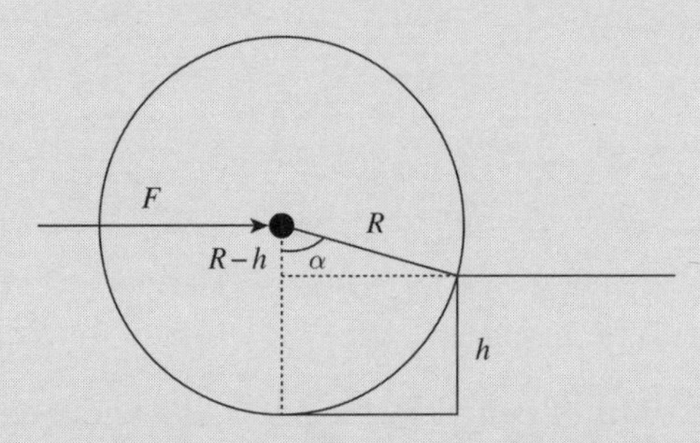

Figure 9.1 A wheel being pushed up a ledge.

This problem is solved most simply by considering the torques about the contact point with the ledge; the ledge exerts some complicated force on the wheel, but we can ignore this entirely because it exerts no torque about the contact point. The wheel will roll up if the torque due to the horizontal force exceeds the torque due to gravity:

$$\tau_g = RMg\sin\alpha = Mg\sqrt{R^2-(R-h)^2},$$
$$\tau_F = RF\sin(\pi/2-\alpha) = RF\cos\alpha = F(R-h),$$
$$\tau_F > \tau_g \implies F > Mg\frac{\sqrt{2Rh-h^2}}{R-h}.$$

Now, let's say we made a mistake calculating $\sin\alpha$, and wrote $\sin\alpha = \dfrac{\sqrt{R^2-h^2}}{R}$. This gives $F > Mg\sqrt{\dfrac{R+h}{R-h}}$. This sort of looks right: it has the right dimensions, and it goes to infinity as $h \to R$, which makes sense (you're never going to be able to push the wheel over the ledge using just a sideways force if the ledge is as high as the radius). However, taking the limiting case of $h \to 0$, we find $F > Mg$. This certainly doesn't make sense: if the ledge disappears, then any force, however small, will allow the wheel to keep rolling. So we know we've made a mistake somewhere.

and you can't remember whether the component of the gravitational force along the ramp is $mg\cos\theta$ or $mg\sin\theta$. Instead of fussing around with similar triangles, just consider what happens when θ is either 0 or $\pi/2$. In the first case, the ramp is horizontal, so the block doesn't slide; in other words, gravity doesn't act at all along the direction of the ramp. In the second case, the ramp is vertical, so the entire force of gravity mg acts downwards and the block just falls straight down. Either of these tell you immediately that the force we're looking for is $mg\sin\theta$.

Checking limiting cases is an extremely powerful strategy if you're running out of time at the end of the test. For sets of answer choices that contain algebraic expressions differing by more than just numerical factors, checking the limiting behavior of the answers usually will let you eliminate some choices. Remember that eliminating even *one* answer choice gives you a positive expected value for that question. If you can quickly identify the relevant limiting cases, and check them against the answer choices, you can often eliminate up to three wrong answers in under a minute. See Example 9.2.

9.4 Numbers and Estimation

Broadly speaking, there are two kinds of physicists: theorists and experimentalists. If you're a theorist, you're probably more comfortable with formulas than numbers, and you might not remember the last time you had to calculate an explicit temperature, energy, or pressure. But a large part of the Physics GRE requires you to think like an experimentalist, estimating rough orders of magnitudes for various physical quantities. Here we'll talk about some strategies for doing so.

First of all, there are some numbers you should just know *cold*. These are the numbers that show up so often in real physics problems that if you haven't already memorized them, you will have after less than a few months of graduate research in the relevant field. Perversely, many of these are *not* the numbers that show up on the Table of Information on the first page of the GRE. Here's the most important example: the binding energy of hydrogen is 13.6 eV. You could memorize the formula for the Bohr energies, $E_n = -\frac{1}{n^2}\frac{m_e e^4}{2(4\pi\epsilon_0)^2\hbar^2}$, plug in all the constants given in the table, and find E_1 after a ton of arithmetic ... or you can memorize this one number.

Actually, this number tells you quite a lot: if you remember that the mass of the electron is about 0.5 MeV$/c^2$ (another number to memorize – see below), this means that you can treat the hydrogen atom nonrelativistically, because the electron's binding energy is much less than its mass. If you know that X-rays have energies of the order of keV, you know that hydrogen atom transition energies are safely below this range, in the ultraviolet. And you know that atoms close to hydrogen in the periodic table will have roughly similar ionization energies: more specifically, the binding energy of each electron increases as the square of the nuclear charge Z, so the ground state energy of helium is about $(13.6)(2^2)(2) \approx 110$ eV, and the binding energy of lithium is about $(13.6)(3^2)(3) \approx 370$ eV. To be clear, these numbers are just approximations – you've probably treated the helium atom using the variational principle in your quantum mechanics class, and you've seen the ground state energy is somewhat less than 110 eV. But these rough estimates are plenty for the GRE – in fact, estimating the binding energy of lithium is a practice question on the Sample Question set released by ETS.

Other important numbers show up as commonly used combinations of fundamental constants. If you're like us, you probably had to memorize the value for h in high school chemistry – but when's the last time you actually had to use the value for h *by itself* in a calculation? If you're calculating anything in quantum mechanics, you use $\hbar$, and if you're doing anything relativistic, you use $\hbar c$. For speed, these combinations are worth memorizing, because they're the ones that you'll actually need. But note that they are currently listed in the GRE formula sheet in case you forget. Similarly, Boltzmann's constant k_B is almost *never* used by itself, but always in combination with temperature. But if you remember that room temperature is about $300\,\text{K} \approx \frac{1}{40}$ eV, you can get the value if you need it. When dealing with combinations of constants, equally important is remembering the units: $\hbar c \approx 200\,\text{MeV}\cdot\text{fm}$ has units of energy $\times$ distance, which tells you that the characteristic distance associated with an object with energy 0.5 MeV is $(200/0.5) \times 10^{-15}$ m, or about 4×10^{-13} m: this is a rough estimate of the Compton wavelength of the electron. (Actually, this is off by a factor of 2π, but who cares? It's good enough as an order of magnitude.)

Based on our experience reviewing past GREs, here is a list of the top five numbers to memorize (in order of importance):

- 13.6 eV – energy of the ground state of hydrogen
- 511 keV – mass of the electron in units of c^2
- 1.22 – coefficient appearing in the Rayleigh criterion, $D \sin\theta = 1.22\lambda$
- 2.9×10^{-3} m·K – Wien's displacement law constant
- 2.7 K – temperature of the cosmic microwave background

You can almost certainly get by with just these numbers. Not included in this list are other numbers you can derive in one or two short steps from numbers given in the Table of Information, such as $\hbar c$ as discussed above.[1]

9.5 Answer Types (What to Remember in a Formula)

The Physics GRE is tricky. Compared to other tests of similar subject matter, such as the AP Physics test or the Physics section of the MCAT, the testmakers throw in answer choices that are deliberately designed to mislead you. Being aware of the common patterns of answer choices can help you avoid these traps, and can often suggest the most efficient approach to a problem. In order of increasing difficulty, here are some patterns you should be aware of.

- **Answer choices with different dimensions.** This was covered in Section 9.2 above, and these problems are some of the easiest because of the possibility of eliminating many answer choices without actually doing any calculations.
- **Order of magnitude.** This was touched on in Section 9.4, and similar to dimensional analysis questions, one can get pretty far just by knowing rough orders of magnitude for common physical situations. See Example 9.3.
- **"Which power of two?"** This pattern is best illustrated by a couple of examples:

[1] Be careful! The most recent GRE included hc on the formula sheet, but often $\hbar c$ is the more useful quantity.

EXAMPLE 9.3

The average intermolecular spacing of air molecules in a room at standard temperature and pressure is closest to

(A) 10^{-12} cm
(B) 10^{-9} cm
(C) 10^{-6} cm
(D) 10^{-3} cm
(E) 1 cm

While you could try to calculate this quantity exactly, using the fact that one mole of gas occupies 22.4 L at STP and so on, it's best just to recognize that A is the scale of nuclear diameters, B is the scale of atomic diameters, and E is macroscopic, which just seems incorrect. So, by common sense, we've narrowed it down to C and D.

1. (A) 2
   (B) 4
   (C) 8
   (D) 16
   (E) 32

2. (A) 0
   (B) $a/3$
   (C) $a/\sqrt{3}$
   (D) a
   (E) $3a$

While the first set is numeric and the second set is symbolic, they're both testing the same thing: do you know the correct power law for a given variable in a certain formula? Often these answer choices will all have the same dimensions, so dimensional analysis won't help you. But the fact that the choices almost always involve nice numbers suggests that memorizing the various constants that accompany formulas is mostly useless: all that matters is the dependence on the various parameters in the problem. As we've emphasized many times, this is especially apparent in the formula for the Bohr energies, where the dependence on reduced mass, nuclear charge, and principal quantum number are all important. On a similar note, if a formula has a simple power-law dependence, such as the Rayleigh formula for small-particle scattering, it's worth simply committing it to memory without asking too many questions about where it came from.

- **Same units, different limiting cases.** This pattern might come from a problem with an angle that can range from 0 to 90°, two unequal masses m and M, or two springs with different spring constants k_1 and k_2. But in any case, while dimensional analysis isn't helpful, taking limiting cases as discussed in Section 9.3 can often help narrow down the answer choices. This pattern lies right on the border between trying to do the problem from the beginning, and forgoing any calculations and just using limiting cases instead. Use your best judgment based on which method you think will be the fastest based on your own strengths and weaknesses.
- **Same units, different numerical factors.** This pattern, which looks like

  (A) $\cos(l/d)$
  (B) $\cos(2l/d)$
  (C) $\cos(l/2d)$
  (D) $\cos(l^2/d^2)$
  (E) $\cos(l^2/2d^2)$

  is tricky, because dimensional analysis is useless, and limiting cases are almost useless. Worse, many of the answer choices only differ by dividing instead of multiplying, increasing the possibility that you land on a trap answer choice by an arithmetic mistake. This pattern is a clue to *slow down*, work through the problem carefully, and try not to refer to the answer choices at any point during your calculation.
- **Random numbers.** Sometimes, you'll have to work out a problem numerically, and all the answer choices will be numbers with no obvious relation to one another. This arises most often in basic kinematics and mechanics problems, where luckily the physics is not an issue – the strategy is just to work slowly and make sure you don't make an arithmetic mistake. Equally as important, many of the wrong answer choices are likely correct answers to an *intermediate* step in the calculation, so, just as mentioned above, try *not* to refer to the answer choices until you're absolutely

through calculating. This reduces the chance you'll get distracted by a trap answer.

9.6 General Test-Taking Strategies

- **Don't practice with a calculator!** You won't be allowed a calculator for the GRE, and the arithmetic required for each problem has been deliberately simplified to avoid messy long division. Resist any temptation to practice with a calculator. We have tried to make the problems in this book as similar as possible to GRE problems, so you should get used to the kinds of arithmetic simplifications you might see on the real test: square roots and factors of π will either be given in the answer choices or be close to whole numbers, and so on.
- **Don't study the night before the exam.** In particular, don't try to cram formulas. Remember, you can derive most of the ones you need from simpler formulas, which should be intimately familiar to you since you've been studying hard for the past several months.
- **Consider working problems symbolically first, only plugging numbers in at the end.** Many GRE problems have extraneous information. If you feel comfortable with algebraic manipulation, it may be a good idea to assign variables to given numbers (if a radius is given as 5 cm, just call it R), find the solution algebraically, and plug in numbers only at the end. Not only will this reduce the amount of arithmetic you have to do, but it may eliminate mistakes because you can check the dimensions of your answer easily. You may also find that you didn't need to use all the information given, or that some factors canceled out.
- **Always guess.** Unlike some other multiple-choice tests, the current version of the GRE does *not* penalize wrong answers. (You may see different instructions in older versions of the test released by ETS, but the most recently released test from 2017 has the new instructions.) Therefore, it is always to your advantage to guess. However, random guessing should be a last resort, and the tips and tricks detailed above will sometimes make it possible to narrow down the answer choices *completely* without ever actually sloving the problem from first principles!
- **Avoid time sinks.** Feel free to take a first pass through the test doing only the problems you feel you know how to answer immediately. On your second pass through the test, you can tackle the more calculation-heavy problems, but try not to spend more than 5 minutes on any particular problem. You should be averaging 1.7 minutes per problem, so it may be worth making a note of the time when you start your second pass to make sure you have enough time to finish the problems that are left. If it looks like you're getting stuck in a time sink, look over the answer choices and see if limiting cases or dimensional analysis can help you narrow down the answer choices. You can always come back to the really tough problems in the last half hour before the test ends.

9.7 Problems: Tips and Tricks

1. Optical phonons in a solid can be excited by infrared light. The typical energy of optical phonons is

   (A) 10^{-4} eV
   (B) 0.1 eV
   (C) 100 eV
   (D) 100 KeV
   (E) 100 MeV

2. A capacitor filled with a dielectric of dielectric constant ϵ is connected to a battery of fixed voltage. If ϵ is doubled, the energy stored in the capacitor is multiplied by a factor of

   (A) 1/4
   (B) 1/2
   (C) 1
   (D) 2
   (E) 4

3. In nuclear magnetic resonance experiments, a nucleus with magnetic moment μ in an external magnetic field B will resonantly absorb radiation of frequency

   (A) $\mu B/\hbar$
   (B) $\mu\hbar/B$
   (C) $\mu/(B\hbar)$
   (D) $\mu\hbar B$
   (E) $B/(\mu\hbar)$

4. Metal A has a Fermi energy of 5 eV and a density of 3 g/cm^3. Metal B has the same number of valence electrons as metal A, but a density of 24 g/cm^3. The Fermi energy of metal B is approximately

   (A) 2.5 eV
   (B) 5 eV
   (C) 10 eV
   (D) 20 eV
   (E) 40 eV

5. A particle of mass m is attached to a spring with spring constant k and feels a frictional force $F = bv$ proportional to its velocity. The particle starts at position x_0 at time $t = 0$ and is observed to undergo oscillatory motion.

Which of the following could describe its position x as a function of time?

(A) $x_0\, e^{-\frac{m}{2b}t}$

(B) $x_0 e^{-\sqrt{\frac{k}{m}}t} \cos\left(\sqrt{\frac{k}{m}}t\right)$

(C) $x_0\, e^{-\frac{b}{2m}t} \sin\left(\sqrt{\frac{m}{k} - \frac{m}{b}t}\right)$

(D) $x_0\, e^{-\frac{b}{2k}t} \cos\left(\sqrt{\frac{b}{m} - \frac{k^2}{2m^2}t}\right)$

(E) $x_0\, e^{-\frac{b}{2m}t} \cos\left(\sqrt{\frac{k}{m} - \frac{b^2}{4m^2}t}\right)$

9.8 Solutions: Tips and Tricks

1. B – Knowing that the binding energy of hydrogen is 13.6 eV (one of your numbers to memorize) tells you that energy transitions in hydrogen are at the eV scale. Since some of the hydrogen transition lines are in the infrared spectrum, this means that infrared photons have energies close to an eV.
2. D – Inserting a dielectric into a capacitor multiplies its capacitance by ϵ. The energy stored in the capacitor can be written as $U = \frac{1}{2}CV^2$, which is useful because we are told the capacitor is at constant voltage. Therefore, the energy scales linearly with C, and hence linearly with ϵ, choice D. This is a classic "which power of 2" problem.
3. A – All the answer choices have different units, so dimensional analysis is the way to go here. Rather than deal with the messy units of electromagnetism, we can remember that the energy of a dipole (anti)aligned with a magnetic field is $U = \mu B$. So μB has units of energy, but so does $\hbar\omega$, which is the energy of a photon of angular frequency ω. This tells us that $\mu B/\hbar$ has units of frequency.
4. D – Since the answer choices resemble a "which power of 2" problem, we know we are not interested in the constants that appear in the formula for the Fermi energy, so this is a good candidate for "derive, don't memorize." Electrons in a metal fill up a Fermi sphere with radius p_F and volume proportional to p_F^3. The total number of electrons is proportional to the density ρ, so $p_F \propto \rho^{1/3}$. The Fermi energy is $E_F = p_F^2/2m_e$, so $E_F \propto \rho^{2/3}$, which is the relation we need to solve this problem. If the density is increased by a factor of 8, the Fermi energy is increased by a factor of $8^{2/3} = 4$, which gives choice D.
5. E – A direct solution of this problem by solving the second-order differential equation for a damped harmonic oscillator would be a time sink. Instead, we can use limiting cases and dimensional analysis. A useful limit to take is $b \to 0$, the case of an undamped oscillator. In that case, we know that the amplitude should be constant and the angular frequency should be $\omega = \sqrt{k/m}$. Only choice E satisfies those criteria. Even if you didn't remember that formula for ω, choices A, C, and D have incorrect units in the exponentials and trig functions. B has correct units, but the fact that it is independent of b is suspicious, since as $b \to 0$ the amplitude should stay constant.

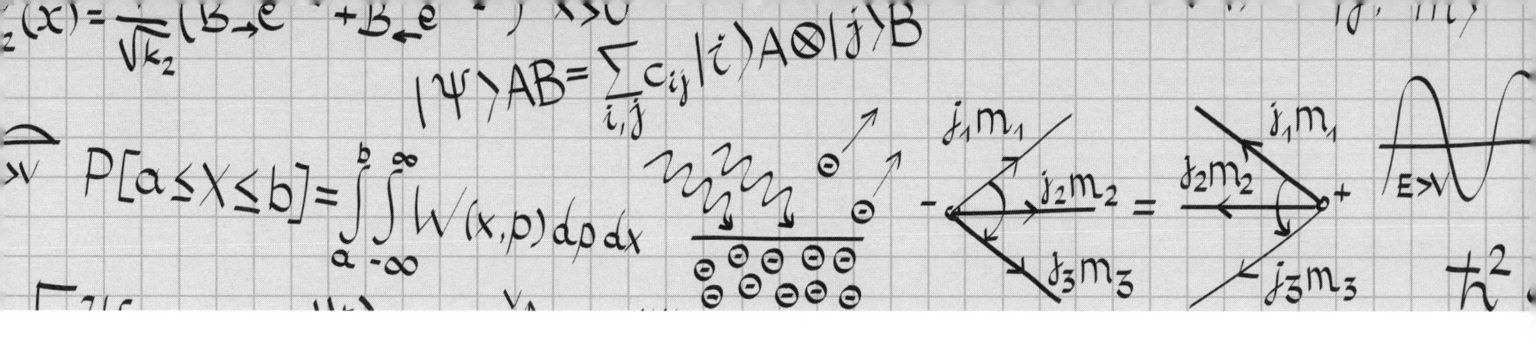

Sample Exams and Solutions

Sample Exam 1

TABLE OF INFORMATION

Rest mass of the electron	m_e	=	9.11×10^{-31} kg
Magnitude of the electron charge	e	=	1.60×10^{-19} C
Avogadro's number	N_A	=	6.02×10^{23}
Universal gas constant	R	=	8.31 J/(mol · K)
Boltzmann's constant	k	=	1.38×10^{-23} J/K
Speed of light	c	=	3.00×10^{8} m/s
Planck's constant	h	=	6.63×10^{-34}J · s = 4.14×10^{-15} eV · s
	$\hbar$	=	$h/2\pi$
	hc	=	1240 eV · nm
Vacuum permittivity	ϵ_0	=	8.85×10^{-12} $C^2/(N \cdot m^2)$
Vacuum permeability	μ_0	=	$4\pi \times 10^{-7}$ T · m/A
Universal gravitational constant	G	=	6.67×10^{-11} $m^3/(kg \cdot s^2)$
Acceleration due to gravity	g	=	9.80 m/s^2
1 atmosphere pressure	1 atm	=	1.0×10^{5} N/m^2 = 1.0×10^{5} Pa
1 angstrom	1 Å	=	1×10^{-10} m = 0.1 nm

Prefixes for Powers of 10

10^{-15}	femto	f
10^{-12}	pico	p
10^{-9}	nano	n
10^{-6}	micro	µ
10^{-3}	milli	m
10^{-2}	centi	c
10^{3}	kilo	k
10^{6}	mega	M
10^{9}	giga	G
10^{12}	tera	T
10^{15}	peta	P

Rotational inertia about center of mass

Rod	$\frac{1}{12}M\ell^2$
Disk	$\frac{1}{2}MR^2$
Sphere	$\frac{2}{5}MR^2$

SAMPLE EXAM 1

Time — 170 minutes

100 questions

Directions: Each of the questions or incomplete statements below is followed by five suggested answers or completions. Select the one that is best in each case and then fill in the corresponding space on the answer sheet.

1. A centrifuge can be used to simulate large gravitational forces. Consider a centrifuge consisting of an arm of length 4 meters, rotating about a fixed pivot at constant speed. What must this speed be to simulate a gravitational acceleration of $9g$? ($\sqrt{|g|} = \sqrt{9.8}$)

(A) $2\sqrt{|g|}$ m/s
(B) $3\sqrt{|g|}$ m/s
(C) $6\sqrt{|g|}$ m/s
(D) $18\sqrt{|g|}$ m/s
(E) $36\sqrt{|g|}$ m/s

2. A block of mass m moving with velocity v collides with a heavier block of mass $4m$, initially at rest. If the collision is perfectly elastic, what is the velocity of the heavier block after the collision?

(A) $4v$
(B) $(1/4)v$
(C) v
(D) $(5/2)v$
(E) $(2/5)v$

3. An LC circuit, consisting of a solenoid and a parallel-plate capacitor, has resonant frequency ω. If the linear dimensions of all circuit elements are doubled, the new resonant frequency is

(A) $\sqrt{2}\omega$
(B) 2ω
(C) ω
(D) $\omega/2$
(E) $\omega/\sqrt{2}$

4. A point dipole with dipole moment $\mathbf{p} = p\hat{\mathbf{z}}$ is placed at the center of a thin spherical conducting shell of radius R. What is the electric field outside the shell?

(A) $\frac{1}{4\pi\epsilon_0}\frac{p}{r^2 R}\hat{\mathbf{r}}$
(B) 0
(C) $\frac{1}{4\pi\epsilon_0}\frac{3(\mathbf{p}\cdot\hat{\mathbf{r}})\hat{\mathbf{r}} - \mathbf{p}}{r^3}$
(D) $-\frac{1}{4\pi\epsilon_0}\frac{p}{r^2 R}\hat{\mathbf{r}}$
(E) $-\frac{1}{4\pi\epsilon_0}\frac{3(\mathbf{p}\cdot\hat{\mathbf{r}})\hat{\mathbf{r}} - \mathbf{p}}{r^3}$

5. The ground state energy of helium is 79 eV. If the ground state wavefunction of helium were a simple product of $1s$ wavefunctions, $\Psi_{100}(\mathbf{r}_1)\Psi_{100}(\mathbf{r}_2)$, the predicted ground state energy would be 108 eV. What is the MAIN factor that accounts for this discrepancy?

(A) Electron–electron Coulomb repulsion
(B) Nonzero orbital angular momentum in the ground state
(C) Spin–spin coupling between the orbital electrons
(D) Spin–spin coupling between the nucleons
(E) None of these

6. The energy of gamma rays from a transition of a nucleus from the first excited state to its ground state is measured. Which of the following is true of the measurement?

(A) Gamma rays from this transition are part of a continuum of gamma rays from the de-excitation of low-lying states.
(B) The measured mean energy must correspond to the energy of a vibrational state of the nucleus.
(C) The measured width of the spectral peak must be $\hbar/(2\tau)$, where τ is the lifetime of the excited state.
(D) The measured mean energy is greater than the true transition energy.
(E) The measured mean energy is less than the true transition energy.

7. A system of electrons is in a box of fixed volume. If the number of electrons in the box is doubled, the Fermi energy is multiplied by a factor of

(A) $2^{-1/2}$
(B) $2^{1/2}$
(C) $2^{2/3}$
(D) 2
(E) $2^{3/2}$

8. A gas of electrons is confined to a two-dimensional surface at $z = 0$ but is otherwise free to move in the x- and y-directions. An external magnetic field is applied so that the electrons feel a harmonic oscillator potential, $U = \frac{1}{2}m\omega^2(x^2 + y^2)$. The temperature of the system is well above the Fermi temperature. What is the specific heat per particle of the electron gas?

(A) $\frac{1}{2}k$
(B) k
(C) $\frac{3}{2}k$
(D) $2k$
(E) $\frac{5}{2}k$

9. A particle with mass m and angular momentum l moves in a constant central potential $U(r) = -k/r$, with $k > 0$. What, if any, is the radius of its stable circular orbit?

(A) The particle has no allowed stable circular orbit.
(B) $\dfrac{l^2}{mk}$
(C) $\dfrac{l^2}{2mk}$
(D) $\dfrac{2l^2}{mk}$
(E) $\dfrac{2l^2}{3mk}$

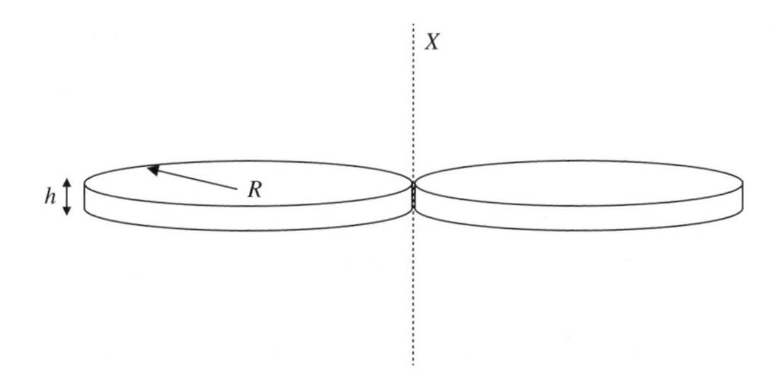

10. Two identical disks shown in the figure above, each of thickness h, radius R, and mass M, are rigidly attached at a point on their edges. What is the moment of inertia of the pair of disks about an axis X, perpendicular to the plane of the disks, which passes through the point where the disks are connected?

(A) MR^2
(B) $\frac{3}{2}MR^2$
(C) $3MR^2$
(D) $6MR^2$
(E) $\frac{3}{2}MRh$

11. A distant galaxy is located at redshift 2. What is the observed wavelength of the 21 cm hyperfine transition line of hydrogen originating from the galaxy?

(A) 7 cm
(B) 10.5 cm
(C) 21 cm
(D) 42 cm
(E) 63 cm

12. A quantum system is prepared in a state $|\Psi\rangle$, which is a superposition of three energy eigenstates. The first two states in $|\Psi\rangle$ have energies E_0 and $2E_0$, and they are measured with probabilities $1/4$ and $1/2$, respectively. If the energy expectation value in the state $|\Psi\rangle$ is $\frac{9}{4}E_0$, what is the energy of the third eigenstate in $|\Psi\rangle$?

(A) $\frac{3}{4}E_0$
(B) E_0
(C) $2E_0$
(D) $4E_0$
(E) $\frac{43}{16}E_0$

13. The Planck mass is given by which of the following expressions?

(A) $\sqrt{\frac{\hbar G}{c^3}}$
(B) $\sqrt{\frac{\hbar G}{c^5}}$
(C) $\sqrt{\frac{\hbar c^3}{G}}$
(D) $\sqrt{\frac{\hbar c^5}{Gk^2}}$
(E) $\sqrt{\frac{\hbar c}{G}}$

14. A spaceship traveling at $0.6c$ towards a planet transmits a signal at 1 GHz to the planet's inhabitants. What frequency is the signal when it is received on the planet?

(A) 1 GHz
(B) 2 GHz
(C) 2.5 GHz
(D) 4 GHz
(E) 8 GHz

15. How much work is required to move a point charge q from infinity to a distance d above an infinite conducting plane?

(A) $-\frac{1}{4\pi\epsilon_0}\frac{q^2}{4d}$
(B) $-\frac{1}{4\pi\epsilon_0}\frac{q^2}{2d}$
(C) $\frac{1}{4\pi\epsilon_0}\frac{q^2}{4d}$
(D) $\frac{1}{4\pi\epsilon_0}\frac{q^2}{2d}$
(E) 0

16. An initially uncharged 10-μF parallel-plate capacitor is charged with a constant current of 1 mA. What is the potential difference between the plates after one second?

(A) 0.01 V
(B) 1 V
(C) 10 V
(D) 100 V
(E) 1000 V

17. A particle in a one-dimensional infinite square well between $x = 0$ and $x = L$ is subject to the following perturbation:

$$\delta V(x) = \begin{cases} V_0, & x < L/2, \\ 0, & \text{otherwise.} \end{cases}$$

What is the leading-order shift in the energy of the first excited state? Recall that the wavefunction for the first excited state is

$$\psi(x) = \sqrt{\frac{2}{L}}\sin\frac{2\pi x}{L}.$$

(A) $-V_0$
(B) V_0
(C) $V_0/4$
(D) 0
(E) $V_0/2$

18. Which of the following is NOT true about the isothermal expansion phase of a Carnot cycle?

(A) The free energy of the gas increases.
(B) The entropy of the gas increases.
(C) The isothermal expansion phase is reversible.
(D) The expansion takes place at the temperature of the "hot" reservoir.
(E) The gas does work on its surroundings.

19. Monochromatic blue light of wavelength 450 nm is shined on a slit of width a. A diffraction pattern is observed on a screen 10 m away. What must a be such that the width of the central diffraction maximum is 100 times the width of the slit?

(A) 45 nm
(B) 450 nm
(C) 0.045 mm
(D) 0.21 mm
(E) 0.30 mm

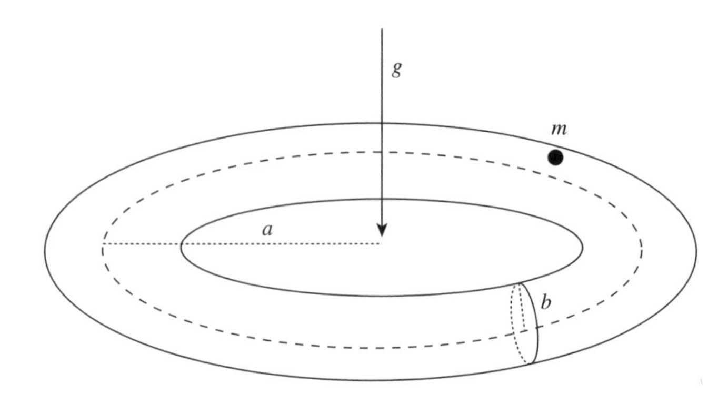

Questions 20 and 21 refer to a particle of mass m, confined to the surface of a torus with central radius a and cross-sectional radius b, oriented such that the Earth's gravitational field points perpendicular to the plane of the circle of radius a. Letting ϕ and θ be the angular coordinates on the circles of radii a and b, respectively, a Lagrangian for this system is

$$L = \frac{1}{2}m(a + b\cos\theta)^2\dot{\phi}^2 + \frac{1}{2}mb^2\dot{\theta}^2 - mgb\sin\theta.$$

20. What is the conjugate momentum to ϕ?

(A) $\frac{1}{2}m\dot{\phi}(a + b\cos\theta)^2$
(B) $m\dot{\phi}(a + b\cos\theta)^2$
(C) $\frac{1}{2}mb^2\dot{\theta}$
(D) $mb^2\dot{\theta}$
(E) $mgb\cos\theta$

21. Which of the following quantities represents the total energy?

(A) L
(B) $L + mgb\sin\theta$
(C) $L - mgb\sin\theta$
(D) $L + 2mgb\sin\theta$
(E) $L - 2mgb\sin\theta$

22. A resistor with resistance R and an inductor with inductance L are in series with a voltage source. For $t < 0$, the voltage is 0. For $t > 0$, the voltage source is V. What time t does it take for the voltage across the inductor to drop to half of its initial level?

(A) $\frac{L\ln 2}{R}$
(B) $\frac{L}{R}$
(C) $\frac{L}{R\ln 2}$
(D) $\frac{2L}{R}$
(E) 0

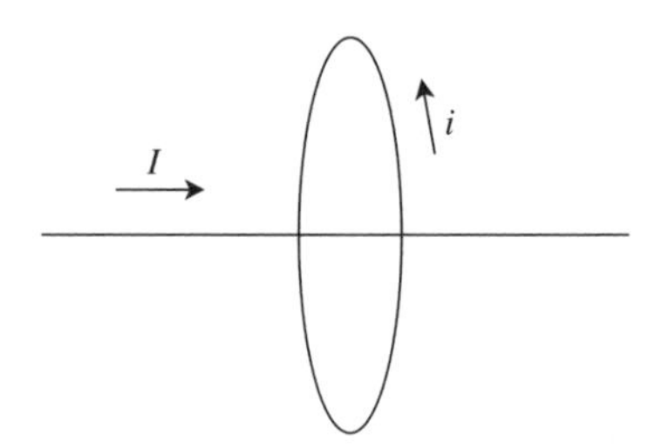

23. A straight wire carrying current I passes through the center of a circular wire carrying current i. If the circular loop of wire has radius R, what is the tension on the circular wire due to the field produced by the straight wire?

(A) $\frac{\mu_0 iI}{2\pi R^2}$
(B) $\frac{\mu_0 I^2}{2\pi R}$
(C) $\frac{\mu_0 i^2}{2\pi R}$
(D) $\frac{\mu_0 iI}{2\pi R}$
(E) 0

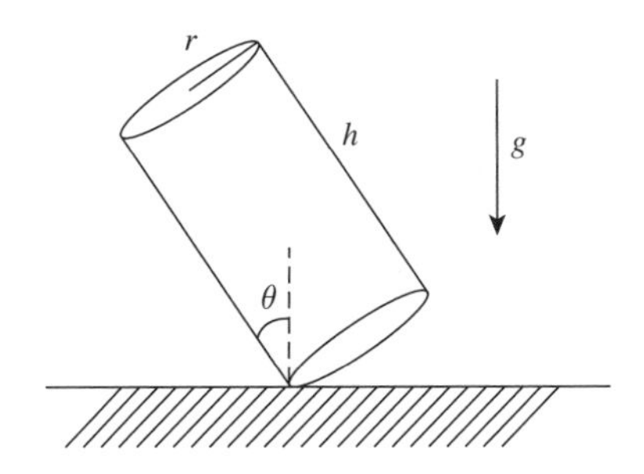

24. A uniform cylinder of height h and radius r is placed on a flat surface and tipped at an angle θ from the vertical. Find θ_0 such that, when the cylinder is released from $\theta > \theta_0$, it falls over.

(A) $\arctan(2r/h)$
(B) $\arctan(r/h)$
(C) $\arctan(r/2h)$
(D) $\arccos(2r/h)$
(E) $\arccos(r/h)$

25. Consider a beam of muons with energy 3 GeV. The muon's mass is approximately 100 MeV$/c^2$, and its lifetime at rest is 2×10^{-6} s. What is the muon lifetime measured by an experimenter in the lab?

(A) 67 ns
(B) 20 μs
(C) 60 μs
(D) 20 ms
(E) 60 ms

26. Graphene, a two-dimensional allotrope of carbon, displays unusual electronic properties. In particular, the dispersion relation for conduction electrons in graphene is

(A) $\omega \propto \sqrt{|k|}$
(B) $\omega \propto |k|$
(C) $\omega \propto |k|^2$
(D) $\omega \propto |k|^3$
(E) $\omega \propto |k|^4$

27. An electron placed in a one-dimensional harmonic oscillator potential $V = \frac{1}{2}kx^2$ is subject to a uniform electric field $\mathbf{E} = E_0\hat{\mathbf{x}}$. For small E_0, the lowest-order, nonzero correction to the ground state energy is

(A) independent of E_0
(B) proportional to E_0
(C) proportional to E_0^2
(D) proportional to E_0^3
(E) proportional to E_0^4

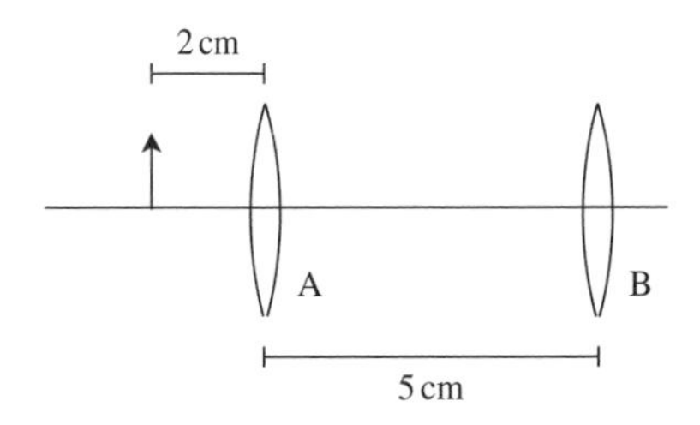

28. In the optical arrangement shown above, converging lenses A and B both have focal length 5 cm. An object is placed 2 cm to the left of lens A. Where is the image of the object located?

(A) 5 cm to the right of B
(B) 6.25 cm to the right B
(C) 12.5 cm to the left of B
(D) 12.5 cm to the right of B
(E) No image is formed.

29. A star of mass m orbits a galaxy of mass M in a circular orbit. The MOND theory postulates that, at small accelerations, Newton's Second Law is replaced by the force law $F = ma^2/a_0$, where a_0 is a constant with dimensions of acceleration. Assuming the MOND force law and Newton's Law of Gravity, what is the relation between the velocity v of the star and the radius of its orbit r?

(A) v is independent of r
(B) v is proportional to $r^{-1/2}$
(C) v is proportional to $r^{1/4}$
(D) v is proportional to r^{-1}
(E) v is proportional to r^2

30. Which values of spin quantum numbers are NOT possible for a system consisting of a spin-1 particle and a spin-2 particle?

(A) $s = 3, m_s = 3$
(B) $s = 1, m_s = 0$
(C) $s = 2, m_s = 1$
(D) $s = 2, m_s = 0$
(E) $s = 0, m_s = 0$

31. The radial wavefunction of the 2p state of hydrogen is

$$R_{21}(r) = \frac{1}{\sqrt{24}} a_0^{-5/2} r \exp(-r/2a_0),$$

where a_0 is the Bohr radius. What is the most probable value of r in this state?

(A) $a_0/2$
(B) a_0
(C) $2a_0$
(D) $4a_0$
(E) $6a_0$

32. Mass spectrometry uses which of the following physical properties of ions to determine the chemical makeup of a substance?

(A) Dipole moment
(B) Nuclear spin
(C) Charge-to-mass ratio
(D) Atomic number
(E) Electronegativity

33. A musician tuning a violin to a tuning fork at 440 Hz hears a beat frequency of 3 Hz. What is the frequency of the note produced by the violin?

(A) 428 Hz
(B) 434 Hz
(C) 437 Hz
(D) 443 Hz
(E) It is impossible to tell from the given information.

34. A spaceship is traveling directly towards a planet at speed $0.5c$. When the ship is a distance of 1 light-hour away from the planet (as measured in the frame of the planet), it fires a missile at the planet with speed $0.5c$. An observer on the planet sees the flash of light from the missile at time t_1, followed by the missile impact at time t_2. What is $t_2 - t_1$?

(A) 0 min
(B) 10 min
(C) 15 min
(D) 30 min
(E) 60 min

35. An ice skater is spinning with arms extended at an angular velocity of 5.0 radians/second. After drawing her arms in, her new angular velocity is 8.0 radians/second. If the skater's moment of inertia with arms extended was I, her moment of inertia with arms drawn in is

(A) I
(B) $3I$
(C) $\frac{8}{5}I$
(D) $\frac{5}{8}I$
(E) $\sqrt{\frac{5}{8}}I$

36. Suppose that a particle in a one-dimensional system has a Lagrangian L with a potential that is constant in time and such that

$$\frac{\partial L}{\partial t} = 0,$$
$$\frac{\partial L}{\partial x} = 0.$$

Which of the following must be true?

I. Energy is conserved.
II. Linear momentum is conserved.
III. The potential is nonzero.
IV. The Euler–Lagrange equations are not satisfied.

(A) I only
(B) II only
(C) I and II
(D) I, II, and III
(E) I, III, and IV

37. A beam of particles with luminosity 10^{22} cm^{-2} s^{-1} is incident upon a target with scattering cross section 10^{-20} cm^2. Assuming a detector has an efficiency of 0.5 for detecting products of the scattering process, how many events will the detector see if the experiment runs for 1 day? Recall that 1 day = 8.64×10^4 seconds.

(A) 4.20×10^2 events
(B) 4.00×10^4 events
(C) 7.20×10^5 events
(D) 4.32×10^6 events
(E) 8.64×10^6 events

38. An electron in a cyclotron moves in a circular orbit at a fixed radius in the presence of a constant magnetic field **B**. If the strength of the magnetic field is tripled, by what factor must the electron's momentum change to keep it orbiting at the same radius?

(A) $\sqrt{3}$
(B) 3
(C) $1/\sqrt{3}$
(D) $1/3$
(E) $3/2$

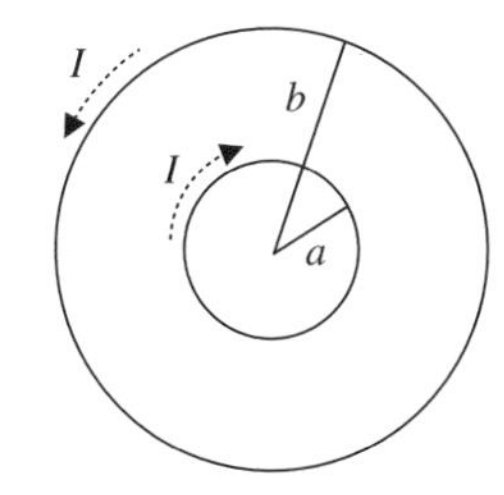

39. Two circular loops of wire of radii a and b are oriented concentrically in the same plane, and they each carry a current I circulating in opposite directions, as shown in the figure above. What is the magnetic field at the center of the loops?

(A) $\frac{\mu_0 I}{2}\left(\frac{1}{a}-\frac{1}{b}\right)$, pointed out of the page
(B) $\frac{\mu_0 I}{2}\left(\frac{1}{a}-\frac{1}{b}\right)$, pointed into the page
(C) $\frac{\mu_0 I}{4}\left(\frac{1}{a}-\frac{1}{b}\right)$, pointed into the page
(D) $\frac{\mu_0 I}{2}\frac{1}{a}$, pointed out of the page
(E) 0

40. Which of the following is true about the total orbital angular momentum operator, L^2, of a particle subjected to an arbitrary force?

I. Always commutes with L_x, L_y, L_z
II. Always commutes with the total angular momentum J^2
III. Always commutes with the Hamiltonian

(A) I only
(B) II only
(C) III only
(D) I and II
(E) I, II, and III

41. A quantum system has a Hamiltonian given by

$$H = \begin{pmatrix} a & 0 & 0 \\ 0 & 0 & -ib \\ 0 & ib & 0 \end{pmatrix},$$

where a, b, c are real positive constants. What are the possible results of a measurement of the energy of the system?

(A) $b, \pm a$
(B) $a, \pm b$
(C) $a, b, a+b$
(D) $a, \pm\sqrt{ab}$
(E) $a, \pm b^2$

42. If magnetic monopoles existed, which of the following expressions would be proportional to the "magnetic charge" of the monopole? You may assume that there are no other sources of electric or magnetic fields present.

(A) $\int(\nabla \cdot \mathbf{E})\, dV$
(B) $\int(\nabla \cdot \mathbf{B})\, dV$
(C) $\int |\mathbf{E}|^2\, dV$
(D) $\int |\mathbf{B}|^2\, dV$
(E) $\int(\mathbf{E} \cdot \mathbf{B})\, dV$

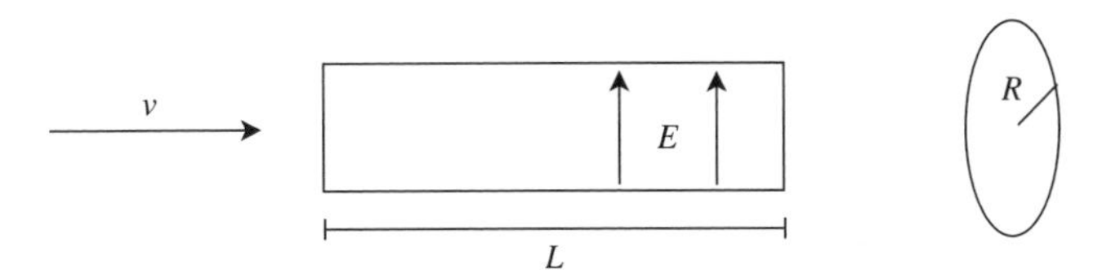

43. A beam of nonrelativistic protons (mass m, charge q) of velocity v enters a region of length L with an electric field E perpendicular to the direction of the beam. At the end of the region of length L is a circular target of radius R. Assuming that the diameter of the beam is much smaller than R, what is the minimum electric field E needed to deflect all protons before they strike the target?

(A) $\dfrac{mLv^2}{2qR^2}$
(B) $\dfrac{2mLv^2}{qR^2}$
(C) $\dfrac{mRv^2}{q^2L^2}$
(D) $\dfrac{2mRv^2}{qL^2}$
(E) $\dfrac{4mLv^2}{qR^2}$

44. Put the following in chronological order, starting with the earliest.

I. Epoch of reionization
II. Nucleosynthesis
III. Inflation

(A) I, II, III
(B) I, III, II
(C) II, I, III
(D) III, I, II
(E) III, II, I

45. For a *monoatomic* ideal gas, which of the following is constant during adiabatic changes of state?

(A) $PV^{1/2}$
(B) PV
(C) $PV^{5/3}$
(D) $PV^{7/5}$
(E) $PV^{9/7}$

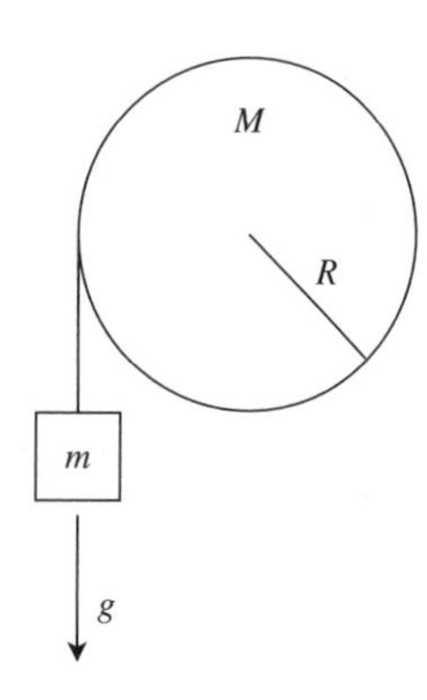

46. A string of length L and negligible mass is completely wound around a solid cylinder of uniform density, of mass M and radius R, and it has a small weight of mass m attached to its end. If the weight is released from rest under the influence of gravity, what is its velocity when the string is entirely unwound?

(A) $\sqrt{\dfrac{4mgL}{M+2m}}$

(B) $\sqrt{\dfrac{2mgL - MR^2}{2m}}$

(C) $\sqrt{2gL}$

(D) $\sqrt{\dfrac{2(m+M)gL}{m}}$

(E) $\sqrt{\dfrac{2mgL - 2MR^2}{m}}$

47. An object is placed at rest in a potential field $U(x, y, z) = x + y^2 - \cos z$. What is the force on the object?

(A) $\mathbf{F}(x, y, z) = -\hat{\mathbf{x}} - 2y\hat{\mathbf{y}} - \sin z\hat{\mathbf{z}}$
(B) $\mathbf{F}(x, y, z) = x\hat{\mathbf{x}} + 2y\hat{\mathbf{y}} - \cos z\hat{\mathbf{z}}$
(C) $\mathbf{F}(x, y, z) = -x\hat{\mathbf{x}} - 2y\hat{\mathbf{y}} + \cos z\hat{\mathbf{z}}$
(D) $\mathbf{F}(x, y, z) = -\hat{\mathbf{x}} - 2y\hat{\mathbf{y}} + \cos z\hat{\mathbf{z}}$
(E) $\mathbf{F}(x, y, z) = \hat{\mathbf{x}} + 2y\hat{\mathbf{y}} + \sin z\hat{\mathbf{z}}$

48. Consider a system with three energy levels $-\epsilon, 0, \epsilon$, and degeneracies $d(-\epsilon) = 2$, $d(0) = 1$, $d(\epsilon) = 3$. What is the energy of the system as $T \to \infty$?

(A) $\epsilon/5$
(B) $\epsilon/6$
(C) $5\epsilon/6$
(D) 0
(E) ϵ

49. In process 1, a monoatomic ideal gas is heated from temperature T to temperature $2T$ reversibly and at constant volume. In process 2, a monoatomic ideal gas freely expands from V to $2V$. Which is the correct relationship between the change in entropy ΔS_1 in process 1 and the change in entropy ΔS_2 in process 2?

(A) $0 < \Delta S_1 < \Delta S_2$
(B) $0 < \Delta S_1 = \Delta S_2$
(C) $0 = \Delta S_1 < \Delta S_2$
(D) $0 < \Delta S_2 < \Delta S_1$
(E) $\Delta S_1 = \Delta S_2 < 0$

50. An electromagnetic wave propagates in vacuum with electric field $E_0 \cos(kx - \omega t)\hat{\mathbf{z}}$. What is the average magnitude of the Poynting vector in SI units, where the average is taken over one period of oscillation?

(A) $\dfrac{4E_0^2}{c\mu_0}$

(B) 0

(C) $\dfrac{E_0^2}{c\mu_0}$

(D) $\dfrac{E_0^2}{2c\mu_0}$

(E) $-\dfrac{E_0^2}{2c\mu_0}$

51. An observation of the reaction $e^+e^- \rightarrow \gamma$ would necessarily violate which of the following conservation laws?

 (A) Lepton number
 (B) Baryon number
 (C) Energy–momentum
 (D) Angular momentum
 (E) Charge conservation

52. The nucleus can be modeled as a degenerate Fermi gas. If the Fermi momentum of nucleons in the carbon nucleus is measured to be 40 MeV/c, which of the following is an approximate lower bound on the nuclear radius?

 (A) 1 nm
 (B) 10 pm
 (C) 100 fm
 (D) 1 fm
 (E) 0.01 fm

53. Which of the following does NOT obey Bose–Einstein statistics?

 (A) Neutrinos
 (B) Photons
 (C) ^{4}He nuclei
 (D) ^{4}He atoms
 (E) Pions

54. The observation of a sharp line of gamma rays of energy 511 keV from the center of our galaxy is most naturally explained by which of the following processes?

 (A) Helium hyperfine transitions
 (B) Hawking radiation
 (C) Ammonia maser transitions
 (D) Electron–positron annihilation
 (E) Supernovae

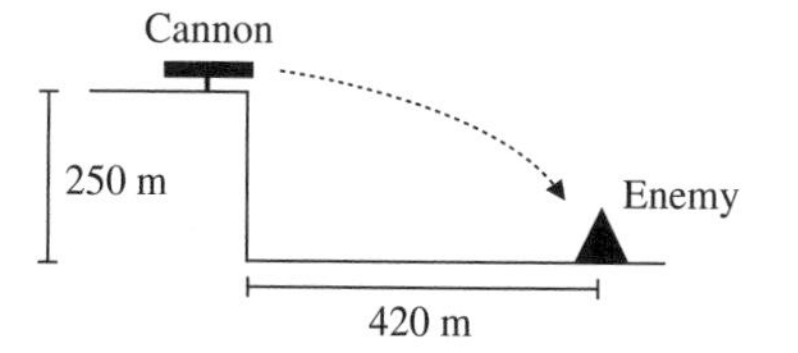

55. A soldier can fire a cannon horizontally from the top of a 250 m cliff. He wants to hit an enemy encampment at a 420 m horizontal distance from the cliff. What must the initial velocity of his cannonball be in order to strike the encampment, neglecting air resistance?

 (A) 22.4 m/s
 (B) 39.6 m/s
 (C) 58.8 m/s
 (D) 94.9 m/s
 (E) 134.2 m/s

56. A rocket traveling at constant speed v in empty space instantaneously expels 10% of its mass in fuel, which is ejected at speed $v/2$ relative to the rocket. What is the final speed of the rocket? You may assume $v \ll c$.

 (A) $v/18$
 (B) $21v/20$
 (C) $19v/18$
 (D) $11v/10$
 (E) $3v/2$

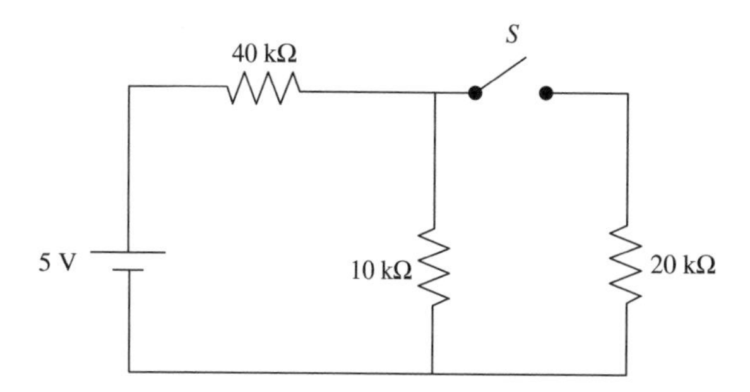

57. Consider the circuit shown in the diagram. When switch S is open, the current through the 10 kΩ resistor is I_1. After switch S is closed, the current through the same resistor is I_2. What is I_2/I_1?

 (A) 1/4
 (B) 5/7
 (C) 1
 (D) 15/14
 (E) 2

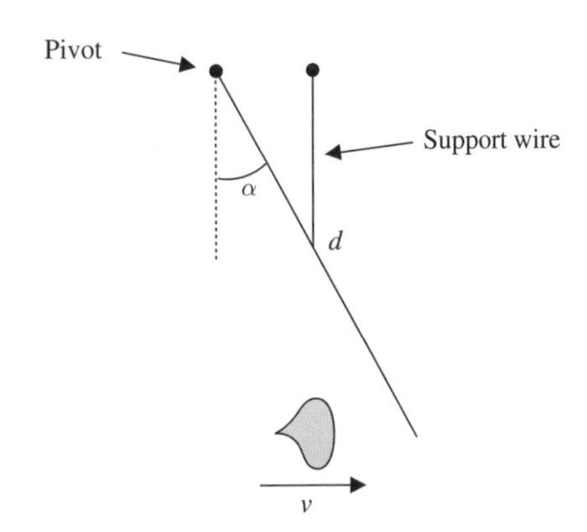

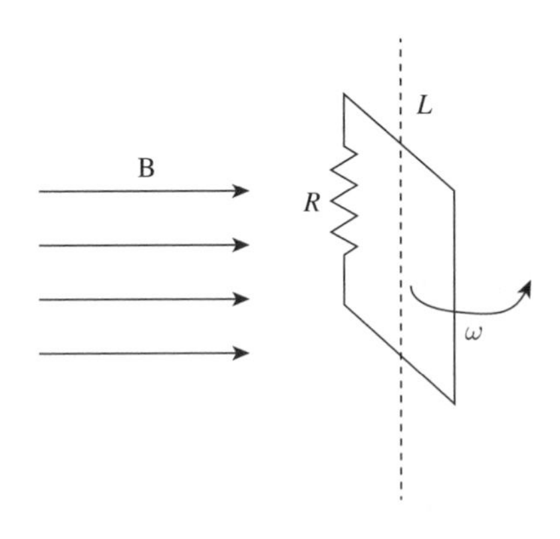

58. A rod of length d and mass M is attached to a pivot and suspended at an angle α from the vertical using a support wire, as shown in the diagram. A lump of clay of mass m is fired at the end of the rod with a velocity v. Just before the clay makes contact with the rod, the wire is cut. Assuming the clay and rod stick together after collision, what is the angular velocity in radians of the rod–clay system? (You may treat the lump of clay as a point mass.)

(A) $\dfrac{mv\cos\alpha}{(M+m)d}$

(B) $\dfrac{3mv\sin\alpha}{(M+m)d}$

(C) $\dfrac{3mv\cos\alpha}{(M+3m)d}$

(D) $\dfrac{3mv}{(M+3m)d}$

(E) $\dfrac{3mv}{Md}$

59. A square loop of wire of side length L, containing a load resistor R, is oriented perpendicular to the xy-plane and rotates about the z-axis at angular frequency ω in the presence of a uniform magnetic field $\mathbf{B} = B_0\hat{\mathbf{x}}$, as shown in the diagram. If $L = 10$ cm, $B_0 = 2$ tesla, and $R = 100.0\ \Omega$, what must ω be so that the average power dissipated in the resistor is 0.5 W?

(A) 25 rad/s
(B) 50 rad/s
(C) 314 rad/s
(D) 354 rad/s
(E) 500 rad/s

60. In calculating the entropy of a microcanonical ensemble, the inverse temperature $\beta = 1/kT$ can be viewed as a Lagrange multiplier enforcing the constraint of fixed total energy. Similarly, the chemical potential μ is related to the Lagrange multiplier for

(A) fermion number
(B) particle number
(C) pressure
(D) volume
(E) magnetization

61. A spin-1/2 particle interacts with a magnetic field $\mathbf{B} = B\hat{\mathbf{z}}$ through a Hamiltonian $H = (-\mu_B g B/2\hbar)\sigma_z$, where μ_B is the Bohr magneton and g is the particle's gyromagnetic ratio. Consider a system of these spin-1/2 particles in equilibrium at temperature T. Let A be the ratio of the number of spin-up particles to spin-down particles. If the strength of the magnetic field is doubled, the new ratio of spin-up to spin-down particles is

(A) A^{-2}
(B) A
(C) A^2
(D) e^A
(E) $A \exp(\mu_B g B/\hbar k T)$

62. Which of the following is equivalent to $\nabla^2(1/r)$?

(A) $-4\pi\delta^3(\mathbf{r})$
(B) $4\pi\delta^3(\mathbf{r})$
(C) 0
(D) 4π
(E) -4π

63. The mass of the proton is 1.67×10^{-27} kg. Which of the following is closest to the Compton wavelength of the proton?

(A) 10^{-15} m
(B) 10^{-13} m
(C) 10^{-12} m
(D) 10^{-10} m
(E) 10^{-9} m

Questions 64 and 65 refer to the following scenario. A K^0 of mass m_K and energy E in the lab frame decays to a π^+ and a π^-, both of mass m_π. The π^+ is observed to be emitted parallel to the K^0 momentum.

64. What is the speed of the π^+ in the K^0 rest frame?

(A) $(1 - 4m_K^2/m_\pi^2)^{1/2}c$
(B) $(1 - 4m_\pi^2/m_K^2)^{1/2}c$
(C) $(1 - m_K^2/m_\pi^2)^{1/2}c$
(D) $(1 - m_\pi^2/m_K^2)^{1/2}c$
(E) $2(m_\pi^2/m_K^2)^{1/2}c$

65. What must be the initial K^0 energy such that the π^- is stationary in the lab frame?

(A) $\dfrac{m_\pi^2 c^2}{2m_K}$
(B) $\dfrac{m_K c^2}{2}$
(C) $\dfrac{m_\pi c^2}{2}$
(D) $\dfrac{(m_K^2 + m_\pi^2)c^2}{2m_\pi}$
(E) $\dfrac{m_K^2 c^2}{2m_\pi}$

66. A clarinet can be treated as a half-open pipe, where sounds are produced by standing pressure waves. For a clarinet of length 0.6 m, which of the following is a possible wavelength of a standing wave?

(A) 0.3 m
(B) 0.6 m
(C) 0.8 m
(D) 1.2 m
(E) 1.5 m

67. A sphere has a polarization of $\mathbf{P}(\mathbf{r}) = Cr^2\hat{\mathbf{r}}$. What is the electric field inside the sphere? (You may find the following fact useful: $\nabla \cdot (v(r)\hat{\mathbf{r}}) = \frac{1}{r^2}\frac{d}{dr}(r^2 v(r))$.)

(A) $-\dfrac{4Cr^2}{\epsilon_0}\hat{\mathbf{r}}$
(B) $\dfrac{2Cr^2}{\epsilon_0}\hat{\mathbf{r}}$
(C) $-\dfrac{Cr^2}{\epsilon_0}\hat{\mathbf{r}}$
(D) $\dfrac{Cr^2}{4\pi\epsilon_0}\hat{\mathbf{r}}$
(E) 0

68. Suppose an electromagnetic plane wave propagating in vacuum in the $+\hat{\mathbf{z}}$-direction has a polarization with the electric field in the $+\hat{\mathbf{x}}$-direction immediately before it strikes a perfect conductor at normal incidence. What are the directions of the **E** and **B** vectors of the transmitted wave?

(A) **E** in $+\hat{\mathbf{x}}$-direction & **B** in $+\hat{\mathbf{y}}$-direction
(B) **E** in $-\hat{\mathbf{x}}$-direction & **B** in $+\hat{\mathbf{y}}$-direction
(C) **E** in $+\hat{\mathbf{x}}$-direction & **B** in $-\hat{\mathbf{y}}$-direction
(D) **E** in $-\hat{\mathbf{x}}$-direction & **B** in $-\hat{\mathbf{y}}$-direction
(E) There is no transmitted wave in a perfect conductor.

69. What is the value of the following commutator?

$$[[[L_x, L_y], L_x], L_x]$$

(A) $-i\hbar^3 L_z$
(B) $i\hbar^3 L_z$
(C) $-i\hbar^3 L_y$
(D) $i\hbar^3 L_y$
(E) $-i\hbar^3 L_x$

70. The vibrational frequency of diatomic oxygen is approximately 5×10^{13} Hz. The temperature at which the vibrational modes of O_2 will begin to be excited is closest to

(A) 20 K
(B) 200 K
(C) 2,000 K
(D) 20,000 K
(E) 2×10^5 K

71. Which of the following does NOT represent a possible observable, written in the position basis, for a free particle in three dimensions?

(A) $-i\hbar\nabla$
(B) $x^2\partial/\partial y$
(C) $x\partial^2/\partial y^2$
(D) $x^2y^2z^2$
(E) xyz

72. The BCS theory of superconductivity explains the superconducting properties of metals at low temperature by supposing that a macroscopic number of metallic electrons all lie in the same ground state. Why does this not violate the Pauli exclusion principle?

(A) BCS theory is incorrect.
(B) Electrons pair off into Cooper pairs, which behave as bosons.
(C) Spin–spin coupling prevents electrons from being in the same state.
(D) The Pauli exclusion principle does not apply to systems at low temperature.
(E) Electrons are not fermions.

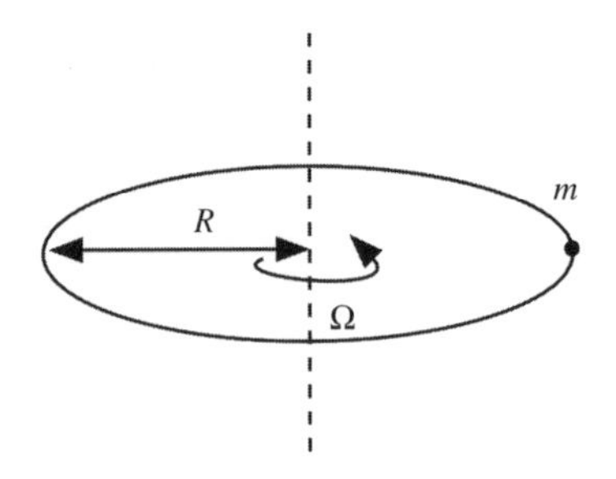

73. A hoop of radius R rotates at constant angular velocity Ω. A small bead of mass m is attached to the hoop, with a frictional force on the bead proportional to the difference in velocity between the bead and edge of the hoop, $F = k(R\Omega - R\omega)$, where ω is the angular velocity of the bead. If the bead begins at angular velocity ω_0, which of the following describes its subsequent motion?

(A) $\omega(t) = \omega_0 e^{-kt/m}$
(B) $\omega(t) = \Omega - \omega_0 e^{-kt/m}$
(C) $\omega(t) = \Omega - \omega_0 e^{-mt/k}$
(D) $\omega(t) = \Omega(1 - e^{-kt/m})$
(E) $\omega(t) = \Omega - (\Omega - \omega_0)e^{-kt/m}$

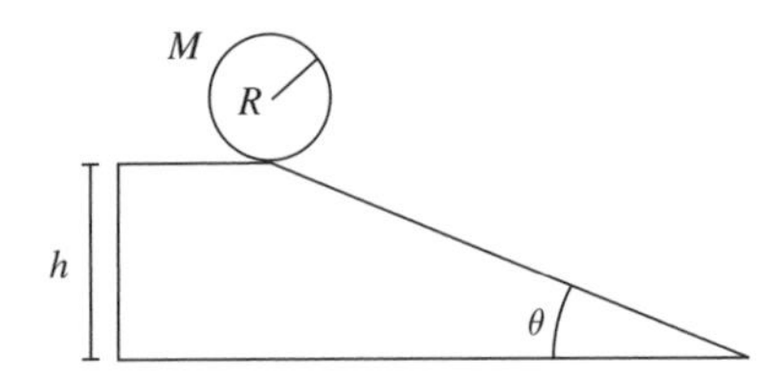

74. Consider a cylinder of radius R, mass M, and density $\rho(r) = Ar^\alpha$ that starts at rest and rolls without slipping down an inclined plane of height h at an angle θ. Assuming no rolling friction, the final velocity of the cylinder at the bottom depends ONLY on

(A) θ and α
(B) M and R
(C) h and M
(D) h and θ
(E) h and α

75. The Δ is a spin-3/2 bound state of three spin-1/2 quarks. The spin part of the wavefunction of the state with $m = +3/2$ is $|\Psi\rangle = |\uparrow\uparrow\uparrow\rangle$. What is the spin part of the wavefunction with definite spin $m = -1/2$?

(A) $|\uparrow\downarrow\downarrow\rangle$
(B) $\frac{1}{\sqrt{3}}(|\uparrow\downarrow\downarrow\rangle + |\downarrow\uparrow\downarrow\rangle - |\downarrow\downarrow\uparrow\rangle)$
(C) $\frac{1}{\sqrt{3}}(-|\uparrow\downarrow\downarrow\rangle + |\downarrow\uparrow\downarrow\rangle - |\downarrow\downarrow\uparrow\rangle)$
(D) $\frac{1}{\sqrt{3}}(|\uparrow\downarrow\downarrow\rangle + |\downarrow\uparrow\downarrow\rangle + |\downarrow\downarrow\uparrow\rangle)$
(E) $|\downarrow\downarrow\downarrow\rangle$

76. What is true of the electromagnetic field at a p-n junction at equilibrium with zero bias voltage applied?

(A) The electric field points toward the p-type semiconductor.
(B) The electric field points toward the n-type semiconductor.
(C) The electric field is parallel to the interface between the p-type and n-type semiconductors.
(D) There is no electromagnetic field.
(E) There is no electric field, but there is a magnetic field pointing toward the n-type semiconductor.

77. In an inertial frame S, two events E_1 and E_2 occur at $(t, x, y, z) = (3, 4, 1, 1)$ and $(1, 3, 0, 1)$, respectively (in units where $c = 1$). In another inertial frame S', which of the following could an observer measure as the spacetime 4-vector between E_1 and E_2?

(A) $(1, 0.5, 1, 1)$
(B) $(2, 1, 0, 0)$
(C) $(3, 2, \sqrt{3}, 0)$
(D) $(2, 0, \sqrt{3}, 0)$
(E) None of these

78. A dark matter experiment takes data for a time T and observes no events. What is the 90% confidence level upper limit that one can place on the event rate in the detector?

(A) One cannot place a limit at the 90% confidence level for this experiment.
(B) $-(1/T)\ln 0.9$
(C) $-(1/T)\ln 0.1$
(D) $(1/T)\ln 0.9$
(E) 0

79. If the proton were a spin-0 particle, which of the following features of the hydrogen energy spectrum would be absent?

(A) Lyman series
(B) Balmer series
(C) 21 cm hyperfine transition
(D) Lamb shift
(E) fine-structure splitting of the $2p$ state

80. An electron neutrino emitted from the Sun may be detected as a tau neutrino on Earth because:

(A) Conservation of lepton number does not apply to tau neutrinos.
(B) Electron neutrinos from the Sun can annihilate and be reemitted as a pair of tau neutrinos.
(C) Electron neutrinos interact with the Earth's magnetic field.
(D) A freely propagating neutrino is a superposition of electron and tau neutrinos.
(E) Electron neutrinos decay faster than tau neutrinos.

81. A pair of electrons is trapped in a "quantum dot." A magnetic field is applied along the z-direction so that the singlet state has energy $-\epsilon$, and the triplet state has energies $-\epsilon/2$, $-\epsilon$, and $-3\epsilon/2$ for spins $+\hbar$, 0, and $-\hbar$ along the z-axis, respectively. What is the probability of finding the electrons in the triplet state, at temperature T?

(A) 0
(B) 1
(C) $\frac{2}{2 + e^{\epsilon/2kT} + e^{-\epsilon/2kT}}$
(D) $\frac{e^{\epsilon/2kT} + e^{-\epsilon/2kT}}{2 + e^{\epsilon/2kT} + e^{-\epsilon/2kT}}$
(E) $\frac{1 + e^{\epsilon/2kT} + e^{-\epsilon/2kT}}{2 + e^{\epsilon/2kT} + e^{-\epsilon/2kT}}$

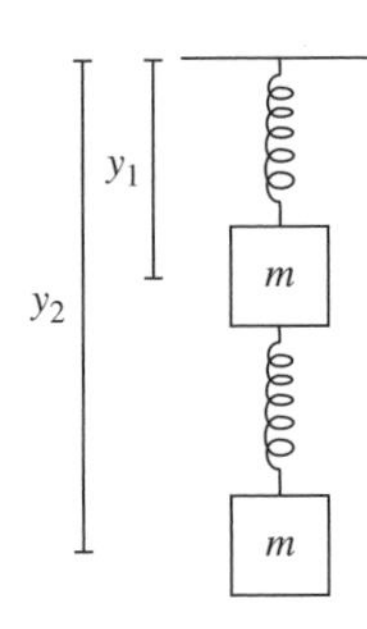

82. The diagram above illustrates a system consisting of a block of mass m hanging from a spring of spring constant k, with another block of mass m hanging from the first block by another spring of spring constant k. What is the total energy of this system?

(A) $\frac{1}{2}m(\dot{y_1}^2 + \dot{y_2}^2) + \frac{1}{2}k(y_1^2 + (y_2 - y_1)^2) - mg(y_1 + y_2)$
(B) $\frac{1}{2}m(\dot{y_1}^2 + \dot{y_2}^2) + \frac{1}{2}k(y_1^2 + (y_2 - y_1)^2) + mg(y_1 + y_2)$
(C) $\frac{1}{2}m(\dot{y_1}^2 + \dot{y_2}^2) - \frac{1}{2}k(y_1^2 + (y_2 - y_1)^2) + mg(y_1 + y_2)$
(D) $\frac{1}{2}m(\dot{y_1}^2 + \dot{y_2}^2) - \frac{1}{2}k(y_1^2 + (y_2 - y_1)^2) - mg(y_1 + y_2)$
(E) $-\frac{1}{2}m(\dot{y_1}^2 + \dot{y_2}^2) - \frac{1}{2}k(y_1^2 + (y_2 - y_1)^2) - mg(y_1 + y_2)$

83. A particle of mass m is in the ground state of an infinite square well of size a, with energy E. The well suddenly expands to size $2a$. What is E'/E, where E' is the expectation value of the energy of the particle after this sudden expansion?

(A) 0
(B) 1
(C) $1/\sqrt{2}$
(D) 1/2
(E) 1/4

84. A particle of mass m and energy E is incident from the left on a delta-function barrier, $V(x) = \alpha\delta(x)$ with $\alpha > 0$. Which of the following gives the coefficient of reflection for the system?

(A) α^2
(B) $\alpha^2 E$
(C) $\frac{\alpha}{\hbar}\sqrt{\frac{m}{2E}}$
(D) $\frac{1}{1 + 2\hbar^2 E/m\alpha^2}$
(E) $\frac{1}{1 + m\alpha^2/2\hbar^2 E}$

85. Which of the following is NOT true about the $2s \to 1s$ transition in the hydrogen atom?

(A) The dominant decay mode is two-photon emission.
(B) It violates $\Delta l = \pm 1$.
(C) It violates $\Delta m = \pm 1$ or 0.
(D) It cannot occur in the electric dipole approximation.
(E) None of these.

86. Measurements of the electric dipole moment of the neutron provide sensitive tests of fundamental physics. If the neutron were found to have a nonzero electric dipole moment, one could directly conclude that which of the following symmetries is violated?

I. Parity
II. Charge conjugation
III. Time reversal

(A) I
(B) II
(C) III
(D) I and II
(E) I and III

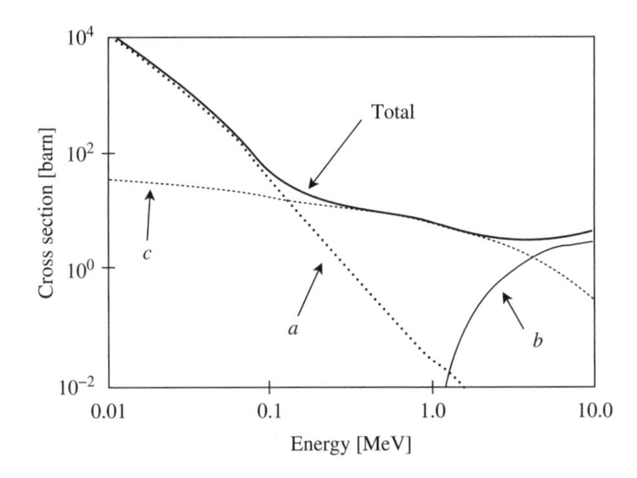

87. The figure above shows the total cross section for photon scattering on a Pb atom as well as the cross sections for several individual process. Why does curve b drop quickly near 1 MeV?

(A) Penetration depth of low-energy photons is small.
(B) Interactions with electrons become significant.
(C) 1.022 MeV threshold for pair production.
(D) Pb has no absorption lines below 1 MeV.
(E) Conservation of angular momentum.

88. Let $f(x) = x$ for $x \in [-\pi, \pi]$. What is the coefficient a_1 of the cos x term in the Fourier series for $f(x)$?

(A) 0
(B) π
(C) 1
(D) 2
(E) 4

89. Suppose that the magnetic field in a region of space is given by $\mathbf{B} = B_0(\hat{\mathbf{x}} + 2x\hat{\mathbf{z}})$. Which of the following could be the vector potential?

(A) $B_0(x\hat{\mathbf{y}} + x^2\hat{\mathbf{z}})$
(B) $-B_0(x\hat{\mathbf{y}} + x^2\hat{\mathbf{z}})$
(C) $-B_0(x^2\hat{\mathbf{y}} + y\hat{\mathbf{z}})$
(D) $B_0(y^2\hat{\mathbf{x}} + z\hat{\mathbf{y}})$
(E) $B_0(x^2\hat{\mathbf{y}} + y\hat{\mathbf{z}})$

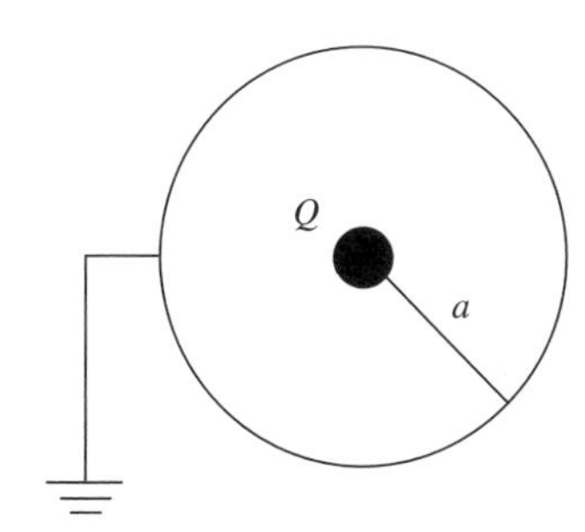

90. Consider a charge configuration consisting of a ball of charge Q surrounded by a thin conducting shell of radius a. The conductor initially has no net charge, but is then connected to ground (the potential at infinity). What is the change in energy of the configuration?

(A) $-\dfrac{Q^2}{4\pi\epsilon_0 a}$

(B) $\dfrac{Q^2}{4\pi\epsilon_0 a}$

(C) $\dfrac{Q^2}{8\pi\epsilon_0^2 a^2}$

(D) $-\dfrac{Q^2}{8\pi\epsilon_0 a}$

(E) $\dfrac{Q^2}{8\pi\epsilon_0 a}$

91. Without the hypothesis of quark color, the quark model would be unable to explain the existence of which of the following spin-3/2 baryons? You may assume the quarks have zero relative orbital angular momentum.

(A) *udd*
(B) *uud*
(C) *uuu*
(D) *uds*
(E) All of the above baryons are allowed.

92. In tabletop atomic spectroscopy experiments using free nuclei, the difference between the frequencies of emitted and absorbed photons driven by the same electronic transition is due to

(A) measurement error
(B) nuclear recoil
(C) gravitational redshift
(D) time dilation
(E) none of these

93. A sequence of NAND gates can create which of the following effective logic gates?

(A) AND
(B) OR
(C) NOT
(D) NOR
(E) all of the above

94. A particle moving in one dimension has the following Lagrangian:

$$L = \frac{1}{2}A\dot{q}^2 - Bq^2.$$

What is the equation of motion of the particle?

(A) $\dot{q} = \dfrac{2B}{A}q$
(B) $\dot{q} = -\dfrac{2B}{A}q$
(C) $\ddot{q} = \dfrac{2B}{A}q$
(D) $\ddot{q} = \dfrac{2A}{B}q$
(E) $\ddot{q} = -\dfrac{2B}{A}q$

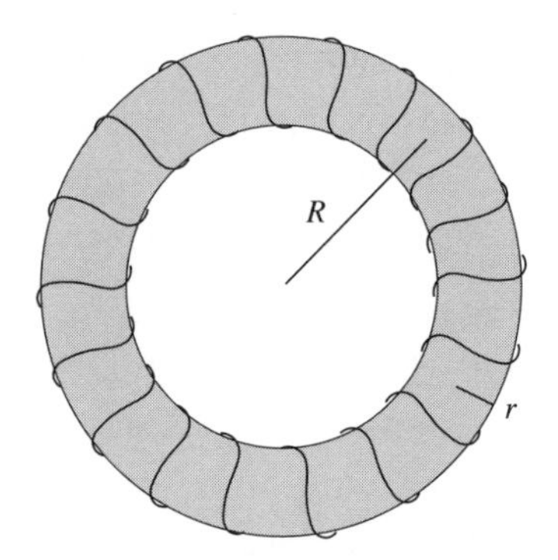

95. A toroidal solenoid of radius R and cross-sectional radius $r \ll R$ has N winds and carries current R. The volume enclosed by the torus is $2\pi^2 Rr^2$. What is the energy stored in the toroidal solenoid?

(A) 0
(B) $\frac{\mu_0 N I^2 r^2}{4\pi R^3}$
(C) $\frac{\mu_0 N^2 I^2 r^2}{4\pi R^3}$
(D) $\frac{\mu_0 N I^2 r^2}{4R}$
(E) $\frac{\mu_0 N^2 I^2 r^2}{4R}$

96. Which of the following is true about a longitudinally polarized wave in three dimensions?

I. There are two linearly independent polarization vectors.
II. The polarization vector(s) is/are perpendicular to the wavevector.
III. The polarization vector(s) is/are parallel to the wavevector.

(A) III only
(B) II only
(C) I only
(D) I and II
(E) I and III

97. Deep water waves obey the dispersion relation $\omega = A\sqrt{k}$, where A is a constant. What is the correct relationship between phase velocity and group velocity for deep water waves?

(A) $v_{\text{phase}} = \frac{1}{2} v_{\text{group}}$
(B) $v_{\text{phase}} = v_{\text{group}}$
(C) $v_{\text{phase}} = 2v_{\text{group}}$
(D) $v_{\text{phase}} v_{\text{group}} = A^4 t^2$
(E) none of these

98. When light of 5000 Å is shined on a thin film of oil ($n = 1.5$) that sits on top of a medium with $n = 2.0$, the intensity of reflected light is minimized. What is the thickness of the oil?

(A) 4×10^{-8} m
(B) 8.33×10^{-8} m
(C) 1.67×10^{-7} m
(D) 1.25×10^{-7} m
(E) 5.0×10^{-7} m

99. Suppose a particle has a normalized wavefunction $\psi(x)$ given by

$$\psi(x) = \begin{cases} \sqrt{3}(1-x), & 0 < x < 1, \\ 0, & \text{otherwise.} \end{cases}$$

What is the expectation value of the position of this particle?

(A) 0
(B) 1
(C) 1/12
(D) 1/4
(E) 1/2

100. What are the energy levels of a quantized system consisting of a massless rigid rod of length a connecting two masses m, where n is a non-negative integer?

(A) $\frac{\hbar^2 n(n+1)}{ma^2}$
(B) $\frac{\hbar^2 n(n+1)}{2ma^2}$
(C) $\frac{\hbar^2 n}{2ma^2}$
(D) $\frac{\hbar^2 n}{ma^2}$
(E) $\frac{\hbar^2 (n+1)}{ma^2}$

Sample Exam 2

TABLE OF INFORMATION

Rest mass of the electron	m_e	=	9.11×10^{-31} kg
Magnitude of the electron charge	e	=	1.60×10^{-19} C
Avogadro's number	N_A	=	6.02×10^{23}
Universal gas constant	R	=	8.31 J/(mol · K)
Boltzmann's constant	k	=	1.38×10^{-23} J/K
Speed of light	c	=	3.00×10^{8} m/s
Planck's constant	h	=	6.63×10^{-34}J · s = 4.14×10^{-15} eV · s
	$\hbar$	=	$h/2\pi$
	hc	=	1240 eV · nm
Vacuum permittivity	ϵ_0	=	8.85×10^{-12} $C^2/(N \cdot m^2)$
Vacuum permeability	μ_0	=	$4\pi \times 10^{-7}$ T · m/A
Universal gravitational constant	G	=	6.67×10^{-11} $m^3/(kg \cdot s^2)$
Acceleration due to gravity	g	=	9.80 m/s^2
1 atmosphere pressure	1 atm	=	1.0×10^{5} N/m^2 = 1.0×10^{5} Pa
1 angstrom	1 Å	=	1×10^{-10} m = 0.1 nm

Prefixes for Powers of 10

10^{-15}	femto	f
10^{-12}	pico	p
10^{-9}	nano	n
10^{-6}	micro	μ
10^{-3}	milli	m
10^{-2}	centi	c
10^{3}	kilo	k
10^{6}	mega	M
10^{9}	giga	G
10^{12}	tera	T
10^{15}	peta	P

Rotational inertia about center of mass

Rod	$\frac{1}{12}M\ell^2$
Disk	$\frac{1}{2}MR^2$
Sphere	$\frac{2}{5}MR^2$

SAMPLE EXAM 2

Time — 170 minutes
100 questions

Directions: Each of the questions or incomplete statements below is followed by five suggested answers or completions. Select the one that is best in each case and then fill in the corresponding space on the answer sheet.

1. A ball of mass m is dropped from a tall building, and experiences a velocity-dependent air resistance force $F = bv$. What is its terminal velocity?

(A) $\frac{b}{mg}$
(B) $\frac{mb}{g}$
(C) $e^{b/m}$
(D) $\frac{mg}{b}$
(E) $\frac{mg}{b}(1 - e^{-b/m})$

2. A charged particle moving in the direction $\hat{\mathbf{n}} = \frac{1}{\sqrt{2}}(\hat{\mathbf{x}} + \hat{\mathbf{y}})$ enters a region of uniform magnetic field $\mathbf{B} = B_0\hat{\mathbf{x}}$. The path of the particle after it enters the field is a

(A) circle
(B) cycloid
(C) helix
(D) straight line
(E) logarithmic spiral

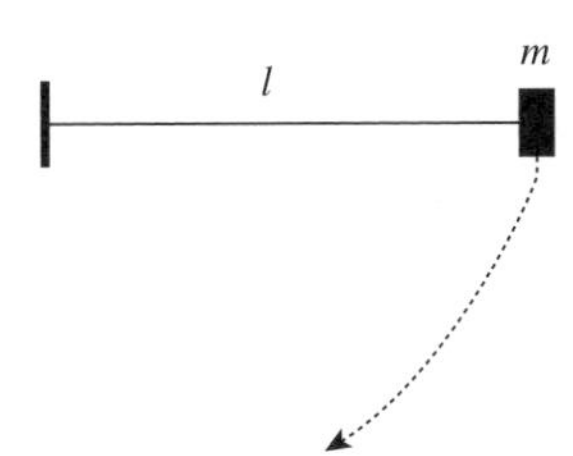

3. A massless rope of length l, attached to a fixed pivot at one end and with a mass m at the other end, is held horizontally and then released, as shown in the diagram. When the mass is at its lowest point, the tension in the rope is

(A) 0
(B) $gl/2$
(C) mg
(D) $2mg$
(E) $3mg$

4. A particle of charge q and mass m is suspended from a massless string. A constant electric field of known magnitude is turned on, perpendicular to the direction of gravity, and the rope forms some angle α with the vertical. A measurement of α determines which of the following quantities?

(A) m
(B) q
(C) q/m
(D) qm
(E) none of the above

5. A hydrogen atom transitions from the $n = 3$ to $n = 2$ states by emitting a photon. What is the wavelength of the photon?

(A) 347 nm
(B) 657 nm
(C) 985 nm
(D) 2.32 μm
(E) 1.34 mm

6. For a quantum operator to represent a physical observable, it must be

(A) Hermitian
(B) positive-definite
(C) finite-dimensional
(D) symmetric
(E) none of the above

7. If the net force on an object is zero, which of the following MUST be true?

I. Its angular momentum is constant.
II. Its velocity is zero.
III. Its acceleration is zero.

(A) I only
(B) II only
(C) III only
(D) I and II
(E) I and III

8. Fluorine is not naturally found as free atoms, but rather in compounds as the ion F^-. The electron configuration of a neutral fluorine atom is

(A) $1s^2\,2s^1$
(B) $1s^2\,2s^2$
(C) $1s^2\,2s^2\,2p^1$
(D) $1s^2\,2s^2\,2p^5$
(E) $1s^2\,2s^2\,2p^6$

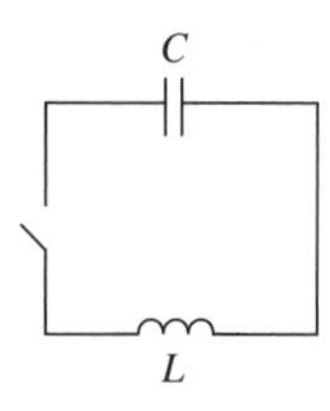

9. In the circuit shown in the diagram, the capacitor is initially charged, and the switch is closed at $t = 0$. Assuming all circuit elements have negligible resistance, the peak magnitude of the current is achieved at

(A) $t = \frac{\pi}{4}\sqrt{\frac{L}{C}}$
(B) $t = \frac{\pi}{\sqrt{LC}}$
(C) $t = 2\pi\sqrt{LC}$
(D) $t = \frac{\pi}{2}\sqrt{LC}$
(E) $t = \frac{\pi}{2\sqrt{LC}}$

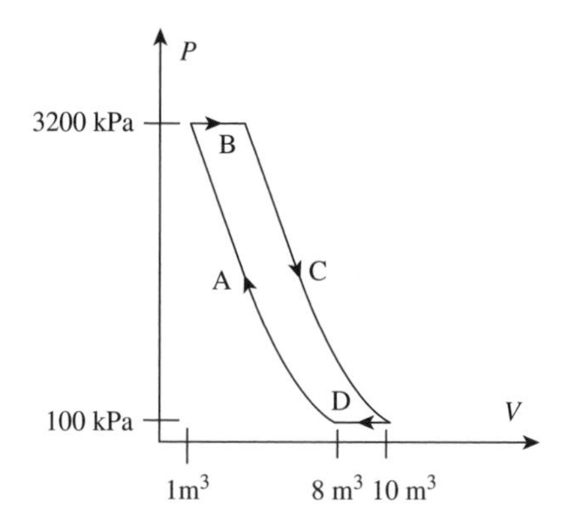

Questions 10 and 11 refer to the P–V diagram of an ideal gas undergoing the Brayton cycle. Steps A and C are isentropic, and steps B and D are isobaric.

10. What is the approximate work done by the gas over one cycle?

(A) −6,200 kJ
(B) −3,100 kJ
(C) 0
(D) 3,100 kJ
(E) 6,200 kJ

11. The gas used in the cycle is most likely

(A) monoatomic
(B) diatomic
(C) triatomic
(D) ionized
(E) heteronuclear

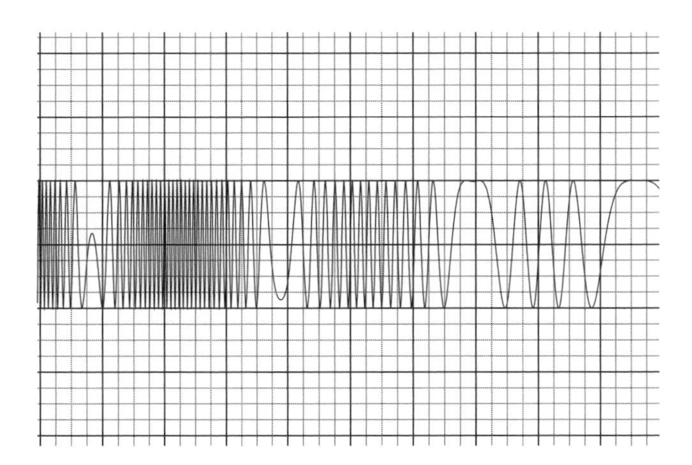

12. The oscilloscope trace shown in the diagram is an example of

(A) frequency modulation
(B) amplitude modulation
(C) pulse-code modulation
(D) single-sideband modulation
(E) clipping

13. Two exoplanets, A and B, are discovered orbiting a star much more massive than either planet. The semimajor axes of the orbits of A and B are found to be a and $a/2$, respectively. What is the ratio of the area enclosed by the orbit of planet A to the area enclosed by the orbit of planet B?

(A) $1/4$
(B) $1/(2\sqrt{2})$
(C) $1/2$
(D) $2\sqrt{2}$
(E) It cannot be determined from the information given.

14. A spin-1/2 particle has the angular wavefunction

$$\psi(\theta,\phi) = \frac{1}{\sqrt{2}}\left(Y_3^0(\theta,\phi) + Y_2^1(\theta,\phi)\right),$$

where $Y_l^m(\theta,\phi)$ are the normalized spherical harmonics. Which of the following is a possible result of measuring the particle's total spin quantum numbers j and m_j?

(A) $j = 3,\ m_j = 0$
(B) $j = 2,\ m_j = 1/2$
(C) $j = 7/2,\ m_j = -1/2$
(D) $j = 7/2,\ m_j = 3/2$
(E) $j = 9/2,\ m_j = -1/2$

15. The normalized energy eigenfunctions of the infinite square well of size L are $\psi_n(x) = \sqrt{\frac{2}{L}}\sin(n\pi x/L)$. The expectation value of energy of the state

$$\Psi = \frac{1}{\sqrt{2}}\psi_2 + \frac{1}{\sqrt{3}}\psi_3 + \frac{1}{\sqrt{6}}\psi_4$$

for a particle of mass m is

(A) $\dfrac{4\pi^2\hbar^2}{3mL^2}$
(B) $\dfrac{8\pi^2\hbar^2}{3mL^2}$
(C) $\dfrac{17\pi^2\hbar^2}{4mL^2}$
(D) $\dfrac{14\pi^2\hbar^2}{3mL^2}$
(E) $\dfrac{23\pi^2\hbar^2}{6mL^2}$

16. An ideal gas is maintained at a temperature of 250 K through contact with a thermal reservoir and is free to expand against a piston. If 5000 J of heat is slowly added to the gas, what is the change in entropy of the gas?

(A) 10 J/K
(B) $10 \ln 2$ J/K
(C) 20 J/K
(D) 40 J/K
(E) 500 J/K

17. The photoelectric effect provides direct experimental evidence for which of the following properties of light?

I. It has two linearly independent polarization states.
II. It carries kinetic energy proportional to its frequency.
III. It travels at a constant speed c in vacuum.

(A) I only
(B) II only
(C) I and II
(D) II and III
(E) I, II, and III

18. The Hamiltonian $H = eE_0z$, describing an atomic electron of charge $-e$ interacting with a uniform electric field in the z-direction, is responsible for

(A) the Zeeman effect
(B) the Lamb shift
(C) hyperfine splitting
(D) the Stark effect
(E) stimulated emission

19. A block of mass m_1 moving with velocity v collides elastically with a block of mass m_2 at rest. If m_1 continues moving in the same direction as it did prior to the collision, one can conclude

(A) $m_1 > m_2$
(B) $m_1 = m_2$
(C) $m_1 < m_2$
(D) momentum was not conserved in this collision
(E) none of the above

20. A 20 cm tall slice of a spherical mirror is oriented such that the image of a 2-meter tall person 1 meter away from the ATM will just fill the surface of the mirror. What must the radius of curvature R and convexity of the mirror be?

(A) $R = 22$ cm, convex
(B) $R = 40$ cm, concave
(C) $R = 80$ cm, concave
(D) $R = 4.5$ m, convex
(E) $R = 18$ m, convex

21. The quantized vibrations of a crystal lattice are called

(A) photons
(B) anyons
(C) phonons
(D) vibrons
(E) rotons

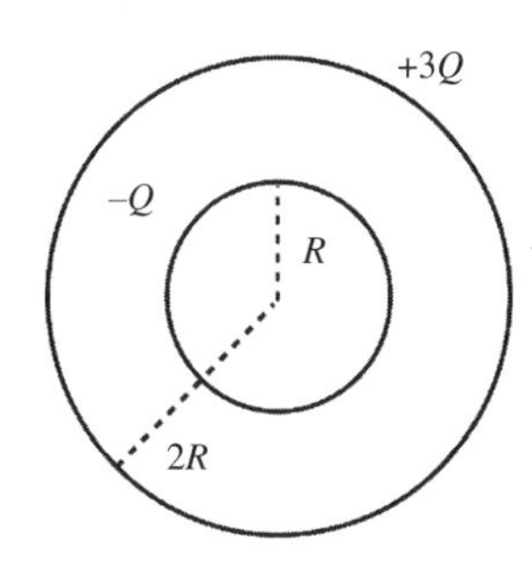

22. Shown in the diagram are two concentric thin spherical shells of radii R and $2R$, the outer one carrying charge $+3Q$ and the inner one carrying charge $-Q$. Setting the electric potential equal to zero at infinity, which of the following graphs best represents the electric potential as a function of r, the distance from the center of the shells?

(A)

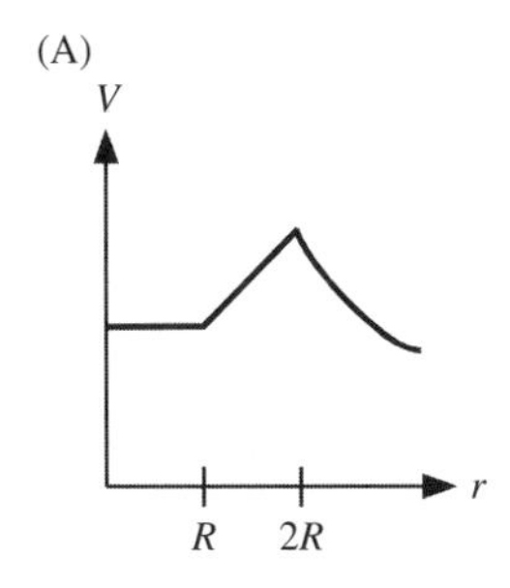

(B)

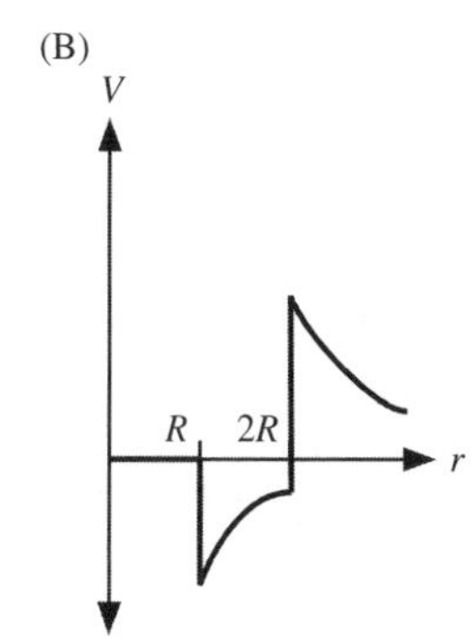

(C)

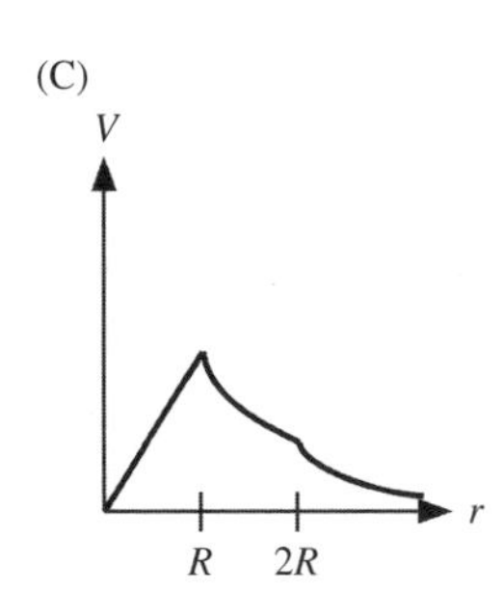

(D)

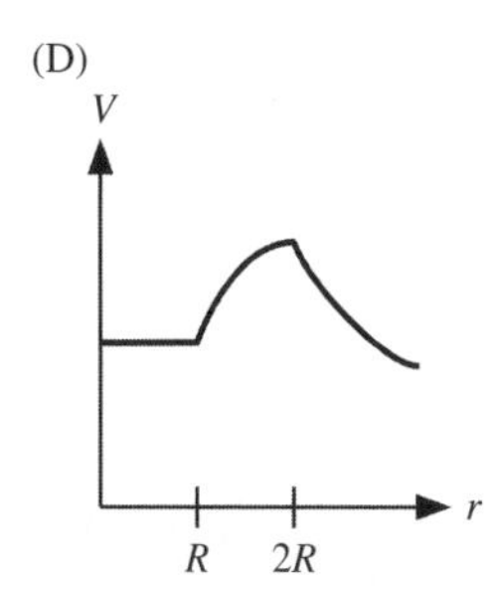

(E)

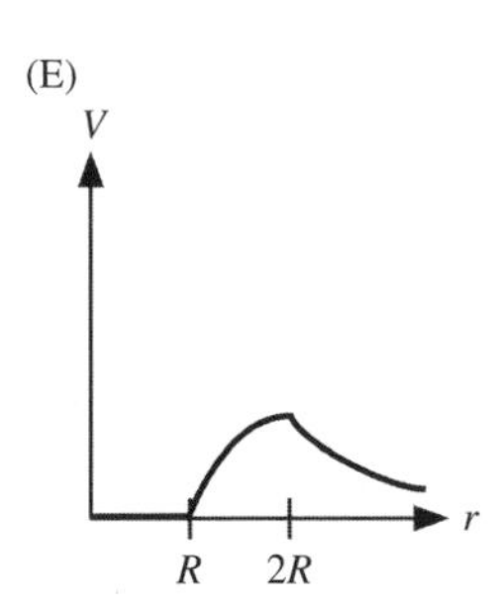

23. The force-carrying particle responsible for binding the quarks in the proton is called the

(A) photon
(B) gluon
(C) W boson
(D) Z boson
(E) Higgs boson

24. A charge $-e$ at the origin is subject to a uniform electric field $\mathbf{E} = -E\hat{\mathbf{x}}$. After traveling to $(3, 4)$, what is the change in potential energy of the charge?

(A) $-3Ee$
(B) $3Ee$
(C) $-5Ee$
(D) $5Ee$
(E) $-7Ee$

25. Which of the following are true statements about Gauss's law for magnetism, $\nabla \cdot \mathbf{B} = 0$?

I. It implies that magnetic monopoles do not exist in nature.
II. It is incompatible with the continuity equation for $\mathbf{J}$.
III. It allows the magnetic field to be written in terms of a vector potential as $\mathbf{B} = \nabla \times \mathbf{A}$.

(A) I only
(B) II only
(C) III only
(D) I and II
(E) I and III

26. A relativistic particle of mass m has momentum $p = mc$. What is the particle's energy?

(A) mc^2
(B) $\sqrt{2}mc^2$
(C) $2mc^2$
(D) $4mc^2$
(E) none of the above

27. The hot, dense gas of electrons and positive ions known as a plasma is capable of supporting charge density waves known as plasma oscillations. Let n_e be the number density of electrons, e the charge of the electrons, and m^* an effective mass of the electrons in the plasma. The frequency of plasma oscillations is

(A) $\omega = \dfrac{\epsilon_0}{n_e e^2 m^*}$

(B) $\omega = \dfrac{m^* e^2}{n_e \epsilon_0}$

(C) $\omega = \dfrac{e^2}{n_e \epsilon_0}$

(D) $\omega = \sqrt{\dfrac{n_e e^2 m^*}{\epsilon_0}}$

(E) $\omega = \sqrt{\dfrac{n_e e^2}{m^* \epsilon_0}}$

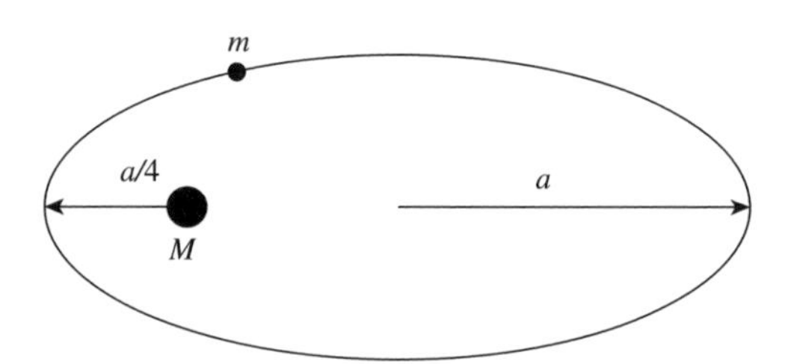

28. A planet of mass m orbits a star of mass M in an elliptical orbit with semimajor axis a, as shown in the diagram. The distance of closest approach to the star is $a/4$. Assuming $m \ll M$, the ratio of the planet's speed at perigee (when the planet is closest to the star) to the planet's speed at apogee (when the planet is furthest away from the star) is

(A) 1/4
(B) 1/3
(C) 4
(D) 7
(E) 16

29. Two identical sailboats race across a lake, starting from rest. Boat 1 reaches the finish line first with velocity v_1, and boat 2 arrives later with velocity $v_2 > v_1$. Let F_{t1} and F_{t2} be the average force per unit *time* on boats 1 and 2, respectively, and let F_{d1} and F_{d2} be the average force per unit *distance* on boats 1 and 2. Which of the following MUST be true?

(A) $F_{t1} > F_{t2}$
(B) $F_{t1} < F_{t2}$
(C) $F_{d1} > F_{d2}$
(D) $F_{d1} < F_{d2}$
(E) none of the above

30. The first two normalized position-space energy eigenfunctions of the harmonic oscillator Hamiltonian $H = \frac{p^2}{2m} + \frac{1}{2}m\omega^2 x^2$ are

$$\psi_0(x) = \left(\frac{m\omega}{\pi\hbar}\right)^{1/4} e^{-m\omega x^2/2\hbar},$$
$$\psi_1(x) = \sqrt{2}\left(\frac{m\omega}{\pi\hbar}\right)^{1/4} xe^{-m\omega x^2/2\hbar}.$$

A delta-function perturbation $V(x) = \epsilon\delta(x)$ is added to the harmonic oscillator Hamiltonian. What are the new energies E_0 and E_1 to first order in perturbation theory?

(A) $E_0 = \hbar\omega/2, E_1 = 3\hbar\omega/2$
(B) $E_0 = \hbar\omega/2 + \epsilon\left(\frac{m\omega}{\pi\hbar}\right)^{1/4}, E_1 = 3\hbar\omega/2$
(C) $E_0 = \hbar\omega/2 + \epsilon\sqrt{\frac{m\omega}{\pi\hbar}}, E_1 = 3\hbar\omega/2$
(D) $E_0 = \hbar\omega/2 + \epsilon\sqrt{\frac{m\omega}{\pi\hbar}}, E_1 = 3\hbar\omega/2 + 2\epsilon\sqrt{\frac{m\omega}{\pi\hbar}}$
(E) $E_0 = \hbar\omega + \epsilon\sqrt{\frac{m\omega}{\pi\hbar}}, E_1 = 2\hbar\omega$

31. At sufficiently high temperature T, which of the following contributes to the total energy of a diatomic molecule?

I. Translational kinetic energy
II. Rotational kinetic energy
III. Vibrational potential energy

(A) I only
(B) II only
(C) I and II
(D) I and III
(E) I, II, and III

32. During the adiabatic expansion phase of a Carnot cycle, one mole of gas expands to twice its original size. The change in entropy of the gas during this process is

(A) $R \ln 2$
(B) $-R \ln 2$
(C) $2R$
(D) $-2R$
(E) 0

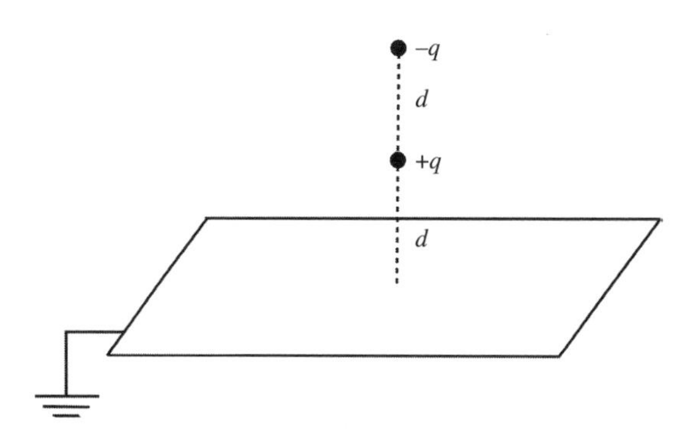

33. A particle of charge $+q$ is placed at the point $(0, 0, d)$, between an infinite grounded conducting plate at $z = 0$ and a stationary charge $-q$ at $(0, 0, 2d)$, as shown in the diagram. What is the force on the charge $+q$?

(A) $-\frac{q^2}{8\pi\epsilon_0 d^2}\hat{\mathbf{z}}$
(B) $-\frac{7q^2}{24\pi\epsilon_0 d^2}\hat{\mathbf{z}}$
(C) $\frac{11q^2}{72\pi\epsilon_0 d^2}\hat{\mathbf{z}}$
(D) $\frac{25q^2}{72\pi\epsilon_0 d^2}\hat{\mathbf{z}}$
(E) $\frac{31q^2}{144\pi\epsilon_0 d^2}\hat{\mathbf{z}}$

Questions 34 and 35 refer to the following scenario. A new star is discovered with an optical telescope, from which it is deduced that the star emits most of its power in the orange region of the visible spectrum, at wavelength approximately 600 nm.

34. Assuming the star behaves as a blackbody and neglecting possible redshift, what is its approximate temperature?

(A) 200 K
(B) 5000 K
(C) 6×10^4 K
(D) 3×10^6 K
(E) 2×10^9 K

35. It is later discovered that the star is receding at a peculiar velocity of $0.2c$. The ratio between the true temperature T_{true} and the measured temperature T_{meas} is

(A) 1/5
(B) 2/3
(C) 3/2
(D) $\sqrt{2/3}$
(E) $\sqrt{3/2}$

36. A one-dimensional system has Lagrangian

$$L(q, \dot{q}, t) = A\dot{q}^2 + \sin(q/L - \omega t)$$

for constants A, L, and ω. What is the Euler–Lagrange equation of motion?

(A) $\dot{q} = \frac{\omega}{2AL} \sin(q/L - \omega t)$
(B) $\dot{q} = -\frac{\omega}{A} \cos(q/L - \omega t)$
(C) $\dot{q} = -\frac{\omega}{2A} \cos(q/L - \omega t)$
(D) $\ddot{q} = \frac{1}{2AL} \cos(q/L - \omega t)$
(E) $\ddot{q} = -\frac{1}{AL} \cos(q/L - \omega t)$

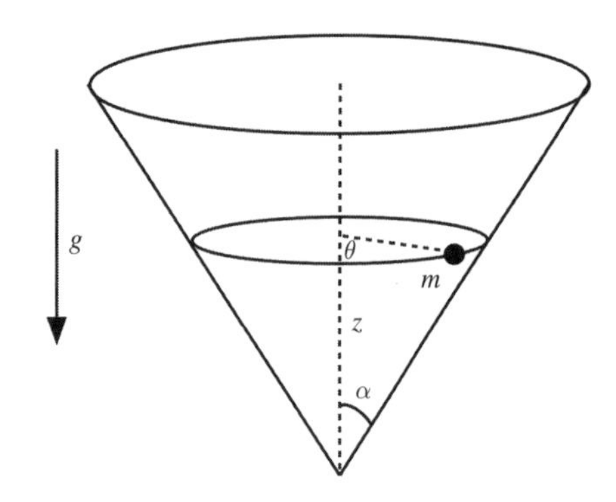

37. A particle of mass m is constrained to move on a cone of opening angle α, oriented as shown in the diagram. The Hamiltonian for this system is given by

$$H = \frac{p_z^2}{2m} \cos^2 \alpha + \frac{p_\theta^2}{2mz^2} \cot^2 \alpha + mgz.$$

What are Hamilton's equations for the coordinate z?

(A) $\dot{p}_z = \frac{p_z}{m} \cos^2 \alpha - mg$; $\dot{z} = \frac{p_z^2}{2mz^2} \cot^2 \alpha$
(B) $\dot{p}_z = \frac{p_\theta^2}{2mz} \cot^2 \alpha - mg$; $\dot{z} = \frac{p_z}{m} \cos^2 \alpha$
(C) $\dot{p}_z = \frac{p_z}{m} \cos^2 \alpha$; $\dot{z} = \frac{p_\theta^2}{mz} \cot^2 \alpha - mg$
(D) $\dot{p}_z = \frac{p_\theta^2}{mz^3} \cot^2 \alpha - mg$; $\dot{z} = \frac{p_z}{m} \cos^2 \alpha$
(E) $\dot{p}_z = -\frac{p_\theta^2}{mz} \cot^2 \alpha + mg$; $\dot{z} = -\frac{p_z}{m} \cos^2 \alpha$

38. The specific heat at constant volume, C_V, of a solid is observed at low temperatures T to follow the Debye law $C_V = AT^3$, with A a constant. What is the internal energy of the solid $U(T)$ as a function of temperature, assuming $U(0) = 0$, in the regime of validity of the Debye law?

(A) $3AT^2$
(B) AT^3
(C) $\frac{1}{3}AT^3$
(D) $\frac{1}{4}AT^4$
(E) AT^4

39. Let $|n\rangle$ denote a set of real, orthonormal energy eigenfunctions of a Hamiltonian $\hat{H}$ in one dimension, with energies E_n. Let $\hat{p}$ denote the momentum operator. Which of the following must be true?

I. $\langle m|n\rangle = m + n$
II. $|n\rangle$ is an eigenfunction of $\hat{p}$
III. $\langle m|\hat{H}|n\rangle = \delta_{mn} E_n$

(A) I only
(B) II only
(C) III only
(D) I and II
(E) II and III

40. The Meissner effect refers to the tendency of superconductors to

(A) develop a surface charge density
(B) expel magnetic fields
(C) acquire a finite resistance at a critical temperature T_c
(D) spontaneously develop an internal electric field
(E) have persistent currents

41. The hydrogen isotope tritium, ^{3}H, contains one proton and two neutrons and has a half-life of approximately 12 years. The binding energy of tritium is closest to

(A) 8.5 eV
(B) 8.5 keV
(C) 8.5 MeV
(D) 8.5 GeV
(E) 8.5 TeV

42. A positive muon stopped by matter can attract an electron to form an exotic bound state known as muonium, where the muon (which has the same charge as the proton) acts as the nucleus. Let m_μ be the mass of the muon, m_e the mass of the electron, and m_p the mass of the proton. What is the Bohr radius of muonium, in terms of the Bohr radius of ordinary hydrogen a_0 and the masses of the particles?

(A) $a_0 \dfrac{m_p}{m_\mu}$
(B) $a_0 \dfrac{m_\mu}{m_p}$
(C) $a_0 \left(\dfrac{m_\mu}{m_p}\right)^2$
(D) $a_0 \dfrac{m_p(m_e + m_\mu)}{m_\mu(m_e + m_p)}$
(E) $a_0 \dfrac{m_p(m_e + m_p)}{m_\mu(m_e + m_\mu)}$

43. Let $\hat{x}$ and $\hat{p}$ be the quantum-mechanical position and momentum operators, respectively. The commutator $[\hat{x}, \hat{p}^2]$ is equivalent to which of the following?

(A) 0
(B) $i\hbar$
(C) $-i\hbar\hat{x}$
(D) $2i\hbar\hat{p}$
(E) $2i\hbar$

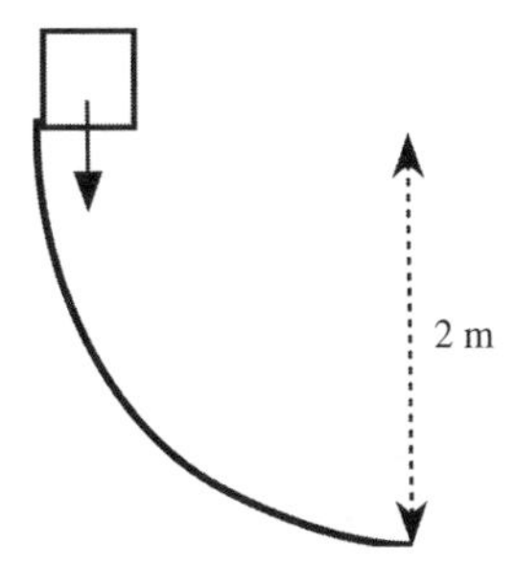

44. A block of mass 2 kg slides down a ramp in the shape of a quarter-circle of radius 2 m, as shown in the diagram. If the block reaches the bottom of the ramp with velocity 4 m/s, then, ignoring air resistance, the work done by friction during the slide down the ramp is most nearly

(A) 0 J
(B) 8 J
(C) 12 J
(D) 24 J
(E) 40 J

45. Consider an infinite charge-carrying wire, of charge per unit length λ. Setting the zero of electric potential at distance a from the wire, what is the electric potential as a function of the distance r from the wire?

(A) $\dfrac{\lambda}{2\pi\epsilon_0} \ln(a/r)$
(B) $\dfrac{\lambda}{4\pi\epsilon_0} \ln(a/r)$
(C) $\dfrac{\lambda}{2\pi\epsilon_0} \ln(r/a)$
(D) $\dfrac{\lambda}{4\pi\epsilon_0 r}$
(E) $\dfrac{\lambda(r-a)}{4\pi\epsilon_0 r}$

46. An ideal beam-splitter is an optical device that lets part of an incident beam of light pass through and reflects the remainder, with no absorption taking place in the beam-splitter. Let the incident beam have complex amplitude E, the reflected beam have amplitude E_r, and the transmitted beam have amplitude E_t. Which of the following MUST be true?

(A) $E_r = E/\sqrt{2}$
(B) $E_r = E/2$
(C) $E = E_r + E_t$
(D) $|E|^2 = |E_r|^2 + |E_t|^2$
(E) $|E|^4 = |E_r|^4 + |E_t|^4$

47. The Hamiltonian operator for a free particle of mass m moving in three dimensions is

(A) $-\dfrac{\hbar^2\nabla^2}{m}$
(B) $-\dfrac{\hbar^2\nabla^2}{2m}$
(C) $-i\hbar\nabla$
(D) $\dfrac{\hbar^2\nabla^2}{2m}$
(E) 0

48. In proton therapy, medium-energy protons are directed at a cancer patient's tumor in order to irradiate it. Which of the following pieces of information would be MOST useful in determining the correct energy and angle with which to fire the protons?

(A) The charge-to-mass ratio of the proton
(B) The distance traveled in human tissue as a function of energy
(C) The cross section for proton scattering on carbon nuclei
(D) The mean lifetime of the proton
(E) The binding energy of the proton

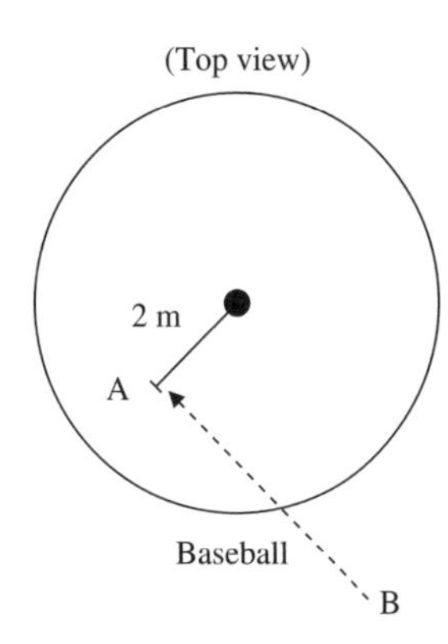

49. Student A, of mass 100 kg, stands 2 meters from the center of a circular platform which is free to rotate on frictionless bearings. Student B, not standing on the platform, tosses student A a baseball of mass 0.09 kg, which reaches student A with a velocity of 20 m/s directed perpendicular to the line joining student A and the center of the platform. If the platform has moment of inertia 200 kg·m^2, what is its approximate angular velocity after student A catches the baseball?

(A) 0.006 rad/s
(B) 0.009 rad/s
(C) 0.018 rad/s
(D) 0.067 rad/s
(E) 0.1 rad/s

50. In Compton scattering, the change in wavelength of the scattered light is given in terms of the electron mass m_e and the scattering angle θ by which of the following?

(A) $\frac{h}{m_e c}(1-\cos\theta)$
(B) $\frac{m_e}{hc}(1+\cos\theta)$
(C) $\frac{h}{m_e c}(1+\sin\theta)$
(D) $\frac{m_e}{hc}\sin^2\theta$
(E) $\frac{1}{hcm_e}(1-\sin\theta)$

51. The binding energy of the electron in the Li^{++} ion is approximately

(A) 1.51 eV
(B) 13.6 eV
(C) 40.8 eV
(D) 122.4 eV
(E) 1102 eV

52. Which of the following statements is NOT consistent with the three laws of thermodynamics?

(A) The entropy of a perfect crystal of a pure substance must approach zero at absolute zero.
(B) The entropy of an isolated system can sometimes decrease.
(C) The ground state degeneracy of a system determines its entropy.
(D) Absolute zero can never be reached in experiments.
(E) The entropy of a system can be nonzero at absolute zero.

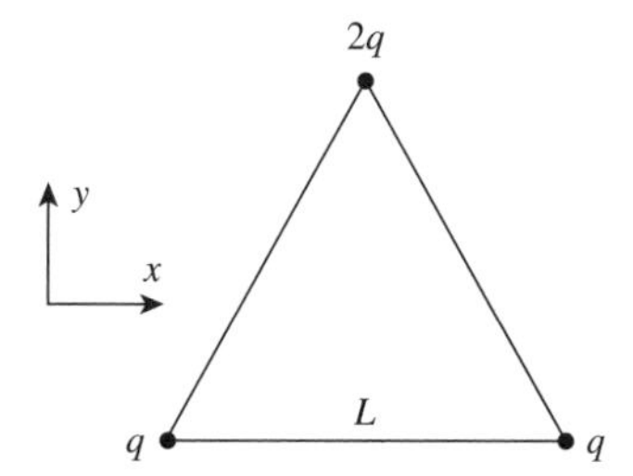

53. Questions 53 and 54 refer to the diagram above, with charges q, q, and $2q$ placed at the corners of an equilateral triangle of side length L, and the $\hat{\mathbf{x}}$- and $\hat{\mathbf{y}}$-axes oriented as shown. What is the electric field at the center of the triangle?

(A) $\frac{q}{4\pi\epsilon_0 L^2}(\hat{\mathbf{x}}+\sqrt{3}\hat{\mathbf{y}})$
(B) $-\frac{3q}{4\pi\epsilon_0 L^2}\hat{\mathbf{y}}$
(C) $\frac{4q}{3\pi\epsilon_0 L^2}\hat{\mathbf{y}}$
(D) $-\frac{q}{\pi\epsilon_0 L\sqrt{3}}\hat{\mathbf{y}}$
(E) 0

54. What is the electric potential at the center of the triangle, relative to infinity?

(A) $\frac{q\sqrt{3}}{\pi\epsilon_0 L}$
(B) $\frac{q\sqrt{3}}{4\pi\epsilon_0 L}$
(C) $\frac{4q\sqrt{3}}{3\pi\epsilon_0 L}$
(D) $\frac{4q}{3\pi\epsilon_0 L^2}$
(E) 0

55. A car engine operates between a cold reservoir at temperature 27 °C and a hot reservoir at 127 °C. What is the minimum amount of heat the engine must absorb from the hot reservoir in a period of 1 minute to obtain a power output of 100 kW?

(A) 1,500 kJ
(B) 6,000 kJ
(C) 8,000 kJ
(D) 18,000 kJ
(E) 24,000 kJ

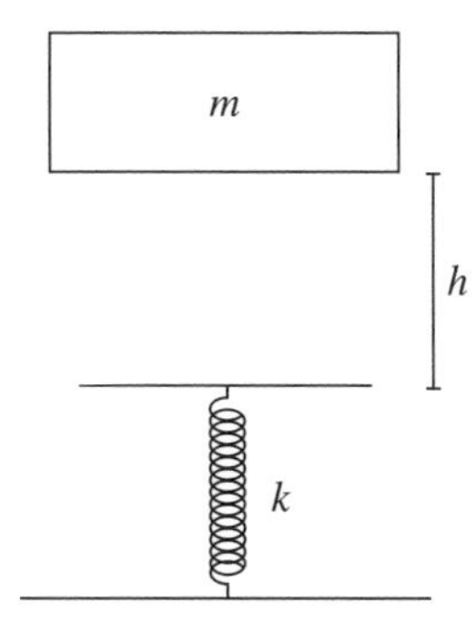

56. A brick of mass m falls onto a massless spring with spring constant k from a height h above it. What is the maximum distance the spring will be compressed from its equilibrium length?

(A) $\frac{mg}{k}$
(B) $\frac{mk}{2gh}$
(C) $\frac{k}{mg}\left(1+\sqrt{\frac{mg}{kh}}\right)$
(D) $\frac{mg}{k}\left(1+\sqrt{\frac{2kh}{mg}}\right)$
(E) $\frac{mg}{k}\left(1+\sqrt{1+\frac{2kh}{mg}}\right)$

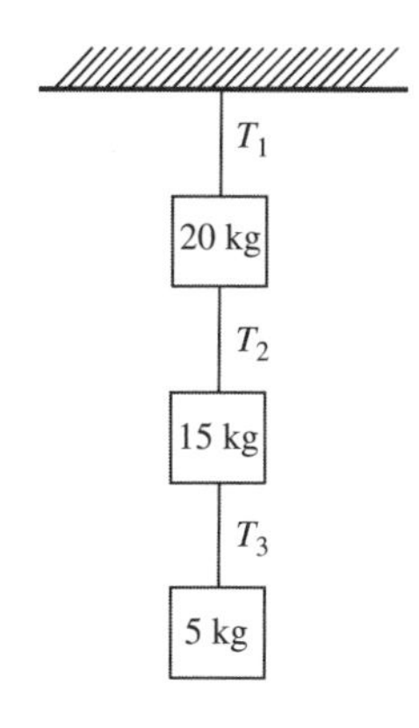

57. Three weights are suspended from a ceiling using massless ropes, as shown in the diagram. The tensions in the ropes are T_1, T_2, and T_3. What is T_1/T_3?

(A) 0.25
(B) 1
(C) 3
(D) 4
(E) 8

58. A particle's normalized spin wavefunction has the form

$$\psi(\theta,\phi) = \sqrt{\frac{3}{2\pi}}\sin\theta\cos 2\phi\sin\phi.$$

What is the expectation value of the particle's z-component of orbital angular momentum L_z?

(A) 0
(B) $-3\hbar/2\pi$
(C) $3\hbar/2\pi$
(D) $-3\hbar/\pi$
(E) $3\hbar/\pi$

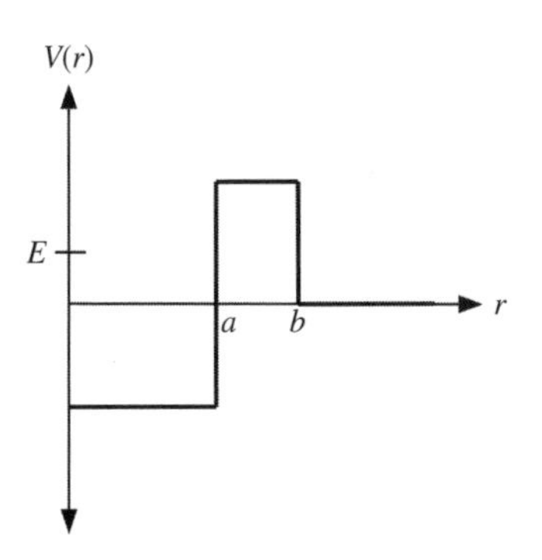

59. The strong nuclear force binding an alpha particle to a nucleus can be modeled by the potential shown in the diagram. Which of the following plots best illustrates the radial wavefunction of an alpha particle with energy E that tunnels out of the nucleus in alpha decay?

(A)

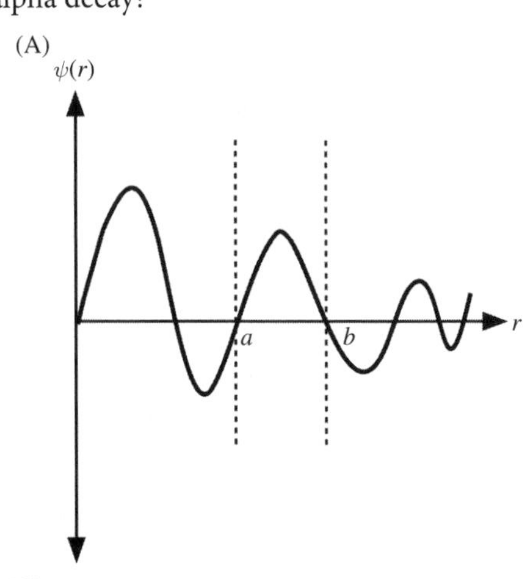

(B)

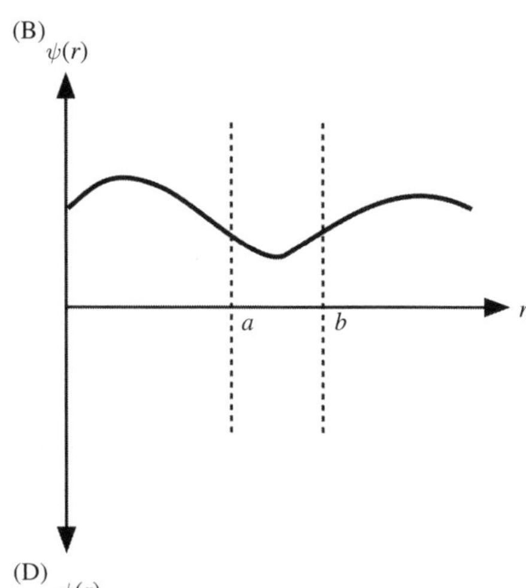

(C)

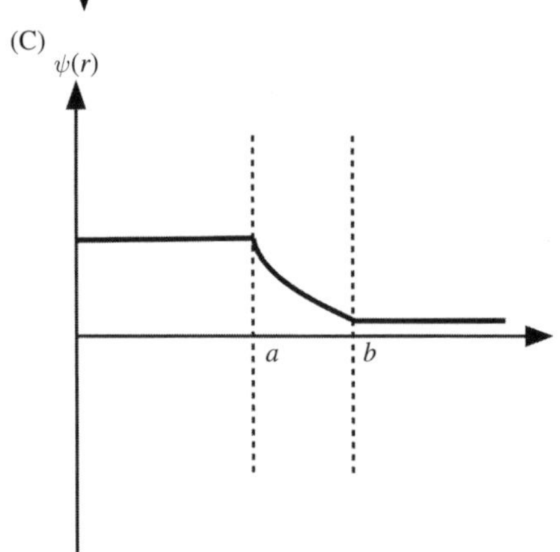

(D)

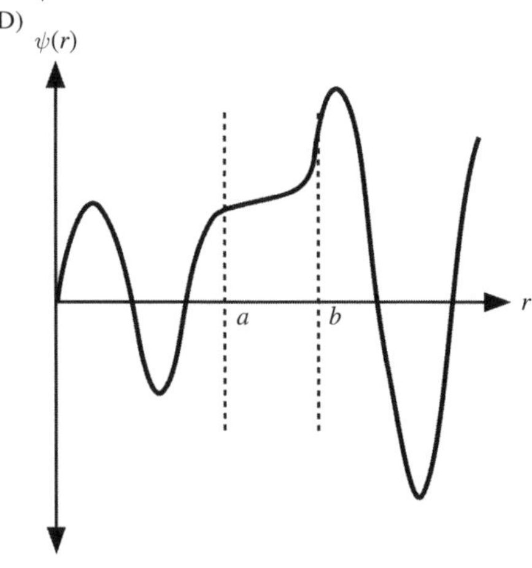

(E)

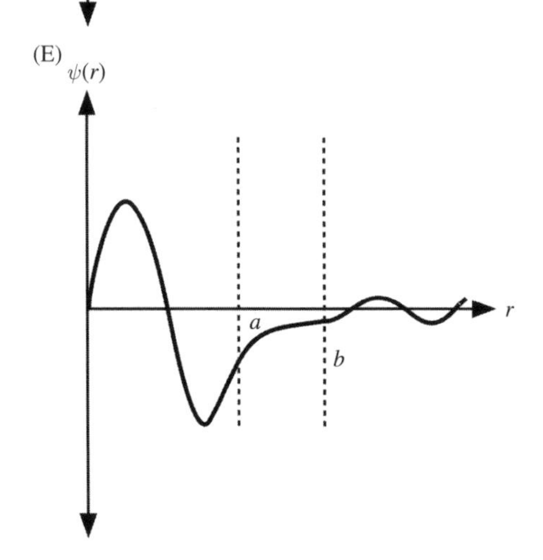

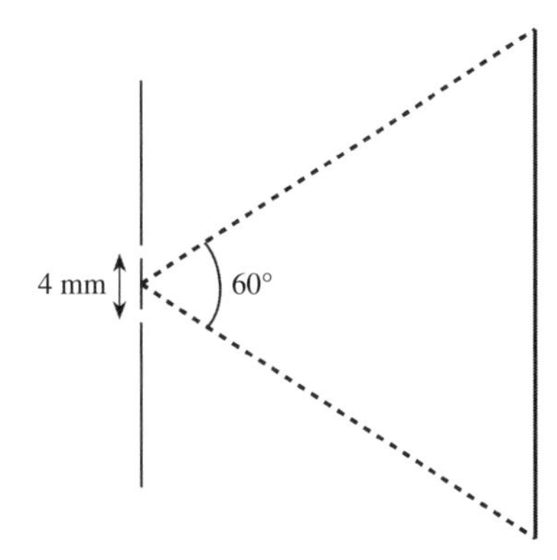

60. A coherent beam of monochromatic light of wavelength 500 nm is directed towards two very thin slits separated by a distance 4 mm. Far behind the slits is a screen covering an angular region of 60°, as shown in the diagram. Approximately how many bright interference bands are visible on the screen?

(A) 0
(B) 4,000
(C) 6,800
(D) 8,000
(E) 13,600

61. A stationary telescope monitoring a rocket ship observes the ship emitting flashes of light at 1-second intervals. If the ship begins moving toward the telescope at speed $0.6c$, with what period does the telescope observe the light flashes?

(A) 0.5 s
(B) 0.8 s
(C) 1 s
(D) 1.2 s
(E) 2 s

62. A particle moving at speed $0.8c$ enters a tube of length 30 m and hits a target at the end of the tube. How far away is the target when the particle enters the tube, in the reference frame of the particle?

(A) 18 m
(B) 24 m
(C) 30 m
(D) 50 m
(E) 60 m

63. The volume of the first Brillouin zone of a simple cubic lattice of lattice spacing a is

(A) a^3
(B) $1/a^3$
(C) $(a/2\pi)^3$
(D) $(2\pi/a)^3$
(E) $a^3/2\pi$

64. Early observations of beta decay of the neutron showed that the emitted electron had a broad energy spectrum, rather than a fixed energy. This was taken as evidence for the existence of the

(A) neutrino
(B) positron
(C) muon
(D) strange quark
(E) pion

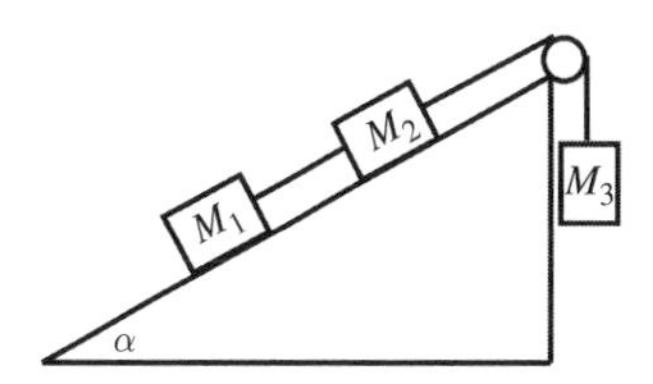

65. Blocks of masses M_1, M_2, and M_3 are arranged on a frictionless inclined plane at angle α as shown in the diagram. The pulley at the top of the plane is frictionless and massless, and the system is in static equilibrium. What is α in terms of M_1, M_2, and M_3?

(A) $\tan^{-1}\left(\frac{M_3}{M_1}\right)$
(B) $\sin^{-1}\left(\frac{M_3}{M_1+M_2}\right)$
(C) $\sin^{-1}\left(\frac{M_1+M_2}{M_3}\right)$
(D) $\cos^{-1}\left(\frac{M_1+M_2}{M_3}\right)$
(E) $\cos^{-1}\left(\frac{M_3}{M_1+M_2}\right)$

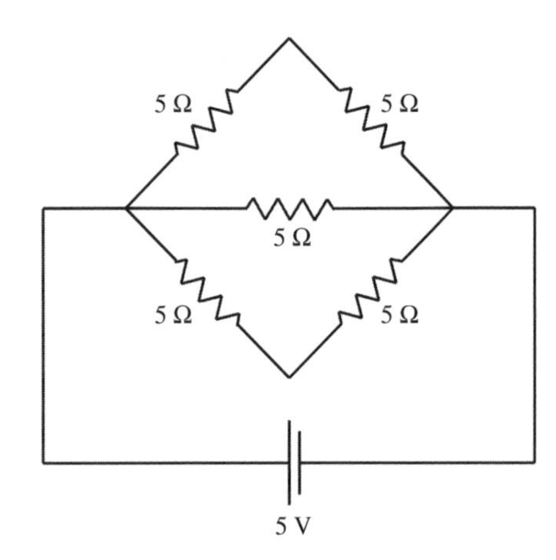

66. A 5 V battery supplies the emf for the circuit shown in the diagram, where all resistors are 5 Ω. What current flows through the circuit? (You may assume the wires are resistanceless and the battery has negligible internal resistance.)

(A) 0.2 A
(B) 0.5 A
(C) 1 A
(D) 2 A
(E) 5 A

67. Questions 67 and 68 refer to the Pauli matrices:

$$\sigma_x = \begin{pmatrix} 0 & 1 \\ 1 & 0 \end{pmatrix}, \quad \sigma_y = \begin{pmatrix} 0 & -i \\ i & 0 \end{pmatrix}, \quad \sigma_z = \begin{pmatrix} 1 & 0 \\ 0 & -1 \end{pmatrix}.$$

What is the determinant of the matrix

$$M = \sigma_x^2 + \sigma_y^2 + \sigma_z^2?$$

(A) 0
(B) 1
(C) 3
(D) 4
(E) 9

68. The state of a spin-1/2 particle is described by the spinor

$$\eta = \mathcal{N}\begin{pmatrix} i \\ 2-i \end{pmatrix}$$

where $\mathcal{N}$ is a normalization constant. What is the expectation value of S_x, the spin projection onto the x-axis, in the state η?

(A) $-\hbar$
(B) $-\hbar/3$
(C) $-\hbar/6$
(D) 0
(E) $\hbar/2$

69. Which of the following could represent the displacement of a standing wave?

(A) $\cos(kx - \omega t)$
(B) $\sin(kx - \omega t)$
(C) $(x - vt)^2$
(D) $\sin kx \cos \omega t$
(E) $\omega t \sin^2 kx$

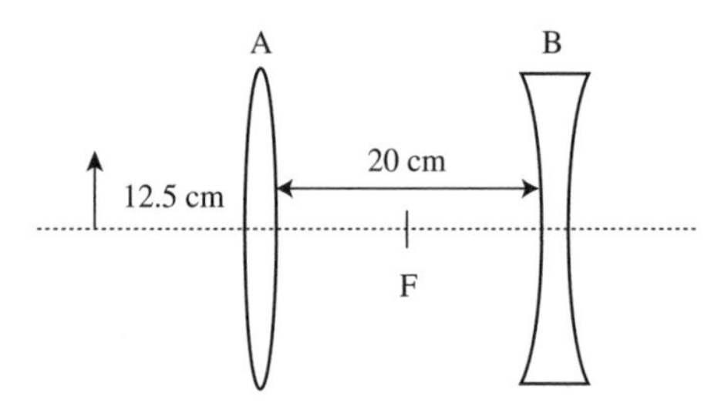

70. A converging lens A and a diverging lens B, both with focal length 10 cm, are arranged so that the midpoint between the lenses F coincides with both lenses' foci. An object is placed 12.5 cm to the left of A. Which of the following gives the correct position and orientation of the image?

(A) 5 cm to the right of A, inverted
(B) 5 cm to the right of A, upright
(C) 7.5 cm to the right of A, inverted
(D) 30 cm to the right of B, inverted
(E) 30 cm to the right of B, upright

71. A parallel-plate capacitor with capacitance C is at rest in frame S, with the plates of the capacitor parallel to the xy-plane. In frame $\overline{S}$, moving in the positive $\hat{\mathbf{x}}$ direction at speed v, what is the new capacitance in terms of C?

(A) C
(B) $C\sqrt{1 - v^2/c^2}$
(C) $\dfrac{C}{\sqrt{1 - v^2/c^2}}$
(D) $C(1 - v^2/c^2)$
(E) $\dfrac{C}{1 - v^2/c^2}$

72. In underground particle detection experiments, the main source of naturally occurring ionizing radiation at energies above 200 MeV is

(A) thermal radiation
(B) cosmic ray muons
(C) solar neutrinos
(D) seismic noise
(E) solar flares

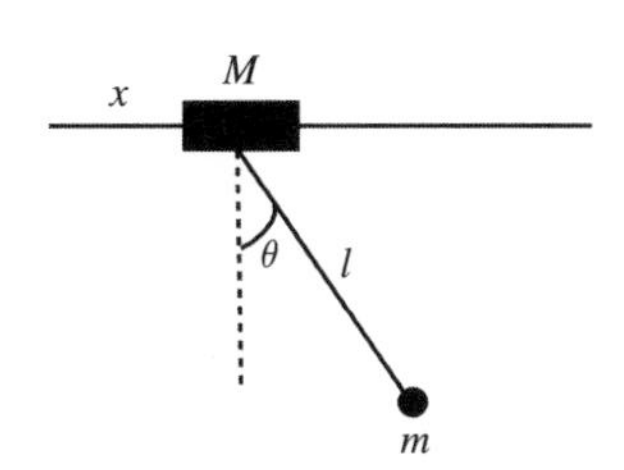

73. A point mass m is attached with a massless rod of length l to a pivot of mass M, which is free to slide along a frictionless bar. Letting x be the position of the pivot and θ the angle of the rod, what is a possible Lagrangian for this system?

(A) $L = \frac{1}{2}M\dot{x}^2 + \frac{1}{2}ml^2\dot{\theta}^2 - mgl\cos\theta$
(B) $L = \frac{1}{2}(M + m)\dot{x}^2 + \frac{1}{2}ml^2\dot{\theta}^2 + mgl\cos\theta$
(C) $L = \frac{1}{2}(M+m)\dot{x}^2 + \frac{1}{2}ml^2\dot{\theta}^2 + ml\dot{x}\dot{\theta}\cos\theta + mgl\cos\theta$
(D) $L = \frac{1}{2}(M+m)\dot{x}^2 + \frac{1}{2}ml^2\dot{\theta}^2 + 2ml\dot{x}\dot{\theta}\sin\theta - mgl\cos\theta$
(E) $L = \frac{1}{2}M\dot{x}^2 + \frac{1}{2}ml^2\dot{\theta}^2\sin^2\theta + ml\dot{x}\dot{\theta}\cos^2\theta - mgl\cos\theta$

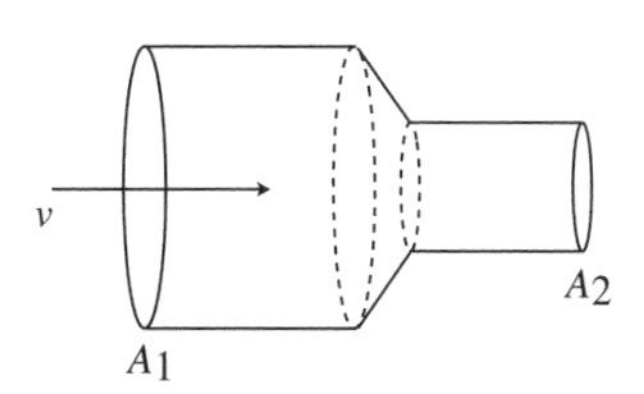

74. A pipe has cross-sectional area A_1 at one point, but subsequently narrows to a cross-sectional area A_2. If the pressure of an incompressible fluid of density ρ flowing toward the narrow end is p in the first region, and its velocity is v, what is the pressure in the second narrow region?

(A) p
(B) $\frac{1}{2}\frac{A_2^2}{A_1^2}\rho v^2$
(C) $p + \frac{1}{2}\left(\frac{A_2^2}{A_1^2} - 1\right)\rho v^2$
(D) $p + \frac{1}{2}\left(1 - \frac{A_1^2}{A_2^2}\right)\rho v^2$
(E) $p + \frac{1}{2}\left(\frac{A_2^2}{A_1^2} + 1\right)\rho v^2$

75. A particle in three dimensions has normalized radial wavefunction

$$\psi(r) = \begin{cases} 0, & 0 \le r \le a \\ \sqrt{a}/r^2, & r > a. \end{cases}$$

What is the probability the particle will be found between $r = a$ and $r = 2a$?

(A) 1/3
(B) 1/2
(C) $1/\sqrt{3}$
(D) $1/\sqrt{2}$
(E) 1

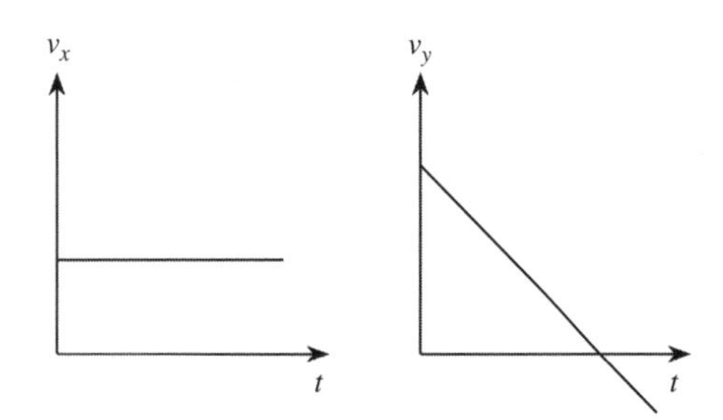

76. The diagram above shows plots of components of velocity v_x and v_y versus time t, with identical scales on both plots. Gravity acts in the $-\hat{\mathbf{y}}$-direction. Ignoring air resistance, these plots could represent which of the following scenarios?

    (A) A ball dropped from the top of a high building
    (B) A rock thrown from a high building at an angle of 45° below the horizontal
    (C) A brick thrown from ground level at an angle of 60° above the horizontal
    (D) A golf ball on an elevated tee struck at an angle of 30° above the horizontal
    (E) A mass attached to a vertical spring which is compressed and then released

77. A boater wearing sunglasses with the polarization axis vertical observes that the intensity of sunlight reflected off the water and transmitted through her sunglasses gradually decreases once the Sun is overhead, and goes to zero when the Sun is 30° above the horizon. What is the index of refraction of the water? You may assume the index of refraction of air is 1.

    (A) $1/2$
    (B) 1
    (C) $\sqrt{2}$
    (D) $\sqrt{3}$
    (E) 2

78. A body of mass m and charge q at the origin is subjected to an electric field $\mathbf{E}(t) = E_0 \sin(\omega t)\hat{\mathbf{x}}$ for a time $T \gg 2\pi/\omega$. Which of the following will cause the average power radiated by the charge to decrease?

    (A) Increasing m
    (B) Increasing q
    (C) Increasing E_0
    (D) Decreasing ω
    (E) Decreasing T

79. A positively charged particle q is traveling at constant (nonrelativistic) velocity in the $+\hat{\mathbf{z}}$-direction and passes through the center of a loop of wire lying in the xy-plane, at $t = 0$. Which of the following plots best illustrates $\Phi(t)$, the electric flux through the loop as a function of t? Assume that the normal to the loop is parallel to the velocity vector of the charge.

(A)

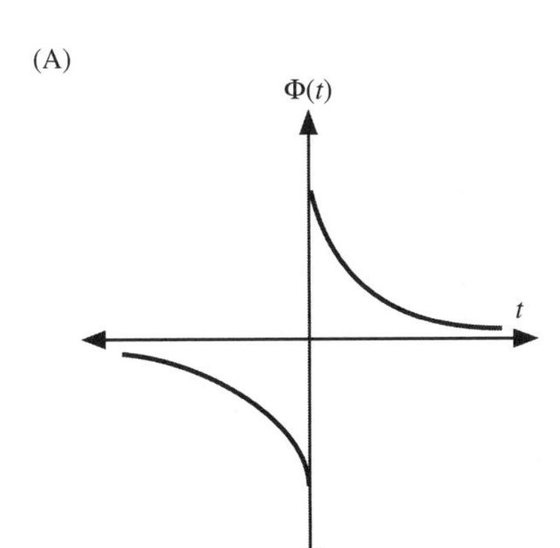

(B)

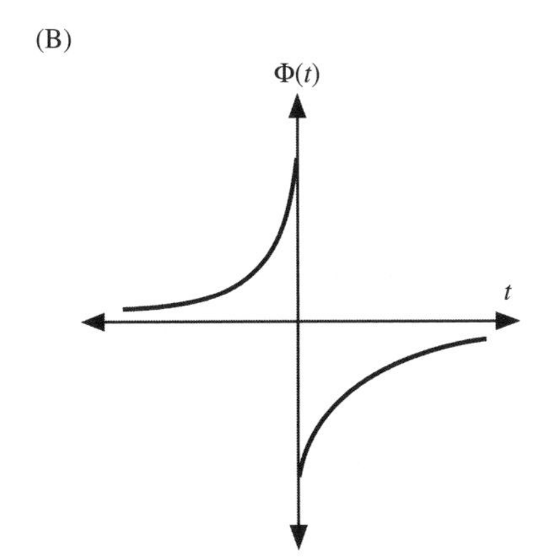

(C)

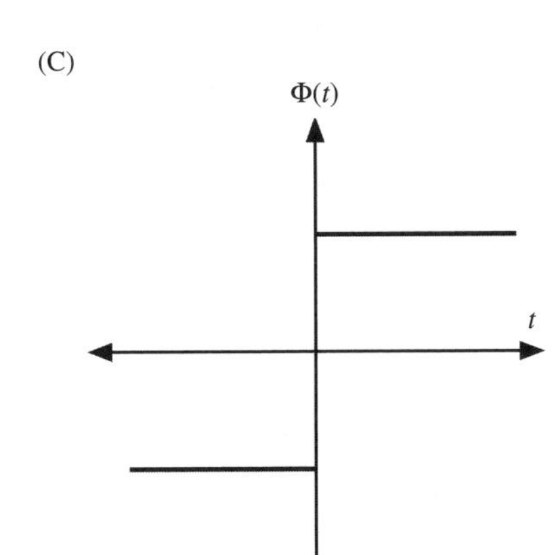

(D)

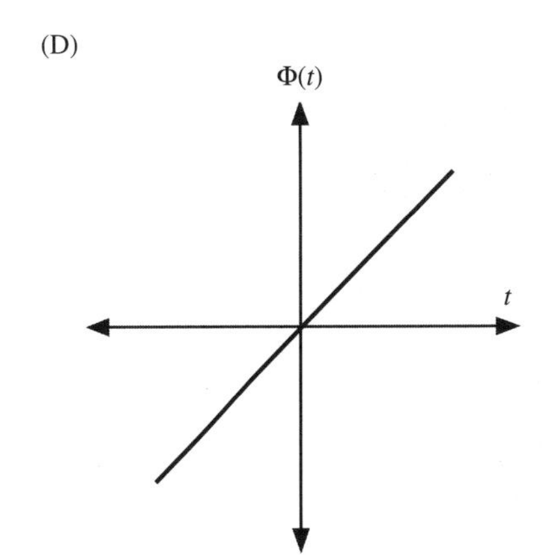

(E)

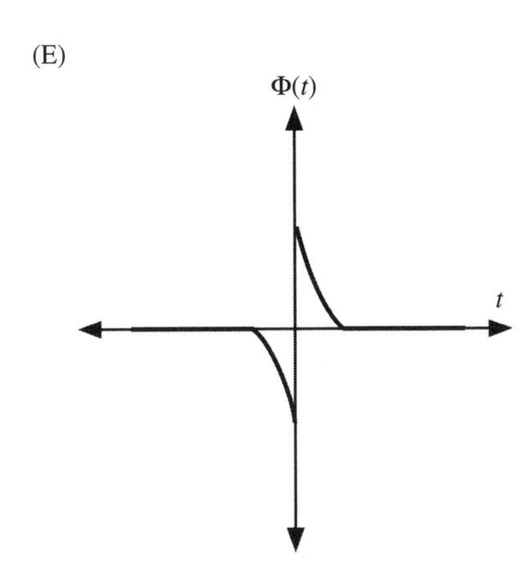

80. For an ideal gas in a container of fixed volume, the most probable speed of the gas molecules as a function of pressure P is proportional to

(A) $P^{-1/2}$
(B) $P^{1/2}$
(C) P
(D) $P^{5/3}$
(E) P^2

81. An ambulance with its siren blaring at constant frequency f drives in a straight line at constant speed directly toward an observer. Which of the following is true of f_{obs}, the siren frequency heard by the observer?

(A) $f_{\text{obs}} = f$.
(B) f_{obs} is constant and greater than f.
(C) f_{obs} is constant and less than f.
(D) f_{obs} continuously increases as the ambulance approaches the observer.
(E) f_{obs} continuously decreases as the ambulance approaches the observer.

82. A photon of energy 25 MeV collides head-on with an electron of energy 50 MeV in the laboratory frame. What is the velocity of the photon in the center-of-momentum frame of the electron and photon?

(A) $0.33c$
(B) $0.5c$
(C) $0.66c$
(D) $0.95c$
(E) c

83. In the quark model, mesons such as the pion are composed of

(A) two quarks
(B) two antiquarks
(C) a quark and an antiquark
(D) three quarks
(E) two baryons

84. Which of the following is the MAIN factor that prevents neutron stars from gravitationally collapsing?

(A) Pauli exclusion principle
(B) Angular momentum
(C) Tidal forces
(D) Spin-down
(E) Strong nuclear force

85. A capacitor C is in an RC circuit with an initial charge Q_0. When the circuit is closed, the energy dissipated in the resistor is used to heat a material of specific heat c_p and mass m, with an efficiency ϵ. Assuming that the material is thermally isolated from everything except the resistor and that the heat capacity of the resistor is negligible compared with the material, what is the change in temperature of the material a long time after the capacitor is discharged?

(A) $\frac{Q_0^2 \epsilon}{2mc_pC}$

(B) $\frac{Q_0^2 \epsilon}{mc_pC}$

(C) $\frac{2Q_0^2 \epsilon}{mc_pC}$

(D) $\frac{Q_0^2}{2\epsilon mc_pC}$

(E) $\frac{Q_0^2}{\epsilon mc_pC}$

86. Two infinite wires a distance d apart carry equal current I in opposite directions. The force per unit length of one wire acting on the other

(A) has magnitude $\frac{\mu_0 I^2}{2\pi d}$ and is attractive
(B) has magnitude $\frac{\mu_0 I^2}{2\pi d}$ and is repulsive
(C) has magnitude $\frac{\mu_0 I^2}{4\pi d^2}$ and is attractive
(D) has magnitude $\frac{\mu_0 I^2}{4\pi d^2}$ and is repulsive
(E) is zero

87. A system of two spin-1/2 particles is subject to the Hamiltonian $H = -A\mathbf{S}_1 \cdot \mathbf{S}_2$, with $A > 0$. What is the degeneracy of the ground state of this system?

(A) 1
(B) 2
(C) 3
(D) 4
(E) This system does not have a ground state.

88. When ultraviolet light of wavelength 350 nm is shined on a container of gas whose molecules have diameter 1 nm, the intensity of the scattered light is I. If the experiment were repeated with the same incident intensity of red light (wavelength 700 nm), the intensity of the scattered light would be

(A) $I/64$
(B) $I/16$
(C) I
(D) $2I$
(E) $16I$

89. The principal decay mode of the 3s state of hydrogen is to

(A) 1s
(B) 2s
(C) 2p
(D) 3p
(E) nothing; the 3s state is stable

90. A new particle's mass is measured in a high-energy physics experiment, and the value reported as 5.43 GeV$\pm$ 0.08 GeV $\pm$ 0.06 GeV, where the first error is systematic and the second is statistical. The total error on the measurement is

(A) 0.0048 GeV
(B) 0.01 GeV
(C) 0.02 GeV
(D) 0.10 GeV
(E) 0.14 GeV

91. A Geiger counter monitoring a radioactive sample records 64 counts in a 1-minute window. The fractional uncertainty on the counting rate is

(A) 1/64
(B) 1/8
(C) 1/4
(D) 1/2
(E) not determinable from the information given

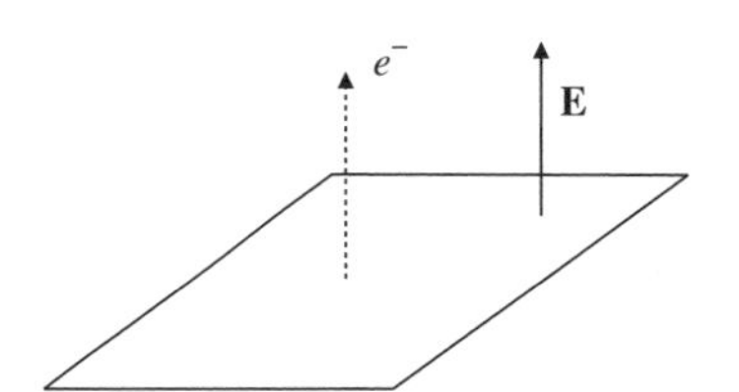

92. An electron is ejected from a metal sheet in the direction normal to the sheet with kinetic energy of 10 eV. A uniform electric field of 100 V/m is applied normal to the sheet, as shown in the diagram. What is the maximum height above the sheet achieved by the electron? You may ignore the effects of gravity.

(A) 1 mm
(B) 1 cm
(C) 10 cm
(D) 1 m
(E) The electron accelerates off to infinity.

93. A bullet of mass 5 g is fired at a block of wood of mass 1 kg, which is hanging from a massless rigid rod of length 0.4 m. The block is thick enough to stop the bullet entirely inside. Which of the following is closest to the minimum velocity of the bullet such that the block makes a complete vertical revolution?

(A) 200 m/s
(B) 400 m/s
(C) 800 m/s
(D) 1000 m/s
(E) 1600 m/s

94. A radio telescope is trained on a binary star system whose angular separation on the sky is 0.061 radians. What is the minimum diameter of the telescope in order to resolve both stars in the binary by observing radio frequency radiation at 200 MHz? (Ignore any atmospheric effects.)

(A) 0.11 m
(B) 1.5 m
(C) 13.3 m
(D) 20 m
(E) 30 m

95. An electron–positron collider creates muons through the reaction

$$e^+ + e^- \to \mu^+ + \mu^-.$$

In the center-of-momentum frame, what is the minimum speed of the electron for this reaction to occur, in terms of the masses m_e and m_μ of the electron and muon?

(A) $(1 - m_e/m_\mu)c$
(B) $\left(\sqrt{1 - m_e/m_\mu}\right) c$
(C) $\left(\sqrt{1 - m_e^2/(2m_\mu^2)}\right) c$
(D) $\left(\sqrt{1 - m_e^2/m_\mu^2}\right) c$
(E) This process can occur at any speed.

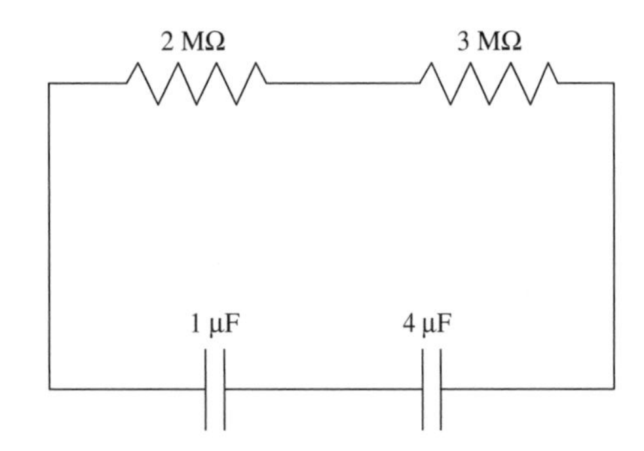

96. What is the time constant of the circuit shown in the diagram?

(A) 2 s
(B) 4 s
(C) 6 s
(D) 10 s
(E) 25 s

97. The self-inductance of an ideal solenoid is L. If the number of coils per unit length is tripled while all other parameters remain the same, the new self-inductance is

(A) $L/9$
(B) $L/3$
(C) $L\sqrt{3}$
(D) $3L$
(E) $9L$

98. Free electron lasers produce coherent light through which of the following mechanisms?

(A) Spontaneous emission
(B) Synchrotron radiation
(C) Population inversion
(D) Optical pumping
(E) Electric dipole transitions

99. What is the magnetic field due to an infinite surface current $\mathbf{K} = K_0\hat{\mathbf{y}}$ flowing along the xy-plane?

(A) $-\frac{\mu_0 K}{2}\hat{\mathbf{z}}$ for $z < 0$, $\frac{\mu_0 K}{2}\hat{\mathbf{z}}$ for $z > 0$
(B) $\frac{\mu_0 K}{2}\hat{\mathbf{z}}$ for $z < 0$, $-\frac{\mu_0 K}{2}\hat{\mathbf{z}}$ for $z > 0$
(C) $-\frac{\mu_0 K}{2}\hat{\mathbf{x}}$ for $z < 0$, $\frac{\mu_0 K}{2}\hat{\mathbf{x}}$ for $z > 0$
(D) $\frac{\mu_0 K}{2}\hat{\mathbf{x}}$ for $z < 0$, $-\frac{\mu_0 K}{2}\hat{\mathbf{x}}$ for $z > 0$
(E) $-\frac{\mu_0 K}{2}\hat{\mathbf{y}}$ for $z < 0$, $\frac{\mu_0 K}{2}\hat{\mathbf{y}}$ for $z > 0$

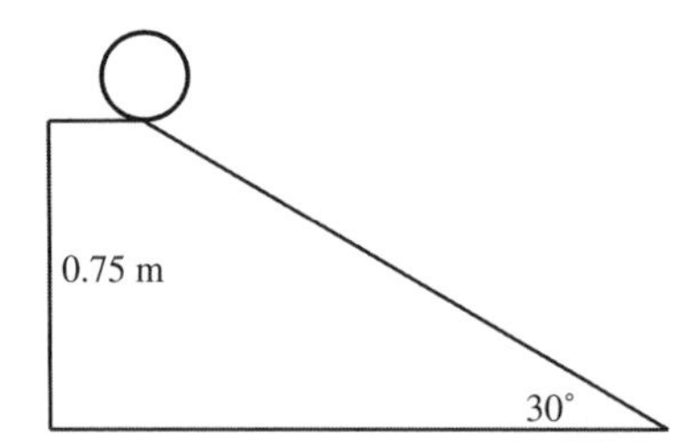

100. A sphere of radius 20 cm and mass 45 g is placed atop a ramp of height 0.75 m and inclination angle 30°. If the ramp were frictionless, the sphere would slide down the ramp in a time t. With friction, the sphere would roll without slipping down the ramp, and reach the bottom in a time t'. What is t'/t?

(A) $\sqrt{2/5}$
(B) $\sqrt{7/10}$
(C) 1
(D) $\sqrt{7/5}$
(E) $\sqrt{3}$

Sample Exam 3

TABLE OF INFORMATION

Rest mass of the electron	m_e	=	9.11×10^{-31} kg
Magnitude of the electron charge	e	=	1.60×10^{-19} C
Avogadro's number	N_A	=	6.02×10^{23}
Universal gas constant	R	=	8.31 J/(mol · K)
Boltzmann's constant	k	=	1.38×10^{-23} J/K
Speed of light	c	=	3.00×10^{8} m/s
Planck's constant	h	=	6.63×10^{-34} J · s = 4.14×10^{-15} eV · s
	$\hbar$	=	$h/2\pi$
	hc	=	1240 eV · nm
Vacuum permittivity	ϵ_0	=	8.85×10^{-12} $C^2/(N \cdot m^2)$
Vacuum permeability	μ_0	=	$4\pi \times 10^{-7}$ T · m/A
Universal gravitational constant	G	=	6.67×10^{-11} $m^3/(kg \cdot s^2)$
Acceleration due to gravity	g	=	9.80 m/s^2
1 atmosphere pressure	1 atm	=	1.0×10^5 N/m^2 = 1.0×10^5 Pa
1 angstrom	1 Å	=	1×10^{-10} m = 0.1 nm

Prefixes for Powers of 10

10^{-15}	femto	f
10^{-12}	pico	p
10^{-9}	nano	n
10^{-6}	micro	μ
10^{-3}	milli	m
10^{-2}	centi	c
10^{3}	kilo	k
10^{6}	mega	M
10^{9}	giga	G
10^{12}	tera	T
10^{15}	peta	P

Rotational inertia about center of mass

Rod	$\frac{1}{12}M\ell^2$
Disk	$\frac{1}{2}MR^2$
Sphere	$\frac{2}{5}MR^2$

SAMPLE EXAM 3

Time — 170 minutes

100 questions

Directions: Each of the questions or incomplete statements below is followed by five suggested answers or completions. Select the one that is best in each case and then fill in the corresponding space on the answer sheet.

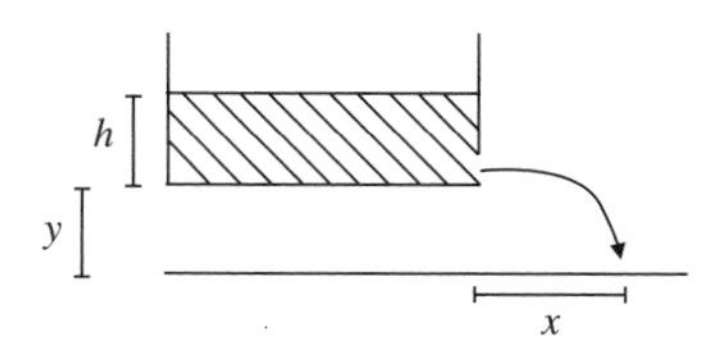

1. A bath of water has a hole in the bottom of one side, as shown in the figure. At what horizontal distance x from the edge of the bath does the draining water land? Neglect effects due to viscosity and surface tension.

(A) $\sqrt{2hy}$
(B) $\frac{\sqrt{hy}}{2}$
(C) $2\sqrt{hy}$
(D) $\frac{h}{2}$
(E) $\frac{2h^2}{y}$

2. When monochromatic blue light of wavelength 450 nm is shined on a sample of hydrogen *atoms*, the intensity of the scattered light is I_0. If the Bohr radius of hydrogen were doubled, what would be the approximate intensity of scattered light?

(A) $I_0/16$
(B) $I_0/4$
(C) $4I_0$
(D) $16I_0$
(E) $64I_0$

3. Two objects in the sky have angular separation 1 arcminute, and emit a broad spectrum of radiation. A telescope with aperture diameter 1 cm could resolve the objects by observing which of the following kinds of radiation? Note that 1 arcminute is approximately 3×10^{-4} rad.

I. Radio
II. Visible
III. X-ray

(A) I only
(B) II only
(C) III only
(D) I and II
(E) II and III

4. A block slides frictionlessly on ice at a constant velocity of 10 m/s. The block suddenly encounters a rough patch where its coefficient of kinetic friction suddenly increases from 0 to 0.5. How far does the block slide before stopping?

(A) 5 m
(B) 10 m
(C) 15 m
(D) 20 m
(E) 100 m

5. The Euler–Lagrange equations are valid for systems with which of the following properties?

I. Systems with time-dependent potentials
II. Systems without rotational symmetry
III. Systems acted on by only conservative forces

(A) II
(B) III
(C) I and II
(D) I and III
(E) I, II, and III

6. Order the following corrections to the Bohr energies of hydrogen from smallest to largest.

   I. Fine structure
   II. Hyperfine splitting
   III. Lamb shift

   (A) I, II, III
   (B) I, III, II
   (C) II, III, I
   (D) III, II, I
   (E) III, I, II

7. What is the difference in energy between the $n = 5$ and the $n = 1$ states of the one-dimensional quantized harmonic oscillator?

   (A) $\hbar\omega$
   (B) $2\hbar\omega$
   (C) $4\hbar\omega$
   (D) $8\hbar\omega$
   (E) $16\hbar\omega$

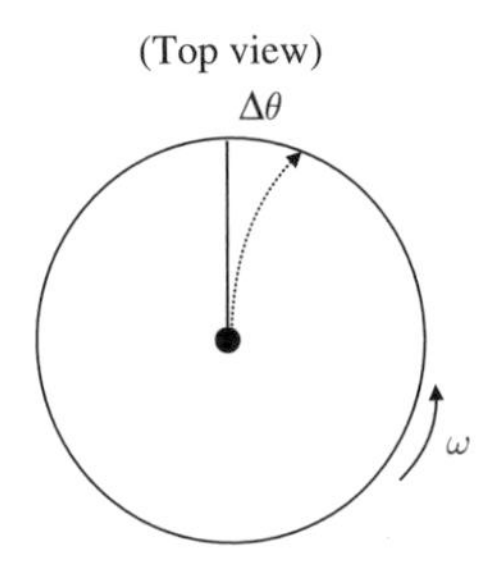

8. A person stands at the center of a frictionless disk of radius R rotating at angular velocity ω, and pushes a puck radially outwards at velocity v. What is the angle $\Delta\theta$ between the point on the edge of the disk where the puck was aimed and the point on the edge of the disk where the puck actually arrives?

   (A) $v/(\omega R)$
   (B) $\omega R/v$
   (C) $2\omega R/v$
   (D) $\omega R/(2v)$
   (E) 0

9. Two parameters x and y were measured with uncorrelated uncertainties Δx and Δy. What is the uncertainty of the quantity $z = x/y$?

   (A) $\sqrt{(\Delta x/x)^2 + (\Delta y/y)^2}$
   (B) $\sqrt{(\Delta x/x)^2 - (\Delta y/y)^2}$
   (C) $z\sqrt{(\Delta x/x)^2 + (\Delta y/y)^2}$
   (D) $z\sqrt{(\Delta x/x)^2 - (\Delta y/y)^2}$
   (E) $z\sqrt{(\Delta x/x)^2 + 2(\Delta y/y)^2}$

10. A circuit made only of which of the following circuit elements may function as a bandpass filter?

    (A) One resistor, one inductor, and one capacitor
    (B) One resistor and one inductor
    (C) One resistor and one capacitor
    (D) Two resistors
    (E) Two capacitors

11. Consider a planet of mass m that orbits a star of mass $M \gg m$. For a fixed orbital angular momentum L, what is the relationship between the energies of the three possible orbit shapes: circular (E_{cir}), elliptical (E_{ell}), or hyperbolic (E_{hyp})?

    (A) $E_{\text{cir}} < E_{\text{ell}} < E_{\text{hyp}}$
    (B) $E_{\text{ell}} < E_{\text{hyp}} < E_{\text{cir}}$
    (C) $E_{\text{hyp}} < E_{\text{cir}} < E_{\text{ell}}$
    (D) $E_{\text{cir}} < E_{\text{hyp}} < E_{\text{ell}}$
    (E) $E_{\text{ell}} < E_{\text{cir}} < E_{\text{hyp}}$

12. For a system of electrons at zero temperature, the energy of the highest occupied quantum state is called the

    (A) zero-point energy
    (B) Einstein energy
    (C) Fermi energy
    (D) Bose energy
    (E) binding energy

13. A beam is made up of particles that have a lifetime of 10^{-8} s at rest. If the beam travels at $0.8c$, at what location down the beamline is there only a fraction $1/e$ of the particles remaining?

    (A) 6.2 m
    (B) 4.0 m
    (C) 3.6 m
    (D) 2.7 m
    (E) 2.2 m

14. An atom has electron configuration $1s^2 2s^2 2p^3$. A measurement of the total orbital angular momentum of the outermost electron in the ground state could return which of the following?

(A) $\hbar$
(B) $\hbar\sqrt{2}$
(C) $2\hbar$
(D) $\hbar\sqrt{6}$
(E) $3\hbar$

15. Which of the following nuclei were produced during Big Bang nucleosynthesis?

I. ^{3}He
II. ^{7}Li
III. ^{55}Fe

(A) I
(B) III
(C) I and II
(D) I and III
(E) I, II, and III

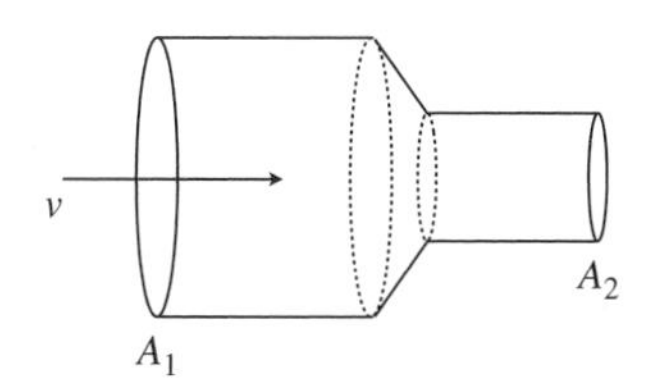

16. A horizontal tube has a wide section of cross-sectional area A_1 and a narrow section with cross-sectional area A_2, as shown in the diagram. If an incompressible fluid of density ρ moves with velocity v through the wide part of the horizontal tube, what is the velocity of the fluid in the narrow section of the tube?

(A) v
(B) $v\frac{A_1}{A_2}$
(C) $v\frac{A_2}{A_1}$
(D) $v\frac{A_1^2}{A_2^2}$
(E) $v\frac{A_2^2}{A_1^2}$

17. A car accelerates from rest at 5 m/s^2. What is its speed after traveling 40 m?

(A) 9.7 m/s
(B) 15 m/s
(C) 20 m/s
(D) 30 m/s
(E) 54.2 m/s

18. A bullet of mass m is fired at velocity v into a block of mass M at rest on a table, where it stops and is embedded. If there is a coefficient of friction μ between the block and the table, how much time does it take for the block to come to rest?

(A) $\dfrac{Mv}{\mu g(m+M)}$
(B) $\dfrac{\mu mv}{g(m+M)}$
(C) $\dfrac{v}{\mu g}$
(D) $\dfrac{mv}{\mu gM}$
(E) $\dfrac{mv}{\mu g(m+M)}$

19. Thermal fluctuations produce voltage fluctuations in all resistors. Which of the following is the spectral density (units of V Hz$^{-1/2}$) of voltage fluctuations in a resistor at temperature T and of resistance R?

(A) $\sqrt{4kT/R}$
(B) $\sqrt{4kR/T}$
(C) $\sqrt{4kTR}$
(D) $\sqrt{kT}$
(E) $\sqrt{4R}$

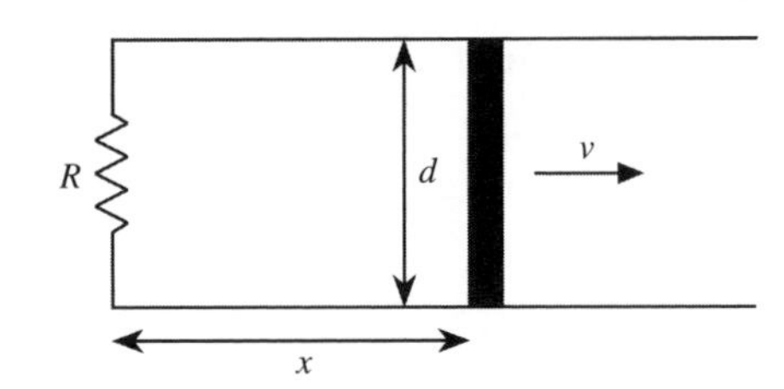

20. A metal bar is pulled at constant velocity $v\hat{\mathbf{x}}$ along two metal rails a distance d apart connected by a resistor of resistance R, as shown in the diagram. There is a constant magnetic field, pointing into the page, of magnitude B. At time T, how much energy has been dissipated in the resistor thus far, as a function of T?

(A) $\dfrac{2BvdT}{R}$

(B) $\dfrac{2(Bvd)^2T}{R}$

(C) $\dfrac{BvdT}{R}$

(D) $\dfrac{(Bvd)^2T}{R}$

(E) $(Bvd)^2TR$

21. If the rms velocity of H_2 gas at 300 K is v, which of the following is closest to the rms velocity of helium gas at the same temperature?

(A) $v/8$
(B) $v/4$
(C) $v/2$
(D) $v/\sqrt{2}$
(E) $2v$

22. A circuit consists of a capacitor C, a resistor R, and an inductor L all in series. The circuit is driven by an AC generator at frequency ω. What is the driving frequency at which the current through the circuit is maximized?

(A) $\sqrt{\dfrac{1}{LC} - \left(\dfrac{R}{2L}\right)^2}$

(B) $\sqrt{\dfrac{1}{LC} - \left(\dfrac{1}{2RC}\right)^2}$

(C) $\sqrt{\dfrac{1}{LC} - \left(\dfrac{R}{L}\right)^2}$

(D) $\sqrt{\dfrac{1}{LC}}$

(E) $\sqrt{\dfrac{1}{LC} - \left(\dfrac{1}{RC}\right)^2}$

23. A pipe with two open ends is 20 cm long. What is the fundamental frequency of the pipe? (You may assume the speed of sound is 343 m/s.)

(A) 1715 Hz
(B) 858 Hz
(C) 563 Hz
(D) 429 Hz
(E) 205 Hz

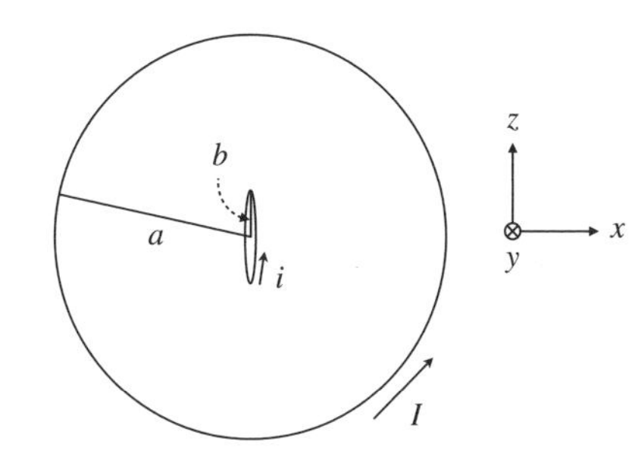

24. A current loop of radius a carrying current I is centered at the origin and lies in the xy-plane. Another loop, carrying current $i \ll I$, and of radius $b \ll a$, is centered at the origin and lies in the xz-plane. What is the torque on the smaller loop about its center?

(A) $\dfrac{\mu_0 \pi i I b^2}{a}$
(B) $\dfrac{\mu_0 \pi i I b^2}{2a}$
(C) $\dfrac{\mu_0 \pi i I b^3}{2a^2}$
(D) $\dfrac{\mu_0 \pi i I b^2}{4a^2}$
(E) $\dfrac{3\mu_0 \pi i I b^2}{2a^2}$

25. A merry-go-round of radius R rotates at an angular velocity of Ω. A ball A is released at radius $R/2$, initially at rest, by a person standing on the merry-go-round. An identical ball B is released at radius R, also at rest. In the noninertial reference frame of the rotating merry-go-round, what is the ratio of the acceleration experienced by ball A to ball B immediately after they are released?

(A) 0
(B) 1/2
(C) 1
(D) 2
(E) 4

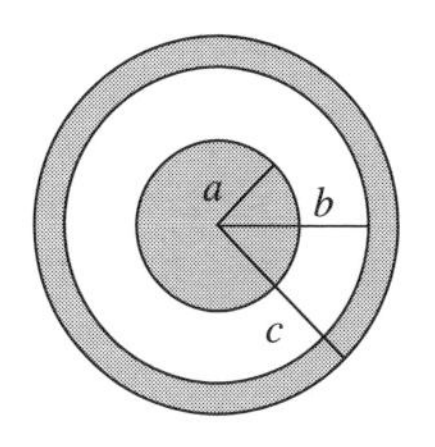

26. A ball of uniform charge density and radius a is surrounded by a conducting shell of inner radius b and outer radius c. Which could be the potential as a function of radius?

(A)

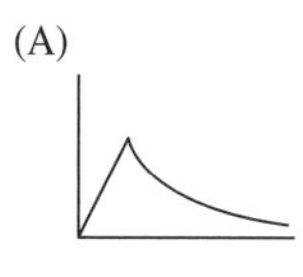

(B)

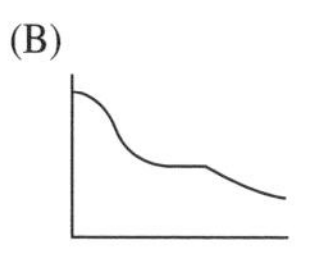

(C)

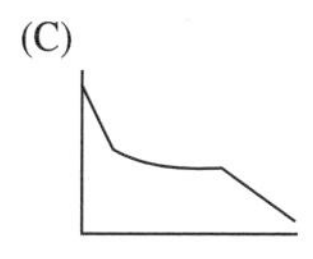

(D)

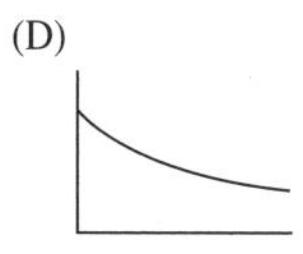

(E)

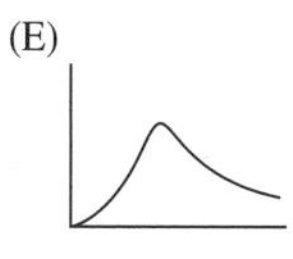

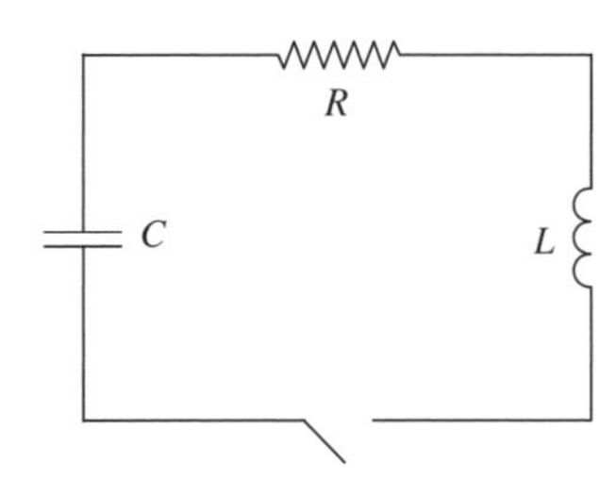

27. A charged capacitor is in series with a resistor and inductor, as in the diagram. Which of the following could be a graph of the current when the switch is closed?

(A)

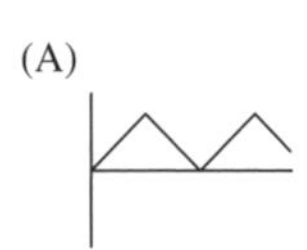

(B)

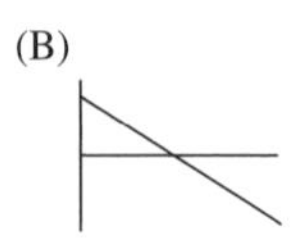

(C)

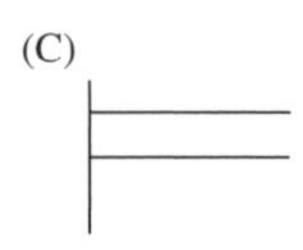

(D)

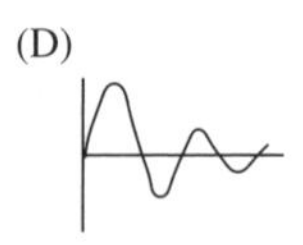

(E)

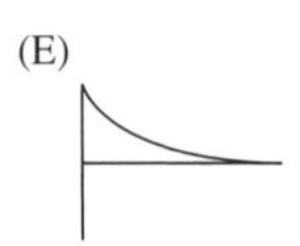

28. A beam of electrons (mass m and charge q) with uniform velocity enters a region of constant magnetic field B perpendicular to the beam direction. Assuming that the electrons are able to follow a circular path completely within the field, how long does it take for the beam to make one complete revolution?

(A) $2\pi m/(qB)$
(B) $\pi m/(qB)$
(C) $m/(2\pi qB)$
(D) $m/(\pi qB)$
(E) $m/(qB)$

29. Photons of wavelength 10 nm are incident on a crystal with interatomic space 80 nm. At what angle is the first-order maximum in the diffraction pattern?

(A) 7.2°
(B) 3.6°
(C) 2.4°
(D) 1.8°
(E) 0.9°

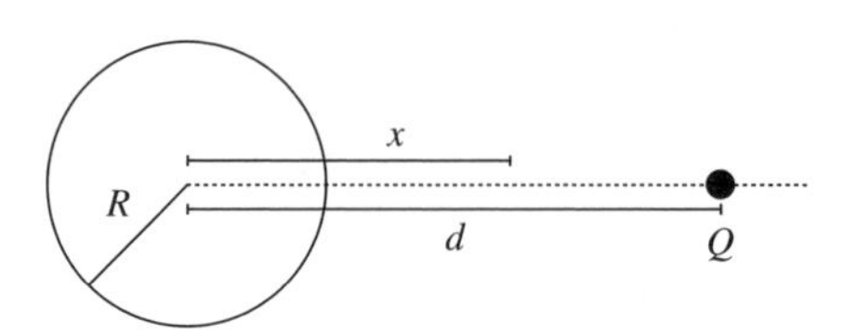

30. A charge Q is brought to a distance d from the center of a grounded conducting sphere of radius R. What is the electric potential at a distance $x > R$ from the center of the sphere along the axis between the charge and the sphere?

(A) $\dfrac{Q}{2\pi\epsilon_0}\left(\dfrac{1}{|x-d|}-\dfrac{d}{|xd-R^2|}\right)$

(B) $\dfrac{Q}{4\pi\epsilon_0}\left(\dfrac{1}{|x-d|}-\dfrac{R}{|xd-R^2|}\right)$

(C) $\dfrac{Q}{4\pi\epsilon_0}\left(\dfrac{1}{|x-d|}-\dfrac{1}{|2x-R|}\right)$

(D) $\dfrac{Q}{4\pi\epsilon_0}\left(\dfrac{1}{|x-d|}-\dfrac{1}{|x|}\right)$

(E) 0

31. What type of lattice is the reciprocal lattice to a simple cubic lattice in three dimensions?

(A) Face-centered cubic
(B) Simple cubic
(C) Body-centered cubic
(D) Simple hexagonal
(E) None of the above

32. The magnetic vector potential in a region of space is given by

$$\mathbf{A}(x, y) = Ay^2\hat{\mathbf{x}} + Cx^2\hat{\mathbf{z}},$$

where A and C are constants. What is the magnetic field in this region?

(A) $-2Cx\hat{\mathbf{y}} - 2Ay\hat{\mathbf{z}}$
(B) $2Cx\hat{\mathbf{y}} + 2Ay\hat{\mathbf{z}}$
(C) 0
(D) $-Cx\hat{\mathbf{y}} - Ay\hat{\mathbf{z}}$
(E) $Cx\hat{\mathbf{y}} + Ay\hat{\mathbf{z}}$

33. The CMB has a temperature of 2.7 K and has a peak intensity at a wavelength of approximately 1 mm. If the CMB were at 5 K, what would be the wavelength with the maximal intensity?

(A) 1 mm
(B) 0.54 mm
(C) 0.32 mm
(D) 5.3 mm
(E) 57 mm

34. Which of the following decays is allowed in the Standard Model?

(A) $\mu^- \to e^- + \nu_e$
(B) $\pi^- \to \gamma$
(C) $\Delta^+ \to p + n$
(D) $\pi^+ \to 2\gamma$
(E) $K^+ \to \mu^+ + \nu_\mu$

35. A nearby star is moving away from the Earth with peculiar velocity $0.1c$. It appears to have an effective blackbody temperature of 10^4 K. What is its true effective blackbody temperature? (Assume a negligible cosmological redshift.)

(A) 0.6×10^4 K
(B) 0.9×10^4 K
(C) 10^4 K
(D) 1.1×10^4 K
(E) 1.4×10^4 K

36. A two-level system has energies $\pm\epsilon$. What is the average energy of the system as the temperature $T \to \infty$?

(A) -2ϵ
(B) $-\epsilon$
(C) 0
(D) ϵ
(E) 2ϵ

37. An ideal gas is confined to half of a rigid box with volume $2V$. If a valve is opened suddenly, letting the gas suddenly fill the full volume of the box, which of the following is unchanged?

I. Internal energy U
II. Temperature T
III. Entropy S

(A) I
(B) III
(C) I and II
(D) I and III
(E) I, II, and III

38. Suppose a heat engine that transfers heat from a warm bath at temperature T_H to a cold bath at T_C has an efficiency

$$e = \frac{T_H + T_C}{T_H + 2T_C}.$$

Which of the following must be violated?

(A) Conservation of energy
(B) First Law of Thermodynamics
(C) Second Law of Thermodynamics
(D) Third Law of Thermodynamics
(E) Postulate of equal *a priori* probabilities

39. Two spaceships pass each other. Spaceship A moves relative to a nearby planet at velocity v_1, while spaceship B moves at velocity v_2 relative to the planet. How fast does spaceship A move relative to spaceship B?

(A) $\dfrac{v_1 - v_2}{1 + v_1 v_2/c^2}$

(B) $\dfrac{|v_1 + v_2|}{1 - v_1 v_2/c^2}$

(C) $|v_1 - v_2|$

(D) $\dfrac{v_1 + v_2}{1 + v_1 v_2/c^2}$

(E) $\dfrac{|v_1 - v_2|}{1 - v_1 v_2/c^2}$

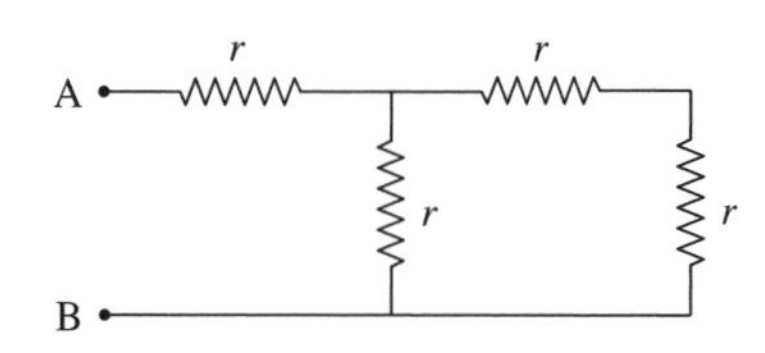

40. Consider the network of resistors with resistance r, shown in the figure. What is the equivalent resistance between terminals A and B?

(A) $r/3$
(B) $2r/3$
(C) r
(D) $4r/3$
(E) $5r/3$

41. An ideal monoatomic gas initially at pressure P undergoes adiabatic expansion from a volume V to a volume $2V$. What is the final pressure of the gas?

(A) P
(B) $P/2$
(C) $P/4$
(D) $2^{-5/3}P$
(E) $2^{-7/3}P$

42. What is the expectation value of L_z for the following wavefunction:

$$\psi(\theta,\phi) = \frac{1}{\sqrt{2}}\left(Y_1^{-1}(\theta,\phi) + Y_1^0(\theta,\phi)\right),$$

where $Y_l^m(\theta,\phi)$ are the spherical harmonics?

(A) $-\hbar$
(B) $-\hbar/2$
(C) 0
(D) $\hbar/2$
(E) $\hbar$

43. Which of the following statements is true in general for one-dimensional spin-0 quantum mechanical systems?

(A) All states are energy eigenstates.
(B) Energies are always quantized.
(C) Eigenvalues of Hermitian operators are real.
(D) All states have real-valued wavefunctions in the x basis.
(E) None of the above.

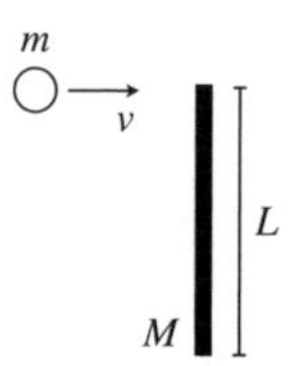

44. A mass m moves at speed v perpendicular to a rod of uniform density, mass M, and length L on a frictionless table. Suppose $m \ll M$. If the mass collides with the end of the rod and sticks to it, at what angular speed does the rod begin to rotate? (You may treat the mass m as a point particle.)

(A) $\dfrac{3mv}{2ML}$

(B) $\dfrac{3mv}{ML}$

(C) $\dfrac{6mv}{ML}$

(D) $\dfrac{12mv}{ML}$

(E) $\dfrac{Mv}{2mL}$

45. An electron is in a magnetic field and has a Hamiltonian $H = \alpha \mathbf{S} \cdot \mathbf{B}$. If the electron is aligned with the magnetic field at $t = 0$, what is its time-dependent wavefunction? ($|+\rangle$ represents a spinor aligned with the magnetic field.)

(A) $\exp(-i\alpha Bt/2)\,|+\rangle$
(B) $\exp(+i\alpha Bt/2)\,|+\rangle$
(C) $\exp(-i\alpha Bt)\,|+\rangle$
(D) $\exp(+i\alpha Bt)\,|+\rangle$
(E) $\exp(-2\pi i\alpha Bt/\hbar)\,|+\rangle$

46. Which of the following is true of observables in quantum mechanics?

 I. They are represented by Hermitian operators.
 II. Multiple observables can never be simultaneously measured.
 III. The operators representing observables must have real eigenvalues.

(A) I only
(B) II only
(C) I and II
(D) I and III
(E) I, II, and III

47. A mass attached to a spring oscillates at frequency f. If the mass attached to the spring triples and the spring constant doubles, what is the new frequency of oscillation of the system?

(A) $(\sqrt{2/3})f$
(B) $(2/3)f$
(C) $(4/9)f$
(D) $(\sqrt{1/2})f$
(E) $(1/2)f$

48. A quantum particle has normalized wavefunction $\psi(x) = \sqrt{5}x^2$ on the interval $[0, 1]$. What is the probability that the particle is found at $0 \leq x \leq 1/2$?

(A) 1/2
(B) 1/4
(C) 1/8
(D) 1/16
(E) 1/32

49. A spin-1/2 particle is in a state

$$|\psi\rangle = \sqrt{\frac{2}{3}}\,|m = 1/2\rangle + \sqrt{\frac{1}{3}}\,|m = -1/2\rangle\,.$$

What is the expectation value of the z-component of its spin?

(A) $3\hbar$
(B) $\hbar$
(C) $\hbar/2$
(D) $\hbar/3$
(E) $\hbar/6$

50. Which of the following physical processes is responsible for producing the photons from a carbon dioxide laser?

(A) Pair annihilation
(B) Bremsstrahlung
(C) Transitions of nuclear energy levels
(D) Transitions between vibrational molecular energy levels
(E) Photoelectric effect

51. A car of mass M pulls a trailer of mass m by a cord. The car's engine exerts a force F on the car. Suppose that the trailer has a coefficient of rolling friction μ, but neglect the coefficient of friction of the car. What is the tension in the cord between the car and trailer?

(A) $\dfrac{m(F - \mu Mg)}{M - m}$
(B) $\dfrac{m(F + \mu Mg)}{M + m}$
(C) $\dfrac{m(F - \mu Mg)}{M + m}$
(D) $\dfrac{\mu m(F - Mg)}{M + m}$
(E) $\dfrac{m(F + \mu Mg)}{\mu(M + m)}$

52. A spin-2 particle has orbital angular momentum $l = 4$. What is the smallest possible value of its total angular momentum quantum number j?

(A) 6
(B) 5
(C) 4
(D) 3
(E) 2

53. An element with a ground state electron configuration of $1s^2 2s^2 2p^6$ is best characterized as a(n)

(A) alkali metal
(B) rare earth metal
(C) semiconductor
(D) halogen
(E) noble gas

54. The ground state wavefunction for the harmonic oscillator is

$$\psi_0(x) = \left(\frac{m\omega}{\pi\hbar}\right)^{1/4} e^{-m\omega x^2/(2\hbar)}.$$

Consider a perturbation to the Hamiltonian given by

$$\delta V(x) = \alpha e^{-\beta x^2}.$$

What is the first-order correction to the ground state energy? Note that

$$\int_{-\infty}^{\infty} e^{-x^2/c^2} = c\sqrt{\pi}.$$

(A) $\alpha\sqrt{\frac{\beta\hbar}{m\omega}}$
(B) $\alpha\sqrt{\frac{m\omega+\beta\hbar}{m\omega}}$
(C) $\alpha\sqrt{\frac{m\omega}{m\omega+\beta\hbar}}$
(D) $\alpha\sqrt{\frac{2m\omega+\beta\hbar}{m\omega}}$
(E) $\alpha\sqrt{\frac{m\omega}{2m\omega+\beta\hbar}}$

55. Let γ be the path in the xy-plane that traverses the square with vertices $(0,0)$, $(1,0)$, $(1,1)$, and $(0,1)$, in that order. What is the line integral $\oint_\gamma \mathbf{f}\cdot d\mathbf{l}$ of the function $\mathbf{f}(x,y) = y\hat{\mathbf{x}} + x\hat{\mathbf{y}}$?

(A) -2
(B) -1
(C) 0
(D) 1
(E) 2

56. Magnetic flux is quantized in type II superconductors. What is the unit of magnetic flux quanta?

(A) $2eh$
(B) $\frac{e}{2h^2}$
(C) $\frac{h}{2e}$
(D) $\frac{h}{2e^2}$
(E) $\frac{e^2}{2h}$

57. Positronium is a bound state of an electron and a positron. Which of the following are true facts about positronium?

I. It obeys Bose–Einstein statistics.
II. Its binding energy is -6.8 eV.
III. It can decay into a single photon.

(A) I only
(B) III only
(C) I and II
(D) I and III
(E) I, II, and III

58. A billiard ball m rolls without slipping with velocity $\mathbf{v}$ on a pool table. After striking the wall of the table in an elastic collision, the ball has velocity $\mathbf{v}'$. The change in the ball's momentum is $\Delta\mathbf{p}$. Which of the following must be true?

I. $|\mathbf{v}'| = |\mathbf{v}|$
II. $\mathbf{v}' = -\mathbf{v}$
III. $|\Delta\mathbf{p}| = m|\mathbf{v}' - \mathbf{v}|$

(A) I only
(B) I and II
(C) I and III
(D) II and III
(E) I, II, and III

59. Which of the following transitions of the hydrogen atom is allowed in the electric dipole approximation? (The entries in parentheses are (n, l, m).)

(A) $(2,1,0) \to (1,0,0)$
(B) $(2,0,0) \to (1,0,0)$
(C) $(3,2,0) \to (1,0,0)$
(D) $(3,2,2) \to (2,1,0)$
(E) $(3,0,0) \to (2,0,0)$

60. Suppose that a satellite orbiting the Sun can be approximated as a perfect blackbody. Assuming the body is in equilibrium with its surroundings, what is the ratio of its blackbody temperature at radius $2R$ from the Sun to the blackbody temperature at radius R from the Sun?

(A) $2^{-3/2}$
(B) 2^{-1}
(C) $2^{-1/2}$
(D) $2^{-1/3}$
(E) $2^{-1/4}$

61. Gas A consists of atoms of mass m, and gas B consists of diatomic molecules where each atom has mass $m/2$. Which of the following best describes the rms speeds v_A and v_B of the particles of gases A and B at temperature T? You may assume T is less than the temperature at which gas B dissociates.

(A) $v_A < v_B$ at low T, $v_A > v_B$ at high T
(B) $v_A > v_B$ at low T, $v_A < v_B$ at high T
(C) $v_A = v_B$ at low T, $v_A > v_B$ at high T
(D) $v_A < v_B$ for all T
(E) $v_A = v_B$ for all T

62. A merry-go-round can be approximated as a disk of uniform density with radius r, and mass M. A merry-go-round is spinning at angular velocity ω before a person steps onto it. What is the change in angular velocity after a person of mass m steps onto the edge of the merry-go-round?

(A) $\dfrac{2\omega m}{M+2m}$
(B) $\dfrac{\omega M}{M+2m}$
(C) $\dfrac{2\omega(m+M)}{M+2m}$
(D) $\dfrac{2\omega m}{M+m}$
(E) $\dfrac{\omega m}{M+m}$

63. The He^+ ion experiences an atomic transition from the $n = 2$ state to the $n = 1$ state. What is the energy of the emitted photon?

(A) 10.2 eV
(B) 13.6 eV
(C) 27.2 eV
(D) 31.4 eV
(E) 40.8 eV

64. A string of mass density $\mu = 1$ g/cm and tension $T = 4 \times 10^3$ N is fixed at both ends. What is the speed of waves on the string?

(A) 2 m/s
(B) 20 m/s
(C) 200 m/s
(D) 2,000 m/s
(E) 2×10^5 m/s

65. Eight charges $+q$ are fixed at the corners of a cube of side length a. A test charge $-q$ is placed at the center of the cube and released. Which of the following MUST be true of this configuration of charges?

I. The force on the test charge due to the six charges $+q$ is zero.
II. The electric potential due to the six charges $+q$ is zero.
III. The test charge is in stable equilibrium.

(A) I only
(B) II only
(C) I and II only
(D) I and III only
(E) I, II, and III

66. A particle of mass m and charge q in a region of homogeneous magnetic field undergoes cyclotron motion at speed $v = 0.6c$ with cyclotron radius R. If the particle's speed is increased to $v = 0.8c$, what is the new cyclotron radius?

(A) $3R/4$
(B) R
(C) $5R/4$
(D) $4R/3$
(E) $16R/9$

67. Unpolarized light is incident on two polarizing filters oriented at 30° to one another. What is the intensity of transmitted light as a fraction of the incident light intensity?

(A) 1/4
(B) 3/8
(C) 1/2
(D) 3/4
(E) 1/8

68. What is the expectation value of the operator

$$\mathcal{O} = |[\hat{x}, \hat{p}]|^2$$

in the ground state of the infinite square well of size L, centered on $x = 0$?

(A) 0
(B) $\hbar^2$
(C) L^2
(D) $\hbar^2 L^2$
(E) $\hbar^2/L^2$

69. What is the inductance of a cylindrical solenoid with N turns, length ℓ, and radius $R \ll \ell$?

(A) $2\mu_0 N^2 R^2/\ell$
(B) $2\mu_0 N^2 \pi R^2/\ell$
(C) $\mu_0 N \pi R^2/\ell$
(D) $\mu_0 N^2 \pi R^2/\ell$
(E) $\mu_0 N^2 R^2/\ell$

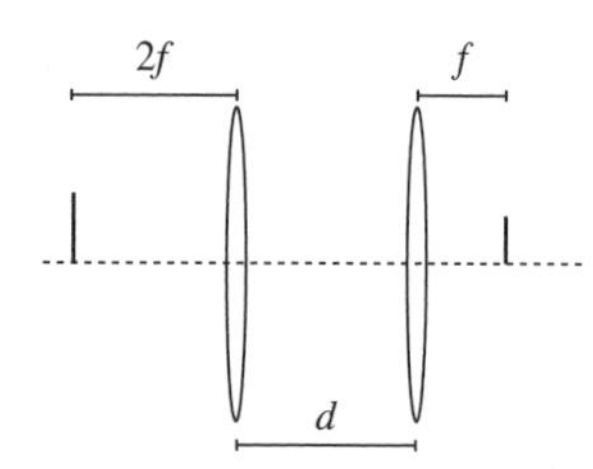

70. Two converging lenses of focal length $f/2$ are placed in series, separated by a distance d. The object is placed a distance $2f$ to the left of the left lens, and the image is upright and located a distance f to the right of the right lens. What is d?

(A) $(2/3)f$
(B) $(5/3)f$
(C) $(7/3)f$
(D) $3f$
(E) $9f$

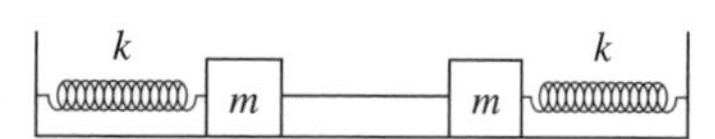

71. Two masses m are connected by springs with spring constants k and a massless rigid rod, as shown in the diagram. What is the frequency of oscillation of the system?

(A) $\sqrt{3k/m}$
(B) $\sqrt{k/m}$
(C) $\sqrt{2k/m}$
(D) $\sqrt{k/2m}$
(E) $2\sqrt{k/m}$

72. Suppose an atomic transition has a lifetime of 3×10^{-10} s. The natural line width of this transition is closest to

(A) 10^{-2} eV
(B) 10^{-4} eV
(C) 10^{-6} eV
(D) 10^{-8} eV
(E) 10^{-10} eV

73. Which of the following quantities change under a general gauge transformation in electromagnetism?

I. Electric potential
II. Electric field
III. Magnetic field

(A) I only
(B) II only
(C) I and II
(D) II and III
(E) I, II, and III

74. The degeneracy of the second excited state of the three-dimensional infinite square well is

(A) 1
(B) 2
(C) 3
(D) 8
(E) 9

75. What is the capacitance of two concentric thin conducting spheres of radii a and $b > a$?

(A) $4\pi\epsilon_0 ab/(b-a)$
(B) $4\pi\epsilon_0 a^2/(b-a)$
(C) $4\pi\epsilon_0 b^2/(b-a)$
(D) $4\pi\epsilon_0 \ln(b/a)$
(E) $4\pi\epsilon_0$

76. How many distinct spin states can be formed by three distinguishable spin-1/2 particles?

(A) 1
(B) 2
(C) 4
(D) 7
(E) 8

77. A thin lens is made of a material with an index of refraction of 1.5. If the radius of curvature of the left side of the lens is 10 cm, and the focal length is 1 m, what is the radius of curvature of the right side of the lens?

(A) 12.5 cm
(B) 6.25 cm
(C) 25 cm
(D) 3.125 cm
(E) 25 cm

78. A car drives at velocity v through a garage of length l, with a front and rear door. The front door is initially open and the rear door is initially closed. In the frame of the garage, the rear door opens when the car is just about to collide with it, and the front door closes at the same time. In the frame of the car, how much time separates the opening of the rear door from the closing of the front door? ($\gamma = 1/\sqrt{1 - v^2/c^2}$.)

(A) 0
(B) $\frac{\gamma v l}{c^2}$
(C) $\frac{vl}{c^2}$
(D) $\frac{\gamma c l}{v^2}$
(E) $\frac{vl}{\gamma c^2}$

79. Consider the change in entropy of an ideal gas in the following situations:

- ΔS_1: temperature doubles, pressure constant
- ΔS_2: temperature doubles, pressure doubles
- ΔS_3: temperature constant, volume doubles

Which of the following is true?

(A) $\Delta S_1 < \Delta S_2 < \Delta S_3$
(B) $\Delta S_1 < \Delta S_3 < \Delta S_2$
(C) $\Delta S_2 < \Delta S_3 < \Delta S_1$
(D) $\Delta S_3 < \Delta S_1 < \Delta S_2$
(E) $\Delta S_3 < \Delta S_2 < \Delta S_1$

80. A mass m attached to a spring of constant k is driven by a force $F(t) = F_0 \sin(\omega t)$. What is the late-time amplitude of the spring oscillations, assuming friction is small but sufficient to damp out transient oscillations?

(A) $\frac{F_0}{k}$
(B) $\frac{F_0}{m\omega^2}$
(C) $\frac{F_0}{|k - 2m\omega^2|}$
(D) $\frac{F_0}{|k - m\omega^2|}$
(E) $\frac{F_0}{|4k - m\omega^2|}$

81. An event occurs at $(t, x, y, z) = (0\text{ s}, 5\text{ m}, 10\text{ m}, 0\text{ m})$ in a reference frame S. At what position x' does the event occur in a reference frame S' with coordinates (t', x', y', z') that is moving at velocity $0.8c$ along the x-axis, relative to S? You may assume that the origin of coordinates coincides in both reference frames.

(A) 15.67 m
(B) 11.33 m
(C) 8.33 m
(D) 6.67 m
(E) 3.00 m

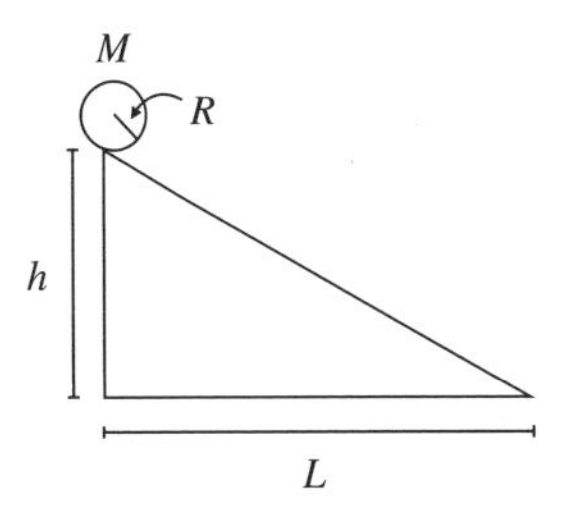

82. A dowel of radius R, mass M, and uniform mass density rolls without slipping down a ramp of length L and height h. What is its speed at the bottom of the ramp?

(A) $\sqrt{2gh}$
(B) $\sqrt{gh}$
(C) $\sqrt{gL}$
(D) $\sqrt{\frac{Rgh}{3L}}$
(E) $2\sqrt{\frac{gh}{3}}$

83. A positively charged particle, initially moving in the $\hat{\mathbf{x}}$-direction, enters a region containing uniform electric and magnetic fields, $\mathbf{E} = E_0(\hat{\mathbf{x}} + 2\hat{\mathbf{y}})$ and $\mathbf{B} = B_0\hat{\mathbf{z}}$. Which of the following is a true statement about the motion of the particle?

(A) The particle moves in a circle.
(B) The particle is confined to the xz-plane.
(C) The particle is confined to the xy-plane.
(D) No work is done on the particle.
(E) The particle moves in a straight line.

84. Dark matter, a hypothetical component of matter in the universe, which feels the gravitational force but does not interact with electromagnetism, can potentially explain all of the following observations EXCEPT:

    (A) Stars at the outskirts of galaxies rotate faster than expected based on the mass inferred from luminous matter.
    (B) Galaxies appear to be receding from us at a rate proportional to their distance.
    (C) Fits to cosmic microwave background anisotropy data require a nonzero energy density which does not couple to photons.
    (D) Gravitational lensing observations of colliding galaxies show that the luminous mass and the gravitational mass are displaced from one another.
    (E) The gravitational potential of baryons alone would not be sufficient to counteract cosmic expansion and form structure such as galaxies.

85. A Geiger counter measures 1,250 events near a radioactive source during 10 seconds, and 350 events during 10 seconds when the source is removed. What is the uncertainty of the rate of events due to the source?

    (A) 3.0 Hz
    (B) 4.0 Hz
    (C) 18.7 Hz
    (D) 35.4 Hz
    (E) 40.0 Hz

86. A signal pulse contains a current that has an exponential risetime of 10 ms and an exponential falltime of 100 ms. Approximately what minimum bandwidth should be used to view the pulse in frequency space on a spectrum analyzer?

    (A) 0.01 Hz
    (B) 1 Hz
    (C) 1 kHz
    (D) 1 MHz
    (E) 1 GHz

87. A fit of a histogram to a model has a χ^2 statistic of 45.6. Which of the following should be used with the χ^2 value to determine whether the model represents the data well?

    I. The number of entries in the histogram
    II. The number of parameters in the model
    III. The number of bins in the histogram

    (A) II only
    (B) III only
    (C) I and III only
    (D) II and III only
    (E) I, II, and III

88. Which of the following is NOT produced in the *pp* cycle of the Sun?

    (A) ^{2}H
    (B) ^{3}He
    (C) ^{4}He
    (D) ^{8}B
    (E) ^{11}C

89. What is the mean energy at temperature T of a two-state system with energy states 0 and ϵ?

    (A) $\epsilon/2$
    (B) $\epsilon/(1+e^{-\epsilon/kT})$
    (C) $\epsilon/(1-e^{\epsilon/kT})$
    (D) $\epsilon/(1+e^{\epsilon/kT})$
    (E) $\epsilon(1+e^{\epsilon/kT})$

90. If the Hubble parameter suddenly were increased to twice its present value, which of the following would change?

    (A) Fundamental particle masses
    (B) Distance between Earth and distant galaxies
    (C) Redshift of distant galaxies
    (D) Cosmological constant
    (E) Gravitational constant G

91. A satellite in orbit around the Earth monitors seismic activity through interferometry, by comparing the phase of a reflected wave before and after an earthquake. If the typical displacement of a point on Earth is 5 cm, which of the following radiation frequencies would be most appropriate to use for this measurement?

(A) 1 Hz
(B) 1 kHz
(C) 1 MHz
(D) 1 GHz
(E) 1 THz

92. Doping a semiconductor does which of the following to the band structure?

(A) No effect on band structure
(B) Eliminates the valence band
(C) Increases the gap energy
(D) Adds additional states between valence and conduction bands
(E) Eliminates the energy gap

93. An infinite conducting cylinder of radius a carries a surface charge density of σ. Assuming that the potential on the surface is 0, what is the potential at a distance $r > a$?

(A) $\frac{1}{\epsilon_0} a\sigma \ln \frac{r}{a}$
(B) $\frac{1}{\epsilon_0} a\sigma \ln \frac{a}{r}$
(C) $\frac{1}{2\epsilon_0} a\sigma \ln \frac{r}{a}$
(D) $\frac{1}{2\epsilon_0} a\sigma \ln \frac{a}{r}$
(E) 0

94. A toy car travels through a vertical loop on a track. If the radius of the loop is 50 cm, what is the minimum initial speed needed by the car at the bottom of the loop to successfully complete the loop without losing contact with the track?

(A) 1.4 m/s
(B) 4.5 m/s
(C) 5.0 m/s
(D) 7.8 m/s
(E) 9.6 m/s

95. Which of the following is equivalent to the commutator $[S_x S_y, S_y]$, where S_x and S_y are quantum-mechanical spin operators?

(A) 0
(B) $i\hbar S_z S_x$
(C) $-i\hbar S_x S_z$
(D) $i\hbar S_y S_x$
(E) $i\hbar S_z S_y$

96. A particle is under the influence of a central potential $U(r)$ with the property that all bound orbits are closed. The particle has an energy E such that $V_{\min} < E < 0$, where $V_{\min}$ is the minimum of the effective potential derived from $U(r)$. Which of the following characterizes the shape of the orbit?

(A) Circular
(B) Elliptical
(C) Parabolic
(D) Hyperbolic
(E) The answer cannot be determined from the information given.

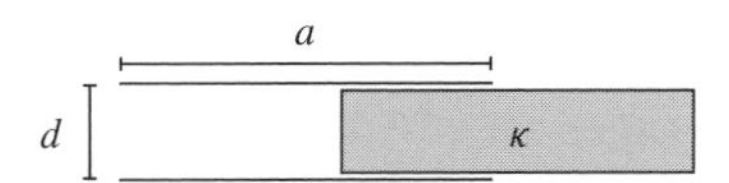

97. A square parallel-plate capacitor has side lengths a and separation d between the plates. A constant voltage V_0 is applied between the plates. A block of dielectric material of dielectric constant κ and the same area and thickness as the capacitor is slowly inserted into the capacitor. What is the change in energy stored in the capacitor by the time the dielectric is fully inserted?

(A) 0
(B) $\frac{(\kappa - 1)\epsilon_0 a^2 V_0^2}{4d}$
(C) $\frac{(\kappa - 1)\epsilon_0 a^2 V_0^2}{d}$
(D) $\frac{(\kappa - 1)\epsilon_0 a^2 V_0^2}{2d}$
(E) $\frac{\kappa \epsilon_0 a^2 V_0^2}{2d}$

98. An object is placed at rest in a potential field $U(x, y, z) = Ax + By^2 - C\cos z$, where A, B, C are constants. What is the force on the object?

(A) $\mathbf{F}(x, y, z) = -A\hat{\mathbf{x}} - 2By\hat{\mathbf{y}} - C\sin z\hat{\mathbf{z}}$
(B) $\mathbf{F}(x, y, z) = Ax\hat{\mathbf{x}} + 2By\hat{\mathbf{y}} - C\cos z\hat{\mathbf{z}}$
(C) $\mathbf{F}(x, y, z) = -Ax\hat{\mathbf{x}} - 2By\hat{\mathbf{y}} + C\cos z\hat{\mathbf{z}}$
(D) $\mathbf{F}(x, y, z) = -A\hat{\mathbf{x}} - 2By\hat{\mathbf{y}} + C\cos z\hat{\mathbf{z}}$
(E) $\mathbf{F}(x, y, z) = A\hat{\mathbf{x}} + 2By\hat{\mathbf{y}} + C\sin z\hat{\mathbf{z}}$

99. Suppose a hydrogen atom is in a uniform external electric field of magnitude E. What is the correction to the ground state energy calculated from first-order perturbation theory? The normalized ground state wavefunction of hydrogen is $\psi(r, \theta, \phi) = \frac{1}{\sqrt{\pi} a_0^{3/2}} e^{-r/a_0}$.

(A) eE
(B) $\frac{1}{2}eE$
(C) $\frac{3}{2}eE$
(D) $\frac{16}{5}eE$
(E) 0

100. In Mössbauer spectroscopy, a source of photons of energy E is moved with velocity $v \ll c$ relative to a target material. The absorption of photons by the target material is then measured, with the Doppler shift from the source velocity producing a small variation in the photon energy. If the absorption peaks of two lines correspond to source velocities of 0 and v, what is the energy splitting between the lines to lowest order in v?

(A) Ev^4/c^4
(B) Ev^3/c^3
(C) Ev^2/c^2
(D) Ev/c
(E) $E\sqrt{v/c}$

Answers to Sample Exam 1

1. C
2. E
3. D
4. B
5. A
6. E
7. C
8. D
9. B
10. C
11. E
12. D
13. E
14. B
15. A
16. D
17. E
18. A
19. E
20. B
21. D
22. A
23. E
24. A
25. C
26. B
27. C
28. D
29. A
30. E
31. D
32. C
33. E
34. C
35. D
36. C
37. D
38. B
39. B
40. D
41. B
42. B
43. D
44. E
45. C
46. A
47. A
48. B
49. D
50. D
51. C
52. D
53. A
54. D
55. C
56. C
57. B
58. C
59. E
60. B
61. C
62. A
63. A
64. B
65. E
66. C
67. C
68. E
69. B
70. C
71. B
72. B
73. E
74. E
75. D
76. A
77. C
78. C
79. C
80. D
81. E
82. A
83. B
84. D
85. C
86. E
87. C
88. A
89. E
90. D
91. C
92. B
93. E
94. E
95. E
96. A
97. C
98. B
99. D
100. A

Answers to Sample Exam 2

1. D
2. C
3. E
4. C
5. B
6. A
7. C
8. D
9. D
10. E
11. A
12. A
13. E
14. C
15. E
16. C
17. B
18. D
19. A
20. A
21. C
22. D
23. B
24. A
25. E
26. B
27. E
28. D
29. D
30. C
31. E
32. E
33. E
34. B
35. E
36. D
37. D
38. D
39. C
40. B
41. C
42. D
43. D
44. D
45. A
46. D
47. B
48. B
49. A
50. A
51. D
52. B
53. B
54. A
55. E
56. E
57. E
58. A
59. E
60. D
61. A
62. A
63. D
64. A
65. B
66. D
67. E
68. C
69. D
70. B
71. B
72. B
73. C
74. D
75. B
76. C
77. D
78. A
79. B
80. B
81. B
82. E
83. C
84. A
85. A
86. B
87. C
88. B
89. C
90. D
91. B
92. C
93. C
94. E
95. D
96. B
97. E
98. B
99. C
100. D

Answers to Sample Exam 3

1. C
2. E
3. E
4. B
5. E
6. C
7. C
8. B
9. C
10. A
11. A
12. C
13. B
14. B
15. C
16. B
17. C
18. E
19. C
20. D
21. D
22. D
23. B
24. B
25. B
26. B
27. D
28. A
29. B
30. B
31. B
32. A
33. B
34. E
35. D
36. C
37. C
38. C
39. E
40. E
41. D
42. B
43. C
44. C
45. A
46. D
47. A
48. E
49. E
50. D
51. B
52. E
53. E
54. C
55. C
56. C
57. C
58. C
59. A
60. C
61. E
62. A
63. E
64. C
65. A
66. E
67. B
68. B
69. D
70. B
71. B
72. C
73. A
74. C
75. A
76. E
77. A
78. B
79. E
80. D
81. C
82. E
83. C
84. B
85. B
86. C
87. D
88. E
89. D
90. C
91. D
92. D
93. B
94. C
95. E
96. B
97. D
98. A
99. E
100. D

Solutions to Sample Exam 1

1. C – This is a simple application of the centripetal acceleration formula:

$$v = \sqrt{ar} = \sqrt{9 \cdot 4g} = 6\sqrt{|g|} \text{ m/s}.$$

2. E – Although we have to just calculate this one, it is good practice for working through basic kinematics problems quickly. If you find yourself getting bogged down in numerical factors, this is a good example of a problem to skip and save for later, as there shouldn't be too many problems of this type on the exam. Recall that elastic collisions conserve both linear momentum and energy. If we call the final velocity of the heavy block v_1 and the final velocity of the light block v_2, and drop all the m's, then we just need to solve the following system of equations:

$$v = 4v_1 + v_2,$$
$$\frac{1}{2}v^2 = \frac{1}{2}4v_1^2 + \frac{1}{2}v_2^2.$$

Plugging the first equation into the second for v, we get

$$(4v_1 + v_2)^2 = 4v_1^2 + v_2^2,$$
$$16v_1^2 + 8v_1v_2 + v_2^2 = 4v_1^2 + v_2^2,$$
$$12v_1^2 + 8v_1v_2 = 0,$$
$$v_2 = -\frac{3}{2}v_1.$$

Then, plugging this into the conservation of momentum equation, we get

$$v = 4v_1 - \frac{3}{2}v_1,$$
$$v_1 = \frac{2}{5}v.$$

3. D – The oscillation frequency of an LC circuit is a good quantity to memorize:

$$\omega = \frac{1}{\sqrt{LC}}.$$

If you (understandably) forget it, it is straightforward to reconstruct it from dimensional analysis by looking for combinations of L and C that give units of s^{-1}. The capacitance scales as $C \sim A/d$, so C increases by a factor of 2 under the doubling. Assuming that our inductor is a solenoid with a fixed number of loops, then the inductance scales just like the solenoid inductance (a 20 second derivation if you forget it) $L \sim A/\ell$. Overall, the frequency drops in half.

4. B – A dipole contains no net charge, and the charge on a conductor will rearrange itself to completely cancel out the dipole field.
5. A – The approximation of a product wavefunction comes from solving the Schrödinger equation by separation of variables, which is only possible when the Coulomb repulsion term is ignored.
6. E – Since the emitted photon carries momentum, the entire atom must recoil slightly to conserve linear momentum. This means that the total energy released in the transition is divided between the gamma and the recoil of the atom. As a result, the photon energy will be slightly less than the true transition energy of the atomic level. Choices A–C simply do not make sense in the context of the problem, and D appears to violate the conservation of energy, so E is the correct choice.
7. C – The Fermi energy is $E_F = \frac{\hbar^2}{2m}(3\pi^2 n)^{2/3}$, where n is the density. Since we are at fixed volume, doubling the number of particles doubles the density, which multiplies E_F by a factor of $2^{2/3}$, choice C.
8. D – Since the gas is well above the Fermi temperature, it is essentially classical. By the equipartition theorem, each quadratic degree of freedom contributes specific heat per particle $k/2$. There are two quadratic degrees of freedom for the kinetic part of the Hamiltonian and two quadratic degrees of freedom for the potential part, so the specific heat is $2k$.
9. B – Recall that the effective potential for radial motion is $V_{\text{eff}}(r) = \frac{l^2}{2mr^2} + U(r)$. The radii of circular orbits are found by solving $\frac{dV_{\text{eff}}}{dr} = 0$, and stability is determined by the sign of $\frac{d^2V_{\text{eff}}}{dr^2}$. Here,

$$\frac{dV_{\text{eff}}}{dr} = -\frac{l^2}{mr^3} + \frac{k}{r^2},$$

and setting this equal to zero and solving for r gives $r = \frac{l^2}{mk}$, choice B. We could check that this is a minimum of V_{eff} by computing the second derivative, but it's easier to just think about the behavior of the potential at $r = 0$ and $r = \infty$. At $r = 0$, the repulsive centrifugal barrier dominates and $V_{\text{eff}} \to +\infty$. As r approaches infinity, the central potential $U = -k/r$ decays more slowly than the centrifugal term, so V_{eff} approaches zero from *below*. Since V_{eff} has only a single critical point, sketching the graph of V shows that this radius is indeed a minimum, and hence an allowed stable circular orbit.

10. C – This is a straightforward application of the parallel axis theorem. Recall from the formula sheet that the

moment of inertia for a single disk about its central axis is $I_{\text{CM}} = \frac{1}{2}MR^2$. From the parallel axis theorem, we can read off the moment of inertia for a single disk about the axis going through the edge:

$$I_{\text{disk}} = I_{\text{CM}} + \frac{1}{2}MR^2 = MR^2 + \frac{1}{2}MR^2 = \frac{3}{2}MR^2.$$

And finally, the moment of inertia of two bodies that are rigidly attached to each other is equal to the sum of the moments of inertia of the individual bodies, so we have the final result

$$I_{\text{total}} = 2\frac{3}{2}MR^2 = 3MR^2.$$

11. E – Redshift z is defined by $1 + z = \lambda_{\text{obs}}/\lambda_{\text{emit}}$. So a galaxy at redshift 2 has wavelengths expanded by a factor of 3, and the 21 cm hydrogen line gets redshifted to 63 cm, choice E. Beware the trap answer D: redshift 2 does *not* mean wavelengths are expanded by a factor of 2!
12. D – The probabilities of states 1 and 2 are $\frac{1}{4}$ and $\frac{1}{2}$, so the probability of measuring state 3 must be $\frac{1}{4}$. We can write the wavefunction as

$$|\Psi\rangle = c_1 |1\rangle + c_2 |2\rangle + c_3 |3\rangle,$$

and the energy expectation value is

$$\langle E \rangle = c_1^2(E_0) + c_2^2(2E_0) + c_3^2 X.$$

The squares of the coefficients are the probabilities of measuring each energy, so we can just plug in values and solve for X:

$$\begin{aligned}
\frac{9}{4}E_0 &= \frac{1}{4}E_0 + \frac{1}{2}2E_0 + \frac{1}{4}X, \\
\frac{1}{4}X &= \frac{9-5}{4}E_0, \\
X &= 4E_0.
\end{aligned}$$

13. E – The Planck mass is the unique combination of fundamental constants $\hbar$, c, and G that has units of mass. Expressing all the constants in terms of mass, length, and time:

$$\begin{aligned}
\hbar &= [M][L]^2[T]^{-1}, \\
c &= [L][T]^{-1}, \\
G &= [M]^{-1}[L]^3[T]^{-2}.
\end{aligned}$$

For a combination $(\hbar)^p(c)^q(G)^r$ to have units of mass, or $[M]^1$, we need:

$$\begin{aligned}
p - r &= 1, \\
2p + q + 3r &= 0, \\
-p - q - 2r &= 0.
\end{aligned}$$

Using your favorite method to solve systems of linear equations, we find $p = 1/2$, $q = 1/2$, $r = -1/2$, so $M_P = \sqrt{\frac{\hbar c}{G}}$, which is E.

14. B – This is a straightforward application of the relativistic Doppler shift formula:

$$\frac{\nu'}{\nu} = \sqrt{\frac{1+\beta}{1-\beta}},$$

where $\beta = v/c$ is the relative velocity of the source with respect to the observer. The easy way to keep track of the signs in the numerator and the denominator is to remember that the observed frequency *increases* when the source is moving towards the observer. Or, if you're partial to astrophysics, when the observer is receding, the signal is *redshifted* (in other words, the frequency decreases). Here the source is approaching, so $\beta = 0.6$ is positive and

$$\frac{\nu'}{\nu} = \sqrt{\frac{1.6}{0.4}} = \sqrt{4} = 2,$$

so the frequency doubles: $\nu' = 2$ GHz, choice B.

15. A – This is a classic method of images problem, with a small twist. To calculate the work, we first need to know the potential energy of the initial charge configuration with the point at infinity. This is just $U_0 = 0$, by definition. The final potential energy can be determined from the image charge configuration, which consists of our charge $+q$ at a distance d above the plane, and an image charge $-q$ at a distance d below the plane. Be sure to remember your signs! One of the charges must be negative because the potential on the conducting plane must, by definition, always be zero (this is the entire point of the method of images). The potential energy of two opposite point charges separated by a distance $2d$ is given by

$$U_{\text{image}} = -\frac{1}{4\pi\epsilon_0}\frac{q^2}{2d}.$$

Now the trick: the total potential energy of the point charge plus plane is half the potential energy of the two point charges. This is because we can think of the total energy of the configuration as

$$U = \int_{\text{all space}} \frac{\epsilon_0}{2}E^2\, dV,$$

but since the plane is a conductor, the electric field $E = 0$ below the plane. By symmetry about the plane, the potential energy of the charge with conductor is half the energy of the two charges. Thus we find

$$W = \Delta U = \frac{U_{\text{image}}}{2} - 0 = -\frac{1}{4\pi\epsilon_0}\frac{q^2}{4d},$$

which is choice A. You could also compute this by integrating the force on the point charge, as discussed in Section 2.1.7.

16. D – The charge after one second is $Q = (10^{-3}\ \text{C}\cdot\text{s}^{-1})(1\ \text{s}) = 10^{-3}$ C. From $V = Q/C$, we get $V = 10^{-3}/10^{-5} = 100$ V.
17. E – The leading-order perturbation is simply

$$\langle 2|\,\delta V\,|2\rangle = \frac{2}{L}\int_0^{L/2} V_0 \sin^2\left(\frac{2\pi x}{L}\right) dx$$
$$= V_0/2.$$

18. A – From $dE = T\,dS - P\,dV$, we do a Legendre transform to get $dF = -S\,dT - P\,dV$. Since the expansion is isothermal, it takes place at constant temperature, so the first term vanishes; the second term is positive because the volume increases, so dF is negative and the free energy decreases, rather than increases. This is consistent with the definition of "free" energy, which is essentially the available energy the gas has to do work on its surroundings. Since it does work in the isothermal expansion phase, at the end of the expansion it has less free energy available to do work. All the other statements are true; see the discussion in Section 4.2.6.
19. E – The angle of the first diffraction minimum is $\sin\theta = \lambda/a$, where λ is the wavelength of the incident light. Since the screen is far away, we approximate $\sin\theta \approx \theta$, so the angular width of the central maximum is 2θ. For a screen a distance L away, the width of the maximum as seen on the screen is $L\tan\theta \approx L\theta$. So we want

$$2L\frac{\lambda}{a} = 100a \implies a^2 = 9\times 10^{-8}\ \text{m}^2$$
$$\implies a = 3\times 10^{-4}\ \text{m} = 0.3\ \text{mm}.$$

20. B – From the definition of conjugate momentum,

$$p_\phi = \frac{\partial L}{\partial\dot\phi} = m(a + b\cos\theta)^2\dot\phi,$$

choice B.

21. D – The Lagrangian isn't explicitly dependent on time, so the first two terms in L represent the kinetic energy T and the third represents the potential energy U. Since $L = T - U$, to get the total energy $T + U$ we have to take $L + 2U$, which is choice D.
22. A – The decay time constant for an RL circuit is $\tau = L/R$. This represents the amount of time required for the voltage across the inductor to drop to $1/e$ of its initial level, or, in other words, $V(t) = V_0e^{-t/\tau}$. Setting $V(t) = V_0/2$, we find that $t = \tau\ln 2 = \frac{L}{R}\ln 2$.
23. E – The straight wire produces a magnetic field that circles azimuthally around in the $\hat{\boldsymbol{\phi}}$-direction. The tension on the circular wire, if any, will be the result of the Lorentz force that the moving charges in the circular wire feel due to the magnetic field established by the straight wire. But the current in the circular wire is also flowing in the $\hat{\boldsymbol{\phi}}$-direction, parallel to the magnetic field. The electrons flowing in the circular wire therefore feel no force and there is no tension. So the correct answer is E.
24. A – Considering the limit of $r \gg h$ eliminates C and D immediately, but we need to calculate to get the exact numerical factors. The cylinder will fall over when it is tipped just past the point where the center of mass is directly above the point of contact of the cylinder with the ground. At that point, the angle θ forms a right triangle with side lengths $h/2$ and r, so we have $\tan\theta = r/(h/2) = 2r/h$, and the angle of the cylinder with the horizontal is A.
25. C – The γ factor is $\gamma = 3000\ \text{MeV}/100\ \text{MeV} = 30$. In the (stationary) lab frame, the lifetime of the muon is $\tau = \gamma\tau_{\text{rest}} = 30\tau_{\text{rest}} = 60\ \mu\text{s}$.
26. B – This problem is a bit of trivia that is difficult to guess. The only way to guess it is to notice that the dispersion relation in B gives energy as proportional to momentum, which is ordinarily true for particles in the extreme relativistic limit (i.e. $E = pc$). Given that the problem refers to unusual electronic properties, it might seem reasonable that this unusual dispersion relation would produce unusual properties. This is indeed what happens in graphene, where electrons and holes behave like massless Dirac fermions, a distinctive feature of a semiconductor with no gap energy and linear dispersion.
27. C – The perturbing potential $V(x) = -qE_0x$ is an odd function, while the ground state wavefunction for the harmonic oscillator is even. Therefore, the first-order perturbation will vanish, and the leading-order correction will come from second-order perturbation theory and be proportional to E_0^2.
28. D – We can use the lens equation twice. For the first lens, we have

$$\frac{1}{5\ \text{cm}} = \frac{1}{2\ \text{cm}} + \frac{1}{s_1'},$$

which implies that $s_1' = -10/3$ cm. The negative sign implies that the image is located to the left of lens A, or 25/3 cm to the left of lens B. Using the lens equation again for the second lens, we find that

$$\frac{1}{5\text{ cm}} = \frac{3}{25\text{ cm}} + \frac{1}{s_2'},$$

which implies that $s_2' = 12.5$ cm. Since the answer is positive, it is located to the right of lens B.

29. A – As is typical for uniform circular motion, we want to set the gravitational force equal to the centripetal force. The acceleration in uniform circular motion is still given by $a = v^2/r$, and the gravitational force is still $F_{\text{grav}} = GMm/r^2$. The MOND force law requires that we set $F_{\text{grav}} = ma^2/a_0$, giving

$$\frac{GMm}{r^2} = m\frac{(v^2/r)^2}{a_0}$$
$$\implies v = (GMa_0)^{1/4}.$$

The factors of r cancel out, and v ends up independent of r.

30. E – When combining spins l_1 and l_2 (where without loss of generality we can take $l_1 < l_2$), we can get all values of l between $l_2 + l_1$ and $l_2 - l_1$ in integer steps. In the present case, with $l_1 = 1$ and $l_2 = 2$, we can get $s = 3, 2, 1$. The only condition on m_s is $|m_s| \leq s$, so choices A–D are all fine. Since $s = 0$ is impossible, the answer is E.

31. D – The radial probability density for the $2p$ state is given by

$$r^2|R_{21}(r)|^2 = \frac{1}{24}a_0^{-5}r^4 \exp(-r/a_0),$$

where the factor of r^2 is from the volume element in spherical coordinates, $dV = r^2 \sin\theta dr\, d\theta\, d\phi$. To find the most probable value, we take the derivative and set to zero (and in doing so, we can ignore all the annoying constants out front):

$$4r^3e^{-r/a_0} - \frac{r^4}{a_0}e^{-r/a_0} = 0$$
$$\implies r = 4a_0,$$

choice D. Forgetting the factor of r^2 from the volume element leads to trap answer C, forgetting to square the wavefunction leads to E, and forgetting both also leads to C. These are all very common mistakes – don't make them!

32. C – This is almost a giveaway, since the "mass" in "mass spectrometry" does indeed refer to the mass of the particle involved. However, it is important to note that only the charge-to-mass ratio can be measured by electromagnetic fields, so the answer can't be D, which can change the mass without affecting the charge.

33. E – Beat frequencies are caused by destructive interference between two closely spaced frequencies, resulting in a modulation with a long enough period that the minimum of each cycle is heard independently. For this problem, all that is relevant is the sum-to-product identity

$$\cos 2\pi at + \cos 2\pi bt = 2\cos\left(2\pi\frac{a+b}{2}t\right)\cos\left(2\pi\frac{a-b}{2}t\right).$$

The second term in the product modulates the wave; we hear beats when its amplitude is zero, which occurs at frequency $a - b$. Note that this is *twice* the apparent frequency of the cosine! If all we know is that this frequency is 3 Hz, and that one of a or b is 440 Hz, it is impossible to determine whether the other frequency is 443 Hz or 437 Hz, since both would give the same beat frequency (the sign of $a - b$ is irrelevant because cosine is even). Hence the correct answer is E.

34. C – This is a relativistic velocity addition problem. In the frame of the planet, the speed of the missile is given by the velocity addition formula:

$$v_{\text{missile}} = \frac{0.5c + 0.5c}{1 + (0.5c)(0.5c)/c^2} = \frac{c}{1.25} = 0.8c.$$

Light travels at c in all frames, so the information that the missile has been fired reaches the planet at $1c$. Since the spaceship is 1 light-hour away ($c \times 1$ h), the observer sees the flash at $t_1 = (c \times 1\text{ h})/1c = 1$ h, and the missile hits at $t_2 = (c \times 1\text{ h})/0.8c = 1.25$ h, so the difference is 0.25 h or 15 min, choice C.

35. D – There are no external torques here, so angular momentum is conserved. Thus $I_0\omega_0 = I_1\omega_1$, and plugging in numbers we arrive at D.

36. C – IV is obviously false since we use the Euler–Lagrange equations to construct the equations of motion for any system with a Lagrangian. III is not always true because we are always free to add a constant to the potential – the fact that L is independent of x only means that the potential must be constant in space, and we can set that constant to zero if we wish.

On the other hand, homogeneity of time *does* imply conservation of energy, and homogeneity of space implies linear momentum conservation. If these statements are unfamiliar to you, then let's prove them. Since L depends on x and $\dot{x}$ only, the total time derivative of the Lagrangian is

$$\frac{dL}{dt} = \frac{\partial L}{\partial x}\dot{x} + \frac{\partial L}{\partial \dot{x}}\ddot{x} + \frac{\partial L}{\partial t}.$$

Using the Euler–Lagrange equations and the fact that $\frac{\partial L}{\partial t} = 0$, we have

$$\begin{aligned} \frac{dL}{dt} &= \dot{x}\frac{d}{dt}\frac{\partial L}{\partial \dot{x}} + \frac{\partial L}{\partial \dot{x}}\ddot{x} \\ &= \frac{d}{dt}\left(\dot{x}\frac{\partial L}{\partial \dot{x}}\right), \\ 0 &= \frac{d}{dt}\left(\dot{x}\frac{\partial L}{\partial \dot{x}} - L\right). \end{aligned}$$

But the quantity in parentheses is precisely the energy of the system that we obtain when we construct the Hamiltonian of the system from the Lagrangian via the Legendre transform.

The situation for momentum is more simple. By Euler–Lagrange, we have

$$\frac{d}{dt}\frac{\partial L}{\partial \dot{x}} = 0,$$

and therefore

$$\frac{\partial L}{\partial \dot{x}} = p,$$

where p is a constant of motion that is defined to be the linear momentum.

37. D – The formula we want is $N = \epsilon \mathcal{L} \sigma T$, where N is the number of events seen, ϵ is the detector efficiency, $\mathcal{L}$ is the luminosity, σ is the cross section, and T is the running time. Note that this just comes straight from dimensional analysis – any possible numerical factors are all absorbed in the definitions of luminosity, cross section, and efficiency. There are 8.64×10^4 seconds in a day, so $N = (0.5)(10^{22})(10^{-20})(8.64 \times 10^4) = 4.32 \times 10^6$, choice D.

38. B – Cyclotron motion is simple enough that it can be derived in a matter of seconds from the Lorentz force law and centripetal force. Rewriting a bit, we find that

$$r = \frac{p}{qB}.$$

Triple the field, and the momentum must also be tripled to maintain constant radius.

39. B – Without calculating anything, we can reason that only choices B and C are allowed. The current is the same in both loops of wire and the field from the smaller loop (pointed into the page, by the right-hand rule) will be stronger than the field from the larger loop (pointed out of the page). So the total field must also point into the page. If you remember the formula for the magnetic field at the center of a single circular loop of wire of radius r,

$$B = \frac{\mu_0 I}{2r},$$

then you can immediately find that the answer is

$$B_{\text{tot}} = \frac{\mu_0 I}{2}\left(\frac{1}{a} - \frac{1}{b}\right),$$

which is choice B.

If you are in a bind and don't remember the formula for the field at the center of a circular loop of wire, it is easy to derive from the Biot–Savart law:

$$\begin{aligned} \mathbf{B} &= \frac{\mu_0}{4\pi}\int \frac{I d\mathbf{l} \times \hat{\mathbf{r}}}{r^2} \\ &= \frac{\mu_0 I}{4\pi r}\int_0^{2\pi} d\phi\, \hat{\mathbf{z}} \\ &= \frac{\mu_0 I}{2r}\hat{\mathbf{z}}. \end{aligned}$$

40. D – Recall that L^2 commutes with each component of L because it commutes with L_z. For the same reason, L^2 commutes with $J^2 = L^2 + S^2 + 2\mathbf{L} \cdot \mathbf{S}$: $\mathbf{L}$ and $\mathbf{S}$ act on different parts of the wavefunction, so all components of $\mathbf{L}$ commute with all components of $\mathbf{S}$, and L^2 commutes with the last term because it commutes with $\mathbf{L}$ as well. However, L^2 only commutes with rotationally symmetric Hamiltonians, and a Hamiltonian need not have rotational symmetry: for example, an atom in a strong magnetic field which picks out a particular direction in space.

41. B – The possible energies are just the eigenvalues of the Hamiltonian matrix, which we can obtain by solving the equation

$$\begin{aligned} \det(H - \lambda I) &= 0, \\ (a - \lambda)(\lambda^2 - b^2) &= 0. \end{aligned}$$

The solutions are clearly $\lambda = a$ and $\lambda = \pm b$, choice B.

42. B – If there were magnetic charges, we would have a Maxwell equation $\nabla \cdot \mathbf{B} \propto \rho_m$ just like we have a Maxwell equation for electric (monopole) charges $\nabla \cdot \mathbf{E} \propto \rho_e$, and Gauss's law for magnetism would give the total charge by integrating over space. This is choice B.

43. D – We could solve this one exactly, but it is clear from the answer choices that we can use dimensional analysis and scaling arguments to rule out answers much more quickly. Choice C does not have the correct units for an electric field (N/C). And choices A, B, and E all have the wrong scaling with the radius of the target.

As the radius of the target increases, the electric field required to deflect the beam should increase. In all of these answers, R appears in the denominator, implying the opposite behavior.

On the other hand, it is not too difficult to solve the problem the normal way, which would be necessary on the exam if a different set of answer choices were given. The perpendicular electric field only provides an acceleration perpendicular to the direction of motion, so we can solve this just like an analogous kinematics problem where gravity provides the perpendicular force. The velocity in the direction of the beam stays v, so the time it takes to strike the target is $t = L/v$. On the other hand, the perpendicular acceleration is $a_{\perp} = qE/m$, so in time t the protons are deflected $\frac{1}{2}a_{\perp}t^2 = \frac{qE}{2m}t^2$ in the perpendicular direction. Setting this equal to R, plugging in $t = L/v$, and solving for E, we obtain $E = \dfrac{2mRv^2}{qL^2}$, choice D.

44. E – In the standard cosmology, the Big Bang is immediately followed by inflation, at the end of which elementary particles are produced in an era known as "reheating." At this stage, quarks and gluons are free particles. Once the universe cools enough to allow quarks to form baryons, we get nucleosynthesis. Even without knowing the exact timeline governing I and II, we can deduce immediately that II precedes I, since (re-)ionization refers to the stripping of an electron from a neutral atom, and to have atoms we must first have nuclei, which are formed during nucleosynthesis. So the correct order is III, II, I, choice E.

45. C – For an adiabatic process, we have $PV^{\gamma} = \text{const}$, where $\gamma \equiv c_P/c_V = (\alpha + 1)/\alpha$ and 2α is the number of degrees of freedom for the system. For an ideal gas $\alpha = 3/2$ and $\gamma = 5/3$.

46. A – A solid cylinder of uniform mass density has moment of inertia $I = \frac{1}{2}MR^2$. The potential energy mgL of the weight when the rope is wound is entirely converted to kinetic energy once the rope is unwound. Note that the kinetic energy has contributions from both the rotational energy of the cylinder and the velocity of the weight. So by conservation of energy,

$$mgL = \frac{1}{2}mv^2 + \frac{1}{4}MR^2\frac{v^2}{R^2} \implies v = \sqrt{\frac{4mgL}{M + 2m}},$$

choice A.

47. A – The force on an object is related to the potential energy by

$$\mathbf{F} = -\nabla U.$$

All we need is to take the gradient of the potential in the question:

$$\begin{aligned}\mathbf{F} &= -\hat{\mathbf{x}}\frac{\partial}{\partial x}(x) - \hat{\mathbf{y}}\frac{\partial}{\partial y}(y^2) + \hat{\mathbf{z}}\frac{\partial}{\partial z}(\cos z)\\ &= -\hat{\mathbf{x}} - 2y\hat{\mathbf{y}} - \sin z\hat{\mathbf{z}}.\end{aligned}$$

48. B – The formula for the energy is

$$E = \frac{\sum_{\epsilon} \epsilon d(\epsilon)e^{-\epsilon/kT}}{\sum_{\epsilon} d(\epsilon)e^{-\epsilon/kT}}.$$

Plugging in the given energies and degeneracies,

$$E = \frac{(-2\epsilon)e^{\epsilon/kT} + (3\epsilon)e^{-\epsilon/kT}}{2e^{\epsilon/kT} + 1 + 3e^{-\epsilon/kT}}.$$

As $T \to \infty$, the exponent goes to zero, so each of the exponential factors collapses to 1. Thus the $T \to \infty$ limit of the energy is

$$E = \frac{-2\epsilon + 3\epsilon}{2 + 1 + 3} = \frac{\epsilon}{6},$$

choice B. This last step is equivalent to saying that, at infinite temperature, each state becomes equally probable, so we can forget about the Boltzmann statistics and just calculate a weighted mean.

49. D – In process 1 we have $\Delta S_1 = \int dQ/T = C\ln 2 = (3/2)Nk\ln 2$. In process 2, we have have $\Delta S_2 = Nk\ln 2$ from the equation for the entropy of an ideal gas. So $\Delta S_1 > \Delta S_2$.

50. D – The Poynting vector is

$$\mathbf{S} = \frac{1}{\mu_0}(\mathbf{E} \times \mathbf{B}),$$

and if the electric field is $\mathbf{E} = E_0\cos(kx - \omega t)\hat{\mathbf{z}}$, then the magnetic field is $\mathbf{B} = -(1/c)E_0\cos(kx - \omega t)\hat{\mathbf{y}}$ for propagation in the $\hat{\mathbf{x}}$-direction. The average magnitude of the Poynting vector is simply

$$\langle \mathbf{S} \rangle = \frac{1}{\mu_0 c}\langle E_0^2\cos^2(kx - \omega t)\rangle = \frac{E_0^2}{2\mu_0 c}.$$

(A very useful fact to remember is that the average of $\sin^2$ or $\cos^2$ over one period is $1/2$.)

51. C – If we consider the reaction in the CM frame of the e^+e^- system, then the γ must be at rest to conserve energy–momentum. But γ always travels at the speed of light, so conservation of energy–momentum is violated.

52. D – This is a straightforward application of the uncertainty principle. We know that

$$\Delta p \Delta x \gtrsim \hbar.$$

So if r is the nuclear radius,

$$r \sim \Delta x \gtrsim \frac{\hbar}{\Delta p}.$$

Using our numerical values, we have

$$r \gtrsim \frac{\hbar c}{40 \text{ MeV}} = \frac{197}{40} \text{ fm} \sim 5 \text{ fm},$$

closest to choice D. Note that we have dropped lots of factors of 2 along the way, which is fine for these "powers of 10" type of problems.

53. A – Choices B–E all are spin-1 objects and therefore are bosons, which obey Bose–Einstein statistics. You should know that the photon has spin-1 from quantum mechanics. The statistics of helium nuclei and atoms follow from the rules of addition of angular momentum: the nucleus has four fermions (two protons and two neutrons), and the atom has two more fermions (the two atomic electrons), and an even number of spin-1/2 particles behaves as a boson with integer spin. Pions are a little tricky, but you might remember that in the quark model they are bound states of a quark and an antiquark, both fermions, so again an even number of fermions gives a boson. Only A, the neutrino, is a spin-1/2 particle and fermion that obeys Fermi statistics.

54. D – The mass of the electron is 511 keV/c^2, which is one of those ubiquitous numbers that should be memorized. So any time the energy scale 511 keV shows up, it must have something to do with electrons. In this case, electrons and positrons at rest in the galactic center can annihilate to two photons, each of energy 511 keV. (Incidentally, the fact that the line is "sharp" means that most of the electrons and positrons in the galactic center are moving slowly, since if annihilation took place between two highly energetic particles, the photons would be boosted in the galactic rest frame. The spectrum observed on Earth would then have a sharp "edge" at 511 keV and a long tail extending to higher energies.)

55. C – From basic kinematics, the cannonball falls a vertical distance of 250 m in time $\frac{1}{2}gt^2$; solving gives $t = \sqrt{500/g}$ seconds. In this time, the cannonball travels a horizontal distance of 420 m, so from $vt = 420$ m, we plug in our previous result for t to get

$$v = 420\sqrt{g/500} \approx 420\sqrt{1/50} \approx 420(1/7) \approx 60 \text{ m/s}.$$

This is closest to choice C, so we choose it and move on. This pattern of answer choices is typical of Physics GRE questions – since they're so widely spaced, it's not at all necessary to do arithmetic to three decimal places. The approximations $g \approx 10$ m/s and $\sqrt{50} \approx 7$ were just fine here.

56. C – Let the mass of the rocket be m. In the frame of the rocket, we balance the momentum of the exhaust with the momentum of the rocket at the moment of expulsion:

$$|\Delta p_{\text{fuel}}| = |\Delta p_{\text{rocket}}|,$$
$$\frac{v}{2}(0.1m) = (0.9m)|\Delta v_{\text{rocket}}|,$$
$$|\Delta v_{\text{rocket}}| = \frac{v}{18}.$$

Transforming to the given frame where the rocket is moving at speed v, the final velocity is $v + v/18 = 19v/18$. The signs are fixed by physical reasoning: if the exhaust is expelled backwards, the rocket's velocity increases.

57. B – Since all we care about is ratios, we can scale all elements of the circuit by some convenient numerical factor: let's divide all the resistances by 10 kΩ to make the numbers easier (so we're working in units of 10^{-4} A whenever we calculate currents). When S is open, the total resistance of the circuit is $4 + 1 = 5$, so the current is $I_1 = 1$. When S is closed, the two resistors in parallel have an effective resistance of $(1 + 1/2)^{-1} = 2/3$. So the total resistance is $4+2/3 = 14/3$, and the total current is $I = 15/14$. The voltage across the 40 kΩ resistor is $IR = 30/7$, so the voltage drop across the 10 kΩ resistor is $5 - 30/7 = 5/7$. Hence the current I_2 is 5/7 in our units; since $I_1 = 1$ in these units, this is also the desired ratio.

58. C – This is a ballistic pendulum problem with two twists: the rod is not massless, and it starts out an angle with respect to the projectile, so we have to be a little careful calculating the angular momentum. The initial angular momentum comes just from the clay and is $L = m\mathbf{v} \times \mathbf{r} = mvd\sin(90° - \alpha) = mvd\cos\alpha$, so by conservation of angular momentum, this is the angular momentum after the collision as well. We get the angular velocity from $L = I\omega$, where the moment of inertia of the clay–rod system is $I = \frac{1}{3}Md^2 + md^2$ (the first term from the moment of inertia of a rod about one end, and the second from the moment of inertia of a point mass). So

$$\omega = \frac{L}{I} = \frac{3mv\cos\alpha}{(M+3m)d},$$

choice C.

59. E – Because the loop rotates, the angle the normal to the loop makes with the magnetic field oscillates sinusoidally, and so does the flux: $\Phi = B_0 A \sin \omega t$. Thus the emf in the loop is $V = d\Phi/dt = B_0 A\omega \cos \omega t$, and the power dissipated in the resistor is $P = V^2/R = B_0^2 A^2 \omega^2 \cos^2 \omega t/R$. Solving for ω, and using the fact that the average of $\cos^2$ is $1/2$, we find

$$\omega = \frac{\sqrt{2PR}}{B_0 A} = 500 \text{ rad/s}.$$

60. B – If you do not recall the mathematical definition of chemical potential, you should at least remember that the chemical potential has something to do with the energy associated with changing the number of particles in a system. This reduces the options to A or B, but there is chemical potential for systems of bosons and fermions, so B is the correct choice.

To be more rigorous, recall that the entropy is $S = \sum_i p_i \ln p_i$. We wish to maximize the entropy with respect to the constraints:

$$0 = \sum_i \epsilon_i p_i - E,$$

$$0 = \sum_i n_i p_i - N,$$

$$0 = \sum_i p_i - 1.$$

To maximize S we introduce the Lagrange multipliers β, μ, and λ, we write

$$S = \sum_i p_i \ln p_i - \beta \left(\sum_i \epsilon_i p_i - E \right) - \mu \left(\sum_i n_i p_i - N \right) - \lambda \left(\sum_i p_i - 1 \right),$$

and solve for

$$\frac{\partial S}{\partial x} = 0.$$

This μ turns out to be the chemical potential and is clearly the Lagrange multiplier that enforces the particle number constraint.

61. C – At first glance, choices D and E do not seem to make much sense. Choice B is a bit suspicious too because it does not contain a factor of 2, implying that the spin distributions are independent of the external magnetic field. Choice A does not make much sense either because the number of spin-up particles should increase, not decrease, with an increase in magnetic field: alignment with the magnetic field is energetically favorable. So C seems to be the best choice.

To decide for sure, we can calculate. Since the eigenvalues of σ_z are ± 1, the possible energies are $E = \pm |H|$. The partition function is

$$Z = e^{E/kT} + e^{-E/kT}.$$

The ratio of spin-up to spin-down particles is just

$$A = \frac{e^{E/kT}}{e^{-E/kT}} = e^{2E/kT}.$$

If we double the magnetic field, we have $E \to 2E$, which implies that $A \to e^{4E/kT} = \left(e^{2E/kT}\right)^2 = A^2$.

62. A – While this is a good fact to memorize, we can get it quickly by recalling Poisson's equation from electromagnetism, in SI units:

$$\nabla^2 V = -\rho/\epsilon_0.$$

Since the potential of a point charge q at the origin is $V = \frac{q}{4\pi\epsilon_0} \frac{1}{r}$, and its charge density is $\rho = q\delta^3(\mathbf{r})$, we can read off $\nabla^2 V = -4\pi\delta^3(\mathbf{r})$.

63. A – The formula for Compton wavelength is

$$\lambda = \frac{h}{mc},$$

and plugging in numbers gives 1.32×10^{-15} m, which is closest to choice A. If you didn't happen to remember the formula for Compton wavelength, you could get it by dimensional analysis. We know the Compton wavelength has something to do with quantum mechanics, so h or $\hbar$ must make an appearance, and the mass of a quantum particle is its only distinguishing characteristic (besides its spin of course, but that has the same units as $\hbar$). To get units of length, we need another dimensionful constant, and c fits the bill. So at worst we would get $\lambda = \frac{\hbar}{mc}$ and be off by a factor of 2π, but happily the answer choices are widely spaced enough that this isn't an issue. Alternatively, we could remember that the defining length scale for nuclear interactions is the fermi, or femtometer, 10^{-15} m, which is approximately the range of the strong force. So it makes sense that the quantum "size" of the proton is close to this value, but certainly not much larger.

64. B – We will work in units where $c = 1$ until the very end of the problem. In its rest frame, the K^0 has energy m_K, and since the decay products have equal mass, the π^+ and π^- each get energy $m_K/2$. The boost factor of

the pion is then $\gamma_{\pi^+} = (m_K/2)/m_\pi$. We now solve for v using $\gamma = 1/\sqrt{1-v^2}$:

$$\frac{1}{\sqrt{1-v^2}} = \frac{m_K}{2m_\pi} \implies v = \sqrt{1 - 4m_\pi^2/m_K^2}.$$

Tacking on a factor of c to give dimensions of velocity gives choice B.

65. E – Again, we will use units where $c = 1$ until the end. The boost factor to go from the rest frame of the K^0 to the lab frame is $\gamma = E/m_K$. In the center-of-momentum frame (the rest frame of the K^0, after it decays), the π^- has an energy $m_K/2$, as found above. To find the energy of the π^- back in the lab frame, we use the γ that we found in the equation above to boost back. Since we want to know when the π^- is at rest, we want to set the energy after the boost equal to the rest energy of the π^-:

$$\gamma m_\pi = \frac{m_K}{2} \implies E = \frac{m_K^2}{2m_\pi}.$$

Since $m_K > 2m_\pi$ for the decay to be kinematically allowed, we have $E > m_K$, which passes a useful sanity check. Restoring a factor of c^2, we conclude that E is correct.

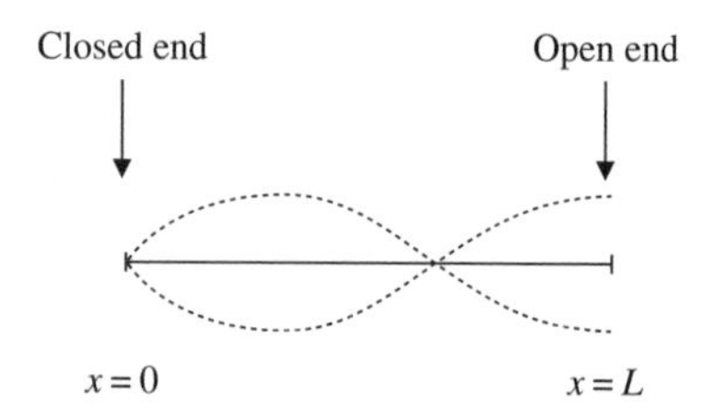

66. C – For a half-open pipe, the open end must be a pressure node, because the air inside and outside the pipe is at atmospheric pressure, hence it is a displacement antinode. On the other hand, the air at the closed end cannot go anywhere, so it is a displacement node. The allowable wavelengths λ in a half-open pipe of length L are then given by the constraint that $L = \lambda\,(1/4 + n/2)$, for non-negative integers n. This implies that

$$\lambda = \frac{4L}{2n+1}.$$

For $n = 1$, we have $\lambda = (4/3)(0.6 \text{ m}) = 0.8$ m. The cartoon above shows the envelope of displacements as a function of distance along the pipe for this λ.

67. C – The field inside the sphere is as if there was a bound charge density of

$$\rho_b = -\nabla \cdot \mathbf{P}.$$

Plugging in the expression for the polarization, we find that

$$\rho_b = -C\frac{1}{r^2}\frac{\partial}{\partial r}\left(r^2 r^2\right) = -4Cr.$$

The electric field can easily be solved using Gauss's law and exploiting the spherical symmetry of the problem:

$$E(r) = \frac{1}{4\pi r^2 \epsilon_0}\int_0^r -4Cr' 4\pi r'^2 dr'$$
$$= \frac{1}{r^2\epsilon_0}\int_0^r -4Cr'^3 dr' = \frac{-Cr^2}{\epsilon_0}.$$

68. E – The electric field inside a perfect conductor is zero, so the magnetic field must vanish as well, and there is no transmitted wave.

69. B – This just involves repeated application of the angular momentum identity

$$\left[L_i, L_j\right] = i\hbar\epsilon_{ijk}L_k.$$

Proceeding step-by-step, we have

$$\left[\left[\left[L_x, L_y\right], L_x\right], L_x\right] = \left[\left[i\hbar L_z, L_x\right], L_x\right]$$
$$= \left[(i\hbar)^2 L_y, L_x\right]$$
$$= -(i\hbar)^3 L_z$$
$$= i\hbar^3 L_z.$$

70. C – The vibrational energies of diatomic molecules are approximately those of a harmonic oscillator, so we solve for T in $kT \simeq \hbar\omega = hf$. To do this quickly, it helps to take advantage of the fact that h is given in the formula sheet in units of eV, and use the mnemonic that kT at room temperature (300 K) is about 1/40 of an eV. We get about 2,400 K, which is closest to choice C. This problem illustrates a fact worth remembering – the vibrational degrees of freedom of light diatomic molecules are "frozen out" at room temperature, and are only unfrozen at temperatures an order of magnitude larger.

71. B – Observables must be Hermitian. The reason for the factor of i in the momentum operator $-i\hbar\nabla$ is precisely to make this single-derivative operator Hermitian – otherwise, we'd pick up an extraneous minus sign during integration by parts. So any operator involving only one derivative that does *not* have a factor of i can't be Hermitian.

72. B – Choices D and E are obviously incorrect. A is incorrect because the chief virtue of the BCS theory is that it *does* give the correct microscopic description of many superconductors. C is interesting, but B is a better answer because it not only escapes the violation of

Pauli exclusion but also gives us a mechanism for superconductivity (bosons can occupy the same state, which produces superconductivity).

73. E – Analyze this system in a frame that rotates with the hoop, and consider only the equation of motion for the tangential component, since the bead is constrained to move on the hoop. There are no Coriolis or centrifugal force terms, so Newton's second law just reads

$$F = -kR\omega = mR\dot{\omega},$$

where ω here means the angular velocity *relative to the hoop*. This has solution $\omega(t) = Ce^{-kt/m}$. Moving back to the stationary reference frame with the replacement $\omega \to \omega + \Omega$, we have $\omega(t) = \Omega + Ce^{-kt/m}$. Imposing the initial condition $\omega(0) = \omega_0$ gives $C = \omega_0 - \Omega$, so

$$\omega(t) = \Omega - (\Omega - \omega_0)e^{-kt/m},$$

which is choice E.

We could also have proceeded by limiting cases. If $\omega_0 = \Omega$, the bead will experience no frictional force and will continue to rotate along with the hoop, $\omega = \Omega$, for all t. This eliminates all but choice E.

74. E – At the top of the hill, the cylinder has only potential energy. At the bottom, it will have purely kinetic energy, which is composed of both translational and rotational kinetic energy. Given the moment of inertia of the cylinder, we can solve the energy conservation equations to find the velocity; since the rotational kinetic energy depends on the moment of inertia, which depends on α, the final velocity must depend on α. Since there is no rolling friction, the difference in potential energy simply depends on the initial and final heights and not on the path taken, so the final velocity depends on h and not θ. This leaves choice E. All energies, kinetic and potential, are proportional to M, so this cancels when using conservation of energy to find the velocity. Similarly, the rolling without slipping condition $v = R\omega$ ensures that both translational and rotational kinetic energy are proportional to v and independent of R.

75. D – The spin-3/2 states are the states with the highest total spin that can be formed from three spin-1/2 particles. The maximal-spin states are always totally symmetric under exchange of the particles. D is the only choice that is totally symmetric. We could also derive this quickly by starting with $|\Psi\rangle$ and applying the lowering operator twice; or even more quickly, by starting with the $m = -3/2$ state $|\downarrow\downarrow\downarrow\rangle$ and applying the *raising* operator *once*.

76. A – A p-type semiconductor has an excess of positive charge carriers or holes, which are empty states in the valence band that electrons can fill. An n-type semiconductor has an excess of electrons. When p- and n-type materials are brought together, the electrons from the n-type material diffuse into the p-type material. This leaves a slight negative charge on the edge of the p-type material and a slight positive charge on the edge of the n-type material. The electric field thus points from n-type to p-type material.

77. C – The two reference frames S and S' must be related by a Lorentz transformation, so with the $(+,-,-,-)$ metric signature, the invariant interval of the position 4-vector in both reference frames must be equal. The separation of E_1 and E_2 in the original frame is (2,1,1,0), which has invariant interval 2. Checking each choice, the invariant interval for A is -1.25, B is 3, C is 2, and D is 1. Choice C is the same as in the original frame, and so is the correct answer.

78. C – With these answer choices, one can make considerable progress using pure dimensional analysis: the answer must have units of $(\text{time})^{-1}$, and must be nonnegative. This leaves only B, C, and E, and E seems rather unreasonable. More formally, assuming that dark matter detection follows a Poisson process with mean rate λ, the probability of seeing zero events is $e^{-\lambda T}$. The 90% confidence level upper limit is the mean rate such that 90% of the time we would see in our experiment a number of events that is inconsistent with our measurement of zero events. Practically, this means that we want to find the mean rate that would produce more than 0 events 90% of the time. In other words, we want to find the rate that gives 0 events only 10% of the time. This is just $0.1 = \exp(-\lambda T)$, or $\lambda = -(1/T)\ln 0.1$.

A limiting-cases analysis also works here. As the confidence level grows and approaches 100%, the upper limit must get weaker (that is, the rate must be larger), since we can never be 100% confident that any finite rate will give zero events in every experiment. The only answer choice that goes to infinity as the confidence level goes to 100% is choice C; indeed, choices B and D both go to zero.

79. C – The 21 cm splitting comes from the hyperfine interaction, which is a spin–spin coupling between the electron and proton spins. Hence, this splitting (between the spin singlet and triplet configurations) is evidence for the proton having spin-1/2.

80. D – Choices A and E are obviously incorrect. B seems initially plausible, but this process would happen just as often with tau neutrinos, so it is far from clear that it would make up the discrepancy. C also seems vaguely plausible, but it does not give an obvious mechanism for production of tau neutrinos. D is the correct answer, with the difference between neutrino mass and flavor eigenstates being the basis of the famous neutrino oscillations. This is the same oscillation effect that takes place in two- and three-state quantum systems.
81. E – By rescaling energies, the partition function can be written as

$$Z = e^{\epsilon/2kT} + e^{-\epsilon/2kT} + 2.$$

The probability of being the triplet state is just

$$P = \frac{e^{\epsilon/2kT} + e^{-\epsilon/2kT} + 1}{e^{\epsilon/2kT} + e^{-\epsilon/2kT} + 2}.$$

82. A – Recalling that the total energy is $H = T + U$, we can solve this problem by a careful consideration of signs in the answer choices. The spring potential energy $U_s = \frac{1}{2}kx^2$, where x is some relative displacement, is always non-negative, so we have only choices A and B. With coordinates y_1 and y_2 as defined in the problem, positive displacements correspond to downward motion, so gravitational potential energy is actually *negative*. Thus we are left with choice A.
83. B – The expectation value of energy is

$$E = \int_0^a \psi^* H \psi \, dx.$$

After a sudden expansion, ψ stays constant. The potential V also stays constant on the interval $[0, a]$, but changes from ∞ to 0 on the interval $[a, 2a]$. Since ψ vanishes on $[a, 2a]$, the kinetic energy operator $T = -\frac{\hbar^2}{2m}\frac{d^2}{dx^2}$ gives zero when acting on ψ. (If you're worried about the fact that the derivative of ψ is discontinuous at $x = a$, meaning that the second derivative is a delta function, note that $\psi(a) = 0$, and zero times a delta function is still zero.) The expectation value of energy after the expansion is therefore

$$\begin{aligned} E' &= \int_0^{2a} \psi^* H' \psi \, dx \\ &= \int_0^a \psi^* H \psi \, dx + \int_a^{2a} \psi^* H' \psi \, dx \\ &= \int_0^a \psi^* H \psi \, dx + 0 \\ &= E. \end{aligned}$$

By the way, this is an application of conservation of energy within the formalism of quantum mechanics – just as the temperature of an ideal gas remains constant during free expansion, the energy of a quantum system remains constant after a sudden change of potential.
84. D – This looks long and complicated, but it's really just a matter of limiting cases. A and B are eliminated by dimensional analysis, since the reflection coefficient must be dimensionless. To eliminate E, note that, as $\alpha \to 0$, the coefficient of reflection must go to zero because the barrier disappears, and the particle continues to propagate to $x > 0$ with probability 1. Choice C looks reasonable at first, but the reflection coefficient must always take a value between 0 and 1, by definition, for all values of parameters in the problem. If α is chosen sufficiently large, then the reflection coefficient of choice C is greater than 1, which is unphysical. This leaves only D.
85. C – The transition $2s \to 1s$ has $\Delta m = 0$ so it does not violate the dipole selection rule.
86. E – The electric dipole moment is a vector quantity, which changes sign under parity transformations, so a nonzero electric dipole moment violates parity. Interestingly, it *also* violates time-reversal invariance. To see this, recall that the neutron *does* have a nonzero magnetic dipole moment. Suppose the magnetic and electric dipole moments were parallel; then, under time-reversal, the magnetic one would change sign but the electric one would remain the same, and the system would not be invariant under time-reversal. So the *relative* orientations of the magnetic and electric dipole moments lead to a violation of time-reversal invariance.
87. C – The sharp drop of curve b is the signature of a process with an energy threshold around 1 MeV. Recalling that the electron mass is about 0.5 MeV, this must be the threshold for pair production.
88. A – $f(x)$ is an odd function on $[-\pi, \pi]$, and cosine is an even function, so all the cosine coefficients in the Fourier series vanish identically.
89. E – We are looking for an $\mathbf{A}$ satisfying $\nabla \times \mathbf{A} = \mathbf{B}$. It's simplest to consider the components of $\mathbf{B}$ one by one. Since $B_x = (\nabla \times \mathbf{A})_x = \partial A_z/\partial y - \partial A_y/\partial z$, with an eye on the answer choices we see that we can only satisfy this by taking $A_z = B_0 y$, since A_y is independent of z in all answer choices. This leaves only C and E. To get $B_z = 2xB_0$, we must have $2xB_0 = (\nabla \times \mathbf{A})_z = \partial A_y/\partial x - \partial A_x/\partial y$. Choice E has the correct sign, and we can check that it also satisfies $(\nabla \times \mathbf{A})_y = 0$.

90. D – The change in energy can be obtained from the change in the electric field energy density outside the sphere. There is no change in energy density inside the sphere, so we will neglect this contribution. Before the grounding, the field outside the sphere has energy

$$\begin{aligned} U_{\text{before}} &= \frac{\epsilon_0}{2}\int_{\text{outside}} \mathbf{E}^2 dV \\ &= \frac{\epsilon_0}{2}\frac{1}{16\pi^2\epsilon_0^2}\int_a^\infty \frac{Q^2}{r^4}4\pi r^2 dr \\ &= \frac{1}{8\pi\epsilon_0}\int_a^\infty \frac{Q^2}{r^2}dr \\ &= \frac{Q^2}{8\pi\epsilon_0 a}. \end{aligned}$$

Afterwards, charge is induced on the conducting sphere to exactly cancel the electric field everywhere outside of the sphere of radius a, so $U_{\text{after}} = 0$. The change is therefore $\Delta U = -\frac{Q^2}{8\pi\epsilon_0 a}$, which is choice D.

91. C – Without color, spin-3/2 baryons with three *identical* quarks and zero orbital angular momentum would have symmetric wavefunctions under the interchange of any two quarks, violating the Pauli exclusion principle.

92. B – By momentum conservation, when a free nucleus emits a photon it must also recoil in the opposite direction, which causes the emitted photon to have a different energy (hence frequency) than it would if the nucleus were held stationary. You may have been thrown by choice C, since the Pound–Rebka experiment used a carefully contrived arrangement of absorbers and emitters in a vertical shaft so that the gravitational redshift was dominant. However, the adjective "tabletop" implies that the experiment takes place at (almost) constant gravitational potential, so gravitational redshift barely contributes.

93. E – It is a useful bit of trivia that a sequence of NAND gates can be combined to create any sequence of basic logical gates. (This is also true of NOR gates.) Even without knowing this, though, we can see fairly easily that A and C must be possible: since NAND is AND followed by NOT, to get an AND we just put two NAND gates in sequence and tie the output of the first to both inputs of the second, and to get NOT we can tie both inputs of a single AND gate together. So if at least two of the answer choices are possible, the answer must be E.

94. E – Applying the Euler–Lagrange equation

$$\frac{d}{dt}\frac{\partial L}{\partial \dot{q}} - \frac{\partial L}{\partial q} = 0$$

to the given Lagrangian, we get $A\ddot{q} - (-2Bq) = 0$. Rearranging gives $\ddot{q} = -\frac{2B}{A}q$, choice E.

95. E – The field of a straight solenoid has uniform magnitude $\mu_0 nI$, where n is the number of turns per unit length; bending this solenoid around into a toroid of radius R sets $n = N/2\pi R$. (This is a hand-wavy argument, but it gives the correct answer, and it's an excellent way to remember the formula without having to rederive it from scratch.) The energy stored in the magnetic field is $U = \frac{1}{2\mu_0}\int \mathbf{B}^2\, dV$, where the integral is taken over all of space. Here the field is only nonzero inside the volume V, and inside this volume the field is approximately uniform (this follows from the statement that $r \ll R$), so

$$U = \frac{V}{2\mu_0}|\mathbf{B}|^2 = \frac{2\pi^2 Rr^2}{2\mu_0}\cdot\left(\frac{\mu_0 NI}{2\pi R}\right)^2 = \frac{\mu_0 N^2 I^2 r^2}{4R},$$

choice E.

96. A – Longitudinally polarized waves propagate in the same direction as the displacement of the wave medium. This means that the polarization vector is forced to be along the direction of propagation, excluding choices I and II. This narrows the answer to A.

97. C – The phase velocity is

$$v_{\text{phase}} = \frac{\omega}{k} = Ak^{-1/2}.$$

The group velocity is

$$v_{\text{group}} = \frac{d\omega}{dk} = \frac{1}{2}Ak^{-1/2},$$

so $v_{\text{phase}} = 2v_{\text{group}}$, choice C.

98. B – The setup $n_1 < n_2 < n_3$ occurs so often that it is probably useful to memorize the result. This is the configuration of an antireflective coating, and the 180° phase shift that occurs at *both* boundaries leads to the condition for destructive interference:

$$2n_{\text{film}}t = \left(m - \frac{1}{2}\right)\lambda,$$

where t is the thickness of the film. (Try to derive this formula if you forgot it.) We conclude that $t_{\text{min}} = \lambda/4n$, which can be memorized with the mnemonic that tn for an antireflective coating is a quarter-wavelength. For the numbers given in this problem, we get $t = 8.33 \times 10^{-8}$ m. So B is correct.

99. D – The wavefunction is already normalized, so using $\langle x\rangle = \int |\Psi(x)|^2 x\, dx$, we have

$$\langle x \rangle = 3\int_0^1 x(1-x)^2\,dx$$
$$= 3\left(\frac{1}{12}\right)$$
$$= \frac{1}{4}.$$

This matches choice D.

100. A – A rigid rod has only rotational degrees of freedom, so its energy is determined by its rotational quantum numbers. The classical formula is $T = L^2/2I$ for the kinetic energy of a rotating body, where I is the moment of inertia about the center of mass, so in the quantum mechanics setting we get $E = \hbar^2 n(n+1)/2I$. The center of mass of this rod is at the center of the rod, so the moment of inertia is $I = 2 \cdot m\left(\frac{a}{2}\right)^2 = \frac{1}{2}ma^2$. Thus,

$$E = \frac{\hbar^2 n(n+1)}{ma^2},$$

choice A.

Note that we could also have done the last step of the problem by calculating the reduced mass of the system, $\mu = \frac{m \cdot m}{m+m} = \frac{m}{2}$, and using the formula $I = \mu r^2$ for a single particle of mass μ.

Solutions to Sample Exam 2

1. D – Rather than solving the equation of motion to find the whole trajectory of the particle, then taking the limit to find the velocity at $t = \infty$, we can simply solve for the velocity v_t at which the force due to gravity balances the air resistance. By Newton's Second Law, there will then be no net force, and the ball will continue to fall at this velocity – this is the terminal velocity. Solving $mg = bv$ for v, we get $v_t = mg/b$. For an alternate solution, note that only choice D has the correct units. Choice E looks like it has the correct units, but you can check that there is no dimensionless combination of the quantities m, g, and b (which are the only ones appearing in the problem), so nothing can appear in an exponential and we are back to choice D.
2. C – The particle has a component of velocity along the direction of the magnetic field. Since the Lorentz force is zero for this component, the particle will continue with the same velocity in the x-direction. However, the perpendicular component of **B** will cause the particle to execute cyclotron motion in the yz-plane. Superimposing these two components of velocity gives a helical path, choice C.
3. E – We first find the velocity at the lowest point, then find the tension needed to supply the required centripetal acceleration. The initial potential energy mgl is converted into kinetic energy $\frac{1}{2}mv^2$ at the bottom, so $v^2 = 2gl$. The centripetal force must be mv^2/l, so $F = m(2gl)/l = 2mg$. However, this is the *net* force: since there is already a force mg due to gravity acting downwards, the tension must provide a force $3mg$ acting upwards for a net force of $2mg$ upwards. Hence choice E.
4. C – Without actually setting up any equations, we can see that C must be correct. The electric field couples to charge (that is, the electric force is qE), but gravity couples to mass (with force mg), so balancing forces gives something like $mg = qE$ with some terms involving α thrown in as well. The only quantity we can get out of this equation, knowing both g and E, is q/m, and a single measurement of α is enough to determine this quantity.
5. B – This is an easy problem, but without a few tricks it is easy to get hung up on the computational details. The Bohr formula gives $E = -E_0/n^2$ for the nth excited state with $E_0 = 13.6$ eV. It's easy to eliminate E by remembering that most of the hydrogen spectrum falls near the visible range, and mm wavelengths correspond to radio waves. Deciding between the rest takes a little bit of computation. Using $hc = 1240$ eV·nm (which is given in the Table of Information on recent tests), we can compute using a couple of numerical approximations to avoid nasty arithmetic:

$$\Delta E = E_0\left(\frac{1}{4} - \frac{1}{9}\right) = 13.6\text{ eV}\left(\frac{5}{36}\right) \approx 2\text{ eV},$$
$$\lambda = \frac{hc}{\Delta E} \approx \frac{1240\text{ eV}\cdot\text{nm}}{2\text{ eV}} \approx 620\text{ nm}.$$

   This is closest to B.
6. A – This follows from the axioms of quantum mechanics: Hermitian observables are guaranteed to have real eigenvalues, and since eigenvalues of operators are results of measurements which must be real numbers, operators must be Hermitian. Some counterexamples for the other choices: the Hamiltonian of a bound state has negative eigenvalues, which violates B; the momentum operator is an operator on infinite-dimensional function spaces, so violates C; the Pauli matrix $\begin{pmatrix} 0 & -i \\ i & 0 \end{pmatrix}$ is Hermitian but not symmetric, so violates D.
7. C – III is the statement of Newton's Second Law. Choice I fails if an extended object is subject to two equal and opposite forces at different locations, creating a net torque, and choice II fails (for example) for an object in empty space moving with some initial velocity.
8. D – You might be tempted to think that this question requires rote memorization of the atomic number of fluorine, but there's an easy way to deduce the correct answer using the information provided and the particular selection of answer choices. The fact that the common form is Fl^- means that fluorine wants to attract an extra electron. This results from an electron shell that is one short of being full, which matches choice D. In fact, E is the electron configuration for both Fl^- and the noble gas neon, with a full $n = 2$ shell. By contrast, A is the electron configuration for lithium, which wants to *lose* an electron to get a full $n = 1$ shell.
9. D – The resonant frequency of an LC circuit is given by $\omega = 1/\sqrt{LC}$. If the capacitor is initially charged, then the charge in the circuit oscillates as $Q(t) = Q_0 \cos\omega t$, and the current is $I = dQ/dt = -Q_0\omega \sin\omega t$. Thus $|I|$ is maximized at $\omega t = \pi/2$, or $t = \frac{\pi}{2}\sqrt{LC}$, choice D.
10. E – The magnitude of the work done by the gas is $\int P\,dV$, which is just the area in the P–V-plane. In this case, the sum of the integrals under each curve is positive, so the total work done by the cycle is also positive. The shape

of the P–V curve is pretty close to a parallelogram, with base 2 m^3 and height 3,100 kPa, so the area is about 6,200 kJ, choice E.

11. A – During an adiabatic process the quantity PV^γ is constant, where $\gamma = C_p/C_v$. Looking at step A, at the endpoint we have $V = 1\,\text{m}^3$, so independent of γ, $PV^\gamma = 3,200$ in the units given. Looking at the start of step A, we solve for γ:

$$(100)(8)^\gamma = 3200 \implies \gamma = 5/3.$$

A monoatomic gas has $\gamma = 5/3$, so choice A is correct.

12. A – The scope trace looks like a carrier wave whose frequency changes as a function of time, which is the defining characteristic of frequency modulation.

13. E – Planetary orbits have two independent parameters corresponding to the two conserved quantities in central-force motion: the energy of the orbit and the angular momentum. The semimajor axis is determined solely by the energy, but the semiminor axis (which is needed to find the area of the orbit) also depends on the angular momentum. Without this additional information, the area of the orbits can't be determined.

14. C – The choices for the orbital angular momentum of the particle are given by the angular wavefunction: we can have either $l = 3$ and $m_l = 0$, or $l = 2$ and $m_l = 1$. Adding these choices to the particle's spin ($s = 1/2$ and $m_s = \pm 1/2$) gives the following possibilities:

$$j = 7/2, m_j = \pm 1/2$$
$$j = 5/2, m_j = \pm 1/2$$
$$j = 5/2, m_j = 1/2, 3/2$$
$$j = 3/2, m_j = 1/2, 3/2$$

Of these, only $j = 7/2$, $m_j = -1/2$ appears in the answer choices, and it is choice C.

15. E – You can either remember the energies of the infinite square well, $E_n = \frac{n^2\pi^2\hbar^2}{2mL^2}$, or derive them from the given wavefunctions using the Hamiltonian operator $-\frac{\hbar^2}{2m^2}\frac{d^2}{dx^2}$. Fortunately the wavefunction is already given to us normalized, so we compute

$$\begin{aligned}\langle E\rangle &= \frac{1}{2}E_2 + \frac{1}{3}E_3 + \frac{1}{6}E_4 = \frac{\pi^2\hbar^2}{2mL^2}\left(\frac{2^2}{2} + \frac{3^2}{3} + \frac{4^2}{6}\right)\\ &= \frac{23\pi^2\hbar^2}{6mL^2},\end{aligned}$$

choice E.

16. C – From the definition of entropy for a reversible process, $\Delta Q = T\Delta S$, so at constant temperature $\Delta S = \Delta Q/T = 20$ J/K, choice C.

17. B – In the photoelectric effect, electrons are ejected from a metal after bombardment of the metal with light of a suitably high energy. The fact that no electrons are ejected for light of low enough frequency, independent of the intensity of the light, implies that the energy of the light depends on the frequency rather than the intensity, and higher frequency means higher energy. Both I and III are true, but are not directly supported by this experiment.

18. D – This is pure fact recall. The splitting of spectral lines by electric fields is called the Stark effect, choice D.

19. A – This is fairly intuitive: we know that if $m_1 = m_2$, then m_1 will stop dead after the collision, and if $m_1 \ll m_2$, then m_1 will just bounce back in the opposite direction, so it's reasonable that if $m_1 > m_2$ the first mass will continue moving in the same direction. More formally, letting v_1 and v_2 be the final velocities of m_1 and m_2 respectively, we solve the conservation of momentum and energy equations $m_1 v = m_1 v_1 + m_2 v_2$ and $\frac{1}{2}m_1 v^2 = \frac{1}{2}m_1 v_1^2 + \frac{1}{2}m_2 v_2^2$ for v_1 to get

$$v_1 = v\frac{m_1 - m_2}{m_1 + m_2},$$

so indeed if $m_1 > m_2$, the numerator is positive and v_1 has the same sign as v.

20. A – We want an upright virtual image with magnification $m = 20\,\text{cm}/2\,\text{m} = 1/10$. This means we need $-s'/s = 1/10$, or $s' = -s/10$. Using the optics equation with $s = 1$ m,

$$\frac{1}{f} = \frac{1}{s} + \frac{1}{s'} = -\frac{9}{s} = -9\,\text{m}^{-1},$$

so $f \approx -0.11$ m and $R = 2f = -22$ cm. In particular, R is negative, so the mirror must be convex.

21. C – You could probably get this from some etymology: vibrations of a lattice are also known as sound, which has the Greek root *phon*, hence "phonons" (choice C).

22. D – The potential must be continuous, which eliminates choice B (which, incidentally, is a graph of the electric *field*). Furthermore, the potential should be constant inside the inner shell since there is no field there, which eliminates C. In the region between the two shells, the potential should look like $-1/r$, since the only field is due to the inner shell, and this eliminates A.

To decide between D and E, we need to do a quick calculation. Depending on the amount of charge on each shell, either D or E could be valid. Call the potential inside the inner shell $V_1(r)$, the potential in between the two shells $V_2(r)$, and the potential outside the outer shell $V_3(r)$. Since the electric field is the *derivative* of the potential, we can add arbitrary constant offsets C_1 and C_2 to the standard Coulomb potentials for the enclosed charge in these regions, so that they match on the boundaries according to the conditions $V_1(R) = V_2(R)$ and $V_2(2R) = V_3(2R)$. When we rewrite the boundary conditions in terms of the Coulomb potentials with the constant offsets, we get the following system of equations:

$$V_1(R) = V_2(R) \iff C_1 = -\frac{q}{4\pi\epsilon_0 R} + C_2,$$

$$V_2(2R) = V_3(2R) \iff -\frac{q}{8\pi\epsilon_0 R} + C_2 = +\frac{2q}{8\pi\epsilon_0 R}.$$

The offset C_1 is clearly positive, so $V_1(r) = C_1 > 0$ and D is the correct answer.

23. B – The quarks in the proton are held together by the strong nuclear force, which is mediated by gluons, choice B. The quarks are also electrically charged, but the electromagnetic force is orders of magnitude weaker than the strong force so is not responsible for binding.
24. A – Potential energy is related to electric potential by $U = qV$. The change in potential is $\Delta V = -\int \mathbf{E} \cdot d\mathbf{l}$, and since $\mathbf{E} \perp \hat{\mathbf{y}}$, only the x-component of the path contributes. This gives $\Delta V = 3E$ since $\mathbf{E}$ is constant, and $U = -3Ee$. To make sure we have the signs right, note that if e is positive, then our particle is negatively charged and moving in the opposite direction from the field lines. This is the direction it would move under the influence of the Lorentz force, so its energy should decrease, which is consistent with the minus sign.
25. E – I is true, as a magnetic monopole would provide a "source term" on the right-hand side of Gauss's law, just as it does for Gauss's law for electricity. In general, by Helmholtz's theorem, we can decompose a vector field such as the magnetic field into the gradient and curl of a scalar and vector potential as $\mathbf{X} = -\nabla\varphi + \nabla \times \mathbf{A}$. Thus III is true as well, since a divergence-free $\mathbf{B}$ implies that the scalar potential can always just be chosen to be zero without changing the gradient or curl of $\mathbf{B}$ in Maxwell's equations. Gauss's law is perfectly compatible with the continuity equation – Ampère's law is the one that needs adjusting, through the addition of Maxwell's displacement current.
26. B – Using $E^2 = p^2c^2 + m^2c^4$, we have $E^2 = 2m^2c^4$, so $E = \sqrt{2}mc^2$, choice B.
27. E – This is pure dimensional analysis. Only E has the correct units, so it is the correct answer. A quick way to see this is to recall the expression for the potential energy of two equal point charges, $W = \frac{e^2}{4\pi\epsilon_0}\frac{1}{r}$, which says that $e^2/\epsilon_0 r$ has units of energy. The only remaining units are nice and familiar powers of mass and length, so we see that E works out to have units of inverse time, or frequency.
28. D – It's easy to get lost in calculations in this problem, computing the total energy of the orbit and so on, but things are drastically simplified because at perigee and apogee the planet's path is perpendicular to the line connecting the planet and the star. Thus, its angular momentum at perigee and apogee is just $mv_p(a/4)$ and $mv_a(7a/4)$, respectively. By conservation of angular momentum in a central force field, we get $v_p/v_a = 7$, choice D.
29. D – The time-averaged force is defined as $\frac{1}{T}\int_{t_i}^{t_f} F\,dt$, where T is the total time. The distance-averaged force is defined similarly, by $\frac{1}{D}\int_{x_i}^{x_f} F\,dx$, where D is the total distance. The quantities inside the integral are also known as total impulse and total work, respectively. But total impulse is change in momentum, and total work is change in kinetic energy. Since boat 2 arrives with a higher speed, we know the impulse must be greater, but the denominator T is greater as well, so we can't definitively say whether F_t increases or decreases with respect to boat 1. However, the *distance* traveled by the two boats is identical, and $v_2 > v_1$ means the kinetic energy of boat 2 is greater than that of boat 1. This implies $F_{d1} < F_{d2}$, choice D.
30. C – The unperturbed energies are $E_0 = \hbar\omega/2$ and $E_1 = 3\hbar\omega/2$. The first-order perturbation theory formula gives

$$\delta E_n = \langle n|V(x)|n\rangle = \epsilon \int_{-\infty}^{\infty} |\psi_n(x)|^2 \delta(x)\,dx = |\psi_n(0)|^2.$$

Since $\psi_1(x)$ vanishes at $x = 0$, it has no first-order shift. $|\psi_0(0)|^2 = \sqrt{m\omega/\pi\hbar}$, so $E_0 = \hbar\omega/2 + \epsilon\sqrt{m\omega/\pi\hbar}$ and $E_1 = 3\hbar\omega/2$, which is C.

31. E – The degrees of freedom associated to choices II and III freeze out at low temperatures, but at sufficiently high temperatures all three are active. Note that if we were dealing with a single atom rather than a diatomic molecule, choices II and III would *not* contribute.
32. E – All steps of the Carnot cycle are reversible, and reversible adiabatic processes take place at constant entropy. This problem illustrates a questionable, but standard, use of GRE terminology: "adiabatic" in the question statement was understood to mean "reversible

adiabatic," not just a process where $\Delta Q = 0$. Indeed, free expansion is also adiabatic, but not reversible, and if the process in question were free expansion, choice A would be the correct answer.

33. E – This is the setup for the method of images, but we must remember there are *two* image charges: $-q$ at distance d below the plate and $+q$ at distance $2d$ below the plate. So the real charge $+q$ feels a force from the real charge $-q$ and both the image charges. The correct prescription is at this point to pretend there is no conductor and just calculate the force directly from all these charges. The force from the closest charge will dominate, so we know that q will be attracted upwards, and thus the force must be in the $+\hat{\mathbf{z}}$-direction, eliminating choices A and B. More precisely,

$$\mathbf{F} = q\mathbf{E} = \frac{q^2}{4\pi\epsilon_0}\left(\frac{1}{d^2} - \frac{1}{(2d)^2} + \frac{1}{(3d)^2}\right)\hat{\mathbf{z}}.$$

At this stage we can make sure all our minus signs are correct: the real charge $-q$ should attract, so the force from the first term should be upwards (positive), whereas the attraction of the image charge $-q$ should be a downwards force (negative), and the repulsion of the image $+q$ should be upwards (positive). Rearranging a bit, we get

$$\mathbf{F} = \frac{q^2}{4\pi\epsilon_0 d^2}\left(1 - \frac{1}{4} + \frac{1}{9}\right) = \frac{31q^2}{144\pi\epsilon_0 d^2},$$

which is E. This problem is rather tricky, because if instead you had tried to get the field by taking the derivative of the potential, you would have been off by a factor of 2 in the middle term (from the image charge associated to our test charge $+q$) and ended up with choice C. This subtlety is discussed in Griffiths *Introduction to Electrodynamics* Sections 3.2.1 and 3.2.3, and is due to the fact that there is no energy cost from moving the image charge along with the test charge. In our opinion, it's better to forget all the complicated reasoning and just memorize the fact that you can pretend that you get forces, rather than potentials, from the method of images.

34. B – Using Wien's displacement law, $T = 3\times10^{-3}\,\text{m}\cdot\text{K}/\lambda$ (since the answer choices are widely separated we can approximate the constant to one significant figure), we find $T \approx 5000$ K.

35. E – A receding star will have its wavelengths redshifted, meaning that $\lambda_{\text{true}} < \lambda_{\text{meas}}$, and hence $T_{\text{true}} > T_{\text{meas}}$. This immediately knocks out choices A, B, and D. To find the exact ratio we use the relativistic Doppler shift formula, which can be applied directly to T rather than λ because of the inverse proportionality between T and λ and the fact that we know which direction the ratio must go.

$$\frac{T_{\text{true}}}{T_{\text{meas}}} = \sqrt{\frac{1+\beta}{1-\beta}} = \sqrt{\frac{1.2}{0.8}} = \sqrt{\frac{3}{2}},$$

so choice E is correct.

36. D – The Euler–Lagrange equation is

$$\frac{d}{dt}\frac{\partial L}{\partial \dot{q}} = \frac{\partial L}{\partial q}.$$

Computing the derivatives, $\partial L/\partial q = \frac{1}{L}\cos(q/L - \omega t)$ and $\partial L/\partial \dot{q} = 2A\dot{q}$, so the Euler–Lagrange equation is

$$2A\ddot{q} = \frac{1}{L}\cos(q/L - \omega t),$$

which becomes choice D after dividing through by $2A$.

37. D – Hamilton's equations for the coordinate z and the conjugate momentum p_z are given by

$$\dot{p}_z = -\frac{\partial H}{\partial z}, \qquad \dot{z} = \frac{\partial H}{\partial p_z}.$$

Plugging in the Hamiltonian to the first equation, we find

$$\begin{aligned}\dot{p}_z &= -\frac{\partial}{\partial z}\left(\frac{p_\theta}{2mz^2}\cot^2\alpha + mgz\right)\\ &= \frac{p_\theta^2}{mz^3}\cot^2\alpha - mg.\end{aligned}$$

Plugging into the second equation, we find

$$\begin{aligned}\dot{z} &= \frac{\partial}{\partial p_z}\left(\frac{p_z^2}{2m}\cos^2\alpha\right)\\ &= \frac{p_z}{m}\cos^2\alpha.\end{aligned}$$

Note that it is easy to confuse the minus signs and coordinates in Hamilton's equations, which make for classic trap answer choices, so it pays to memorize these, or rederive them quickly by relating $H = p^2/2m + U(x)$ to Newton's law $F = -U'(x)$.

38. D – C_V is defined as $\partial U/\partial T$, so all we have to do is integrate C_V with respect to T. $\int AT^3\,dT = \frac{1}{4}AT^4$, so choice D is correct. We could have narrowed down the answer choices first, by dimensional analysis: since C_V has units of energy/temperature, $U(T)$ could only have been D or E.

39. C – Since the states $|n\rangle$ are orthonormal, $\langle m|n\rangle = \delta_{mn}$, so I is false. For II, recall that two operators commute if and only if they share the same eigenfunctions. Since $\hat{x}$

does not commute with $\hat{p}$, then $[\hat{H}, \hat{p}] \neq 0$ for Hamiltonians whose potential term depends on x, thus the eigenfunctions of the Hamiltonian and the momentum operator are not shared in general. III is true, and follows from both orthonormality and the fact that the states $|n\rangle$ are energy eigenfunctions with eigenvalues E_n. So C is the correct answer.

40. B – As is typical for specialized topics problems, this is pure fact recall. Expulsion of magnetic fields is an interesting characteristic of superconductors. For the record, A and D are false, and C and E are true but irrelevant.

41. C – Energy scales of MeV are typical of nuclear binding energies and reactions, so C is the only choice that has roughly the correct order of magnitude. One way of remembering this is that $\hbar c \approx 200\,\text{MeV} \cdot \text{fm}$, and nuclear dimensions are on the order of several femtometers. But we could get pretty far just by a process of elimination. The binding energy certainly can't be more than three times the mass of a nucleon (in units of c^2), or else tritium would have negative mass. Since nucleons have masses of about 1 GeV/c^2, choices D and E are impossible. Furthermore, energies of eV are typical of chemical reactions, rather than nuclear reactions, so the only remaining reasonable choices are B and C.

42. D – If we keep the nucleus with charge +1 and the orbiting particle with charge -1, the Bohr radius goes as $a \sim 1/\mu$, where μ is the reduced mass of the system. The reduced mass of the hydrogen atom is $\frac{m_e m_p}{m_e + m_p}$, and the reduced mass of muonium is $\frac{m_e m_\mu}{m_e + m_\mu}$, so taking the correct ratio gives

$$a = a_0 \frac{m_e m_p / m_e + m_p}{m_e m_\mu / m_e + m_\mu} = a_0 \frac{m_p(m_e + m_\mu)}{m_\mu(m_e + m_p)},$$

which is D. For an alternate solution, we can use limiting cases: in the limit $m_e \to 0$, we must have $a = a_0$ since the nucleus becomes infinitely heavy compared to the electron, and the reduced mass is then independent of the nucleus mass. Choice D is the only one that satisfies this property.

43. D – Just apply the canonical commutation relation $[\hat{x}, \hat{p}] = i\hbar$ repeatedly. Using the commutator identity $[A, BC] = [A, B]C + B[A, C]$, we have

$$[\hat{x}, \hat{p}^2] = [\hat{x}, \hat{p}]\hat{p} + \hat{p}[\hat{x}, \hat{p}] = (i\hbar)\hat{p} + \hat{p}(i\hbar) = 2i\hbar\hat{p},$$

which is D.

44. D – The work–energy theorem tells us that the difference between the initial and final energies must be the work done by friction; in the absence of friction, energy would be conserved. In symbols, let h be the radius of the quarter-circle ramp, v the block's velocity at the bottom, and m its mass.

$$\begin{aligned} W &= mgh - \frac{1}{2}mv^2 \\ &\approx (2\text{ kg})(10\text{ m/s}^2)(2\text{ m}) - (0.5)(2\text{ kg})(4\text{ m/s})^2 \\ &= (40 - 16)\text{ J} \\ &= 24\text{ J}, \end{aligned}$$

where as usual we have approximated $g \approx 10\text{ m/s}^2$.

45. A – Potentials of line sources are logarithmic, which eliminates D and E. Furthermore, the potential should blow up to $+\infty$ at $r = 0$, which eliminates C. To decide between A and B, we have to go back to Gauss's law to find the field. Taking a Gaussian cylinder of height h and radius r centered on the wire, the charge enclosed is $Q = \lambda h$. The field should be radial, so there is no field parallel to the endcaps of the cylinder, and the fact that the field is constant at constant radius r gives $E(2\pi r h) = \lambda h/\epsilon_0$, so $E = \lambda/2\pi r\epsilon_0$. Computing V we find

$$V = -\int_a^r E(r)\,dr = -\frac{\lambda}{2\pi\epsilon_0}(\ln r - \ln a) = \frac{\lambda}{2\pi\epsilon_0}\ln(a/r),$$

matching choice A.

46. D – If there is no absorption, the sum of the intensities of the outgoing beams is equal to the incident intensity. Since the intensity is proportional to the square of the amplitude, choice D is correct. Choice A is true for a 50/50 beam-splitter, which divides an incoming beam into two equal-intensity beams, but this is a special case and not true in general.

47. B – The classical free-particle Hamiltonian is $p^2/2m$, so to get the quantum operator we make the replacement $p \to -i\hbar\nabla$ to get

$$H = (-i\hbar\nabla)^2/2m = -\hbar^2\nabla^2/2m,$$

matching choice B.

48. B – This is mostly common sense. The penetration depth of the protons as a function of energy would be the most useful piece of information, since this would allow the proton depth to be matched with the tumor position. Note that protons of a fixed energy traveling through matter tend to penetrate to a relatively constant depth, losing the majority of their energy near the end of their track (a phenomenon usually referred to as the "Bragg peak" of their dE/dx curve). This makes them particularly suitable for cancer treatment because the physical location of a dose from a proton beam is well localized.

This is in contrast to neutrons, for example, which tend to diffuse much more broadly as they lose their energy.

49. A – First using conservation of *angular* momentum, the angular momentum of the baseball is $L = (0.09 \text{ kg})(20 \text{ m/s})(2 \text{ m}) = 3.6 \text{ kg} \cdot \text{m}^2/\text{s}$, so this must be the angular momentum of the student–ball–platform system after it starts rotating. The total moment of inertia of the system (ignoring the ball, whose mass is negligible compared to the student) is $I = m_{\text{student}} r^2 + I_{\text{platform}} = (100 \text{ kg})(2 \text{ m}^2) + 200 \text{ kg} \cdot \text{m}^2 = 600 \text{ kg} \cdot \text{m}^2$. Using $L = I\omega$, we find $\omega = 0.006$ rad/s. Note that forgetting the contribution of the platform's moment of inertia would lead to choice B, and forgetting the student would lead to choice C.

50. A – You might think you needed to memorize the formula for Compton scattering, but really all you need here is to take the correct limits. As $\theta \to 0$, there should be no change in wavelength since no scattering takes place: this eliminates B, C, and E. However, D has the wrong units, which leaves only choice A. You could also observe that there should be a nonzero wavelength shift at $\theta = \pi$, where the incident light is reflected directly back off the electron accompanied by a suitable change in kinetic energy; this also eliminates D.

51. D – Li^{++} is a hydrogen-like ion with one electron orbiting a nucleus of charge $+3$. Recalling that the binding energy of hydrogen-like systems is proportional to the square of the nuclear charge Z, the binding energy of Li^{++} is $3^2 = 9$ times the binding energy of hydrogen. 9×13.6 eV = 122.4 eV, so D is correct.

52. B – Choice B is false because it is forbidden by the Second Law of Thermodynamics. The Third Law comes in various forms, but all of them require that the entropy approaches a constant at absolute zero. A, C, and E are all true by the Boltzmann definition of entropy $S = k \ln \Omega$, where Ω is the degeneracy of the system. A perfect crystal of a pure substance has a nondegenerate ground state, so $\Omega = 1$ and $S = 0$. But if the ground state were degenerate, then S could conceivably be nonzero at absolute zero when the system is in the ground state. Since a system at absolute zero is in its ground state, the Boltzmann definition of entropy implies that the entropy of a system approaches a constant at absolute zero. The fact that D follows from the Third Law is somewhat less obvious, but also true and proven in many textbooks.

53. B – By symmetry, if the triangle consisted of equal charges $+q$ at each vertex, the field at the center would be zero. So the field at the center is entirely due to the surplus of charge $+q$ at the top vertex compared to the other two vertices. The distance to the center of the triangle from a vertex is $L/\sqrt{3}$, so

$$\mathbf{E} = -\frac{q}{4\pi\epsilon_0}\frac{1}{(L/\sqrt{3})^2}\hat{\mathbf{y}} = -\frac{3q}{4\pi\epsilon_0 L^2}\hat{\mathbf{y}},$$

choice B.

54. A – We just use $V = \frac{q}{4\pi\epsilon_0}\frac{1}{r}$ with $r = L/\sqrt{3}$, and sum over all charges. In fact, since all charges are equidistant from the center, we can do the sum immediately:

$$V = \frac{2q + q + q}{4\pi\epsilon_0}\frac{1}{L/\sqrt{3}} = \frac{q\sqrt{3}}{\pi\epsilon_0 L},$$

which is choice A.

55. E – The thermodynamic limit for heat engine efficiency is $W/Q = 1 - T_C/T_H$, where W is the work output of the engine in one cycle, Q is the heat provided by the hot reservoir in one cycle, and T_C and T_H are the cold and hot reservoir temperatures, respectively. In the given problem $T_C = 300$ K and $T_H = 400$ K, so the maximum efficiency is $W/Q = 0.25$, and $Q_{\text{min}} = 4W$. A power output of 100 kW corresponds to 100 kJ/s, so in 1 minute the work done is 6,000 kJ, and $Q_{\text{min}} = 24,000$ kJ, choice E.

56. E – Doing the usual conservation of energy routine, we set the zero of potential energy at the top of the equilibrium position of the spring. The brick starts with potential energy mgh, and at maximum compression x the spring has potential energy $\frac{1}{2}kx^2$. However, there is an additional contribution $-mgx$ to potential energy from the compression of the spring below its equilibrium position. We have the following quadratic equation for x:

$$\frac{1}{2}kx^2 - mgx = mgh$$
$$\implies x = \frac{mg \pm \sqrt{m^2g^2 + 2mghk}}{k}$$
$$= \frac{mg}{k}\left(1 \pm \sqrt{1 + \frac{2kh}{mg}}\right).$$

Discarding the spurious negative solution, we get choice E. We could also have done this by dimensional analysis and limiting cases. The only answers with correct units are A, D, and E, and in the limit $h \to 0$ (where the block is just sitting on top of the spring and then released), only E gives the correct compression $2mg/k$.

57. E – Solving this by free-body diagrams is straightforward but rather time consuming. Just using a little physics

intuition, T_1 must support the mass of all the blocks, a total of 40 kg, while T_3 only has to support the lightest mass of 5 kg. So $T_1/T_3 = 8$, choice E.

58. A – Recall that the z-component of orbital angular momentum has the operator $\hat{L}_z = -i\hbar\partial/\partial\phi$. The hard way is to compute the expectation value from the definition,

$$\langle L_z \rangle = \int_0^{\pi} \int_0^{2\pi} d\phi\, d\theta\, \psi^*(\theta,\phi) \left(-i\hbar\frac{\partial}{\partial\phi}\right) \psi(\theta,\phi) = 0,$$

but the easy way is to note that this wavefunction is purely real, and so acting on it by $\hat{L}_z$ will give something imaginary. Expectation values of observables must be real, so the expectation value must be zero. Yet another way to see this is to remember that the ϕ dependence in spherical harmonics is carried by $e^{im\phi}$, and that sines and cosines are odd and even linear combinations of these exponentials with m values of opposite signs. So positive and negative L_z contribute equally and cancel in the end for any real orbital wavefunction.

59. E – The potential given is a finite square barrier, and at the given energy E, the barrier is a classically forbidden region for the alpha particle. So its wavefunction in that region must be an exponential, rather than a sinusoid. Furthermore, the tunneling probability is less than 1, so the amplitude of the wave outside the barrier must be less than the initial amplitude. Only choice E matches this description. Note that choice C is incorrect both because the wavefunction is not sinusoidal outside the barrier, and because the derivative of the wavefunction is discontinuous at both the walls; discontinuities in the derivative only occur in *infinite* potential wells, not finite ones.

60. D – Apply the double-slit relation $d\sin\theta = m\lambda$ to find m, with $d = 4\,\text{mm}$, $\lambda = 500\,\text{nm}$, and $\theta = 30°$, since θ measures the angle from the *center* of the arrangement. This gives $m = 4{,}000$, and we must multiply by 2 to cover both halves of the screen. So approximately 8,000 fringes are visible, choice D. (We say "approximately" because there is also a fringe at $\theta = 0$, but also the very last fringes at $\theta = 30°$ may or may not be visible at the edges of the screen.)

61. A – This is the setup for the relativistic Doppler shift. Since the ship is approaching the telescope, the period should decrease relative to the emitted period, since the light pulses are being received more often. This fixes the signs in the numerator and denominator:

$$\frac{T_{\text{obs}}}{T_{\text{emit}}} = \sqrt{\frac{1-\beta}{1+\beta}} = \sqrt{\frac{0.4}{1.6}} = \sqrt{\frac{1}{4}} = \frac{1}{2},$$

so $T_{\text{obs}} = 0.5$ s, choice A.

62. A – This is an example of length contraction: in the rest frame of the particle, the tube is approaching at speed $0.8c$, and so has its length contracted by a factor

$$\frac{1}{\gamma} = \sqrt{1 - v^2/c^2} = \sqrt{1 - 0.64} = 0.6.$$

So, in the particle's rest frame, the tube is $(30\,\text{m})(0.6) = 18$ m long, which is choice A.

63. D – The reciprocal lattice to a simple cubic lattice is also a simple cubic lattice, but with side length $2\pi/a$. The first Brillouin zone is again a cube, centered at one of the points on the reciprocal lattice, which bisects the reciprocal lattice vectors and hence also has side length $2\pi/a$. Thus its volume is $(2\pi/a)^3$, which is D.

64. A – Unfortunately this is just memorization. Beta decay is $n \to p + e + \nu$ (actually, the ν should be an antineutrino, but this distinction is irrelevant for the problem), and the third particle present in the final state is responsible for the broad energy spectrum of the remaining two. It is possible to narrow down the answer choices slightly using conservation laws: the positron, muon, and strange quark are all charged, so their presence in beta decay would violate charge conservation.

65. B – This one is simplest by limiting cases. Because we can always replace M_1 and M_2 along with the rope connecting them by a single block of mass $M_1 + M_2$, the answer must involve only $M_1 + M_2$ and not M_1 or M_2 individually. Furthermore, as $M_3 \to 0$, α must approach 0° or else M_1 and M_2 would slide down the ramp. Finally, if $M_1 + M_2 = M_3$, we must have $\alpha = 90°$. The only choice matching these limits is B.

66. D – The equivalent resistance of the top two resistors is $10\,\Omega$, as well as for the bottom two. Thus we have equivalent resistances of $10\,\Omega$, $5\,\Omega$, and $10\,\Omega$ in parallel, with total equivalent resistance of $(1/10 + 1/5 + 1/10)^{-1} = 2.5\,\Omega$. From $V = IR$, the current is $I = 2$ A, choice D.

67. E – There are several ways to do this, but the quickest is to remember that each of the Pauli matrices squares to the 2×2 identity matrix. You can of course check this by direct matrix multiplication, but it's a very useful fact to remember. Then $M = 3I_{2\times 2}$, which has determinant 9, choice E.

68. C – The expectation value of S_x is given by

$$\langle S_x \rangle = \eta^\dagger \hat{S}_x \eta = \frac{\hbar}{2}\eta^\dagger \sigma_x \eta.$$

The normalization constant works out to be $\mathcal{N} = 1/\sqrt{6}$, so

$$\langle S_x \rangle = \frac{1}{6} \cdot \frac{\hbar}{2} \left((-i \;\; 2+i) \begin{pmatrix} 0 & 1 \\ 1 & 0 \end{pmatrix} \begin{pmatrix} i \\ 2-i \end{pmatrix} \right) = -\hbar/6,$$

choice C. As it must, the expectation value comes out to be real: if it doesn't, you know you've made an arithmetic mistake.

69. D – A standing wave should be of the form $A(x)B(t)$ with B periodic in time. Only choice D matches this form.
70. B – First find the image formed by the lens A. Here the focal length is positive, so

$$\frac{1}{12.5\text{ cm}} + \frac{1}{s_1'} = \frac{1}{10\text{ cm}} \implies s_1' = 50\text{ cm},$$

and the image is real and 50 cm to the right of A, so 30 cm to the right of B. The diverging lens acts next, and because the image is on the opposite side from the incoming light rays, it acts as a virtual object for B:

$$-\frac{1}{30\text{ cm}} + \frac{1}{s_2'} = -\frac{1}{10\text{ cm}} \implies s_2' = -15\text{ cm},$$

so the image is virtual and 15 cm to the left of B, or 5 cm to the right of A. Since $s_1' > 0$, the first magnification is $m_1 = -s_1'/s_1 < 0$, so the first image is inverted. But both the second object and image distances s_2 and s_2' are negative, so $m_2 < 0$, and the final image is upright.

71. B – You could solve this by simply remembering the field transformations under a Lorentz transformation, but it's also instructive to reason geometrically. In $\overline{S}$, there will be a length contraction along the direction of motion, reducing the area of the parallel plates by a factor of $1/\gamma = \sqrt{1 - v^2/c^2}$, but no change in the plate separation. The capacitance $C = \epsilon_0 A/d$ will then become $C/\gamma = C\sqrt{1 - v^2/c^2}$, choice B.
72. B – Muons have a mass of about 100 MeV, and cosmic ray muons are usually produced with energies substantially higher than this, so B is at least plausible. E might have high enough energy, but charged particles created by solar flares are usually captured by the Earth's magnetic field. Thermal radiation, seismic noise, and solar neutrinos have energies orders of magnitude too small, so B is the best choice.
73. C – The kinetic energy of the pivot is $\frac{1}{2}M\dot{x}^2$, and the pivot has no potential energy. Setting the zero of potential at the height of the bar, the potential energy of the mass m is $U = -mgl\cos\theta$. The position of m is $(X, Y) = (x + l\sin\theta, -l\cos\theta)$, so taking time derivatives,

$$(\dot{X}, \dot{Y}) = (\dot{x} + l\cos\theta\dot{\theta}, l\sin\theta\dot{\theta}).$$

The kinetic energy is

$$\begin{aligned} T &= \frac{1}{2}m(\dot{X}^2 + \dot{Y}^2) \\ &= \frac{1}{2}m(\dot{x}^2 + 2l\cos\theta\dot{x}\dot{\theta} + l^2\cos^2\theta\dot{\theta}^2 + l^2\sin^2\theta\dot{\theta}^2) \\ &= \frac{1}{2}m(\dot{x}^2 + l^2\dot{\theta}^2 + 2l\cos\theta\dot{x}\dot{\theta}), \end{aligned}$$

and taking $L = T - U$ gives choice C. Note that we had to start with the expression for kinetic energy in terms of Cartesian coordinates, rather than immediately jumping to polar coordinates, which would have neglected the contribution to the kinetic energy of the hanging mass coming from the motion of the pivot.

74. D – The relevant equations are the Bernoulli equation for an incompressible fluid,

$$\frac{1}{2}\rho v_1^2 + p_1 = \frac{1}{2}\rho v_2^2 + p_2,$$

and the continuity equation, which implies

$$v_1 A_1 = v_2 A_2.$$

Setting $p_1 = p$ and $v_1 = v$, we solve for the relevant variables and find

$$p_2 = p + \frac{1}{2}\rho v^2 \left(1 - \frac{A_1^2}{A_2^2}\right),$$

choice D. Note that this is consistent with Bernoulli's principle (in this context also known as the Venturi effect), where a decrease in cross-sectional area is accompanied by a decrease in pressure. For a partial alternate solution, we could have examined limiting cases: for $A_1 = A_2$, we must have $p_2 = p$, which eliminates B and E.

75. B – Calculating the probability (remembering the r^2 factor in the volume element),

$$\begin{aligned} P &= \int_a^{2a} |\psi(r)|^2 r^2\, dr = a \int_a^{2a} \frac{1}{r^4} r^2\, dr \\ &= a\left(\frac{1}{a} - \frac{1}{2a}\right) = \frac{1}{2}, \end{aligned}$$

which is B. Note that we don't have to worry about the angular part of the wavefunction, which is assumed to be normalized on its own, and so we only have to do the radial part of the volume integral.

76. C – The v_x graph tells us that v_x is constant and positive, and v_y tells us that v_y is initially positive but decreases linearly, which means constant negative acceleration. In other words, something is falling, but started with positive velocity in both the x- and y-directions. And $v_y(0) \approx 2v_x(0)$, which means that the initial angle of the

velocity vector was about 60°, which is consistent with choice C. In particular, $v_y(0) > v_x(0)$, so choice D is not a possibility.

77. D – This is the setup for Brewster's angle: when the Sun is at Brewster's angle with respect to the normal to the ocean, the reflected wave is completely polarized parallel to the ocean. Since the polarization axis of the sunglasses is perpendicular to the ocean, no transmitted light passes through the polarizer. The information given in the problem tells us that Brewster's angle is 60°, so $n_2 = \tan 60° = \sqrt{3}$, choice D. Incidentally, choice A can be eliminated immediately since indices of refraction can't be less than 1.
78. A – A charge in a sinusoidally oscillating field is just a dipole oscillator, which radiates power according to the Larmor formula

$$P = \frac{q^2 a^2}{6\pi \epsilon_0 c^3},$$

where a is the acceleration of the charge. For the purposes of this problem, you don't need to remember all the constants and numerical factors: the crucial part is the dependence on q and a. The acceleration is obtained from $F = ma = qE$, giving $a = qE/m$, and (dropping the irrelevant constants)

$$P \propto q^2(qE/m)^2 = \frac{q^4 E_0^2}{m^2} \cos^2 \omega t.$$

Taking the time average, $\langle \cos^2 \omega t \rangle = 1/2$, which is just a numerical factor (in particular, it's independent of ω). So we have

$$\langle P \rangle \propto \frac{q^4 E_0^2}{m^2},$$

which shows that if m is increased, $\langle P \rangle$ decreases. Increasing q and E_0 will increase $\langle P \rangle$. Changing T won't change the average power: since power is instantaneous energy per unit time, changing the measurement time can never change the power. Curiously, changing ω won't change the average power either, in contrast to the case of a dipole antenna where the emitted power scales as ω^4.
79. B – While actually calculating the flux as a function of time would require doing some tricky integrals, we can see that, because $q > 0$, the flux is positive from $t = -\infty$ to $t = 0$, and negative from $t = 0$ to $t = +\infty$. In particular, the flux must go to zero as $t \to \pm\infty$ because the charge is infinitely far away and the field dies off as $1/r^2$. Furthermore, because the strength of the electric field at $z = 0$ increases as the charge approaches, the flux does not simply increase at a constant rate, but accelerates. Finally, there is a discontinuity at $t = 0$ where the flux jumps from positive to negative, because at t slightly negative, all the field lines have components parallel to the normal to the loop, but at t slightly positive, all the field lines have components antiparallel. B is the only plot that satisfies all of these conditions.
80. B – The most probable speed of particles following the Maxwell distribution is $v = \sqrt{2kT/m}$. From the ideal gas law, P is directly proportional to T at fixed volume, so the most probable speed goes as $v \propto \sqrt{P}$, choice B.
81. B – The Doppler shift formula for a source approaching an observer in a straight line gives a constant observed frequency as the source approaches, and a different constant observed frequency as the source recedes because the sign of the relative velocity changes. The wavefronts are being compressed as the ambulance travels in the direction of the emitted signal, so the frequency is higher, hence choice B. This is a bit counterintuitive because the usual situation observed in real life is choice E, which corresponds to an ambulance passing by the observer with a nonzero distance of closest approach b. There, the reason for the sliding frequency is that the direction between the source and observer is constantly changing, and the Doppler shift only occurs in the direction of the source's motion, so there is an angle-dependent factor which changes continuously as the source moves past the observer.
82. E – Without doing any calculations, we know the answer must be E because the speed of a photon in *any* inertial reference frame is c. This is one of the axioms of special relativity: the speed of light is constant in any inertial reference frame.
83. C – Mesons are defined as bound states of a quark and an antiquark. Choice D describes baryons, not mesons, and none of the other choices are found in nature as bound states (except choice E, but that describes a nucleus).
84. A – Neutron stars are essentially giant spheres of neutrons. Since neutrons are fermions, they cannot all collapse to be in the same position state by the Pauli exclusion principle. All of the other choices could possibly be argued in vaguely plausible ways, but A is clearly the correct answer.
85. A – This can be quickly determined from energy conservation. The total energy stored in the capacitor is

$$E = \frac{Q_0^2}{2C}.$$

Assuming that all of this energy is dissipated by the resistor, the heat transferred to the material is $\Delta Q = \frac{Q_0^2 \epsilon}{2C}$, and the change in temperature can be found from $\Delta Q = mc_p \Delta T$:

$$\Delta T = \frac{\Delta Q}{mc_p} = \frac{Q_0^2 \epsilon}{2mc_p C}.$$

86. B – We can get the field of an infinite wire from Ampère's law. Taking an Amperian loop around the wire at radius r, we get $2\pi rB = \mu_0 I$, so $B = \mu_0 I/2\pi r$, where we have assumed the field is circumferential. Setting $r = d$ and using $\mathbf{F} = I\, d\mathbf{l} \times \mathbf{B}$ gives a force per unit length $|\mathbf{F}|/dl = \mu_0 I^2/2\pi d$. The direction comes from the right-hand rule, or by remembering that wires behave oppositely from charges: like currents attract while opposite currents repel. Hence choice B.
87. C – Writing

$$\mathbf{S}_1 \cdot \mathbf{S}_2 = \frac{1}{2}((\mathbf{S}_1 + \mathbf{S}_2)^2 - \mathbf{S}_1^2 - \mathbf{S}_2^2) = \frac{1}{2}\left((\mathbf{S}_1 + \mathbf{S}_2)^2 - \frac{3}{2}\hbar^2\right),$$

we see the possible values of H depend on the total spin $S_{\text{tot}}^2 = (\mathbf{S}_1 + \mathbf{S}_2)^2$; in other words, whether the two particles are in the singlet or triplet state. In the singlet state with spin 0, $H = -A(-\frac{3}{4}\hbar^2) = 3A\hbar^2/4$, and in the triplet state where $S_{\text{tot}}^2 = (1)(2)\hbar^2$, $H = -A\hbar^2/4$. However, there are three linearly independent total spin states in the triplet, so the triplet has a degeneracy of 3. Since $A > 0$, the triplet is the lowest-energy state, hence choice C.
88. B – The diameter of the molecules is only relevant in that they are much smaller than the wavelength of incident light. This puts us in the regime of Rayleigh scattering, where intensity is proportional to λ^{-4}. So doubling the wavelength multiplies the scattered intensity by 1/16, which is choice B. Incidentally, this is the same phenomenon that is responsible for the blue color of the sky, since solar light scattering off air and water molecules has a much higher intensity toward the blue end of the spectrum.
89. C – By the selection rules, we need $\Delta l = \pm 1$, so $3s$ must decay to $2p$ in the electric dipole approximation. All other decay modes will be suppressed relative to this "allowed" transition, hence choice C. (The selection rules for m will be satisfied if the $2p$ state also has $m = 0$.) Note that, depending on the total spin, the $3p$ state is at best degenerate with $3s$, up to hyperfine-structure corrections. This is a tiny energy splitting, so the transition would still prefer to go to $2p$.
90. D – Statistical and systematic errors are independent so they add in quadrature: $\sigma_{\text{tot}} = \sqrt{(0.08\text{ GeV})^2 + (0.06\text{ GeV})^2} = 0.10$ GeV.
91. B – Radioactive decays follow a Poisson probability distribution, whose variance is equal to N, the number of counts. Hence the fractional error is $\sigma/N = \sqrt{N}/N = 1/\sqrt{N}$, which for $N = 64$ gives 1/8, choice B.
92. C – This is a simple kinematics problem dressed up with some unfamiliar units. While we could convert the electron kinetic energy to joules and use $F = qE$ with the electron charge q in coulombs, it's much simpler to recall the definition of eV ("electron-volt") as the potential energy gained by an electron traveling through a potential difference of 1 V. The electron reaches its maximum height when its kinetic energy is converted entirely to potential energy, which occurs when the potential relative to the sheet is 10 V. 100 V/m = 1 V/cm, so the 10 V potential occurs at a height of 10 cm, choice C. (The electron is negatively charged, so an electric field directed along its direction of motion will cause it to decelerate – trap answer E would be correct for a positively charged particle.)
93. C – Let ℓ be the length of the rod, v be the initial velocity of the bullet, V be the velocity of the block immediately after the collision, and m and M be the masses of the bullet and block, respectively. Momentum is conserved in the collision with the block, so we can solve for V:

$$mv = (M+m)V$$
$$\implies V = \frac{m}{M+m}v \approx \frac{m}{M}v,$$

where the last approximation comes from the fact that the bullet is much lighter than the block. For the block to just barely make a complete revolution, the block must reach the very top of its circular path with zero velocity. There, it will have potential energy $(M+m)g(2\ell)$ with respect to its initial position, so we solve for V by conservation of energy:

$$\frac{1}{2}(M+m)V^2 = 2(M+m)g\ell$$
$$\implies V = 2\sqrt{g\ell}.$$

Plugging in for V and inserting the numbers in the problem gives

$$v = 2\frac{M}{m}\sqrt{g\ell} \approx 800 \text{ m/s},$$

choice C.

94. E – This is an application of the Rayleigh criterion, $\sin\theta = 1.22\lambda/D$. Solving for D gives $D = 1.22\lambda/\sin\theta$. Since $\theta = 0.061$ rad is very small, we can approximate $\sin\theta \approx \theta$. From $\lambda f = c$, we have $\lambda = 1.5$ m, and $D = 30$ m, choice E.
95. D – In the center-of-momentum frame, the electron and the positron each have energy γm_e where $\gamma = 1/\sqrt{1 - v^2/c^2}$ is the Lorentz factor. The total energy in this frame is $2\gamma m_e$, which must be at least the energy of the two muons at rest, $2m_\mu$. Solving $2\gamma m_e = 2m_\mu$ for v, we find choice D.
96. B – The time constant of an RC circuit is simply RC. We have two resistors in series, so the effective resistance is 2 MΩ + 3 MΩ = 5 MΩ, and two capacitors in series whose effective capacitance is $(1/1\,\mu\text{F} + 1/4\,\mu\text{F})^{-1} = 0.8\,\mu\text{F}$. Multiplying these gives a time constant of 4 s, which is B.
97. E – The field in the solenoid is proportional to the number of coils per unit length, n. But the flux through the solenoid is proportional to n^2, since we have to sum the fluxes through each of the n coils. From the definition of inductance, $L = \Phi/I$, L is also proportional to n^2, so tripling n increases L by a factor of 9, as in E.
98. B – In a free electron laser, electrons are forced in a sinusoidal path, and the resulting acceleration produces coherent synchrotron radiation, choice B. All the other choices are steps in the operation of normal lasers, but they all require electronic transitions in atoms – the key description "*free* electron" should at least eliminate choices A and E.
99. C – If we imagine the surface current as being made up of a bunch of parallel wires all pointing in the $\hat{\mathbf{y}}$-direction, we see that there are equal and opposite contributions to a field in the $\hat{\mathbf{z}}$-direction, from wires at $+x$ and $-x$ for any value of x. So there is no field in the $\hat{\mathbf{z}}$-direction. Moreover, there can't be any field parallel to the wires, so $\mathbf{B}$ must point in the $\pm\hat{\mathbf{x}}$-direction. This leaves only C and D. To get the sign right, apply the right-hand rule to all the wires, which shows that above the plane, the field from every wire is in the $+\hat{\mathbf{x}}$-direction. Since C and D only differ by a sign, we're done: we don't actually have to use Ampère's law to calculate the magnitude. Always be on the lookout for shortcuts like this on the GRE!
100. D – It turns out that *all* of the numerical values in the question are irrelevant. This is not an uncommon occurrence on the GRE, so it pays to work out everything in terms of variables until the end of the problem. Let m be the mass of the sphere, R its radius, h the height of the ramp, and v its velocity at the bottom. Two observations make this problem quite simple. First, while the forces on the sphere are different whether or not friction is present, they must be *constant* throughout the fall down the ramp. So the acceleration down the ramp is constant as well. Second, from kinematics we know $v^2 = 2a\Delta x$ for constant acceleration a, so combining this with $v = at$ gives $t = 2\Delta x/v$; the time is inversely proportional to the velocity at the bottom of the ramp. The distance traveled Δx will cancel out in the ratio t'/t.

 To find the velocity at the bottom, use conservation of energy. Without friction, $mgh = \frac{1}{2}mv^2$, so $v = \sqrt{2gh}$. With friction, there is an additional rotational kinetic energy $\frac{1}{2}I\omega^2 = \frac{1}{2}(\frac{2}{5}mR^2)(\frac{v'^2}{R^2}) = \frac{1}{5}mv'^2$. Now we have

$$mgh = \left(\frac{1}{2} + \frac{1}{5}\right) mv'^2 \implies v' = \sqrt{\frac{10}{7}gh},$$

 so $t'/t = v/v' = \sqrt{7/5}$, choice D. We could have narrowed the choices down to D or E from the beginning, since the additional rotational kinetic energy implies the velocity at the bottom will be less than in the frictionless case. From the constant acceleration argument, this means that $t' > t$, which eliminates choices A–C.

Solutions to Sample Exam 3

1. C – Whenever a small amount of liquid with mass m is emitted from the bath, the potential energy released by the level of the water dropping is $\Delta U = mgh$. Neglecting small effects such as surface tension and viscosity, all of this energy is converted into kinetic energy with a velocity directed parallel to the ground. The velocity parallel to the ground is therefore

$$\frac{1}{2}mv^2 = mgh,$$
$$v = \sqrt{2gh}.$$

The same result can also be derived from Bernoulli's principle applied to the top and bottom of the bath, noting that atmospheric pressure is the same at both locations and cancels. In fact, it's common enough that it has its own name, "Torricelli's law," and may be worth memorizing so you don't have to repeat the derivation.

The time required for free fall in the vertical direction follows from basic kinematics:

$$t = \sqrt{2y/g}.$$

Substituting this into $x = vt$, we find that

$$x = 2\sqrt{hy}.$$

2. E – Hydrogen atoms are much smaller than 450 nm so the small-particle Rayleigh formula applies. If the Bohr radius were doubled by the a^6 dependence in the Rayleigh formula we get intensity $2^6 I_0 = 64I_0$, choice E.

3. E – This is a straightforward application of the Rayleigh criterion. The minimum aperture diameter D needed to resolve objects with separation $\Delta\theta$ emitting light of wavelength λ is

$$D = \frac{1.22\lambda}{\Delta\theta},$$

where we have assumed $\Delta\theta$ is small so we can use the small-angle approximation. For the purposes of this problem, which is really an order-of-magnitude question, the coefficient 1.22 isn't even necessary. Radio waves have wavelengths on the mm to km scale, visible light has wavelengths between 400 and 700 nm, and x-rays have wavelengths of approximately 0.01 to 10 nm. Applying the criterion, we can estimate

$$D_{\text{radio}} \geq \frac{1.22(10^{-3}\text{ m})}{3 \times 10^{-4}\text{ rad}} \approx 5\text{ m},$$

$$D_{\text{vis}} \geq \frac{1.22(500 \times 10^{-9}\text{ m})}{3 \times 10^{-4}\text{ rad}} \approx 2\text{ mm}.$$

At this point we don't even have to do the calculation for x-rays (nor do we really have to know exactly what their wavelengths are), since we know that a 1 cm aperture is large enough to resolve visible light, so it must be large enough to resolve x-ray light, which has an even smaller wavelength. So both II and III are possible, giving choice E.

4. B – The force due to friction is $F = \mu mg$, so the acceleration due to friction is given by $ma = \mu mg$, or

$$a = \mu g,$$

opposite to the direction of motion. We can obtain the distance traveled by the useful kinematic identity

$$v_f^2 - v_i^2 = 2a\Delta x.$$

Plugging in numbers, we find that

$$\Delta x = \frac{v_i^2}{2a} = \frac{v_i^2}{2\mu g} = \frac{(10\text{ m/s})^2}{10\text{ m/s}^2} = 10\text{ m}.$$

Just as easily, we can use the work–energy theorem: the work done by friction to bring the block to rest must be equal to its initial kinetic energy, so

$$\mu mg\Delta x = \frac{1}{2}mv_i^2 \implies \Delta x = 10\text{ m}$$

as before. Note that the mass cancels out in both solutions.

5. E – The fastest way to solve this problem is to just think of some examples where you have used Lagrangians to solve a problem. Clearly, we use Lagrangians all the time to solve problems with conservative forces; for example, a particle in a gravitational field. There's also nothing special about rotational symmetry; for example, again, we can use Lagrangians to solve for motion in a uniform gravitational field at the surface of the Earth, which has cylindrical but not full rotational symmetry. Finally, one rarely deals with time-dependent potentials, but the fact that the Euler–Lagrange equations include time derivatives suggests that they do not assume anything to be constant with time.

6. C – The hyperfine structure is the smallest correction, orders of magnitude smaller than the fine-structure effects. It is produced by the interaction of the nuclear dipole moment with the magnetic field produced by the orbit of the electrons. The next smallest is the Lamb

shift, which arises due to interactions between the electron and the vacuum and causes a shift in the relative magnitude of the s and p orbital energies for the $n = 2$ states. The fine-structure corrections arise due to spin–orbit coupling, relativistic effects, and the so-called Darwin term. The combined fine-structure effects are the largest contribution. The easiest way to remember the hierarchy is to remember the factors of the fine-structure constant α: fine-structure corrections are proportional to α^4, the Lamb shift is proportional to α^5, and hyperfine-structure corrections are proportional to $\alpha^4(m_e/m_p) \ll \alpha^5$.

7. C – The energies of the harmonic oscillator are just $E_n = \hbar\omega(n + 1/2)$. The difference between the $n = 5$ and $n = 1$ states is therefore $4\hbar\omega$.
8. B – While it is tempting to think of this problem as an application of the Coriolis force, it is much easier to solve in the nonrotating frame. It takes the puck a time $t = R/v$ to reach the edge of the disk. By this time, the disk has rotated through an angle $\Delta\theta = \omega t = \omega R/v$.
9. C – The answer is a common enough formula that you may have it memorized. If you do not, you can derive it from the usual rules for propagation of uncertainties. The uncertainty of the function $z = x/y$ is

$$(\Delta z)^2 = \left(\frac{\partial z}{\partial x}\right)^2 (\Delta x)^2 + \left(\frac{\partial z}{\partial y}\right)^2 (\Delta y)^2$$
$$= \frac{1}{y^2}(\Delta x)^2 + \frac{x^2}{y^4}(\Delta y)^2$$
$$= z^2 \left(\left(\frac{\Delta x}{x}\right)^2 + \left(\frac{\Delta y}{y}\right)^2\right).$$

Taking square roots to get Δz gives choice C.

10. A – A bandpass filter only allows signals to propagate that are between two frequencies (not to be confused with high-pass or low-pass filter, which are only one-sided). Intuitively, inductors suppress high-frequency signals because voltages are high for fast oscillations and capacitors suppress low-frequency ones because of excessive charge buildup. It therefore is most reasonable that a bandpass filter would require both an inductor and a capacitor. Choice A is the only option that has both of these circuit elements. (Note however that both high-pass and low-pass filters can be made with either RL or RC circuits, and stringing two of these circuits together would give a bandpass filter, but this requires more circuit elements than allowed by the answer choices.)

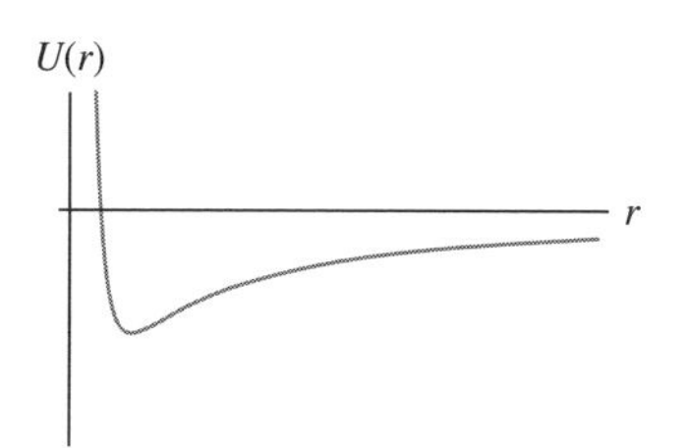

11. A – The effective potential is shown in the figure above. Since a circular orbit is at a fixed radius, E_{cir} corresponds to an energy at the minimum of the effective potential. An elliptical orbit is at a variable radius but is still bound, so E_{ell} corresponds to an energy above the minimum of the effective potential but less than zero. Finally, a hyperbolic orbit is unbound at all radii (even at $r \to \infty$), so $E_{\text{hyp}} > 0$. Putting it all together, we find that $E_{\text{cir}} < E_{\text{ell}} < E_{\text{hyp}}$.
12. C – Electrons are fermions, and so at zero temperature they cannot all collect in the ground state. Instead, they fill out the so-called Fermi sphere, and the energy of electrons at the boundary is called the Fermi energy, choice C. In a pinch, if you only remembered that electrons are fermions, you might be able to guess that the answer had something to do with Fermi.
13. B – The time dilation changes the apparent lifetime in the rest frame by a factor of $\gamma = 1/\sqrt{1 - 0.64} = 1/0.6 = 1.67$. The distance traveled during this time is $0.8c \times (1.67 \times 10^{-8}\text{ s})$, or about 4 m.
14. B – The outermost electron is in the p orbital, and so has $l = 1$. This means $L^2 = l(l+1)\hbar^2 = 2\hbar^2$, so $|\mathbf{L}| = \hbar\sqrt{2}$.
15. C – Superficially this may seem like a rather technical question from early universe cosmology. On the other hand, it is clear that I and II are light nuclei, while III is fairly heavy. Your intuition should tell you that light nuclei were probably the first elements produced after the Big Bang, leading you to guess C. Indeed this is correct. No elements heavier than beryllium were produced in the early universe before it cooled below the temperature needed for nucleosynthesis. Heavier elements were not produced until the first stars formed and combined the lighter elements into heavier ones via nuclear fusion.
16. B – The fluid conservation equation implies that

$$vA_1 = v_2A_2,$$

so B is the correct choice. This follows immediately from conservation of mass and the fact that the fluid is incompressible.

17. C – We can use the equation from basic kinematics:

$$v^2 = 2a\Delta x.$$

Plugging in numbers, we find that the velocity is 20 m/s.

18. E – While this problem can be analyzed to a large extent using limiting cases, it is also simple to work out explicitly. The velocity of the block–bullet system after the collision is obtained by conservation of momentum:

$$v' = \frac{mv}{m+M}.$$

The force due to friction is simply constant and given by

$$F = (m+M)a = -\mu(m+M)g.$$

From basic kinematics at constant acceleration, we know that

$$\Delta t = \frac{\Delta v}{a} = \frac{mv}{\mu g(m+M)}.$$

Alternatively, you could eliminate all choices except E by considering the limiting behavior of the masses and coefficient of friction.

19. C – This question refers to Johnson noise, the voltage fluctuations that arise in resistors due to thermal fluctuations that induce resistance fluctuations. The derivation of this formula is not difficult, but it may be unfamiliar to you. Luckily, we can solve this one by pure dimensional analysis. A spectral density of voltage fluctuations has units of V $\text{Hz}^{-1/2}$. The funny units are chosen because rms values of noise sum in quadrature. It therefore is most natural to write a spectrum of square power over frequency because then noise simply adds. Choice C is the only option that has the correct units. Note that *almost always*, k and T show up in the combination kT which has units of energy. Here, it's most convenient to write energy as $[V][Q]$, or voltage times charge. From $V = IR$, resistance R has units of $V/I = [V][Q]^{-1}[T]$. To get $\text{Hz}^{-1/2}$ or $[T]^{1/2}$, we want $\sqrt{R}$ somewhere, so computing the units,

$$\sqrt{kTR} \sim \sqrt{[V][Q][V][Q]^{-1}[T]} \sim [V][T]^{1/2},$$

and C has the right units as promised.

20. D – At time t the bar has traveled a distance $x = vt$. Since the magnetic field is perpendicular to the loop, the flux through the loop is

$$\Phi = \int \mathbf{B} \cdot d\mathbf{A} = Bxd.$$

By Faraday's law, the emf in the circuit is $\mathcal{E} = d\Phi/dt = Bvd$, so the power is $P = \mathcal{E}^2/R = (Bvd)^2/R$, and integrating this from $t = 0$ to $t = T$ gives the total energy $(Bvd)^2T/R$, choice D.

21. D – Recall that the rms velocity of an ideal gas is

$$v_{\text{rms}} = \sqrt{\frac{3kT}{m}}.$$

Helium is a monoatomic gas since noble gases don't form chemical bonds under standard conditions, so helium gas, with a mass of approximately 4 amu, is approximately two times heavier than that of molecular hydrogen, with a mass of approximately 2 amu. Thus, its rms velocity is $v/\sqrt{2}$, choice D.

22. D – The resonant frequency of an LC circuit is

$$\omega_0 = \sqrt{\frac{1}{LC}},$$

which is also the frequency at which the current through an RLC circuit is maximized, so the answer is D. If you don't remember this fact, just remember that the resonant frequency is defined as the frequency where the imaginary part of the total impedance vanishes. Adding a series resistance only changes the *real* part of the impedance, so the resonant frequency is unchanged.

23. B – An open pipe has pressure nodes at both ends, so the wavelength of the fundamental vibration is $2L$. The frequency is

$$f = \frac{c}{\lambda} = \frac{343 \text{ m s}^{-1}}{0.4 \text{ m}} = 858 \text{ Hz}.$$

24. B – Since the two loops are perpendicular to each other, the torque on the smaller loop is given by

$$N = mB,$$

where $m = \pi i b^2$ is the magnetic moment of the smaller loop, and B is the magnitude of the magnetic field at the center of the larger loop. This can be readily obtained from the Biot–Savart law, which reduces to

$$B = \frac{\mu_0}{4\pi}\frac{2\pi a I}{a^2} = \frac{\mu_0 I}{2a}.$$

Putting the pieces together, we find that the torque is

$$N = \frac{\mu_0 \pi I i b^2}{2a}.$$

25. B – When ball A is released, it experiences a centrifugal acceleration in its reference frame:

$$a = \frac{v^2}{r} = \frac{\Omega^2 R}{2}.$$

Ball B experiences an acceleration given by

$$a = \frac{v^2}{r} = \Omega^2 R,$$

so the ratio is 1/2.

26. B – The electric field inside the sphere can be quickly obtained from Gauss's law, and is found to be $\propto r$. By integration, we conclude that the potential is $\propto -r^2$. In the region between the charged sphere and the conductor, there is the usual $\propto r^{-1}$ potential. Inside the conductor, the potential is constant. Finally, outside the conductor, the potential is again $\propto r^{-1}$. Only choice B satisfies these requirements.
27. D – Only choices D and E are behaviors that are even remotely possible for an RLC circuit. E is almost a correct plot of an overdamped RLC circuit (the case with no visible oscillation from the inductor), but the current starts out at a nonzero value. This should not happen because the inductor should oppose the sudden increase in current, producing a more gradual increase from zero current to some maximum, as seen in choice D.
28. A – The cyclotron radius is

$$r = \frac{mv}{qB}.$$

If the electrons travel at velocity v, then we have

$$v = \frac{2\pi r}{T},$$

and so

$$T = \frac{2\pi m}{qB}.$$

29. B – This is classic Bragg diffraction. We can simply use the well-known formula for first-order diffraction:

$$2d \sin\theta = \lambda.$$

Solving for the angle in the small-angle approximation, we have

$$\theta \approx \frac{\lambda}{2d}.$$

Plugging in numbers, and approximating π by 3 when converting from radians to degrees,

$$\theta \approx \frac{10}{2 \cdot 80} \times \frac{180}{\pi} \approx \frac{30}{8} \approx 3.75^\circ,$$

which is closest to 3.6°, choice B.

30. B – If this strikes you as an insanely difficult problem for the GRE, realize that it can be solved without any calculation by considering limiting cases. The most useful limit here is $x \to R$: since the sphere is grounded and conducting, we know that the potential must vanish when $x = R$. Choices B and E are the only options that satisfy this basic requirement. Choice E is obviously wrong, because as $R \to 0$ we should still see some dependence on x from the potential of the charge Q by itself.
31. B – It is a useful fact that the simple cubic lattice is self-dual; that is, the lattice is equal to its reciprocal lattice. In case this isn't obvious, it should be clear to you from the definition of the reciprocal lattice. If $\mathbf{r} = la\hat{\mathbf{x}} + ma\hat{\mathbf{y}} + na\hat{\mathbf{z}}$ with $l, m, n \in \mathbb{Z}$ are the lattice sites of the simple cubic lattice, then the reciprocal lattice vectors are those vectors $\mathbf{k}$ satisfying the relation

$$e^{i\mathbf{k}\cdot\mathbf{r}} = 1.$$

It is clear that this relation will hold as long as $\mathbf{k} = \frac{2\pi}{a}(p\hat{\mathbf{x}} + q\hat{\mathbf{y}} + r\hat{\mathbf{z}})$, where $p, q, r \in \mathbb{Z}$. But these $\mathbf{k}$ vectors are just the points of another simple cubic lattice. So the reciprocal lattice of a simple cubic lattice is itself cubic.

32. A – This is a straightforward computation using the definition of the magnetic field in terms of the vector potential. Recall that we have

$$\mathbf{B} = \nabla \times \mathbf{A}.$$

Direct substitution of the vector potential into this equation gives choice A.

33. B – This is a simple application of Wien's displacement law, which holds for blackbody radiation. The CMB is one of nature's most perfect blackbodies, so Wien's law is applicable. It states that the wavelength of maximum intensity emitted by a blackbody is

$$\lambda = \frac{b}{T},$$

where b is a constant. We just set

$$\lambda T = \lambda' T',$$

so that

$$\lambda' = \frac{\lambda T}{T'} = \frac{1 \text{ mm} \cdot 2.7 \text{ K}}{5 \text{ K}} \approx 0.5 \text{ mm},$$

which is closest to B.

34. E – The best way to solve this problem is by a process of elimination. B and D are trivially forbidden by charge conservation. A is forbidden by conservation of lepton number (you cannot get two leptons from one). C is a bit more subtle, but it is forbidden because the Δ^+ is

a low-lying excitation of the proton, which is not sufficiently massive to decay to a proton and neutron. In addition, this decay would violate baryon number since Δ^+, p, and n each have baryon number equal to 1. This leaves E: note that lepton number is conserved because μ^+, the antimuon, has mu-lepton number -1.

35. D – This is another application of Wien's displacement law. Relating observed to true quantities, we have

$$\frac{b}{T_{\text{obs}}} = \lambda_{\text{obs}} = \sqrt{\frac{1+\beta}{1-\beta}}\lambda_{\text{true}} = \sqrt{\frac{1+\beta}{1-\beta}}\frac{b}{T_{\text{true}}}.$$

The true temperature is therefore

$$T_{\text{true}} = \sqrt{\frac{1+\beta}{1-\beta}}T_{\text{obs}}.$$

Plugging in numerical values, we obtain about 1.1×10^4 K.

36. C – Intuitively, one might expect the energy to go to the larger of the two states, but this does not turn out to be the answer. The partition function is given by

$$Z = e^{-\beta\epsilon} + e^{\beta\epsilon}.$$

The average energy is given by

$$\langle E \rangle = \frac{\epsilon e^{-\beta\epsilon} - \epsilon e^{\beta\epsilon}}{e^{-\beta\epsilon} + e^{\beta\epsilon}}.$$

As $T \to \infty$, we have $\beta \to 0$, so

$$\langle E \rangle \to 0.$$

We can understand this result because $T \to \infty$ means $kT \gg \epsilon$, so the thermal fluctuations overpower the small energy splitting due to ϵ, and each state is equally populated.

37. C – If the gas expands, then clearly energy is conserved as long as the box is reasonably thermally isolated from its surroundings. Somewhat less obviously, the temperature is also unchanged. During free expansion of an ideal gas, we have $PV = P'V'$, and thus temperature must remain constant. Alternatively, you can see that the temperature is unchanged since the temperature is related to the internal energy of the gas via $U = (3/2)NkT$. The entropy of an ideal gas, on the other hand is given by

$$S = Nk\left[\ln\left(\frac{V}{N}\left(\frac{4\pi mU}{3Nh^2}\right)^{3/2}\right) + \frac{5}{2}\right],$$

which has a clear volume dependence. Another formula that is probably more familiar to you and easier to remember for the exam is that the entropy change during free expansion of an ideal gas is

$$\Delta S = Nk\ln\left(\frac{V'}{V}\right).$$

This follows directly from the full expression for the entropy above, but it is quite a bit more practical for GRE-style questions.

38. C – Energy conservation is not necessarily violated because $e < 1$ is still possible. The First Law of Thermodynamics is essentially a restatement of conservation of energy, so this should be a good clue that neither of the first two choices are correct. The Third Law of Thermodynamics is the statement that objects cannot be cooled to absolute zero, which has nothing to do with the situation at hand. Finally, the postulate of equal *a priori* probabilities is a fundamental assumption of statistical mechanics, which states that all of the microstates corresponding to each macrostate of a system are equally probable. This clearly has nothing to do with the heat engine at hand. This leaves the Second Law of Thermodynamics as the correct answer.

We could also have seen this right away since, as $T_H \to T_C$, $e \to 2/3$. This is a clear violation of the Second Law since it would imply that useful work could be done between two reservoirs at the same temperature; in particular, it violates the Carnot bound $e = 1 - \frac{T_C}{T_H}$.

39. E – The addition of velocities formula is generally written in the form

$$s = \frac{u+v}{1+\frac{uv}{c^2}}.$$

It is critical to understand the notation here. This equation holds for a body A that is traveling at velocity v with respect to another body B that is traveling at velocity u with respect to a reference frame S. In this notation, s is the velocity of A with respect to the reference frame S. In the problem, we can identify v as the velocity v_1 of spaceship A with respect to the planet. We can identify s as the velocity v_2 of spaceship B with respect to the planet. And we want to solve for the absolute value of u, the speed of spaceship A relative to spaceship B. Making these substitutions and solving, we find that the speed is as given in the solution.

Equivalently, we can think of everything in the reference frame of B. Then u is the velocity of the planet relative to B, v is the velocity of A relative to the planet. and s is the velocity of A relative to B. Making the

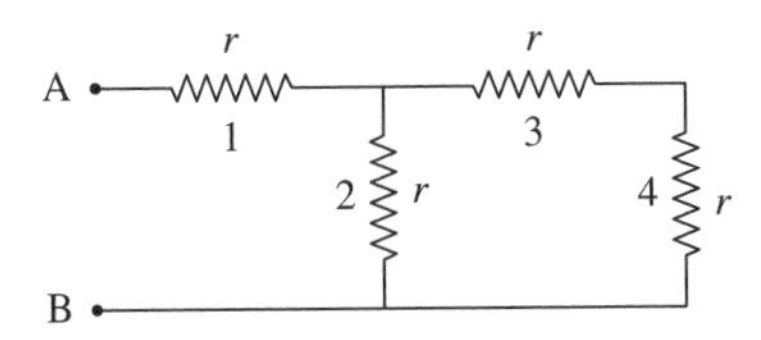

identification $u = -v_2$, $v = v_1$, we obtain the same result.

40. E – This is a straightforward equivalent resistance problem. Labeling the resistors as shown above, resistors 3 and 4 are in series, with equivalent resistance $2r$. This equivalent resistance is in parallel with resistor 2, giving

$$R_{234} = \left(\frac{1}{2r} + \frac{1}{r}\right)^{-1} = \frac{2r}{3}.$$

Adding resistor 1 in series, we get a total resistance of $5r/3$, or choice E.

41. D – The invariant of a gas undergoing a reversible adiabatic change is

$$PV^{\gamma} = \text{const},$$

where γ is the ratio of specific heat at constant pressure to specific heat at constant volume. For a monoatomic ideal gas, we have $\gamma = 5/3$. Thus we have

$$P = P_f \times 2^{5/3},$$

and

$$P_f = 2^{-5/3}P.$$

42. B – To compute this, we simply need to recognize that $L_z Y_l^m(\theta, \phi) = \hbar m Y_l^m(\theta, \phi)$. Using this fact, we have

$$\langle L_z \rangle = \frac{1}{2}\int d\Omega \left(Y_1^{*-1}(\theta,\phi) + Y_1^{*0}(\theta,\phi)\right)$$
$$L_z\left(Y_1^{-1}(\theta,\phi) + Y_1^{0}(\theta,\phi)\right) = -\frac{\hbar}{2}.$$

Even this setup is a little overkill here: since Y_1^{-1} has eigenvalue $-\hbar$ and Y_1^0 has eigenvalue 0, and they appear in the wavefunction with equal relative coefficients, the average must be $\frac{1}{2}(-\hbar + 0) = -\hbar/2$.

43. C – Choice A is clearly incorrect. Superpositions of energy states are permissible states, and they are certainly not energy eigenstates. Choice B is also incorrect. The simplest counterexample is the free particle (i.e. the solution of the Schrödinger equation with $V = 0$), which can have any positive energy. The wavefunction for this case $\psi(x) = Ae^{ikx}$ also is a simple counterexample for choice D. Choice C is, of course, true for quantum systems in any number of dimensions. You may remember that observables are represented by Hermitian operators, so it stands to reason that their eigenvalues should be real.

44. C – We cannot easily guess away the solutions, so we must solve this problem explicitly. We just need to use conservation of angular momentum. In order to do this, we need to know the moment of inertia of the mass–rod system. Since the mass is a point particle, the total moment of inertia is the sum of the point particle moment of inertia and the rod moment of inertia. And, since $m \ll M$, the center of mass of the new system is approximately the center of mass of the rod, so we compute both moments about the center of the rod:

$$I = \frac{1}{12}ML^2 + m\left(\frac{L}{2}\right)^2.$$

The angular momentum before the collision is equal to the angular momentum afterwards:

$$mv\left(\frac{L}{2}\right) = \left(\frac{1}{12}ML^2 + \frac{1}{4}mL^2\right)\omega.$$

So the solution is

$$\omega = \frac{6mv}{(M+3m)L}.$$

Under the approximation $m \ll M$, we can drop the $3m$ term in the denominator, which just gives

$$\omega = \frac{6mv}{ML},$$

choice C.

45. A – The time dependence of the wavefunction is given by $e^{-iHt/\hbar}$. For an eigenstate we have $e^{-iEt/\hbar}$, where $E = \alpha\hbar B/2$ for the $|+\rangle$ state. The wavefunction for this state is thus $\exp(-i\alpha Bt/2)|+\rangle$.

46. D – Observables in quantum mechanics clearly must have real eigenvalues because we measure real numbers in physical experiments. We also require that observable operators be Hermitian, partially in order to ensure that the eigenvalues are real. So, I and III are correct. II is false because multiple operators can commute with the Hamiltonian and therefore can have simultaneous eigenstates.

47. A – The oscillation frequency of a mass on a spring is

$$f_0 = \frac{1}{2\pi}\sqrt{\frac{k}{m}}.$$

While you should definitely know this expression and how to derive it, you can easily derive it from dimensional analysis in a pinch. The equation of motion for this system only involves k and m, and there is only one combination of these constants that gives units of frequency. The prefactor is unimportant for this problem, since we're only interested in the fractional change of a quantity. That said, if we increase the mass by a factor of 3 and increase the spring constant by a factor of 2, then the new oscillation frequency is just

$$f = \frac{1}{2\pi}\sqrt{\frac{2k}{3m}} = \sqrt{\frac{2}{3}}f_0.$$

48. E – The probability of finding the particle on $[0, 1/2]$ is the integral over this interval of $|\psi|^2$:

$$P = 5\int_0^{1/2} x^4\,dx = \frac{1}{32}.$$

49. E – The spin-up and spin-down components of the wave function are orthogonal, so we simply have

$$\langle\psi|\,S_z\,|\psi\rangle = \frac{2}{3}\left(\frac{\hbar}{2}\right) + \frac{1}{3}\left(\frac{-\hbar}{2}\right) = \frac{\hbar}{6}.$$

50. D – Carbon dioxide lasers produce photons when electron impacts on nitrogen excite vibrational modes of the molecule. Collisions between nitrogen and CO_2 then excite vibration modes of the molecule, and the de-excitation produces the laser light.

51. B – This can be quickly solved by writing the equations of motion for the car and the trailer. For the car we have

$$Ma = F - T.$$

For the trailer, we have

$$ma = T - \mu mg.$$

Solving for T by substituting for a, we arrive at

$$T = \frac{m(F + \mu Mg)}{M + m}.$$

52. E – The total angular momentum obtainable when summing together angular momenta l and s are integral values between $|l + s|$ and $|l - s|$. The smallest total angular momentum is therefore $j = 2$.

53. E – Recall that noble gases such as helium, neon, and argon have full electron shells, which is the case here: each shell has $2n^2$ states, so there are two electrons in the $n = 1$ shell and eight electrons in the $n = 2$ shell. (This happens to be the electronic configuration for neon.) Alkali metals and rare earth metals have one or two additional electrons over a full shell, respectively, and halogens are one electron short of a full shell. Semiconductors are not easily identified from their electronic configurations.

54. C – This seems like a nasty problem, but considering limiting cases makes it easy. As $\beta \to 0$, the perturbation becomes constant, and thus the energy shift must also be the constant α: this eliminates A, D, and E. As $\beta \to \infty$, the perturbation disappears entirely, which leaves only choice C.

For completeness, here's the exact solution. The energy shift of the ground state is given by taking the expectation value of the perturbation with respect to the ground state wavefunction:

$$\langle 0|\,\delta V\,|0\rangle = \alpha\left(\frac{m\omega}{\pi\hbar}\right)^{1/2}\int_{-\infty}^{\infty} e^{-m\omega x^2/\hbar}e^{-\beta x^2}\,dx$$
$$= \alpha\left(\frac{m\omega}{m\omega + \beta\hbar}\right)^{1/2}.$$

55. C – By Stokes's theorem, we can relate the line integral of $\mathbf{f}$ around a closed curve to the surface integral of $\nabla \times \mathbf{f}$ over a surface bounded by that curve. But $\nabla \times \mathbf{f} = 0$ so the line integral must vanish. We can also easily do the integral explicitly. The integral over the first segment is zero because only the $\hat{\mathbf{x}}$ part contributes, and $y = 0$; the same is true for the last segment. The second segment contributes 1, but the third segment contributes -1 since it is traversed in the $-\hat{\mathbf{x}}$-direction, so the total is zero as before.

56. C – All answers except C can be eliminated with dimensional analysis. The dimensions of magnetic flux are a bit tricky: it's magnetic field strength times area, but we can relate this to electrical charges by Faraday's law:

$$(B\text{-field})(\text{area}) = (\text{voltage})(\text{time})$$
$$= (\text{energy}\cdot\text{time})/\text{charge},$$

which has units of $\text{J}\cdot\text{s/C}$.

57. C – I is true: electrons and positrons are both fermions, and a system consisting of an even number of fermions behaves as a boson and obeys Bose–Einstein statistics. This can be shown more rigorously by the rules for addition of angular momentum: two spin-1/2 particles can have a total spin of 0 or 1, both of which are integers rather than half-integers. II is also true, since the binding energy of a two-particle bound state is proportional to the reduced mass of the system. In the case of the hydrogen atom, the reduced mass is $m_p m_e/(m_e + m_p) \simeq m_e$ and the ground state energy is given by -13.6 eV. In the case of positronium, the reduced mass is $m_e^2/(2m_e) = m_e/2$, so the ground state energy must be about half that of hydrogen. III is false: going to the rest frame of positronium, a massive bound state cannot decay into

a single massless photon without violating conservation of relativistic energy–momentum.

58. C – Energy is conserved in an elastic collision, and since the wall of the pool table doesn't move, the ball must have the same energy before and after the collision. Thus $\frac{1}{2}m|\mathbf{v}|^2 = \frac{1}{2}m|\mathbf{v}'|^2$, and $|\mathbf{v}| = |\mathbf{v}'|$, so I is true. II is only true if $\mathbf{v}$ is normal to the wall. III is true by the definition of momentum.

59. A – The only transition that is allowed is A. The others are forbidden by the selection rules $\Delta l = \pm 1$ and $\Delta m = 0, \pm 1$.

60. C – If the power of radiation emitted by the satellite is P, then the blackbody temperature is T such that $P = \sigma T^4$. If the radius of the orbit doubles, then the amount of radiation received from the Sun decreases by a factor of 4. In equilibrium, power emitted also decreases by a factor of 4. This means that the blackbody temperature must change by a factor of $4^{-1/4} = 2^{-1/2}$ in order to satisfy $P = \sigma T^4$.

61. E – The particles of gas A and gas B have the same total mass. By the equipartition theorem, the rms velocity *only* depends on the motion of the center of mass of the molecule, and thus should be the same for gases A and B. Gas B has rotational degrees of freedom at low temperatures and vibrational degrees of freedom at higher temperatures, but these only affect the specific heat and not the rms speed.

62. A – Since angular momentum is always conserved, we can use it to solve this problem quickly. The moment of inertia of the disk in the problem about the axis of rotation is

$$I_{\text{disk}} = \frac{1}{2}Mr^2.$$

The moment of inertia of the person at the edge of the disk is

$$I_{\text{person}} = mr^2.$$

The conservation equation is

$$I_{\text{disk}}\omega = (I_{\text{disk}} + I_{\text{person}})\omega',$$

so the change in angular velocity is

$$\Delta\omega = |\omega' - \omega| = \left|\frac{I_{\text{disk}}}{I_{\text{disk}} + I_{\text{person}}} - 1\right|\omega = \frac{2\omega m}{M + 2m}.$$

63. E – The equation for the energy levels of a hydrogen-like atom with Z protons is

$$E_n = \frac{Z^2(13.6\text{ eV})}{n^2}.$$

Calculating the change in energy between the two states with $Z = 2$ for helium, we find that the emitted photon must have energy

$$\Delta E = 4(13.6\text{ eV})\left(1 - \frac{1}{4}\right) = 40.8\text{ eV}.$$

64. C – The given information suggests that the speed of a wave on a vibrating string only depends of μ and T, so by dimensional analysis, the quantity with the correct units is

$$c = \sqrt{\frac{T}{\mu}}.$$

In fact, this is the correct answer even up to dimensionless factors, but because the answer choices are of the "numbers and estimation" type, any such factors are irrelevant for getting the correct answer. Plugging in the given numbers,

$$c = \sqrt{\frac{T}{\mu}} = \sqrt{\frac{4 \times 10^3\text{ N}}{0.1\text{ kg/m}}} = 200\text{ m/s},$$

choice C.

65. A – By symmetry, the force on the test charge at the center is zero, so I is clearly true. II is false because the potential from each charge at the corners adds and is nonzero. III is false for a subtle, but important reason. Earnshaw's theorem states that any configuration of electrostatic charges cannot be in stable equilibrium. The charge at the center is in an unstable equilibrium: the force vanishes, but if the charge is perturbed in certain directions, it will move away from the center instead of returning to its original position.

66. E – The cyclotron radius is given by $R = \gamma mv/(qB)$, where B is the magnetic field strength and γ is the Lorentz factor. It's easy to remember this formula because it's identical to the nonrelativistic version, except with the momentum $p = mv$ replaced by the relativistic momentum γmv. For $v = 0.6c$, we have $\gamma = 5/4$ and $\gamma v = 3c/4$. For $v = 0.8c$, $\gamma = 5/3$ and $\gamma v = 4c/3$. All other factors cancel out in the ratio, so the new cyclotron radius is $\frac{4c/3}{3c/4}R = 16R/9$.

67. B – Let the initial intensity be I_0. The intensity of unpolarized light is attenuated through a single linear polarizer by a factor of 2. The emitted light leaving the first polarizer is now linearly polarized, with intensity $I' = I_0/2$. Malus's law gives the intensity I'' going through the second filter as

$$I'' = I'\cos^2\theta = \frac{I_0}{2}\cos^2\theta.$$

For $\theta = 30°$, we have

$$I'' = \frac{I_0}{2}\left(\frac{3}{4}\right) = \frac{3}{8}I_0.$$

The total attenuation is therefore 3/8.

68. B – Recalling the fundamental commutation relation $[\hat{x}, \hat{p}] = i\hbar$, we see that

$$\mathcal{O} = |i\hbar|^2 = \hbar^2,$$

so $\mathcal{O}$ is just a constant operator. The expectation value of a constant in any state is just the value of that constant, so $\langle \mathcal{O} \rangle = \hbar^2$.

69. D – Recall that the inductance L is defined through

$$\Phi_B = LI.$$

We obtain the magnetic field through the solenoid by Ampère's law:

$$B\ell = \mu_0 NI.$$

The magnetic flux through the solenoid interior is

$$\Phi_B = \frac{\pi R^2 \mu_0 NI}{\ell},$$

so the inductance of *one* wind of the solenoid is

$$L = \frac{\pi R^2 \mu_0 N}{\ell}.$$

Since inductance adds in series, the inductance of N winds is

$$L = \frac{\pi R^2 \mu_0 N^2}{\ell}.$$

70. B – This is a simple application of the thin lens equation. The first lens satisfies

$$\frac{2}{f} = \frac{1}{2f} + \frac{1}{q},$$

where q is some undetermined position of the image from the first lens. The second lens satisfies

$$\frac{2}{f} = \frac{1}{d-q} + \frac{1}{f}.$$

Solving for d, we obtain

$$d = \frac{5}{3}f.$$

Note that we implicitly assumed that $d > q$, otherwise the sign conventions would be different; since the first equation gives $q = 2f/3$, this assumption is self-consistent.

71. B – We can replace the masses connected by the rod with an effective mass $2m$, coupled to a spring of effective spring constant $2k$ (one spring pulls while the other pushes, so the forces add, and are equivalent to a single spring with constant $2k$). This has an identical equation of motion to a single mass attached to a single spring, with the usual frequency $\omega = \sqrt{2k/2m} = \sqrt{k/m}$.

72. C – The natural line width is given by the energy–time uncertainty relation:

$$\Delta E \sim \frac{\hbar}{\Delta t}.$$

Plugging in numbers, we arrive at $\Delta E = 2.2 \times 10^{-6}$ eV, which is closest to C.

73. A – By definition, a general gauge transformation corresponds to changes of the scalar and vector potentials that leave the physical electric and magnetic fields unchanged. This information alone is sufficient to answer the question. In case you were wondering, however, a general gauge transformation corresponds to a change of the scalar potential by

$$V \to V - \frac{\partial f}{\partial t},$$

and a change of the vector potential by

$$\mathbf{A} \to \mathbf{A} + \nabla f,$$

where $f(\mathbf{r}, t)$ is some real-valued function.

74. C – The energies of the three-dimensional infinite square well are proportional to $n_x^2 + n_y^2 + n_z^2$, where n_x, n_y, and n_z are *positive* integers (if $n = 0$, the wavefunction is identically zero). The ground state has all of the n's equal to 1. The first excited state occurs when one of the n's is 2, and the others are 1. The second excited state has two of the n's equal to 2 and the third equal to 1. (Note that since $2^2 + 2^2 + 1^2 < 3^2 + 1^2 + 1^2$, this state has less energy than the state where one of the n's is 3.) There are three such combinations: $(1, 2, 2)$, $(2, 1, 2)$, and $(2, 2, 1)$.

75. A – Suppose that there is a charge $+Q$ on the surface of the inner sphere, and a charge of $-Q$ on the surface of the outer sphere. The potential in the region between the two spheres is just the potential of a point charge:

$$V(r) = \frac{1}{4\pi\epsilon_0}\frac{Q}{r}.$$

The potential difference between the two spheres is therefore

$$\Delta V = \frac{Q}{4\pi\epsilon_0}\left(\frac{1}{a} - \frac{1}{b}\right).$$

Because $Q = CV$, we can read off the capacitance as

$$C = \frac{4\pi\epsilon_0 ab}{b-a}.$$

76. E – This can be solved just by dimension counting. The Hilbert space of a spin-1/2 particle has dimension 2, so the total Hilbert space of three such distinguishable particles has dimension $2^3 = 8$. This is true even if we rearrange the states into linear combinations with definite total spin, which for this problem happens to be completely unnecessary. Note that the adjective "distinguishable" is *crucial* here: without it, we would have to worry about symmetry or antisymmetry of the spatial and spin wavefunctions to satisfy Fermi–Dirac statistics.
77. A – This is a simple application of the lensmaker's equation:

$$\frac{1}{f} = (n-1)\left(\frac{1}{R_1} - \frac{1}{R_2}\right).$$

Solving for R_2, we obtain

$$R_2 = \left(\frac{1}{R_1} - \frac{1}{f(n-1)}\right)^{-1}.$$

Plugging in numbers, we find that $R_2 = 12.5$ cm. Incidentally, since R_2 is positive, the corresponding surface is concave, and we have an example of a convex–concave lens which happens to have positive curvature (as opposed to the more common case of a convex–convex converging lens, which always has positive curvature).
78. B – Let's define coordinates by setting the time at which the doors open and close in the garage's frame to be $t = 0$, and letting x be the distance from the front door. In the frame of the car, the first equation of the Lorentz transformation gives the time at which the doors open and close:

$$t' = \gamma\left(t - \frac{vx}{c^2}\right).$$

The front door therefore closes at time $t_1' = 0$. The rear door opens at time $t_2' = \gamma(0 - vl/(c^2)) = -\gamma vl/c^2$. Notice, of course, that even though the doors move at the same instant in the frame of the garage, they do not move at the same time in the frame of the moving car. This is a prototypical example of the relativity of simultaneity.
79. E – This problem requires remembering the basic scaling law for the entropy of an ideal gas,

$$S = Nk \ln \frac{VT^{3/2}}{N} + \text{constants},$$

in light of the equation of state for an ideal gas $PV = NkT$. If the temperature doubles but pressure remains constant, then volume must also double. So $\Delta S_1 = (5/2)Nk \ln 2$. If temperature doubles but pressure also doubles, then volume must be constant. The entropy only changes by $\Delta S_2 = (3/2)Nk \ln 2$. Finally, if the temperature is constant, but the volume doubles, then the entropy only changes by $\Delta S_3 = Nk \ln 2$. The correct order is choice E.
80. D – The equation of motion for a forced oscillator is

$$m\ddot{x} = -kx + F_0 \sin(\omega t).$$

This inhomogeneous ODE may look a little daunting at first, but it is clear that it is solved by a simple solution of the form

$$x(t) = A \sin(\omega t),$$

which allows us to cancel the sine in the driving force. This is the so-called "particular solution" to the ODE; there will be a general solution piece as well, but the assumptions in the problem allow us to ignore this "transient" term. Plugging in this *ansatz*, we obtain

$$-Am\omega^2 = -Ak + F_0,$$

and therefore

$$|A| = \frac{F_0}{|k - m\omega^2|}.$$

Alternatively, just use limiting cases and physical intuition: if the driving force is applied at the resonant frequency $\sqrt{k/m}$ of the system, the amplitude should blow up to infinity. Only choice D satisfies this condition.
81. C – There is no motion along the y- or z-directions, so these coordinates remain unchanged: $y' = y$ and $z' = z$. The x'-coordinate, on the other hand, can be found directly from the Lorentz transformation equations:

$$x' = \gamma\,(x - vt) = \frac{1}{\sqrt{1 - 0.8^2}}(5\text{ m}) = 8.33\text{ m}.$$

Choice E may be tempting since it is reminiscent of Lorentz contraction, but remember that lengths must be measured using simultaneous events in S', while the event in the problem does *not* occur at $t' = 0$.
82. E – This problem involves conservation of energy, which is pure potential at the top of the ramp but is a combination of translational and rotational energy at the bottom. To obtain the rotational energy, recall (from the formula sheet given at the beginning of the test) that the moment of inertia of a cylinder rotating about its axis is

$$I = \frac{1}{2}MR^2.$$

The potential energy at the top of the ramp must be equal to the sum of the rotational and translational energy at the bottom of the ramp. Using the rolling-without-slipping condition $v = R\omega$, we have

$$\begin{aligned} Mgh &= \frac{1}{2}Mv^2 + \frac{1}{2}I\omega^2 \\ &= \frac{1}{2}Mv^2 + \frac{1}{4}MR^2\left(\frac{v}{R}\right)^2, \\ gh &= \frac{3}{4}v^2, \\ v &= 2\sqrt{\frac{gh}{3}}. \end{aligned}$$

83. C – The electric field is confined to the xy-plane, so since the particle starts out in the xy-plane, the electric force on the particle will certainly keep it in that plane. The magnetic force will be perpendicular to both **B** and the particle's path, but **B** is *already* perpendicular to the xy-plane, so the magnetic force will act along the vector perpendicular to the path, which also lies in the xy-plane. So C is correct. A would be true if there was only a magnetic field, D is false because the electric field does work, and E is false because of the magnetic contribution.
84. B – Without knowing too much about the details of dark matter, this question can be answered by recognizing that dark matter, like any other matter, is gravitationally *attractive*. Choice B is related to the expansion of the universe, which requires an energy density that is effectively *repulsive*: this is dark energy. Choices A, C, D, and E are all seen as compelling evidence in favor of the existence of dark matter.
85. B – Both the counting rates follow Poisson statistics, where the uncertainty on N counts is $\sqrt{N}$. The first count represents the number of signal + background events, while the second count with the source removed represents the number of background-only events. The uncertainty in the number of signal + background events is $\sqrt{1,250}$ events, and the uncertainty in the number of background events is $\sqrt{350}$. The number of signal events is obtained by subtracting the background, and the uncertainty on this count is given by

$$\Delta N = \sqrt{1,250 + 350} = \sqrt{1,600} = 40.$$

The rate uncertainty is therefore 4.0 Hz. Note that the errors *add* in quadrature, even though we are *subtracting* a background, because the two experiments are independent: forgetting this would lead to trap answer A.

86. C – While a precise answer to this question can be calculated, it is too laborious to do so during the exam. Just note that, since the fastest part of the pulse has a time constant of about 10 ms, it may produce some signal around $(0.01\text{ s})^{-1} = 100$ Hz. The slow part of the pulse may produce some signal around 10 Hz. We therefore need a bandwidth of at least 100 Hz, which is closest to 1 kHz, choice C.
87. D – In order to compute a p-value from a χ^2 statistic, we need to know the number of degrees of freedom. The χ^2 distribution looks different for each number of degrees of freedom. I is not relevant for a χ^2 test. II and III are ingredients that together determine the number of degrees of freedom. Neither alone is sufficient, however.
88. E – The *pp* cycle of the Sun consists of nine different nuclear reactions. It is rather fascinating and elegant, but chances are low that you would need to remember the details for an exam. Simple logic can lead to the right answer here, though. As the name suggests, the *pp* cycle starts with fusion of protons into heavier elements. The most natural guess for the nucleus not produced by the cycle would be the nucleus of highest atomic number, ^{11}C. This turns out to be the correct answer; the nucleus with the largest number of protons produced in the *pp* cycle of the Sun is ^{8}B.
89. D – The partition function of this two-state system is

$$Z = 1 + e^{-\beta\epsilon},$$

where $\beta = 1/kT$. It would be perfectly valid to compute the mean energy using $\langle E\rangle = -\partial \ln Z/\partial\beta$, but this problem is simple enough that, for GRE purposes, direct computation is faster:

$$\begin{aligned} \langle E\rangle &= \frac{1}{Z}\sum_i \epsilon_i e^{-\epsilon_i/kT} \\ &= \frac{\epsilon e^{-\epsilon/kT}}{1 + e^{-\epsilon/kT}} \\ &= \frac{\epsilon}{1 + e^{\epsilon/kT}}, \end{aligned}$$

where the last step follows from multiplying numerator and denominator by $e^{\epsilon/kT}$.

90. C – The Hubble parameter relates the velocity of receding objects in space to their distance from us through Hubble's law, $v = H_0 d$. So if H_0 were suddenly doubled, our distance to the objects would be unchanged and the velocity would have to double. Since redshift is determined by the velocity at which distant objects travel with

respect to us, the redshift of these objects would have to change.

91. D – If the radiation used for interferometry has wavelength λ, the phase difference caused by a displacement Δx is $\Delta\phi = 2k\Delta x = 4\pi\Delta x/\lambda$ (the factor of 2 is from traversing the path twice, once from the satellite to the ground and again on the return, but since this is an order of magnitude problem, it won't really matter). This suggests that we want λ the same order of magnitude as Δx, but slightly larger so that the phase shift does not exceed 2π. If $\lambda = 5$ cm, the frequency is $\omega = c/\lambda = 6$ GHz, which is closest to choice D.

92. D – Doping a semiconductor adds additional unbound electrons or holes into the semiconductor by acting as donors or acceptors. This clearly influences the band structure somehow, so A is incorrect. B and E are incorrect because adding a small amount of dopant should not turn the semiconductor into a perfect insulator (B) or a perfect conductor with no band gap (E). C seems suspect because adding more free charge carriers should, if anything, decrease the energy needed for electrons to excite into the valence band.

93. B – We can get the radial electric field outside the cylinder from Gauss's law, using a Gaussian cylinder of radius r and length L. This just gives $E(2\pi rL) = \sigma(2\pi aL)/\epsilon_0$, or

$$E = \frac{a\sigma}{\epsilon_0 r}.$$

Integrating from a to some radius r, we obtain a potential

$$V(r) = -\int_a^r E(r')\,dr' = -\frac{a\sigma}{\epsilon_0}(\ln r - \ln a) = \frac{a\sigma}{\epsilon_0}\ln\frac{a}{r}.$$

Be careful with signs! Forgetting the minus sign in the potential is easy and leads to trap answer A.

94. C – This problem has two short steps: we need to find the speed of the car at the top of the loop in terms of the speed at the bottom, and then find what speed is needed in order for gravity to provide the required centripetal force. If the speed at the bottom is v_b, then the speed v_t at the top of the loop is given by

$$\frac{1}{2}mv_b^2 = 2mgR + \frac{1}{2}mv_t^2.$$

Canceling the mass and simplifying, we have

$$v_t^2 = v_b^2 - 4gR.$$

Setting the gravitational force equal to the centripetal force at the top of the loop, we have

$$\begin{aligned} mg &= \frac{mv_t^2}{R}, \\ gR &= v_b^2 - 4gR, \\ v_b &= \sqrt{5gR} \\ &\simeq \sqrt{25\ \text{m}^2/\text{s}^2} \\ &\simeq 5.0\ \text{m/s}. \end{aligned}$$

If v_b is smaller than this critical value, then gravity provides an extra radial force which causes the car to accelerate radially; in other words, to fall off the track. If v_b is larger than the critical value, the extra centripetal force must be provided by the normal force of the track.

95. E – Expanding out the commutator and factoring, we have

$$\begin{aligned} [S_xS_y, S_y] &= (S_xS_y)S_y - S_y(S_xS_y) \\ &= (S_xS_y)S_y - (S_yS_x)S_y = [S_x, S_y]S_y. \end{aligned}$$

(You could also have used the formula for commutators of products, $[AB, C] = A[B, C] + [A, C]B$.) Now applying the commutation algebra for spin angular momentum, we have $[S_x, S_y] = i\hbar S_z$, so the whole commutator is $i\hbar S_zS_y$.

96. B – Since we are not given the form of the potential, choice E may be tempting. However, Bertrand's theorem states that the only central potentials that produce closed noncircular orbits are the Kepler potential and the harmonic oscillator potential. Any central potential will have circular orbits at $E = V_{\text{min}}$, but the problem tells us that $E > V_{\text{min}}$ so the orbit cannot be circular. We already know that bound orbits of the Kepler potential with $E > V_{\text{min}}$ are elliptical. For the harmonic oscillator, since $r^2 = x^2 + y^2 + z^2$, the coordinates decouple and the motion is a superposition of three harmonic oscillators in the three coordinate directions. The motion must lie in a plane, per the usual arguments for a central potential, so we can set the amplitude of the z oscillation to zero and just consider oscillators in the x- and y-directions. These are sinusoidal, but if they have different amplitudes and/or phases, they will produce ellipses. So the nonminimum energy orbits of a harmonic oscillator are also ellipses, and choice B is correct.

97. D – The energy stored in a capacitor of capacitance C at constant voltage is $E = \frac{1}{2}CV_0^2$. The capacitance with the dielectric inserted is $C = \kappa\epsilon_0 a^2/d$. The "bare"

capacitance without the dielectric is $C = \epsilon_0 a^2/d$. The difference in energy is therefore

$$\Delta E = \frac{(\kappa - 1)\epsilon_0 a^2 V_0^2}{2d}.$$

98. A – To obtain the force from the potential energy, we just take the negative of the gradient:

$$\mathbf{F}(x, y, z) = -\nabla U = -\hat{\mathbf{x}}\frac{\partial}{\partial x}(Ax) - \hat{\mathbf{y}}\frac{\partial}{\partial y}(By^2)$$

$$+\hat{\mathbf{z}}\frac{\partial}{\partial z}(C\cos z) = -A\hat{\mathbf{x}} - 2By\hat{\mathbf{y}} - C\sin z\hat{\mathbf{z}}.$$

99. E – The ground state wavefunction of hydrogen is spherically symmetric, so we will denote it by $\psi(r)$ (as we will see, we don't actually need the functional form, symmetry is enough to solve this problem as stated). Defining the z-direction along the direction of the electric field, the perturbation Hamiltonian is $\Delta H = eE_0 z = eE_0 r\cos\theta$, and the first-order correction is given by

$$\int_0^\infty \int_0^{2\pi} \int_0^\pi \psi^*(r)\psi(r)(eE_0 r\cos\theta)r^2 \sin\theta \, d\theta \, d\phi \, dr = 0.$$

However, the θ integral vanishes:

$$\int_0^\pi \cos\theta \sin\theta \, d\theta = \frac{1}{2}\int_0^\pi \sin 2\theta \, d\theta$$
$$= -\frac{1}{4}(\cos 2\pi - \cos 0) = 0.$$

Thus the first-order correction vanishes.

100. D – We want the nonrelativistic limit of the relativistic Doppler shift, so we Taylor expand for $\beta \ll 1$:

$$\frac{\lambda'}{\lambda} = \sqrt{\frac{1+\beta}{1-\beta}} = (1+\beta)^{1/2}(1-\beta)^{-1/2}$$
$$\approx (1+\beta/2)(1+\beta/2) \approx 1+\beta.$$

Thus the Doppler-shifted energy is

$$E' = \frac{hc}{\lambda'} = \frac{hc}{\lambda(1+v/c)} \approx \frac{hc}{\lambda}(1 - v/c) = E(1 - v/c),$$

so the energy shift is Ev/c. In fact, this is what we might have expected from using the nonrelativistic Doppler shift expression with wave speed c. Note that the sign of β or v doesn't matter here since all we care about is the magnitude of the energy difference.

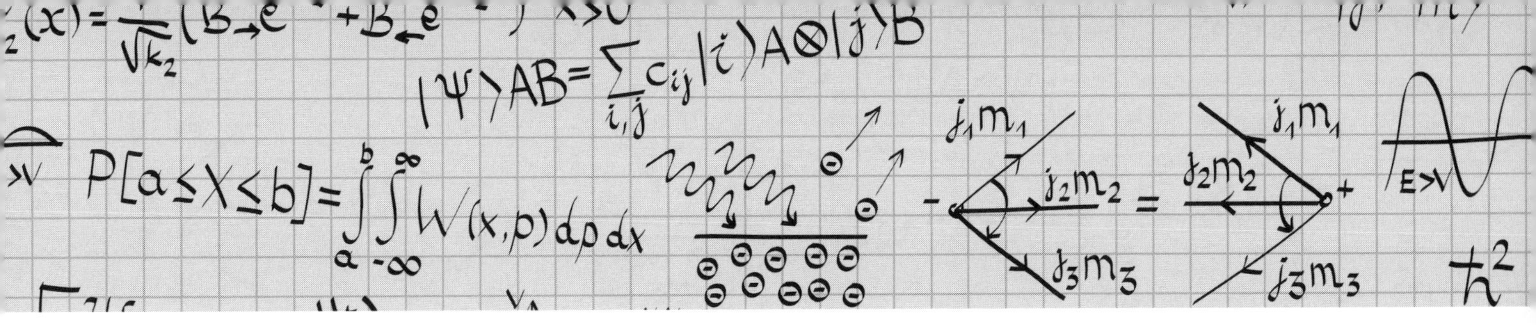

REFERENCES

N.W. Ashcroft and N.D. Mermin, *Solid State Physics* (New York: Saunders College Publishing/Harcourt Brace College Publishers, 1976).

E. Fermi, *Thermodynamics* (New York: Dover Publications, 1956).

D.J. Griffiths, *Introduction to Electrodynamics*, 4th edn (San Francisco: Pearson, 2013; Cambridge: Cambridge University Press, 2017).

D.J. Griffiths, *Introduction to Quantum Mechanics*, 2nd edn (San Francisco: Pearson, 2005; Cambridge: Cambridge University Press, 2016).

D.J. Griffiths, *Introduction to Elementary Particles*, 2nd revised edn (Weinheim: Wiley-VCH, 2008).

P. Horowitz and W. Hill, *The Art of Electronics*, 3rd edn (Cambridge: Cambridge University Press, 2015).

L.A. Kirkby, *Physics: A Student Companion* (Banbury: Scion Publishing, 2011).

C. Kittel and H. Kromer, *Thermal Physics*, 2nd edn (New York: W.H. Freeman, 1980).

C. Kittel, *Elementary Statistical Physics* (New York: John Wiley & Sons, 1958; Mineola: Dover Publications, 2004).

C. Kittel, *Introduction to Solid State Physics*, 8th edn (Hoboken: John Wiley & Sons, 2005).

G. Knoll, *Radiation Detection and Measurement*, 4th edn (Hoboken: John Wiley & Sons, 2010).

F. Mandl, *Statistical Physics*, 2nd edn (New York: John Wiley & Sons, 1988).

E.M. Purcell and D.J. Morin, *Electricity and Magnetism*, 3rd edn (New York: Cambridge University Press, 2013).

F. Reif, *Fundamentals of Statistical and Thermal Physics* (Long Grove: Waveland Press, 2009).

O. Svelto, *Principles of Lasers*, 5th edn (New York: Springer, 2010).

S.T. Thornton and J.B. Marion, *Classical Dynamics of Particles and Systems*, 5th edn (Belmont: Brooks/Cole-Thomson Learning, 2004).

R.K. Wangsness, *Electromangetic Fields*, 2nd edn (Hoboken: John Wiley & Sons, 1986).

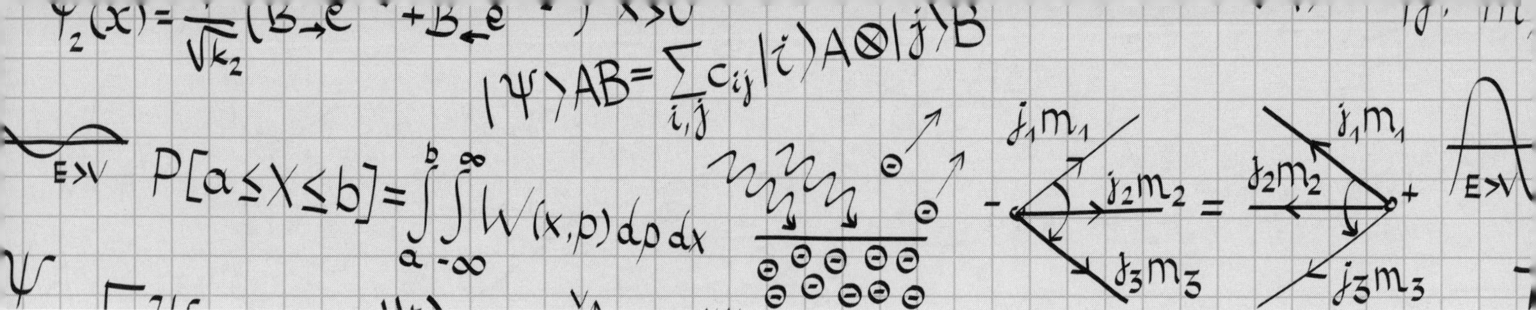

EQUATION INDEX

Classical Mechanics

(1.1) $x(t) = v_{0x}t + x_0, \quad y(t) = -\frac{1}{2}gt^2 + v_{0y}t + y_0$ (p. 5)

(1.2) $v_f^2 - v_i^2 = 2a\Delta x$ (p. 5)

(1.3) $a = \frac{v^2}{r}$ (p. 5)

(1.4) $F = \frac{mv^2}{r}$ (p. 5)

(1.5) Translational kinetic energy: $\frac{1}{2}mv^2$ (p. 7)

(1.6) Rotational kinetic energy: $\frac{1}{2}I\omega^2$ (p. 7)

(1.7) Gravitational potential energy on Earth: mgh (p. 7)

(1.8) Spring potential energy: $\frac{1}{2}kx^2$ (p. 7)

(1.9) $\Delta U = -\int_a^b \mathbf{F} \cdot d\mathbf{l}$ (p. 8)

(1.10) $\mathbf{F}_{\text{grav}} = \frac{Gm_1m_2}{r^2}\hat{\mathbf{r}}$ (p. 8)

(1.11) $\mathbf{F} = -\nabla U$ (p. 8)

(1.12) $v = R\omega$ (p. 9)

(1.13) $E_{\text{initial}} + W_{\text{other}} = E_{\text{final}}$ (p. 11)

(1.14) $W = \Delta\text{KE}$ (p. 11)

(1.15) $W = \int \mathbf{F} \cdot d\mathbf{l}$ (p. 11)

(1.16) $\mathbf{L} = \mathbf{r} \times \mathbf{p}$ (p. 13)

(1.17) $\mathbf{L} = I\boldsymbol{\omega}$ (p. 13)

(1.18) $\boldsymbol{\tau} = \mathbf{r} \times \mathbf{F}$ (p. 13)

(1.19) $L = I\omega$ (p. 13)

(1.20) $\tau = \frac{dL}{dt}$ (p. 13)

(1.21) $F_{\text{centrifugal}} = -m\Omega^2 r$ (p. 14)

(1.22) $F_{\text{Coriolis}} = -2m\boldsymbol{\Omega} \times \mathbf{v}$ (p. 14)

(1.23) $I = mr^2$ (p. 14)

(1.24) $I = \int r^2\, dm$ (p. 14)

(1.25) $I = I_{\text{CM}} + Mr^2$ (p. 14)

(1.26) $\mathbf{r}_{\text{CM}} = \frac{\int \mathbf{r}\, dm}{M}$ (p. 15)

(1.27) $\mathbf{r}_{\text{CM}} = \frac{\sum_i \mathbf{r}_i m_i}{M}$ (p. 15)

(1.28) $L(q, \dot{q}, t) = T - U$ (p. 16)

(1.29) $\frac{d}{dt}\frac{\partial L}{\partial \dot{q}} = \frac{\partial L}{\partial q}$ (p. 17)

(1.30) $p_i \equiv \frac{\partial L}{\partial \dot{q}}$: momentum conjugate to q (p. 18)

(1.31) $H(p, q) = \sum_i p_i\dot{q}_i - L$ (p. 18)

(1.32) $H = T + U$ (if U does not depend explicitly on velocities or time) (p. 18)

(1.33) $\dot{p} = -\frac{\partial H}{\partial q}, \quad \dot{q} = \frac{\partial H}{\partial p}$ (p. 18)

(1.34) $L = \frac{1}{2}m\dot{r}^2 + \frac{1}{2}mr^2\dot{\phi}^2 - U(r)$ (p. 20)

(1.35) $l = mr^2\dot{\phi}$ (p. 20)

(1.36) $V(r) = \frac{l^2}{2mr^2} + U(r)$ (p. 20)

(1.37) $\mu = \frac{m_1m_2}{m_1 + m_2}$ (p. 20)

(1.38) $E = T + V = \frac{1}{2}m\dot{r}^2 + \frac{l^2}{2mr^2} + U(r)$ (p. 20)

(1.39) $F = m\ddot{x} = -kx$ (p. 22)

(1.40) $\omega = \sqrt{\frac{k}{m}}$ (p. 22)

Electricity and Magnetism

(2.42) $\mathcal{E} = -\frac{d\Phi_B}{dt}$ (p. 50)

(2.43) $\Phi_{21} = M_{12}I_1$ (p. 50)

(2.44) $\Phi_B = LI$ (p. 50)

(2.45) $\mathcal{E} = -L\frac{dI}{dt}$ (p. 51)

(2.46) $L = \frac{\mu_0 N^2 A}{l}$ (solenoid) (p. 51)

(2.47) $U_L = \frac{1}{2}LI^2$ (p. 51)

(2.48) $\mathbf{p} = q\mathbf{r}_1 - q\mathbf{r}_2 = q\mathbf{d}$ (p. 52)

(2.49) $\mathbf{p} = \int \mathbf{r}\rho(\mathbf{r})d^3\mathbf{r}$ (p. 52)

(2.50) $V(\mathbf{r}) = \frac{1}{4\pi\epsilon_0}\frac{\mathbf{p}\cdot\hat{\mathbf{r}}}{r^2}$ (p. 52)

(2.51) $\mathbf{N} = \mathbf{p} \times \mathbf{E}$ (p. 52)

(2.52) $U = -\mathbf{p}\cdot\mathbf{E}$ (p. 52)

(2.53) $\mathbf{m} = I\mathbf{A}$ (p. 52)

(2.54) $\mathbf{N} = \mathbf{m} \times \mathbf{B}$ (p. 53)

(2.55) $U = -\mathbf{m}\cdot\mathbf{B}$ (p. 53)

(2.56) $\sigma_b = \mathbf{P}\cdot\hat{\mathbf{n}}$ (p. 54)

(2.57) $\rho_b = -\nabla\cdot\mathbf{P}$ (p. 54)

(2.58) $\epsilon_0 \mapsto \epsilon = \kappa\epsilon_0$ (p. 54)

(2.59) $C = \frac{\epsilon A}{d} = \kappa\frac{\epsilon_0 A}{d}$ (p. 54)

(2.60) $c = 1/\sqrt{\epsilon_0\mu_0}$ (p. 55)

(2.61) $\tilde{\mathbf{E}}(\mathbf{r}) = \tilde{E}_0 e^{i(\mathbf{k}\cdot\mathbf{r}-\omega t)}\hat{\mathbf{n}}$ (p. 55)

(2.62) $\tilde{\mathbf{B}}(\mathbf{r}) = \frac{1}{c}\tilde{E}_0 e^{i(\mathbf{k}\cdot\mathbf{r}-\omega t)}(\hat{\mathbf{k}}\times\hat{\mathbf{n}})$ (p. 55)

(2.63) $\mathbf{S} = \frac{1}{\mu_0}(\mathbf{E}\times\mathbf{B})$ (p. 55)

(2.64) $\mathbf{S} = \frac{1}{2\mu_0}\mathrm{Re}(\tilde{\mathbf{E}}\times\tilde{\mathbf{B}}^*)$ (p. 55)

(2.65) $I = \langle S\rangle = \frac{1}{2}c\epsilon_0 E_0^2$ (p. 55)

(2.66) $P = \frac{q^2a^2}{6\pi\epsilon_0 c^3} = \frac{\mu_0 q^2 a^2}{6\pi c}$ (p. 56)

(2.67) $\langle S\rangle = \left(\frac{\mu_0 p_0^2\omega^4}{32\pi^2 c}\right)\frac{\sin^2\theta}{r^2}$ (p. 56)

(2.68) $\langle P\rangle_E = \frac{\mu_0 p_0^2\omega^4}{12\pi c}$ (p. 56)

(2.69) $\langle P\rangle_B = \frac{\mu_0 m_0^2\omega^4}{12\pi c^3}$ (p. 56)

(2.70) $V_R = IR$ (p. 57)

(2.71) $V_C = \frac{Q}{C}$ (p. 57)

(2.72) $V_L = L\frac{dI}{dt}$ (p. 57)

(2.73) $R_{\text{eq}} = \sum_i R_i$ (series) (p. 57)

(2.74) $\frac{1}{C_{\text{eq}}} = \sum_i \frac{1}{C_i}$ (series) (p. 57)

(2.75) $L_{\text{eq}} = \sum_i L_i$ (series) (p. 57)

(2.76) $\frac{1}{R_{\text{eq}}} = \sum_i \frac{1}{R_i}$ (parallel) (p. 57)

(2.77) $C_{\text{eq}} = \sum_i C_i$ (parallel) (p. 57)

(2.78) $\frac{1}{L_{\text{eq}}} = \sum_i \frac{1}{L_i}$ (parallel) (p. 57)

(2.79) $R = \frac{\rho\ell}{A}$ (p. 57)

(2.80) $\sum_k I_k = 0$ (p. 57)

(2.81) $\sum_k V_k = 0$ (p. 57)

(2.82) $P = IV = \frac{V^2}{R} = I^2R$ (p. 57)

(2.83) $\tau_{RL} = L/R$ (p. 58)

(2.84) $\tau_{RC} = RC$ (p. 58)

(2.85) $\omega_0 = \frac{1}{\sqrt{LC}}$ (p. 58)

Optics and Waves

(3.1) $\frac{\partial^2 f}{\partial t^2} = v^2\frac{\partial^2 f}{\partial x^2}$ (p. 63)

(3.2) $f(x,t) = A\cos(kx - \omega t + \delta)$ (p. 63)

(3.3) $\lambda = \frac{2\pi}{k}, \quad T = \frac{2\pi}{\omega}, \quad \omega = 2\pi f$ (p. 64)

(3.4) $\omega = vk$ (p. 65)

(3.5) Phase velocity: $\frac{\omega}{k}$ (p. 65)

(3.6) Group velocity: $\frac{d\omega}{dk}$ (p. 65)

(3.7) $v = \sqrt{\frac{T}{\mu}}$ (p. 65)

(3.8) $\omega/k = c/n$ (for light waves) (p. 65)

(3.9) $\lambda \to \frac{\lambda}{n}$ (p. 65)

(3.10) $I = I_0 \cos^2\theta$ (p. 66)

(3.11) $\theta_B = \arctan\left(\frac{n_2}{n_1}\right)$ (p. 66)

(3.12) Constructive interference $\Longleftrightarrow$ phase difference of $2m\pi$ (p. 67)

(3.13) Destructive interference $\Longleftrightarrow$ phase difference of $(2m+1)\pi$ (p. 67)

(3.14) $\delta = k\Delta x$ (p. 67)

(3.15) Maxima: $d\sin\theta = m\lambda$ (p. 67)

(3.16) Minima: $d\sin\theta = (m+1/2)\lambda$ (p. 67)

(3.17) $a\sin\theta = m\lambda, \quad m = 1, 2, \ldots$ (p. 68)

(3.18) $\Delta x = nd$ (optical path length) (p. 69)

(3.19) $n_2 > n_1$: phase shift of π (p. 69)

(3.20) $n_2 < n_1$: *no* phase shift (p. 69)

(3.21) First circular diffraction minimum: $D\sin\theta = 1.22\lambda$ (p. 70)

(3.22) Maxima: $d\sin\theta = n\lambda/2$ (p. 70)

(3.23) Reflection: $\theta_i = \theta_r$ (angle of incidence equals angle of reflection) (p. 71)

(3.24) Refraction: $n_1\sin\theta_1 = n_2\sin\theta_2$ (Snell's law) (p. 71)

(3.25) $c_2\sin\theta_1 = c_1\sin\theta_2$ (p. 71)

(3.26) $\frac{1}{s} + \frac{1}{s'} = \frac{1}{f}$ (p. 71)

(3.27) $f = R/2$ (p. 71)

(3.28) $\frac{1}{f} = (n-1)\left(\frac{1}{R_1} - \frac{1}{R_2}\right)$ (p. 71)

(3.29) Positive distances $\Longleftrightarrow$ same side as light rays (incoming for s, outgoing for s') (p. 71)

(3.30) Negative distances $\Longleftrightarrow$ opposite side as light rays (incoming for s, outgoing for s') (p. 71)

(3.31) $I \propto I_0\,\lambda^{-4}\,a^6$ (p. 72)

(3.32) $f = \left(\frac{v+v_r}{v-v_s}\right) f_0$ (p. 73)

Thermodynamics and Statistical Mechanics

(4.1) $p_i = \frac{e^{-\beta E_i}}{\sum_j e^{-\beta E_j}}$ (p. 78)

(4.2) $\beta = \frac{1}{k_B T}$ (p. 78)

(4.3) $\langle \mathcal{O} \rangle = \sum_i p_i \mathcal{O}_i$ (p. 79)

(4.4) $p_i = \frac{e^{-\beta E_i}}{Z}$ (p. 79)

(4.5) $Z = \sum_j e^{-\beta E_j}$ (p. 79)

(4.6) $\langle E \rangle = \sum_i p_i E_i = \frac{\sum_i E_i e^{-\beta E_i}}{Z} = -\frac{\partial}{\partial\beta}\ln Z$ (p. 79)

(4.7) $S = k_B \ln\Omega$ (p. 79)

(4.8) $S = -k_B \sum_i p_i \ln p_i = \frac{\partial}{\partial T}(k_B T\ln Z)$ (p. 79)

(4.9) $S = Nk_B\left(\ln\frac{V}{N} + \frac{3}{2}\ln T + \frac{5}{2} + \frac{3}{2}\ln\frac{2\pi m k_B}{h^2}\right)$ (p. 80)

(4.10) $S = Nk_B \ln\frac{VT^{3/2}}{N} + \text{constants}$ (p. 80)

(4.11) $Z_N = \frac{1}{N!h^{3N}}\int e^{-\beta H(\mathbf{p}_1,\ldots\mathbf{p}_n;\mathbf{x}_1,\ldots\mathbf{x}_n)} d^3\mathbf{p}_1 \cdots d^3\mathbf{p}_n d^3\mathbf{x}_1 \cdots d^3\mathbf{x}_n$ (p. 80)

(4.12) $\binom{N}{M} = \frac{N!}{(N-M)!M!}$ (p. 80)

(4.13) $\ln(n!) \approx n\ln n - n$ (p. 80)

(4.14) $\Delta U = Q - W$ (p. 81)

(4.15) $\Delta S \geq \int \frac{\delta Q}{T}$ (p. 82)

(4.16) $PV = Nk_B T$ (p. 82)

(4.17) $\delta W = P\,dV$ (reversible) (p. 83)

(4.18) $\delta Q = T\,dS$ (reversible) (p. 83)

(4.19) $\Delta S = \int \frac{\delta Q}{T}$ (p. 83)

(4.20) $PV^\gamma = \text{constant}$ (p. 83)

Quantum Mechanics and Atomic Physics

Special Relativity

(6.8) $w = \dfrac{v+u}{1+vu/c^2}$ (p. 125)

(6.9) $x^0 = ct, \quad x^1 = x, \quad x^2 = y, \quad x^3 = z$ (p. 125)

(6.10) $x^\mu = (x^0, x^1, x^2, x^3) = (ct, x, y, z)$ (p. 125)

(6.11) $\beta = v/c$ (p. 125)

(6.12) $\begin{pmatrix} x^{0'} \\ x^{1'} \\ x^{2'} \\ x^{3'} \end{pmatrix} = \begin{pmatrix} \gamma & -\gamma\beta & 0 & 0 \\ -\gamma\beta & \gamma & 0 & 0 \\ 0 & 0 & 1 & 0 \\ 0 & 0 & 0 & 1 \end{pmatrix} \begin{pmatrix} x^0 \\ x^1 \\ x^2 \\ x^3 \end{pmatrix}$ (p. 125)

(6.13) Energy–momentum: $p^\mu = (E/c, \mathbf{p})$ (p. 126)

(6.14) Current density: $j^\mu = (c\rho, \mathbf{J})$ (p. 126)

(6.15) Wavevector: $k^\mu = (\omega/c, \mathbf{k})$ (p. 126)

(6.16) $\mathbf{p} = \gamma m \mathbf{v} = \dfrac{m\mathbf{v}}{\sqrt{1-|\mathbf{v}|^2/c^2}}$ (p. 126)

(6.17) $E_0 = mc^2$ (p. 126)

(6.18) $T = E - mc^2$ (p. 126)

(6.19) $E = \gamma mc^2$ (p. 126)

(6.20) $T = (\gamma - 1)mc^2$ (p. 126)

(6.21) $a \cdot b \equiv a^0b^0 - a^1b^1 - a^2b^2 - a^3b^3$ (p. 126)

(6.22) Timelike: $(\Delta x)^2 > 0$ (p. 127)

(6.23) Spacelike: $(\Delta x)^2 < 0$ (p. 127)

(6.24) Lightlike or null: $(\Delta x)^2 = 0$ (p. 127)

(6.25) $E^2 = \mathbf{p}^2c^2 + m^2c^4$ (p. 127)

(6.26) $\sum_i p_i^\mu = \sum_f p_f^\mu$ (p. 127)

(6.27) $\dfrac{\lambda'}{\lambda} = \sqrt{\dfrac{1+\beta}{1-\beta}}$ (p. 129)

(6.28) $\beta = 0.6 \implies \gamma = 1.25$ (p. 129)

(6.29) $\beta = 0.8 \implies \gamma = 5/3$ (p. 129)

Laboratory Methods

(7.1) $\sigma_S^2 = \dfrac{1}{n-1}\sum_{i=1}^{n}(x_i - \bar{x})^2$ (p. 135)

(7.2) $\sigma_{\text{tot}} = \sqrt{\sigma_{\text{stat}}^2 + \sigma_{\text{sys}}^2}$ (p. 135)

(7.3) $\sigma_z^2 = \sum_{i=1}^{n}\left(\dfrac{\partial z}{\partial x_i}\right)^2 \sigma_{x_i}^2$ (p. 135)

(7.4) $X = \dfrac{x/\sigma_x^2 + y/\sigma_y^2}{1/\sigma_x^2 + 1/\sigma_y^2}$ (p. 136)

(7.5) $\sigma_{\text{tot}}^2 = \dfrac{1}{1/\sigma_x^2 + 1/\sigma_y^2}$ (p. 136)

(7.6) $P(n) = \dfrac{\lambda^n e^{-\lambda}}{n!}$ (p. 136)

(7.7) Capacitor: $Z = \dfrac{1}{i\omega C}$ (p. 137)

(7.8) Inductor: $Z = i\omega L$ (p. 137)

(7.9) Resistor: $Z = R$ (p. 137)

(7.10) Series: $Z_{\text{tot}} = Z_1 + Z_2 + \cdots Z_n$ (p. 137)

(7.11) Parallel: $Z_{\text{tot}}^{-1} = Z_1^{-1} + Z_2^{-1} + \cdots Z_n^{-1}$ (p. 137)

(7.12) $\overline{A \cdot B} = \overline{A} + \overline{B}$ (p. 139)

(7.13) $\overline{A + B} = \overline{A} \cdot \overline{B}$ (p. 139)

(7.14) $E_{\text{max}} = E_\gamma - \phi$ (p. 140)

(7.15) $\lambda = \dfrac{h}{mc}$ (p. 140)

(7.16) $\Delta\lambda = \dfrac{h}{mc}(1 - \cos\theta)$ (p. 140)

(7.17) $N = N_0 e^{-t/\tau}$ (p. 141)

(7.18) $t_{1/2} = \tau \ln 2$ (p. 141)

(7.19) $\dfrac{1}{\tau} = \dfrac{1}{\tau_1} + \dfrac{1}{\tau_2} + \cdots$ (p. 141)

Specialized Topics

(8.1) $k_F = (3\pi^2 n)^{1/3}$ (p. 151)

(8.2) $E_F = \dfrac{\hbar^2}{2m}(3\pi^2 n)^{2/3}$ (p. 151)

(8.3) $\rho(E) = \dfrac{V\sqrt{2}}{\pi^2\hbar^3} m^{3/2}\sqrt{E}$ (p. 151)

(8.4) $N = \int_0^{E_F} \rho(E)\, dE$ (p. 151)

(8.5) $\rho(E_F) = \dfrac{3}{2}\dfrac{N}{E_F}$ (p. 151)

(8.6) $N_C \approx \rho(E_F)(k_B T) \sim N\frac{k_B T}{E_F}$ (p. 151)

(8.7) $\frac{\lambda_0}{\lambda_T} = \frac{a(\text{today})}{a(T)}$ (p. 152)

(8.8) $v = H_0 D$ (p. 152)

(8.9) $z(T) = \frac{\lambda_0}{\lambda_T} - 1$ (p. 152)

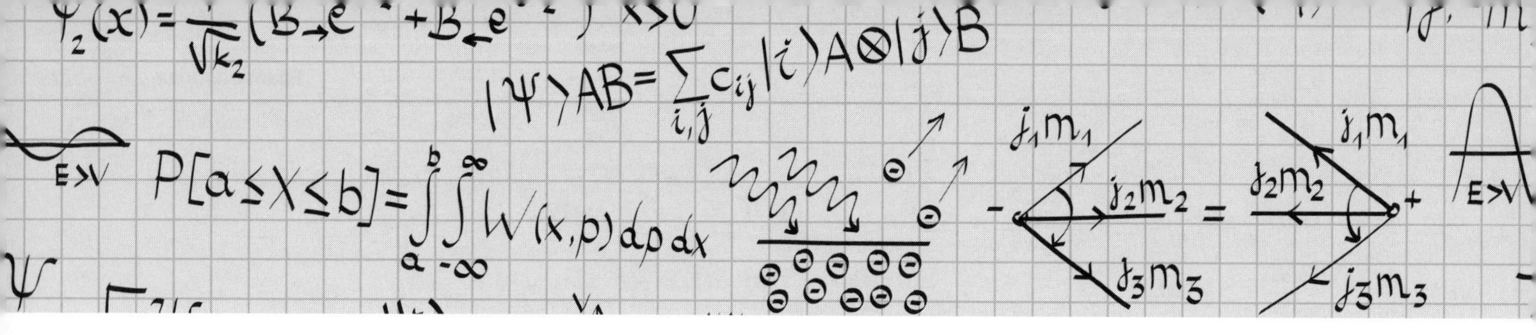

SUBJECT INDEX

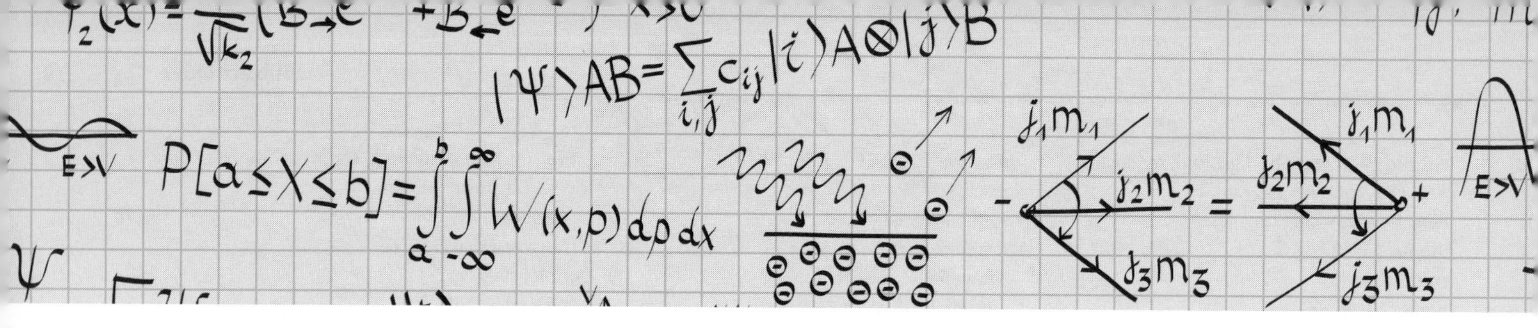

PROBLEMS INDEX